General
Virology

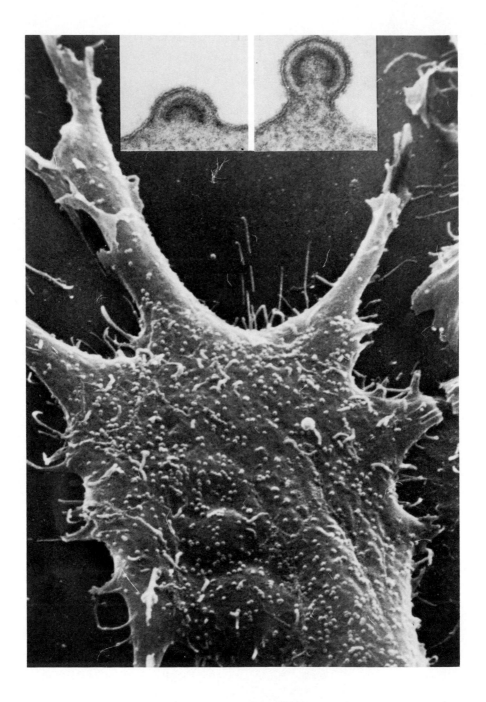

Scanning electron micrograph showing numerous murine
leukemia viruses at surface of an infected cell. Inset
shows section through "budding" virus particles.
(Photographs courtesy of Dr. P. K. Y. Wong)

General Virology

3rd edition

S. E. LURIA
Massachusetts Institute of Technology

JAMES E. DARNELL, JR.
Rockefeller University

DAVID BALTIMORE
Massachusetts Institute of Technology

ALLAN CAMPBELL
Stanford University

JOHN WILEY & SONS

New York • Chichester • Brisbane • Toronto

Library of Congress Cataloging in Publication Data:

Main entry under title:

General virology.

First ed. by S. E. Luria; 2d ed. by S. E. Luria
and J. E. Darnell.
Bibliography
Includes index.
1. Virology. I. Luria, Salvador Edward.
II. Luria, Salvador Edward. General virology.
QR360.L7 1977   576'.64   77-9498
ISBN 0-471-55640-8

Printed in the United States of America

10 9 8 7 6 5

TB 4/6/81

Preface

The third edition of this book appears at a time when virology as a biological science risks having its identity blurred as a result of its own successes. The growth of molecular biology, out of the twin roots of structural biochemistry and cellular genetics, both of which have fed largely on virus research, tends to obscure the limits between the basic biological disciplines.

Like all sciences that come to maturity, such as physics since Planck and Bohr or chemistry since Fischer and Pauling, biology since Watson and Crick has become unified at a finer level of specification than that provided by Darwin and Mendel. The borders between branches of biology are being overrun by the rising tide of modern cell biology. As a result, the teaching of virology as a biological science, flourishing in the 1950s and 1960s, is being dispersed among more catholic courses. Yet the very fact that virology has played so central a role in the development of molecular biology and remains a most vital field of biological and biomedical research justifies the present effort to maintain the unity of the field within a book whose previous editions have, we trust, helped catalyze the integration of virology into cell biology.

The growing sophistication of courses in general biology, genetics, and biochemistry that provide the basic preparation for biology students has made it possible to restrict the present edition of *General Virology* more precisely to its own subject by deemphasizing summaries and surveys of ancillary disciplines. This has allowed us to expand and bring up to date the strictly virological topics without increasing substantially the size of the book.

The methodological advances in the last 10 years, especially in physical biochemistry and electron microscopy, have been only peripheral to the enormous development of the field of virology in the same period. The most important advances have been, on the one hand, the flourishing of bacteriophage research as a branch of molecular genetics and, on the other hand, the role of tumor viruses and other viruses in the study of cellular biochemistry. The present picture of viral biochemistry and viral genetics promises wide advances not only in cell biology but in the study of molecular evolution.

These developments have helped to attract to virology a large influx of enthusiastic and productive researchers. The field has flourished and expanded beyond the territory that can be presented in detail within a book of this size. Fortunately the intellectual framework of the field, while expanding, has remained intact, so that it is still feasible to present an integrated and coherent overall picture. The wealth of virological literature has made it desirable for us to use as references a number of valuable and readily available monographs as well as recent reviews of specific fields. We have retained the principle of direct reference to original articles when these represent either historically significant landmarks, recent important contributions, or sources of specific methodologies and insights. As in previous editions, the choice has been dictated in large part by the inevitably fragmentary knowledge and recollections of a group of authors who are practitioners rather than chroniclers of virus research.

Collaboration between the four authors of the present edition has been made easy and fruitful by the fact that they have been associated with one another in various ways in the course of their careers. As a result they share

v

certain views and biases as well as a common enthusiasm for virology as a frontier in the study of life. The authors are grateful to the many colleagues who provided copy for illustrations. They are especially indebted to Professor Lindsay M. Black who advised them on recent developments in the field of plant virus research and provided criticism of the parts of this volume dealing with plant viruses.

S. E. Luria

James E. Darnell, Jr.

David Baltimore

Allan Campbell

Contents

14/Animal Virus Multiplication: DNA Viruses and Retroviruses   343

15/Effects of Animal Viruses on Host Cells and Organisms   391

Introduction: The Science of Virology

1

Virology as a Biological Science

Virology has become a basic biological science in its own right. Just as bacteriology emerged early this century as a biological science so did virology become, around the middle of the century, a body of knowledge and generalizations with its own perspectives and its own internal development. From its start as a branch of pathology—human and animal pathology on the one hand, plant pathology on the other—the science of virology has developed to a point where progress is dictated as much by the logic of its internal development as by the demands of applied fields.

A major stimulus to unification was the growth of knowledge about bacterial viruses or bacteriophages and the integration of bacteriophage research with bacterial genetics. A conscious effort was made by animal and plant virologists to use phage-work methods and concepts in attacking a variety of viral problems. At the same time there has emerged a new discipline, *molecular biology*, which unifies the various approaches to the structure, function, and organization of the macromolecules in which biological specificity is embodied. Virology has become an integral part of molecular biology because it deals with subcellular entities, the viruses, whose structure and organization belong in the macromolecular domain.

DEFINITIONS OF VIRUSES

The subject matter of virology, the viruses, cannot be defined by the common-sense criteria applied to animals or plants. The definition of viruses is itself somewhat arbitrary, and in fact many definitions have been proposed, an indication of the difficulty of the task.

Lwoff (1957) proposed to define viruses as "strictly intracellular and

1

potentially pathogenic entities with an infectious phase and (1) possessing only one type of nucleic acid, (2) multiplying in the form of their genetic material, (3) unable to grow and to undergo binary fission, and (4) devoid of a Lipmann system" (that is, a system of enzymes for energy production). This definition, whose implications will become clearer in later chapters, stresses the noncellular nature of viruses, their dependence on host-cell metabolism, and the fact that at some stage of its reproductive cycle the specific material of a virus is reduced to an element of genetic material, a nucleic acid.

Another definition considered viruses to be "elements of genetic material that can determine, in the cells where they reproduce, the biosynthesis of a specific apparatus for their own transfer into other cells" (Luria, 1959). This definition stresses the independence of the viral genome, its reproduction, and its specialization for cell-to-cell transfer rather than the lack of metabolic self-sufficiency.

In the first edition of this book (Luria, 1953a) at a time when the nature of genetic materials was uncertain and the events of their replication and function were still obscure, it seemed desirable to employ a strictly operational definition; that is, one that would limit itself to providing certain criteria for inclusion or exclusion of given objects into the category of viruses. The definition then adopted was: "Viruses are submicroscopic entities capable of being introduced into specific living cells and of reproducing inside such cells only." A more satisfactory definition (modified from Luria and Darnell, 1967) is a composite one: *"Viruses are entities whose genomes are elements of nucleic acid that replicate inside living cells using the cellular synthetic machinery and causing the synthesis of specialized elements that can transfer the viral genome to other cells."*

This definition conveys the two essential qualities of viruses: first, the possession of specific genetic materials that utilize the biochemical machinery of the host cells; and, second, the possession of an extracellular infective phase represented by specialized objects or *virions*, which are produced under the genetic control of the virus and serve as vehicles for introducing the viral genome into other cells.

The first quality defines intracellular parasitism. This property is shared with viruses by other classes of parasites, including some bacteria, fungi, and protozoa. The definition of viruses, however, stresses the intimate nature of viral parasitism, which may be called *parasitism at the genetic level.* Parasites such as the malarial plasmodia, the leprosy bacillus, the rickettsiae, and the chlamidiae have cellular organization—a set of chromosomal genes, a ribosomal apparatus, and a mitochondrial apparatus or its equivalent, with more or less complete systems for energy release and utilization. Whether their cellular organization is that of eucaryotic cells or of procaryotic cells, these parasites are cellular organisms. Their obligatory intracellular parasitism presumably reflects requirements for exogenous supply of special nutrients or of metabolic intermediates rather than a need for a host-cell machinery through which the genome of the parasite can express itself.

The fact that virus reproduction is strictly dependent on host cells separates viruses from free-living, extracellular submicroscopic parasites, such as the mycoplasmas, which possess cellular organization and are probably stripped-down bacteria without rigid cell wall. The mutational loss of cell-wall production in common bacteria yields the so-called L forms (Smith, 1964) and provides a model of how such regressive evolution may have occurred.

The distinction between the intracellular and the extracellular forms of a virus implied in the above definition is essential. Within a host cell, disappearance of the virion as a structural entity and release of the viral genome are essential steps of the process that leads to the integration of the viral genome into the metabolic machinery of the cell and finally to replication of the viral genome. The preeminent role of viral nucleic acid in the process of virus infection has been confirmed by the fact that isolated nucleic acid molecules of certain viruses can initiate infection, as first shown for tobacco mosaic virus (Gierer and Schramm, 1956).

In this light the production of mature, complete virions has come to be considered as the culmination of a morphogenetic process directed by the viral genome. Certain genes of the virus direct the production of proteins that generate a shell or *capsid,* into which a viral nucleic acid element becomes enclosed (or "encapsidated"). Other viral proteins may complicate the form of the capsid by adding spikes or tails, as described in Chapter 3. The capsid may finally become surrounded by an *envelope* related to cell membranes but containing some virus-specific proteins. Viruses which by mutation lose the ability to produce a complete virion may persist in nature as *defective* viruses either by transmission from host to host in the form of infectious nucleic acid or by vertical spread, that is, by propagation from a virus-carrying cell to its daughter cells and even from generation to generation through the germ cells.

Even though the small size of all virions is not specifically mentioned in the definition of viruses (there is no theoretical reason why a virus could not have virions of the same order of size as the cells it parasitizes) the small size of the particles of all known viruses has played an important historical role in determining the technology of virology. Filtration has traditionally been used to separate the infectious virus particles from contaminating bacteria and to prove the nonbacterial etiology of viral diseases. Hence the traditional designation of *filtrable virus,* which replaced the earlier term of *ultravirus* and later gave way to the present usage of the word *virus.*

VIRUSES AS ORGANISMS

Are viruses living? Are they organisms? When it was found that the virions of a virus, purified and concentrated from extracts of infected cells, possess uniform size, shape, and chemical composition and can even crystallize (Stanley, 1935), the question arose of reconciling the "molecular" nature of these particles with their ability to reproduce. As usual in cases of this kind, the difficulties were semantic rather than substantial. Words such as "organisms" and "living" have unambiguous meanings only when applied to the objects for which they were originally coined: a frog is an organism, a dog that runs and barks is living. But what is it that makes a frog or a dog an organism?

According to Lwoff (1957) an organism is an "independent unit of integrated and interdependent structures and functions." A frog is such a unit; the individual cells of a frog are also units of integrated and interdependent structures and functions, but are not independent in the usual sense of the word. (Even the frog itself is not reproductively independent because of sexual reproduction.) In unicellular organisms the cell becomes the independent unit, hence the organism. Cellular organelles—mitochondria, chromosomes, chloroplasts—are not organisms because they lack independence.

According to this definition, a virus would not be an organism because it is not independent: it depends on a living cell for the expression and replication of its genetic material. If the criterion of independence is accepted, then viruses are not more independent than the individual elements of genetic material of cells, which also depend on cellular integrity for their expression. A gene or a chromosome is not an organism, runs the argument, hence a virus is not an organism.

We may consider a different definition of organism, one that stresses individuality, historical continuity, and evolutionary rather than functional independence. A cellular gene or gene complex has individuality and historical continuity, but its evolution is confined within the limits of the cell lineage; hence, it is not an organism. In multicellular species, whether animals or plants, individual cell lines do not evolve independently. Hence their cells are not organisms. Change, to be evolutionarily significant, must be tested through the generation of new individuals. According to this reasoning, *an organism is the unit element of a continuous lineage with an individual evolutionary history*.

A virus attains a relatively independent evolutionary history by its peculiar adaptation to transfer from host to host. It can survive the death of the cell and organism that it parasitizes; in fact, it often exploits it. It can test its range of variants in a series of situations that are not confined within any one lineage of host organisms. A virus can move between hosts of different species, genera, or even phyla. It may even move from a plant to an insect and multiply in the cells of both. It can, if suitably adapted, explore evolutionary niches of the most unrelated kinds. Thus a virus has definitely more independence than any cellular organelle; it is more an organism, evolutionarily speaking, than a chromosome or even a cell of a metazoon, even though functionally it is much less independent than any such cell. In Chapter 19 the question of the evolutionary independence of viruses will be examined more closely in the light of current knowledge.

Similar considerations may help us out of the quandary about the living or nonliving nature of viruses. Lwoff (1957) defined life as "a property, a manifestation, or a state of cells and organisms," considered as independent units of structure and function. Hence he regarded viruses as nonliving because he considered them as nonorganisms. In an earlier edition of this book (Luria, 1953), it was stated that "A material is living if, after isolation, it retains a specific configuration that can be reintegrated into the cycle of genetic matter." This made life identical with the possession of an independent, specific, self-replicating pattern of organization. Accordingly, a protein is not living since its specific structure, its amino acid sequence, cannot serve as template to be copied by any cell. The specific base sequence of the nucleic acid of a gene, however, can be copied; it is part of the informational library of a living organism.

Is, then, genetic nucleic acid alive? The foregoing definition makes this attribute subject to the test of extraction and reintegration into a different lineage of cell and organisms. Thus a virus is considered alive and so is any other bit of genetic material that can withstand extraction, enter a living cell, be copied within it, and become at least temporarily a part of its heredity. In bacteria of many different genera fragments of purified DNA can undergo reintegration if they are introduced into suitable living bacteria. This is the so-called "transformation" phenomenon (Avery et al., 1944). Hence any fragment of DNA from these bacteria might be considered as living.

Yet there is an important difference between the transferability of bacterial DNA and that of a viral genome. The transfer of DNA fragments among bacteria, even though it does occasionally take place in nature, is but an accidental event without major evolutionary significance, whereas the transfer of a viral genome is an essential part of viral existence, the outcome of a selective specialization.

The nucleic acid of some viruses, extracted either from the virions or from infected cells, can enter other cells, replicate within them, and produce virions. The efficiency of such successful infection by naked viral nucleic acids is usually lower by orders of magnitude than the efficiency of infection by virions. Specialization for transfer, therefore, may be a reasonable basis for considering viruses "more living" than other fragments of genetic material, in the same way as it is a reasonable basis for considering them "more organisms" than any cellular organelles, including chromosomes and genes. The problem of the so-called *viroids*, minute elements supposedly responsible for certain diseases of plants, will discussed in Chapter 17.

Origins of Virological Knowledge

Illnesses now known to be caused by viruses have been recognized for thousands of years. A Chinese description of a pestilence dating from the tenth century B.C. sounds very much like smallpox. Yellow fever, known for centuries in tropical Africa and as a scourge of ships in the African trade, was probably responsible for the legends of cursed ships, the Flying Dutchman and the Ancient Mariner (Burnet, 1945). Plant virus diseases such as potato leaf roll and the ornamental variegation known as tulip break have been known for centuries (Bawden, 1964).

The mosaic disease of tobacco has played a distinct role. Its transmissibility by mechanical inoculation with sap of infected plants was demonstrated by Mayer in 1886. In 1892 Iwanowsky reported the transmission of tobacco mosaic by sap filtered through bacteria-proof filters. His report went unnoticed, but in 1899 Beijerinck succeeded in proving the serial transmission of tobacco mosaic by bacteria-free filtrates in which no microscopic organism could be detected. Impressed by this unexpected finding, Beijerinck described the agent of tobacco mosaic as a *contagium vivum fluidum*, meaning by this an infective agent which reproduced, and therefore had life, but which was in a state of dispersion different from that of organisms.

Beijerinck's proof of the filterability and serial transmissibility of the agent of tobacco mosaic was the starting point of virus research, which we now call virology. Similar findings were soon made with several animal viruses and the existence of discrete particles or *elementary bodies* as the infective agents was demonstrated microscopically.

After 1930 the perfecting of electron microscopy revealed the structural details of virions. The rod-shaped virions of tobacco mosaic were successfully crystallized by Stanley (1935). The ability of purified viral nucleic acid to initiate infection, proved first with the RNA of tobacco mosaic virus by Gierer and Schramm, was confirmed with the RNA or DNA of many other viruses.

In 1908 Ellerman and Bang found that a leukosis of chicken was transmissible by cell-free filtrates, and in 1911 Rous discovered a virus that produced malignant tumors in chickens. The generalization of this discovery to other fowl cancers opened the way to the recognition of viruses as a major group of agents of neoplastic transformations. The discovery of bacteriophages or bacterial viruses by Twort and d'Herelle provided a powerful stimulus to the integration of virology into a unified science.

Major Groups of Viruses

It is customary to subdivide viruses, according to the nature of their hosts, into plant viruses, animal viruses, and bacterial viruses or bacteriophages. Even such broad subdivisions create ambiguities, as with plant viruses that can multiply in their insect vectors. Each virus has a range of hosts, more or less related organisms in which it can reproduce. Since viruses are discovered as pathogens it is logical to classify them in terms of their "major host," the host, that is, whose response to the viruses first came to the attention of man.

BACTERIAL VIRUSES

There is hardly a group of bacteria for which bacteriophages have not yet been found. Bacteria for which no phage has yet been reported are those about which our knowledge is still quite limited. Bacteriophages of mycoplasmas have been described (Gourlay et al., 1973).

The host range of phages does not cut across well-established taxonomic boundaries between bacterial groups. Thus phages active on micrococci will not multiply in streptococci, and phages of enteric bacteria do not usually multiply in pseudomonads. Phage specificity may be broader or narrower than the rather flimsy classification boundaries that separate bacterial genera and species. For example, a phage may multiply only on a certain strain of *Escherichia coli,* whereas another phage reproduces in many strains of *E. coli* and the closely allied genus *Shigella*. Since bacteria can acquire phage resistance by discrete mutational steps a strain sensitive to several phages may produce a series of stable mutants resistant to one or more of the phages.

Phages that attack blue-green algae have been discovered (Safferman and Morris, 1963). This is an additional confirmation of the widely recognized biochemical and taxonomic relationship between the *Cyanophyceae* and the bacteria (see Stanier, 1964).

ANIMAL VIRUSES

Table 1.1 presents a partial list of important animal viruses. The only invertebrates in which virus diseases have been reported are the insects, which are, of course, the economically most important and therefore the most thoroughly studied class of invertebrates.

Virus diseases are known in fish (carp pox, infectious tumors) and in amphibia (kidney tumor of the leopard frog; Lucké, 1938). In birds one finds many virus diseases, some of them economically very important, for example, Newcastle disease and laryngotracheitis. Neoplastic viral diseases of fowl, including sarcomas and leukoses, are a favorite material in the study of the relation of viruses to tumors.

Virus diseases have been recognized in most domestic mammals as well as in many wild ones. Virus diseases of humans include such major epidemiological problems as smallpox, yellow fever, poliomyelitis, measles, mumps, rabies, and various types of encephalitis. Table 1-1 includes a representative list of important mammalian viruses and the diseases they produce in their major hosts.

PLANT VIRUSES

A list of important plant viruses is presented in Table 1-2. There are relatively few known viruses in gymnosperms, ferns, fungi or algae. The flowering plants on the other hand, are hosts to many types of viruses. In fact, viruses rank next to fungi in causing plant diseases of economic importance, including diseases of potatoes, beans, beets, tobacco, sugarcane, cocoa, and fruit crops.

NOMENCLATURE AND CLASSIFICATION OF VIRUSES

Virus nomenclature and classification are a troublesome area of virology. The reasons should be evident from the above discussion of the nature of viruses. The purpose of classification is to group into meaningful categories those organisms that are more closely related, using morphological and physiological criteria. The ideal goal is to establish groups that reveal evolutionary and phylogenetic relationships in addition to providing a convenient system of nomenclature. At present, however, there is no reason to believe that viruses form a single group of organisms having a common ancestry and a common evolutionary history. A relation of viruses to cellular elements is as plausible as a relation of viruses to each other, and a case might be made for an origin of new viruses from cellular elements. Any classification of viruses is bound to be mainly a "determinative key" of practical value for use by workers in the field.

Traditionally, the nomenclature used for animal or plant viruses consisted of giving the name of the disease produced in the major host followed by the word "virus." Bacteriophages were named by code symbols (letters and numbers, such as T1, C16, S13, ϕX174) derived from laboratory practice. This haphazard method of nomenclature served fairly well the day-by-day needs of the investigator. But as more information about viruses became available the desirability of a nomenclature that grouped viruses of similar properties became recognized. As soon as one attempts to classify viruses, however, one faces the problem of the choice of criteria: the nature of the major hosts, or the type of disease produced, or the properties of the virions, or the features of

TABLE 1.1 THE ANIMAL VIRUSES: CHARACTERISTICS AND MEMBERS OF THE MAJOR GROUPS

Family	General characteristics	Typical agents	Specific characteristics
RNA VIRUSES			
Picornavirus	Icosahedral, naked, 27 nm capsids with single-stranded infectious RNA of 2.7×10^6 daltons.	Human: Poliovirus Coxsackie Rhinovirus Foot and mouth disease virus of cattle	3 serotypes can cause paralysis Cause variety of symptoms especially myositis Over 100 serotypes, acid labile, "common cold" viruses Economically very important
Togavirus	Enveloped, icosahedral 50–70 nm capsids with single-stranded infectious RNA of 4×10^6 daltons; formerly known as "arboviruses"; infect animals, birds, humans, insects.	Type A (Alphavirus) such as Eastern, Western encephalitis Type B (Flavivirus) such as dengue, yellow fever, St. Louis encephalitis	Mosquito-borne, about 20 viruses which show serological cross-reactivity; cause encephalitis in humans, other mammals, can be fatal Some mosquito, some tick-borne several dozen cross-reacting species; cause encephalitis as well as other serious systemic illnesses
Paramyxovirus	Enveloped, overall particle shape pleomorphic ~150 nm in diameter, helical nucleocapsids with diameter 18 nm and length of 1000 nm contain single-stranded "negative strand" RNA of 7×10^6 daltons plus virion polymerase.	Mumps, Newcastle disease virus of chickens Measles, respiratory syncytial virus of humans, distemper of dogs, and others	Contain neuraminidase and hemagglutinin in single protein; infectious for many tissues Lack neuraminidase, have hemagglutinin; infectious for many tissues
Orthomyxovirus	Enveloped, pleomorphic 80–120 nm particles, helicai nucleocapsid 6–9 nm diameter, varies in length;	Influenza A: humans, swine, birds, cattle Influenza B: humans Influenza C: humans	Contain neuraminidase and hemagglutinin in separate proteins; A strain most important for

Group	Properties	Examples	Disease/Host
	contains segmented, negative-strand RNA in total of ~5-6 × 10^6 daltons, possesses virion polymerase.		human disease; undergoes constant antigenic variation
Rhabdovirus	Enveloped bullet-shaped particles, 70 × 175 nm, contain helical nucleocapsid with single-stranded negative strand RNA of 4 × 10^6 daltons and virion mRNA polymerases. Infect mammals, insects, plants (see lettuce necrotic yellow virus e.g.).	Vertebrate: Vesicular stomatitis virus Rabies Invertebrate: Sigma virus of *Drosophilia*	Infects insects and mammals (cattle) Humans, bats, dogs and other mammals subject to neurological destruction Causes sensitivity to CO_2
Reovirus	Icosahedral, naked, double-shelled virions, outer 75-80 nm, inner 45 nm, contain 10 or more double-stranded RNA molecules ranging from 0.4-2.8 × 10^6 daltons and virion transcriptase. Plant viruses also in this group.	Reovirus of humans, many other mammals, birds, cause mild illness of respiratory, GI tract Cytoplasmic polyhedrosis virus of insects.	Mammalian and avian groups serologically distinguishable but common antigen among mammalian group
Retrovirus	Enveloped, roundish particles about 100 nm diameter with helical nucleocapsid containing 6 × 10^6 daltons of RNA consisting of 2 identical molecules 3 × 10^6 each, hydrogen bonded into "70S" RNA plus small RNA from host. Contain reverse transcriptase and multiply by integration into DNA of host. Cause leukemia, sarcoma, various other malignances.	Leukosis group in birds (Rous sarcoma virus) Mouse viruses and other mammalian viruses Visna virus Foamy agents	Numerous antigenic subgroups; not all tumorigenic No definite human representative but cause leukemias in cats, mice, and cattle Causes slow infection in sheep resulting in demyelination in CNS Dogs, cats, others; no known pathology; do not transform cells

TABLE 1.1 (*Continued*)

Family	General Characteristics	Typical Agents	Specific Characteristics
RNA-VIRUSES			
Bunyavirus	Enveloped RNA containing viruses 90–100 nm diameter with segmented negative strand genomes $3–4 \times 10^6$ in helical nucleocapsids; arthropod borne.	Bunyawera viruses	Hundreds of members serologically distinguishable; can cause encephalitis; only recently recognized as different from togaviruses
Coronavirus	Enveloped, 80–120 nm particles with helical nucleocapsid, distinctive morphology with bulbous peripheral spikes	Human strains	Common colds and possible GI disease
Arenavirus	Enveloped, pleomorphic 100–300 nm with helical nucleocapsid	Lassa virus	Rare serious generalized infection in humans spread from rodents
		Lymphocytic choriomeningitis viruses	Causes chronic infections in rodents
DNA-VIRUSES			
Parvoviruses	Icosahedral, naked ~20 nm capsid with single-stranded DNA of $1.2–1.8 \times 10^6$ daltons; virions contain either of 2 duplex strands; both defective and infectious types exist	Minute virus of mice (nondefective) Adeno-associated virus (AAV, defective)	All defectives depend on coinfection with adenovirus for growth
Papovavirus	Icosahedral, naked capsid 45–55 nm with closed circular double-stranded DNA from $3–5 \times 10^6$ daltons; cause cell transformation and tumors in animals	Polyoma virus of mice	Wide distribution in nature, natural tumors unknown
		Simian virus 40	Wide distribution in nature, natural tumors unknown
		Shope papilloma (Rabbits; many other papilloma viruses known) Human wart virus	Occur frequently in nature in association with papillomas of skin
Adenovirus	Icosahedral, naked 80 nm capsid with linear DNA $20–30 \times 10^6$	Human adenovirus	31 serotypes known, cause upper respiratory disease; 12,18,31

Group	Properties	Virus	Effects/Disease
	daltons; highly developed virions with spikes; can transform cells; all types share common antigen; types can be distinguished by several antigens	Mammalian adenovirus subgroups	most oncogenic in laboratory. Infect cattle, dogs, mice, monkeys
		Celovirus	Formerly "chicken orphan virus"
Herpesvirus	Enveloped, icosahedral 100 nm nucleocapsids with double-stranded linear DNA of 10^8 daltons; grow in nucleus, bud through nuclear membrane; cause "latent" infections	Human herpes simplex, types 1,2	Type 1 causes "fever blisters"; Type 2 causes genital herpes (?carcinoma, cervix uteri)
		Epstein-Barr	Causes infectious mononucleosis. associated with Burkitt's lymphoma
		Zoster (varicella)	Causes "shingles" (chronic infection of neural ganglia) in adults and "chicken pox" in children
		Pseudorabies	Causative agent of "mad itch" in swine and cattle
		Cytomegalovirus	Infect humans and many other species, each host specific; associated with fetal damage
		Lucké	Causes frog adenocarcinoma
		Marek's disease	Causes tumors in birds
Poxvirus	Enveloped complex structure 160 × 200 nm, characteristic "brick-shape" with helical nucleocapsid and associated "lateral bodies"; DNA is double-stranded 160 × 10^6 daltons; virions contain many enzymes and at least 30 proteins including virion RNA polymerase; host range includes insects to mammals	Variola (human)	Causes smallpox
		Vaccinia (human)	Provides immunity to smallpox
		Mammalian pox viruses for many species	
		Myxoma-Fibroma	Causes systemic disease and tumors in rabbits

TABLE 1.2 SOME MAJOR GROUPS OF PLANT VIRUSES*

Type virus species of group	Other viruses in group	Virion measurements ~nm	How type species transmitted in nature	Remarks on type species
ssRNA, elongated virions				
Tobacco mosaic	Cucumber green mottle mosaic, odontoglossum ringspot, ribgrass mosaic, sunn-hemp mosaic	300 × 18	Mechanical wounds	First virus discovered; object of many classical studies
Potato X	Cassava common mosaic, clover yellow mosaic, cymbidium mosaic, narcissus mosaic	515 × 13	Mechanical wounds	Latent in many potato varieties
Potato Y	Lettuce mosaic, sugarcane mosaic, tobacco etch, tulip break	730 × 11	Aphids, nonpersistently	At least 30 viruses in group
Beet yellows	Carnation necrotic fleck, citrus tristeza	1250 × 10	Aphids, semipersistently	
Potato S	Chrysanthemum B, pea streak, poplar mosaic, red clover vein mosaic	650 × 12	Aphids (?)	Latent in many potato varieties
Tobacco rattle	Pea early browning	190 × 250 45–115 × 25	*Trichondorus* nematodes	Two particles†
ssRNA, isometric virions				
Brome mosaic	Broad bean mottle, cowpea chlorotic mottle	25		Three particles‡
Cowpea mottle	Bean pod mottle, radish mosaic, red clover mottle, squash mosaic	28	Beetles	Two particles†
Cucumber mosaic	Peanut stunt, tomato aspermy	30	Aphids, nonpersistently	Three particles†
Prunus necrotic ringspot	Apple mosaic, elm mottle, tobacco streak	23	Pollen	Three particles†

12

Barley yellow	Beet western yellow ? potato leaf roll	22	Aphids, circulative	Transcapsidation occurs between strains in plants but not in vector
Tobacco ringspot	Cherry leaf roll, grapevine fanleaf, peach rosette mosaic raspberry ringspot	28	Xiphinema nematodes	Two particles.† Chronic infectious stage may be virtually symptomless.
Tomato bushy stunt	Turnip crinkle	30		Very stable virions
Turnip yellow mosaic	Cacao yellow mosaic, eggplant mosaic, okra mosaic, wild cucumber mosaic	28	Flea beetles	Purified preparations contain particles both with and without RNA
ssRNA, bacilliform virions				
Potato yellow dwarf	Lettuce necrotic yellows, maize mosaic, rice transitory yellowing, wheat American striate mosaic	380 × 75	Leafhopper‡	Multiplies in vector. Virion contains ssRNA, proteins, and lipid. One virion infects.
dsRNA, isometric virions				
Wound tumor	Maize rough dwarf, maize wallaby ear disease, rice dwarf, sugarcane Fiji disease	70	Leafhopper	Multiplies in vector, dsRNA in 12 segments, one virion infects.
dsDNA				
Cauliflower mosaic	Dahlia mosaic	50	Aphids, nonpersistently	Prominent inclusions in infected cells

Other viruses that are probably type species for additional groups include alfalfa mosaic (three particles†), cacao swollen shoot (transmitted by mealy bugs), tobacco necrosis (transmitted by the fungus *Olpidum*), satellite (replicates only in the presence of tobacco necrosis), tomato spotted wilt (contains lipid, transmitted by thrips).

* Based mostly on information from *Descriptions of plant viruses* by Commonwealth Mycological Institute and Association of Applied Biologists (A. J. Gibbs, B. D. Harrison, and A. F. Murant, Eds.) Sets 1–9; 1970 to 1975; A. Gibbs, in *Adv. Virus Research* 14:263 (1969); E. M. J. Jaspars, in *Adv. Virus Research* 19:37 (1974).

† Different specific parts of genome are encapsidated in separate particles all needed to produce one infection.

‡ There are at least 16 members of the group. About half of them multiply in and are transmitted by aphids instead of leafhoppers.

the reproductive cycle. A good system should employ several or all of these criteria, with either equal or hierarchical prominence.

Plant pathologists used to favor systems of viral classification analogous to the Linnaen system of Latin binomials. In 1948 Holmes proposed a binomial system in which the viruses formed the order *Virales*, with three suborders, *Phagineae* (bacteriophages), *Phytophagineae* (plant viruses), and *Zoophagineae* (animal viruses), further subdivided into families, genera, and species. This system has found little favor outside plant pathology, and even there it has come into disuse.

Progress of electron microscopy in the last 30 years has provided ample data about the size and morphology of the particles of most known viruses. Purification of virions has identified the type of nucleic acid they contain and has made it possible to compare the nucleotide sequences in the nucleic acids of different viruses. Serological tests have characterized specific virion proteins, whose cross reactions are indices of chemical and genetic relatedness. Hence it became clear that morphological and serological properties of virions and the properties of their nucleic acid are convenient and meaningful criteria of classification.

Such a system, proposed by a group of virologists (Lwoff et al., 1962) is presented with modifications in Table 1-3. Both the properties of the nucleic acids and the geometry of viral capsids (Caspar and Klug, 1962; see Chapter 3) are used as the criteria for grouping viruses into classes.

The International Committee on Nomenclature of Viruses has proposed a dual system of nomenclature consisting, on the one hand, of generic names ending in *-virus* for individual virus groups and, on the other hand, of eight-digit cryptograms purported to describe each virus according to a conventional key (Wildy, 1971). Despite its prestigious sponsorship, this system of classification appears not to have found favor with virologists.

General Methodology of Virus Research

The presence of a virus in a host organism is recognized by the occurrence of some abnormal manifestation, which may be either a natural disease or an experimental infection. Whenever the presence of a virus is suspected one must find a set of conditions—suitable host, suitable route of infection—such that the virus will produce recognizable changes in an organism. Thus much of the practice of virology consists of setting up experimental infections.

To prove that a disease is caused by a certain infectious agent one must fulfill Koch's postulates: (1) demonstration that the agent is regularly found in diseased hosts; (2) cultivation on suitable media; (3) reproduction of the disease in experimental organisms by means of cultures of the agent; (4) reisolation of the agent from artificially infected hosts. The same postulates apply, *mutatis mutandis*, to the diagnosis of viral diseases. The postulates become (Rivers, 1937): (1) isolation of virus from diseased hosts; (2) cultivation in experimental hosts or host cells; (3) proof of filterability (to exclude larger pathogens); (4) production of a comparable disease in the original host species or in related ones; and (5) reisolation of the virus.

Cultivation and identification of viruses constitute the main body of the

TABLE 1.3 CRITERIA FOR CLASSIFICATION OF VIRUSES*

Nucleic acid	Capsid symmetry	Presence of envelope	Diameter of helical capsid or number of morphological units†	Typical members	Comments or groupings
RNA	Helical	−	100–130 Å 170–200 Å 250 Å	Potato X TMV Barley stripe mosaic	
	Helical	+	100 Å 170 Å	Influenza; fowl plague NDV; measles	Myxovirus group
	Cubic	−	32 units	RNA phages; turnip yellow mosaic	
			60 units 92 units	Polio; bushy stunt Reovirus; wound tumor	2-stranded RNA
DNA	Helical	+	90–100 Å?	Vaccinia	Pox group
	Cubic	−	12 units 42 units 252 units 812 units	φX174 Polyoma, papilloma Adeno Tipula iridescent	
		+	162 units	Herpes, pseudorabies	Herpes group
	Cubic, tailed	−		T-even and other tailed phages	

* Modified from Lwoff et al., *Cold Spring Harbor Symp.* **27**, 51 (1962).

† The morphological units or capsomeres are those visible in electron micrographs (see Chapter 3 for their relation to structural protein subunits).

art of virology as practiced by the diagnostician. Material suspected of containing a virus—a bacterial lysate, a tissue extract, or a sample of a body fluid—is converted, if necessary, into a liquid form by grinding or homogenizing under controlled conditions. Centrifugation and filtration are employed to remove large cellular fragments as well as contaminating microorganisms. The suspension is then inoculated into suitable hosts or host-cell suspensions or onto cell monolayers. On susceptible cell layers, such as bacteria growing on an agar plate or animal cells growing on glass surfaces, one may observe local lesions or *plaques,* whose appearance may be characteristic for a given virus. The plaques are due to local cycles of virus infection and multiplication accompanied by complete or partial cell destruction. Even viruses whose multiplication produces no recognizable discrete plaques may be detected and characterized by the appearance of damage to the cell layer or by other tests.

When the material to be tested is not placed on cell layers but is injected into a host organism the experimenter watches for *general reactions,* that is, for indications of successful infection: death, disease, or presence of specific reactions such as antibody formation.

If no symptoms result either on cell layers or in the inoculated organism virologists use "blind passages," that is, repeated transfer of material which often enhances the virulence or the titer of a virus.

BACTERIOPHAGE PLAQUES

The main symptom of the action of phage on susceptible bacteria is *lysis,* that is, a more-or-less complete dissolution of bacterial cells with release

FIGURE 1-1. *Plaques produced by six different phages on a common host,* E. coli *strain B. Reduced to about one-third size. From Demerec and Fano,* Genetics **30,** *119 (1945).*

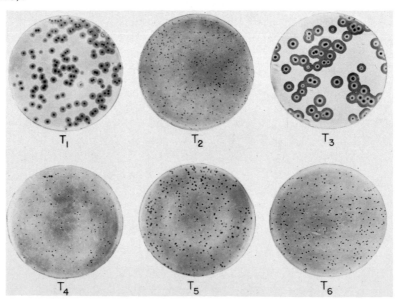

of phage virions. Incomplete lysis may be due to (1) termination of bacterial growth by nutrient limitation since phage can grow only in actively metabolizing bacteria; (2) presence of cells that are genetically resistant to the phage; or (3) establishment of lysogeny, a carrier state in which the cells do not lyse, retain the phage genome, and become immune to lysis if reinfected with that phage.

If proper amounts of a phage are placed on the surface of a nutrient agar plate with enough bacteria to produce a continious layer of growth, one observes localized phage colonies or plaques (Figure 1-1). Phage and bacteria are either spread directly on the surface of the agar plate or, preferably, are mixed together in a soft agar medium and poured as a top-agar layer on the agar plate (Gratia, 1936).

A phage plaque may contain 10^7 to 10^9 virions, descendents of a single one, and can serve as a source of a pure phage line. Plaque size depends on the dimension of the phage virions because smaller particles diffuse faster through the agar, as well as on the rates of phage replication, release, and adsorption. Plaque morphology is generally characteristic for a given phage. Rings within the plaque may be present because of waves of changing bacterial growth rate. Halos around the plaque proper are due to production of enzymes that attack bacterial capsules (Adams and Park, 1956).

ANIMAL VIRUS LESIONS

The skin of animals is impermeable to viruses, so that injection or instillation, e.g. onto the nasal or conjunctival membrane is required to bring virus to susceptible cells in the animal organism. Symptoms observed at the level of the organism, the tissue, or the cells, and especially the immunological responses, serve to identify the viral agent.

Before cell monolayers became available a useful test system was inoculation onto the chorioallantoic membrane of the chick embryo (Beveridge and Burnet, 1946) (Figure 1-2). On the moist uniform cell layer the virus that multiplies at one site spreads only by contiguity forming lesions as *pocks* that are fairly characteristic for specific viruses. Cultured cell monolayers are much

FIGURE 1-2. Chorioallantoic membrane lesions produced by myxoma virus (left) and Murray Valley encephalitis virus (right). Courtesy Dr. F. M. Burnet.

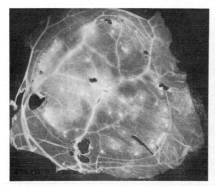 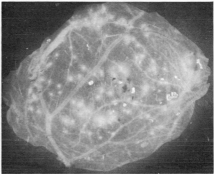

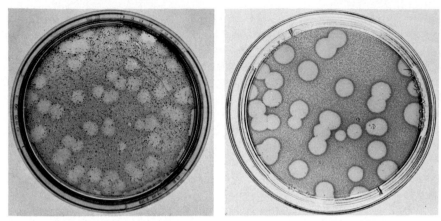

FIGURE 1-3. Plaques produced by Western equine encephalomyelitis virus on chick embryo fibroblasts (left, courtesy Dr. R. Dulbecco) and by poliovirus on HeLa cells (right).

preferable, produced either from suspensions of cells taken directly from animal tissues (primary cultures) or from established cell lines (permanent cultures). As the cells are covered with a thin agar layer the virus can spread only to contiguous cells so that discrete plaques will result (Dulbecco and Vogt, 1953; see Figure 1-3). Observation of plaques is facilitated by treating the cell cultures with vital dyes, such as neutral red, that differentiate live cells from dead ones. The character of the local response depends on virus and cell properties. On cell monolayers, for example, tumor viruses produce foci of cell proliferation rather than plaques (Manaker and Groupé, 1956). The cells used may be either genetically or developmentally specific in their virus susceptibility. The genetic basis of virus sensitivity is best understood for animals such as mice whose inbred strains make it possible to do precise genetic analysis. For example, resistance to yellow fever virus has been attributed to a single dominant gene (Sabin, 1952). The genetic basis of response to tumor viruses will be discussed in Chapter 16.

PLANT VIRUS LESIONS

Direct mechanical inoculation of virus suspensions by rubbing on leaves can lead to formation of local lesions and to generalized infection. Rubbing must break cell walls to bring the virus in contact with the protoplasts; an abrasive such as carborundum may be used. The most common type of local response is the *necrotic lesion,* a spot of characteristic appearance often with rings or halos (see Figure 1-4). The lesions are limited because the infected cells die so rapidly that necrosis remains localized near the point of entry because the virus can not spread and the plant survives.

The necrotic response of a plant to a virus is genetically controlled. In pepper, for example, the response to tobacco mosaic virus (= TMV) is controlled by a gene with three alleles: *L* (dominant) causes a local response; *l'*

(a) (b)

(c) (d)

FIGURE 1-4. *Local necrotic lesions produced by plant viruses.* (a)*Tobacco mosaic virus on* Nicotiana glutinosa; (b) *tomato mosaic virus on* N. glutinosa; (c) *tomato bushy stunt on* N. glutinosa; (d) *tobacco ringspot on* N. tabacum. *Courtesy Dr. K. M. Smith.*

if homozygous causes loss of the leaf before general infection can occur; *l* causes a generalized infection (Holmes, 1937).

Local foci of virus infection can often be revealed, even when no necrosis occurs, by staining with iodine after removal of chlorophyll (Holmes, 1931). When local lesions are not seen a virus infection may manifest itself by a variety of symptoms: chlorosis, mosaic, streak, yellows, ringspot, flower break, and leaf roll—terms used to describe the peculiar abnormalities observed in diseased plants.

FIGURE 1-5. Tumors on roots of sweet clover infected with wound tumor virus. Courtesy Dr. L. M. Black.

Some plant viruses cause specific proliferative diseases. Wound tumor virus produces multiple tumors that arise at sites where the plants have been wounded (Black, 1945; see Figure 1-5).

Many viruses cannot be transmitted by mechanical inoculation of infective materials onto leaves. Some can be transmitted only by arthropod vectors, which become contaminated with the virus while feeding on infected plants and can subsequently inoculate it into susceptible plant tissues (see Chapter 17).

The most complex problems are those of plant diseases that can be transmitted only by grafting a diseased shoot onto a healthy stock. Even if virionlike objects are found in the diseased plants it may be hard to prove that they are the agents of disease.

Titration
2 of Viruses

The Infectious Unit

Virus activity is measured by determining the amount of virus-containing material necessary to obtain a specific response in a host. The response may be an all-or-none one—presence or absence of infection—or a quantitative one, for example, the time required for infection to appear or the number of lesions in a layer of susceptible cells. The quantitative determination of viral activity is called *titration*.

Titration of viruses resembles a viable count of bacteria rather than a chemical titration. In chemical titration what is determined is the amount of a given reagent with which an unknown amount of the chemical can react. In virus titration, the reproduction of virion results in an *amplification* of the effects of very small amounts of virus, so that the manifestations produced by small or large amounts of virus may be similar. In other words, one is dealing with "all-or-none" effects, each resulting from the action of a single element—one virion or one molecule of viral nucleic acid.

The smallest amount of virus capable of producing a reaction is called an *infectious unit* and the titer of a virus suspension is usually stated in terms of the number of infectious units per unit volume. For example, if the smallest amount of a suspension of influenza virus that causes pneumonia in mice upon nasal instillation is 0.1 ml of a $1/10^6$ dilution of an infected lung emulsion, the titer of the emulsion may be given as 10^7 infectious units per ml: $1/(10^{-1} \times 10^{-6}) = 10^7$. In the following section it will be seen that this definition of the infectious unit needs refinement to acquire a precise meaning.

Titration procedures require methods by which an inoculation can be scored as positive or negative. For animal viruses, for example, this may be the presence or absence of visible damage in a cell culture, inflammatory reaction at the site of skin or corneal inoculation, or paralytic symptoms after intracere-bral inoculation. Sometimes one can detect successful virus reproduction even in the absence of visible host reactions, for example, by the appearance of hemagglutinating capacity in the allantoic fluid of eggs inoculated with myxoviruses.

The titration methods of choice are those based on counts of discrete lesions produced on a cell layer that has been inoculated with a measured amount of virus. When the numbers of susceptible cells available to virus can be considered as practically infinite, as in a bacterial layer on a nutrient agar dish or in a confluent monolayer of animal cells, the viral lesions or plaques are colonies of virus comparable to colonies of bacteria on a solid medium. Like bacterial colony counts, the plaque counts are directly proportional to the amounts of virus added to the monolayers.

Theory

There is an important difference between bacterial colony counts and viral plaque counts. For bacteria it is easy to show by microscopic observation that one cell can produce a colony. In fact for some bacteria practically every cell in a culture can form a colony. No such direct verification is easily feasible for viruses. In most titrations an infectious or plaque-forming unit (= pfu) corresponds to many virions. The question arises: does initiation of infection require a cooperation between several virions, or is one virion capable of initiating infection, but only with a probability less than unity (for example, because it becomes inactivated or fails to find a susceptible cell)?

The choice is easily made for plaque counts, which are generally linear functions of the amount of virus tested (Figure 2-1). This linearity can be

FIGURE 2-1. *The proportionality between plaque counts and phage concentrations for two bacteriophage preparations. From Ellis and Delbrück,* J. Gen. Physiol. **22,** *365 (1939).*

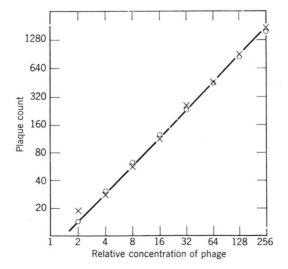

accounted for only by assuming that each plaque stems from the action of a single material element, that is, a virion or any infectious viral component. If a number n of elements greater than one were needed to initiate a plaque, the number of plaques should increase with amount of virus, not linearly but with the nth power of the inoculum.

To look at this matter more closely, assume that in a virus suspension the virions are present as isolated material particles, like bacteria in a liquid culture. The distribution of virions into individual samples will follow the statistical equation of the Poisson distribution,

$$p_r = \frac{s^r e^{-s}}{r!} \qquad [2\text{-}1]$$

where s is the average number of particles per sample; r the actual number in a given sample; $r!$ ("factorial" of r) is the product $r \times (r - 1) \times (r - 2) \times \ldots \times 3 \times 2$; and p_r is the probability of having exactly r particles in a given sample (and also, therefore, the expected frequency of samples containing exactly r particles). For $r = 0$, $p_0 = e^{-s}$ (since $0! = 1$). This is the frequency of samples without particles. The frequency of samples with one particle is $p_1 = se^{-s}$; with two particles, $p_2 = (s^2/2)e^{-s}$, and so on. The frequency distribution of plaques on a series of plates inoculated with equal amounts of several dilutions of a virus suspension agrees very well with the distribution expected from Equation 2-1.

This means that one plaque can stem from the action of one virion. If two virions were needed to produce a plaque (for example, by infecting the same cell on the monolayer) doubling the amount of virus in the inoculum should quadruple the number of plaques. More precisely, let us derive from Equation 2-1 the probabilities that there are at least one, at least two, at least three elements per sample when the average number of elements per sample is s:

$$p_{r>0} = 1 - p_0 = 1 - e^{-s} \qquad [2\text{-}2]$$

$$p_{r>1} = 1 - (p_1 + p_0) = 1 - (s + 1)e^{-s} \qquad [2\text{-}3]$$

$$p_{r>2} = 1 - (p_2 + p_1 + p_0) = 1 - \left[\left(\frac{s^2}{2}\right) + s + 1\right]e^{-s} \qquad [2\text{-}4]$$

These equations also represent the frequency of samples that receive *at least* one, or at least two, or at least three elements in a series of random samples of equal volume. Taking Equations 2-2 and 2-3, expanding e^{-s} in series of powers, and neglecting cubic and higher terms, we obtain:

$$p_{r>0} = 1 - (1 - s) = s \qquad [2\text{-}2']$$

$$p_{r>1} = 1 - s + s^2 \ldots - 1 + s - \frac{s^2}{2} \ldots = \frac{s^2}{2} \qquad [2\text{-}3']$$

The probability that a sample contains at least one element is (obviously) proportional to the average number, whereas the probability of having at least two elements increases with the square of the average number of elements.

Similar considerations apply to certain viral assays in which the response is not the formation of countable plaques but the presence or absence of mass manifestation of infection or disease as discussed below. In such cases what is scored is, for example, the frequency of positive responses in a series of animals receiving equal volumes of one or more dilutions of a virus suspension. Again Equations 2-2, 2-3, 2-4 apply. A convenient way to use them is shown in Figure 2-2, which shows how the frequency of "positive" responses depends on the concentration of virus in the inoculum according to the assumptions of the different Equations 2-2, 2-3 and so on. Data from phage and from animal viruses are in excellent accord with the curve corresponding to Equation 2-2, that is, with curve 1 in Figure 2-2 (Feemster and Wells, 1933; Lauffer and Price, 1945).

It is important to keep in mind that tests of the type we have just discussed do not prove that every virion present does succeed in producing an infection. Even though one virion can cause an infection, some virions will not, either because of intrinsic differences or because of different events during the test. For example, some may be inactivated by inhibitors or may attach to dead host cells. The statistical analysis tells us only that infection *can* be initiated by a single particle and does not require collaboration among several.

In dealing with virus infectivity, Equations 2-1 and 2-2 should conveniently be rewritten to take into account that each virion may have less than 100% chance of being successful and becoming an infective unit:

$$p_i = \frac{(qs)^r e^{-qs}}{r!} \qquad [2\text{-}1a]$$

$$p_{i>0} = 1 - (qs + 1)e^{-qs} \qquad [2\text{-}2a]$$

in which p_i is the probability of having exactly i "successful" virions in a sample, $p_{i>0}$ is the probability of having at least one such successful virion, and q is the fraction of virions that will be successful in the test. Hence qs is the average number of successful virions per sample.

Viral titers can be compared with the numbers of virions counted by

FIGURE 2-2. *The theoretical dependence of the frequency of positive samples (samples with at least* n *particles) on the concentration of particles in an inoculum. Curves 1, 2, 3, 4 represent the expected frequencies for* n = 1 *or* 2 *or* 3 *or* 4. *Abscissa: logarithm of concentration (empirical scale). Ordinate: frequency of positive samples. Note the increasing steepness of the response-concentration curve for increasing values of the minimum number of particles required. From Lauffer and Price,* Arch. Biochem. Biophys. **8,** 449 (1945).

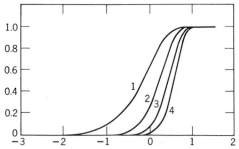

electron microscopy or estimated indirectly. For some bacteriophages the agreement is excellent, the number of virions being between 1 and 2.5 times the number of plaque-forming units produced on the most sensitive host bacteria under optimal conditions (Luria et al., 1951). This means that at least 40% and sometimes nearly 100% of the particles are infectious and actually succeed in forming plaques ($q = 0.4$ to 1.0).

The correspondence is seldom so good for plant or animal viruses. For animal viruses the ratios of virions to infective units vary over a wide range. Some preparations of vaccinia virus contain only four times as many virions as infective units; with mouse encephalitis ratios as low as 2.6 have been reported (Franklin et al., 1959). With purified poliovirus the ratios of virions to plaque-forming units have occasionally been less than 100:1 (Schwerdt and Fogh, 1957), although in most preparations the ratio is higher.

Infectious nucleic acids, whether from phage, plant, or animal viruses, generally have lower titration efficiencies (that is, higher ratios of molecules to infective units) than the corresponding virions. Even with free nucleic acid, however, the titration curve follows the course expected from Equation 2-2, indicating that one nucleic acid molecule can initiate infection (Franklin et al., 1959). As a rule, only intact DNA or RNA molecules are infectious. Interesting deviations, confirming the general rule, are encountered in multiple infection of cells with nucleic acid molecules that contain less than the full genetic complement of one virion (Okubo et al., 1964; Mosig, 1963; see Chapter 7).

In summary, the titer of a virus suspension, expressed in infectious units per unit volume, corresponds to the number of virions (or viral nucleic acid molecules) that succeed in producing infection under the particular conditions employed in the test. The numerical relation between successful virions and total virions cannot be decided by titration alone but require direct counts of virions. These counts, in turn, cannot distinguish between active and inactive virions, nor between those that are differentially infectious for different hosts or by different routes of inoculation.

Bacteriophage Assay

Measured samples of dilutions of a phage suspension are plated on suitable nutrient agar plates together with an excess of bacteria, either by spreading or in a surface layer of soft nutrient agar. The number of plaques that appear on the continuous bacterial layer after incubation is directly proportional to the amount of phage plated (Figure 2-1). If y plaques are produced by plating a volume v (in milliliters) of a dilution x, the titer n of plaque-forming units (pfu) per ml is $n = y/vx$. For example, 0.1 ml of a dilution $1:10^5$ gives an average of 230 plaques per plate; the titer is $230/(0.1 \times 10^{-5}) = 2.3 \times 10^8$ pfu per ml.

Using a series of N plates and assuming a random distribution (sampling errors only), the standard deviation σ should be equal to the square root of the mean number $\bar{n}$ of plaques per plate. The experimental standard deviation $\sigma' = \sqrt{\Sigma(n - \bar{n})^2/(N - 1)}$, compared with σ, provides a test of the randomness of the distribution of particles. The larger the value of $\bar{n}$, the smaller will be the *coefficient of variation* $\sigma'/\bar{n}$. Of course, when the plaques become too numerous, they are confluent and uncountable. The standard error of the mean $\bar{n}$ is given by $\sigma'\sqrt{N}$; hence, $\bar{n}$ becomes a better estimate of the true mean as the number of plates used increases.

In all virus titrations a *different pipette should be used for each successive dilution*. This requirement is very critical since a pipette that has contained billions of virus particles can hardly be rinsed free of them all.

Titration of Animal Viruses

Various approaches have long been used to devise methods analogous to the phage plaque count for titrating animal viruses. The discrete lesions or pocks produced on the chorioallantoic membrane of the chick embryo (Beveridge and Burnet, 1946) provided a relatively convenient system for some viruses. A major step forward was the introduction of monolayers of cells growing *in vitro* (Dulbecco and Vogt, 1953). Cytopathic viruses, that is, viruses that cause cell destruction, form plaques whose number is proportional to the amount of virus plated. Some tumor viruses give foci of proliferation that can be counted to yield titers in foci-forming units (Rubin, 1960).

Viruses that produce hemagglutinins may give rise, on a monolayer of sensitive cells, to localized and countable areas of virus-containing cells recognizable by *hemadsorption*; that is, by the ability to fix and retain red blood cells when the monolayer is flooded with a red cell suspension and then washed (Shelokov et al., 1958).

Some animal viruses do not form plaques or foci on cell monolayers and for them the only titrations available are end-point titration methods. That is, samples of successive virus dilutions are inoculated into susceptible animals and the *dilution end-point* is determined by scoring each inoculation as positive or negative. The titer, in infectious units per milliliter, is calculated by assuming that the last positive dilution contains at least 1 unit. Serial dilutions by factors of 10, $\sqrt{10}$, or 2 are used; the smaller the factor, of course, the more precise is the titer obtained. Titrations by dilution end-point are more

TABLE 2-1   VIRUS TITRATION BY THE 50% END-POINT METHOD

		Calculation of LD_{50} (50% lethal dose)		
Dilution	Amount inoculated	Positive responses (death)	Negative responses (survival)	% Positive
10^{-3}	0.1 ml	4	0	100
10^{-4}	0.1 ml	4	0	100
10^{-5}	0.1 ml	3	1	75
10^{-6}	0.1 ml	1	3	25
10^{-7}	0.1 ml	0	4	0

50% end point $= 10^{-5.5}$.
0.1 ml of a $10^{-5.5}$ dilution $= 0.1 \times 3.16 \times 10^{-6}$ ml $= 1$ LD_{50}.
Titer in LD_{50} per ml $= 1/(0.1 \times 3.16 \times 10^{-6}) = 3.17 \times 10^6$.

accurate when several samples of each dilution are tested. Around the end-point, one or more dilutions will give some positive and some negative responses. A common practice is to estimate by interpolation the dilution that would give 50% positive and 50% negative responses and to express the titers in multiples of the 50% infective dose or ID_{50} (Table 2-1).

Titration of Plant Viruses

In recent years it has become possible to titrate certain plant viruses such as potato yellow dwarf on monolayer cultures of cells taken from insects that are vectors of the viruses and in which the virus multiplies (Grace, 1969). Unfortunately, only a few viruses can as yet be titrated in this way. Viruses like tobacco mosaic (TMV), which do not grow in insect tissues, have traditionally been titrated by counting necrotic lesions produced on plant leaves inoculated directly (Holmes, 1929). The virus is applied to cheesecloth or mixed with some abrasive and then rubbed into the leaf because only those cells whose wall has been broken can become infected. Specialists have refined this technique to improve the reliability of the results.

It is important to realize that titration of a plant virus on a rubbed leaf is not equivalent to titration of viruses on continuous cell monolayer. The abraded leaf surface offers a limited number of entry sites rather than an unlimited number as on a monolayer. The best curves of plant virus titration on leaves (Figure 2-3) saturate when all the infectable sites are infected. The equation corresponding to this curve is: $y/N = 1 - e^{-s}$, where y is the actual number of lesions produced, N is the maximum number of lesions that could be produced by a large excess of virus, and s is the number of virions in the sample tested (Bald, 1937; Lauffer and Price, 1945). Useful titrations fall on the left side, nearly linear portion of the curve. For plant viruses the efficiency q, or probability that a virion produces a lesion, is quite low, less than 10^{-3} for TMV virions and much lower still for titration of naked TMV nucleic acid.

FIGURE 2-3. *The relation between the count of necrotic lesions produced by tobacco mosaic virus on* Nicotiana glutinosa *leaves and the concentration of virus in the inoculum. The virus concentration is given in units of virus antigen as determined by precipitin tests. From Beale,* Contrib. Boyce Thompson Institute **6,** *407 (1934).*

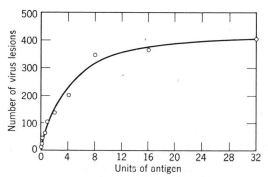

Titration and Loss of Infectivity

Accurate titration methods are important in the study of virus growth as well as in a variety of experimental procedures that require that viral materials be purified or subjected to fractionation. Accurate titration is especially useful in testing and evaluating the response of viruses to chemical or physical agents.

Figure 2-4 shows the loss of infectivity, that is, the decrease in plaque counts of two bacteriophages as a function of time after mixing them with an antiphage serum containing a neutralizing antibody. The curves illustrate several facts:

1. The course of inactivation is exponential, according to the equation $P_t/P_o = e^{-Kt}$; where P_t is the active phage titer, and at time t, P_o is the original titer, and K is a constant. (Note the analogy with Equation 2-1: there an infection required at least one successful virion; here a killing requires at least one "hit.")

2. The rate of the inactivation reaction is approximately proportional to the concentration of the serum: $P_t/P_o = e^{-kCt}$, where k is the reaction time constant. A linear dependence of inactivation rate on serum concentration

FIGURE 2-4. *Neutralization of bacteriophages by homologous or heterologous antiserum. The fraction* P/P_0 *of residual infectious phage is plotted as a function of time of exposure to antiserum at 37°C.* ●: *Phage T2 + anti-T2 serum diluted 1:64,000.* ○: *Same, diluted 1:16,000.* □: *Same, diluted 1:4000.* ×: *Phage T4 + anti-T2 serum diluted 1:4000.*

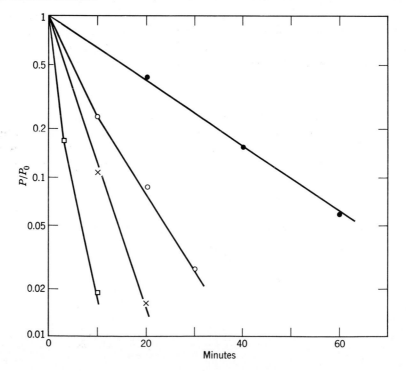

indicates that a virion can be inactivated by the action of a single antibody molecule.

3. For a given serum, the reaction rate constant differs for the homologous and the heterologous virus: anti-T2 serum at a given concentration inactivates phage T2 faster than the related phage T4.

Comparable studies can be done on many viruses and with a variety of agents. Considering the effects of antiviral sera somewhat more closely, it turns out that results are not always as simple as those illustrated in Figure 2-4. These were obtained with hyperimmune antiphage serum; that is, serum from an animal that had received repeated injections of the antigen. Early sera against phage, animal, or plant viruses contain mainly weakly binding antibody, so that antigen-antibody complexes are more or less readily resolved upon dilution. A typical experimental finding with an early serum is presented in Figure 2-5.

An interesting technique, applicable to any accurately assayable virus, is

FIGURE 2-5. *Reactivation of reversibly inactivated phage T4 by dilution. The steep curve ("Inactivation") shows the inactivation of T4 phage in an early serum. The upper curve ("Reactivation") shows the maximal reactivation obtained on incubation after dilution of the reaction mixture. The "Inactivation" data are obtained by the* decision *method, which consists of exposing some of the phage-serum mixture to concentrated bacteria before diluting it, then adding an excess of hyperimmune serum that kills all those phages that have failed to attach to cells (inactivated phages). A large portion of these are reactivated upon dilution. From Jerne and Avegno,* J. Immunol. **76,** *200 (1956).*

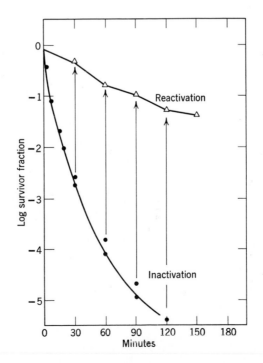

TABLE 2-2 THE SERUM-BLOCKING POWER OF PHAGE AND ULTRAFILTRATE OF PHAGE LYSATE

Preincubation of antiphage serum with ultraviolet-inactivated homologous purified phage or with a phage-free ultrafiltrate of a lysate of the homologous phage reduces the phage-neutralizing ability of the serum.

Tube number	Contents	4 hr at 48° C	Add test phage T2r	4 hr at 48° C	Assay of test phage	Percent survival of test phage
1	Broth				7.7×10^5	100
2	Serum*				2.2×10^4	3
3	Serum + 7.0×10^9 T2 (UV inactivated)				7.7×10^5	100
4	Serum + 2.3×10^9 T2 (UV inactivated)				4.8×10^5	62
5	Serum + 7.7×10^8 T2 (UV inactivated)				1.8×10^5	23
6	Serum + 2.6×10^8 T2 (UV inactivated)				4.3×10^4	6
7	Serum + 8.7×10^7 T2 (UV inactivated)				1.6×10^4	2
8	Serum + ultrafiltrate diluted 1:3				1.7×10^5	22
9	Serum + ultrafiltrate diluted 1:9				5.3×10^4	7
10	Serum + ultrafiltrate diluted 1:27				2.5×10^4	3

From DeMars et al., *Ann. Inst. Pasteur* **84**, 113, 1953.
* Serum anti-T2 1:40,000.
Conclusion: The serum ultrafiltrate contained an amount of serum-blocking power corresponding to that of about 2×10^9 UV-irradiated T2 particles.

the test for serum-blocking power (Table 2-2). This test measures the amount of materials present that can compete with the virus for neutralizing antibody. It is a particularly useful test to detect specific viral material in cases of abortive infections.

Precise titrations, including the assay of a few infectious virions among an excess of inactive ones, are critically important for the verification of noninfectivity of antiviral vaccines consisting of various inactivated virions. A small portion of virions may be more resistant than the majority to an inactivating agent, either genetically, or by association with protective substances, or by being altered in the course of the inactivating treatment. Special methods have to be developed for each situation.

In general, the sensitivity of the virions of a given virus to inactivation depends on specific properties of the proteins of that virus, so that valid rules apply only to closely related viruses. One exception is the sensitivity of viruses to X-ray, which is roughly predictable from the amount and type of nucleic acid in the virions (see Table 2-3). The reason for this regularity is that X-rays produce molecular breaks, and one break in the nucleic acid genome of a virion has a high probability of causing inactivation. The course of inactivation is generally exponential, conforming to equation

$$V_D = V_o e^{-kD} \tag{2-5}$$

where V_o and V_D are the initial virus titer and the titer after receiving a dose D of X-rays. In Equation 2-5 kD is the average number of "hits" per virion. Since

TABLE 2-3 THE RELATION BETWEEN THE SENSITIVITY OF VIRIONS TO X-RAYS (DIRECT EFFECT) AND THE TYPE AND AMOUNT OF THEIR NUCLEIC ACID

Virus	Nucleic acid		Mean lethal dose $D_{1/e}$, rads	Efficiency° = inactivating hits per ionization
	Type	Amount N, daltons		
Phage R17	RNA	9 $\times 10^5$	8.4 $\times 10^5$	0.5
Tobacco ringspot	RNA	1.5 $\times 10^6$	4.6 $\times 10^5$	0.5
Tomato bushy stunt	RNA	1.5 $\times 10^6$	4.5 $\times 10^5$	0.5
Tobacco mosaic virus	RNA	2.06 $\times 10^6$	2.0 $\times 10^5$	0.9
Rous sarcoma	RNA	1.2 $\times 10^7$	2 $\times 10^5$	0.15
Newcastle disease	RNA	~1.0 $\times 10^7$	4 $\times 10^4$	~0.9
Phage ϕX174	DNA, 1 strand	1.6 $\times 10^6$	3.8 $\times 10^5$	0.6
Shope papilloma	DNA	5 $\times 10^6$	4.4 $\times 10^5$	0.17
Phage T7	DNA	2.4 $\times 10^7$	1.3 $\times 10^5$	0.12
Phage λ	DNA	3.3 $\times 10^7$	1.0 $\times 10^5$	0.11
Adeno type 5	DNA	2.3 $\times 10^7$	7.0 $\times 10^4$?	0.23
Phages T2, T4, T6	DNA	1.3 $\times 10^8$	4.0 $\times 10^4$	0.07
Vaccinia	DNA	1.6 $\times 10^8$	~5 $\times 10^4$	0.05

Modified from Kaplan and Moses, *Science* **145,** 21, 1964, and Terzi, *J. Theor. Biol.* **8,** 233, 1965.
° The efficiency e is calculated as follows: $e = 3.7 \times 10^{11}/ND_{1/e}$ because a dose of 1 rad produces 2×10^{12} ionizations per gram of water, corresponding to about 3.7×10^{11} ionizations in 3.7×10^{11} daltons of matter with the average atomic weight of nucleic acids.

the dose D can be measured physically in numbers of ionizations produced per virion (or per unit mass of virion-like material) k becomes a measure of the efficiency of inactivation in hits per ionization. Table 2-3 shows that the efficiency is high for the smaller viruses, which have a higher ratio of nucleic acid to protein in their virions (see Chapter 3), and is lower for the largest, protein-rich virions.

Properties
3 of Virions

Purification of Virions

A purified virus preparation is a suspension or a crystal of virions that is reasonably free from contaminating materials. The virions of a given virus appear to be homogeneous in a variety of tests, much more homogeneous in chemical and physical properties than bacterial or yeast cells from a pure culture. The homogeneity of viral particles is of the same order as that of protein molecules; each molecule of a protein or each virion of a virus has, at least at first approximation, the same size, shape, and composition as all the others. This homogeneity is characteristic of things that are not produced by growth in size followed by division but are built or assembled from smaller subunits in a prescribed pattern—amino acids covalently joined into proteins or nucleic acid and protein molecules noncovalently assembled into virions.

The virions, as will be made clear in following chapters, are produced as the end-products of a process that includes the replication of the viral genome, the synthesis (and sometimes chemical processing) of viral proteins, and the assembly of these proteins with the viral genome into complete virions. The ordering forces that cause the folding and assembling are those that make any given system assume a configuration of minimal energy, as in crystallization or in formation of oriented gels. The homogeneity of virions of a given virus reflects then the uniqueness of the most probable configuration for a given set of viral materials. The homogeneity of virions is generally less precise for viruses with enveloped virions than for those with naked capsids. Envelopes, like all membrane elements, are formed by the closing up of phospholipid bilayers into vesicles. This process is intrinsically less precise than the assembly of proteins because closure can occur equally well at different points of a bilayer.

The complete virions are seldom the only product of virus biosynthesis.

Virus-infected cells contain not only complete virions but also virion-related materials, some of which may be precursors, others by-products, of virion synthesis. Purification of virions involves the removal of these virus-related materials, whose recognition, however, is informative about the steps of virus synthesis.

METHODS OF PURIFICATION

Concentration and separation of virions are accomplished by applying fractionation techniques to a starting material. The starting material consists of a culture fluid or an extract prepared by disrupting cells or tissues with ultrasonic vibration or by grinding alone or with alumina, sand, or glass powder. Purification is monitored by infectivity titration or other specific tests on separated fractions. Differential centrifugation, as described below, is a powerful tool in fractionation and virion purification. Salting out with ammonium sulfate, isoelectric precipitation, and other methods of protein fractionation is particularly useful in concentrating virions from large batches of crude materials.

Filtration through bacteriological filters removes particles larger than virions. The classical bacteriological filters (Berkefeld and Chamberlain-type) have been replaced by asbestos, fritted glass, and especially cellulose membrane filters (Millipore-type). Filters that retain particles whose linear dimensions are 500 nm or more will let most virions through because the dimensions of virions range from 10 to 300 nm.[1] Some virions may be lost on bacteriological filters because of electrostatic attractions. Many phages, for example, become irreversibly trapped in Seitz-type asbestos pads.

Ultrafilter membranes of special construction and graded porosity (for example, the "Gradacol" membranes; Elford, 1938) were widely used to estimate virion size before the electron microscope came into general and convenient use but have rarely been used in recent years.

The design of purification schemes for viruses with different physico-chemical characteristics is discussed in Chapter 4.

The Morphology of Virions

ELECTRON MICROSCOPY

Electron microscopy with a resolving power of the order of 0.4 to 1 nm (4 to 10 Å) has revealed the details of the structures of the particles of many viruses. In its simplest form electron microscopic examination produces photographs of material dried from a volatile medium onto a specimen holder

---

[1] Units commonly employed in virus work: a micrometer (μm) is 10^{-4} cm; a nanometer (nm), 10^{-7} cm; an angstrom (Å), 10^{-8} cm. Recall for comparison that the distance between the centers of two carbon atoms in the ethane molecule is 1.5 Å; the distance between successive amino acids in the polypeptide chain of a fibrous protein is 3.5 Å; the distance between pairs of bases in DNA along the axis of the double helix is 3.4 Å. The unit of molecular weight or dalton is $\frac{1}{12}$ the mass of the ^{12}C carbon atom or 1.66×10^{-24} g, which is the inverse of the Avogadro number 6.02×10^{23}.

consisting of a support of low electron-absorbing materials (collodion or pure carbon). To avoid distortions the objects are either dried from the frozen state (Williams, 1953) or are prepared by the critical-point method (Anderson, 1952). Improved resolution of images of objects with rotational symmetry can be achieved by rotational filtration of electron micrographic images (Crowther and Amos, 1971; see also Figure 3-12).

Interaction with the electrons in the specimen scatters the electrons from the incident beam; therefore, the electron micrographic image reveals local differences in amount of material on the path of the beam: differences in thickness and density, or the presence of heavy atoms. Electron opacity can be increased by staining with heavy metal compounds such as osmium tetroxide.

In the *shadow-casting* technique (Williams and Wyckoff, 1946) an opaque material—gold, palladium, uranium—is vaporized at a known angle to the plane of the specimen. The "shadow" of an object reveals its height and shape. Figure 3-1 shows a micrograph in which a double shadow-casting technique clearly reveals the icosahedral shape of the virions of the *Tipula* virus. Finer details are revealed by the *replica* technique, in which a mold or replica of the object is made by pouring a solution of a plastic material over the object; when the plastic is dry it is stripped off and photographed. The ridges and valleys on the replica reproduce the surface details.

Negative staining (Brenner and Horne, 1959) consists of drying a suspension of virions in a solution of phosphotungstate or uranyl acetate, which at neutral pH do not combine with proteins. Upon drying, objects such as virions stand out on the electron-opaque background. Since the opaque

FIGURE 3-1. *A virion of Tipula iridescent virus, doubly shadowed to show the icosahedral shape. Courtesy Dr. R. C. Williams.*

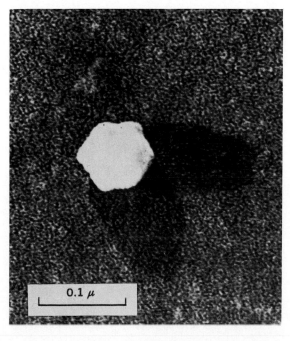

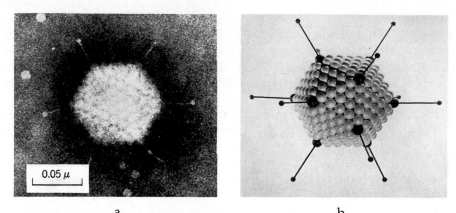

a   b

FIGURE 3-2. A virion of adenovirus (a) and a model of its capsid structure (b). From Valentine and Pereira, J. Mol. Biol. **13,** 13 (1965). Courtesy Dr. R. C. Valentine.

substances can enter the nooks and crannies between adjacent molecular components of the virion's surface the micrographs reveal the finest structural details, as seen for example in Figure 3-2. Empty virion shells are easily distinguished from full virions because they become filled with the opaque stain (Figure 3-3). Other refinements of electron microscopy will be mentioned

FIGURE 3-3. Virions of Shope rabbit papilloma virus, showing arrangement of hollow capsomeres on particles of different shapes and diameters. From Finch and Klug, J. Mol. Biol. **13,** 1 (1965). Courtesy Dr. J. T. Finch.

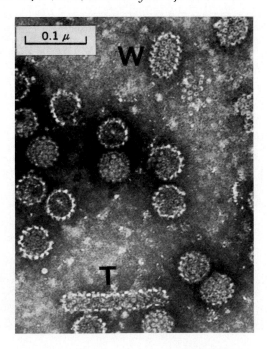

in later sections. All methods of specimen preparation for electron microscopy appear to produce some shrinkage and/or flattening, hence volume space within virion capsids should be calculated cautiously.

X-RAY DIFFRACTION

Purified preparations of virions that form either true crystals (for example, poliovirus and tomato bushy stunt virus) or paracrystals, that is, oriented gels of rod-shaped particles like those of TMV, lend themselves to structural analysis by means of X-ray diffraction, which yields information on the shape and internal structure of virions. The goal is to uncover the symmetry elements in the distribution of the structural units that make up the particles. This method was the first to reveal that a virion is not an amorphous object, but contains structures made up of regularly assembled subunits (Bernal and Fankuchen, 1941). It is the arrangement of these subunits that gives a virion its intrinsic features of molecular symmetry.

ORGANIZATION OF VIRIONS

CAPSID STRUCTURE.   The basic structure of all virions is the capsid, which encloses the nucleic acid. Capsids are made up of protein subunits assembled according to relatively simple geometrical principles based on elementary physical considerations. Thus completely unrelated viruses such as a phage, an animal virus, and a plant virus may have capsids that are similarly built and almost indistinguishable morphologically.

Crick and Watson (1957) pointed out that capsids must be built of many identical subunits since viruses do not have in their nucleic acid enough genetic information to code for sufficient different proteins to make a capsid. There are two ways in which identical asymmetrical subunits such as protein molecules can be assembled to build stable, regular capsids: helical assemblies and closed shells. Correspondingly there are only two basic types of capsids, helical and isometric (or quasi-spherical). Each of these arrangements is reached by capsid proteins by a process of self-assembly that must be energetically favorable. That is, the form of the capsid must be a form of minimal free energy for the specific proteins of a given virus. The specific shape of the protein molecules that serve as subunits for the capsids, and the pattern of bonds that these subunits can make with each other, determines the actual shape and size of the capsid. The stability of the final assembly depends on the number and strength of the weak bonds between protein molecules in the capsid. The greater the free energy released in assembly the more stable will the assembly be.

Helical Capsids.   The virions of many plant viruses and several phages have naked helical capsids, without envelopes (Franklin et al., 1957; Klug and Caspar, 1960). Electron microscopy of TMV, the best known virus of this group, shows rods 15 to 17 nm thick (Figure 3-4). The rods seen in micrographs may show some variation in length, but only those rods that are 3000 Å long are infectious. Fragments of TMV rods photographed along the

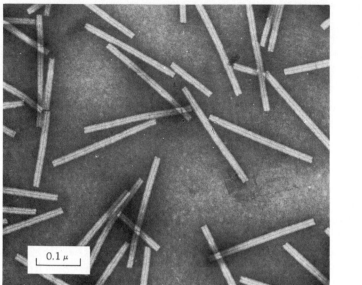

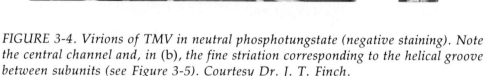

FIGURE 3-4. Virions of TMV in neutral phosphotungstate (negative staining). Note the central channel and, in (b), the fine striation corresponding to the helical groove between subunits (see Figure 3-5). Courtesy Dr. J. T. Finch.

long axis of the virion have hexagonal contour with a central hole 40 Å in diameter.

X-ray diffraction analysis on wet or dried gels of TMV shows an axial repeat of 69 Å, reflecting a left-handed helical assembly of 49 protein subunits for every three turns (Figure 3-5). A virion contains 2130 (±2%) identical protein subunits, each a single protein molecule 17,400 in molecular weight. The protein subunits are tapered on the outside forming an external groove (Figure 3-5). The amino acid sequence of the protein molecules is fully known (see Chapter 4). The virion has a molecular weight 39×10^6 (±3%) and contains a single RNA molecule weighing 2.06×10^6 daltons.

Concentrated solutions of TMV spontaneously form paracrystals (liquid crystals), that is, arrangements in which adjacent particles are arranged in orderly fashion together side-by-side but not lengthwise. In a wet gel the distance between centers of particles is 175 Å, about equal to the distance between centers measured to the outer edge of the subunits. When the paracrystals are dried, the distance between centers of particles becomes 152 Å. The RNA strand is spirally wound in a groove formed by the protein subunits about 40 Å from the central axis, as shown in Figure 3-5.

TMV treated with dilute alkali releases its protein either as monomers or in a partly aggregated form, the so-called A protein. The protein can reaggregate with TMV-RNA to form infectious virions (Fraenkel-Conrat and Williams, 1955). The TMV protein can also reaggregate with RNA from other viruses (Sugiyama, 1966) or with synthetic polynucleotides. Even without RNA, the A protein tends to aggregate by itself to rods; the X-ray diffraction

FIGURE 3-5. *Schematic representation of the structure of TMV. The six-turn segment shown corresponds to about 1/30 the length of the full virion. The RNA strand is coiled between the turns of the 23 Å-pitch helix of protein subunits; the diameter of the coil is 80 Å. The central hole has a diameter of 40 Å. From Caspar, in:* Horsfall and Tamm, Viral and Rickettsial Infections of Man, *New York: Lippincott, 1965, Courtesy Dr. D. Caspar.*

pattern of the rods shows the absence of the set of reflections corresponding to the RNA strand. The reaggregation of TMV protein *in vitro* is a clear illustration that the assembly of viral protein subunits to form helical capsids is a crystallization-like phenomenon that reflects the intrinsic properties of the subunits themselves.

The TMV protein can also aggregate in the form of flat disks of 17 subunits each (Markham et al., 1964). The flat disk is a more open structure than the helical assembly (Champness et al., 1976) and may be the favored assembly type *in vivo*. It appears that disks serve as primers to initiate the assembly of the virion (see Chapter 17). Upon assembly the disks undergo a transition from the flat shape to a lock-washer shape, corresponding to about one turn of the TMV helix (Figure 3-6). The shift, which is accompanied by displacement of carboxylic groups within each subunit (Caspar, 1956), is caused by a conformational change of the protein subunits upon interaction with TMV-RNA (Klug, 1972).

The TMV capsids are rather rigid rods. At least one phage has rigid capsids like those of TMV (Gibbs et al., 1975). The capsids of other plant viruses, such as sugar beet yellows (Figure 3-7) or potato virus X, are helical but flexible rods and so are the capsids of many animal viruses inside their envelopes (Figure 3-8). Flexibility indicates that the subunits are held together by less strong, more displaceable bonds than those present in TMV-like rods.

Isometric (Quasi-Spherical) Capsids. Many viruses have quasi-spherical capsids, which in the electron microscope reveal regular features that show them to be polyhedra rather than spheres. These are called isometric because

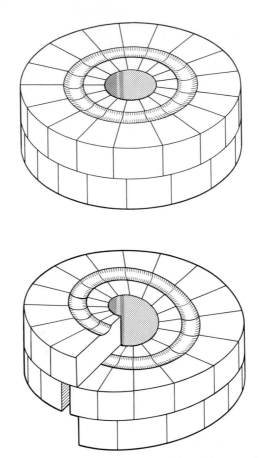

FIGURE 3-6. The TMV protein disk and its helical counterpart. See also Figure 17-4.

FIGURE 3-7. Virions of sugar beet yellows virus. Courtesy Dr. R. W. Horne.

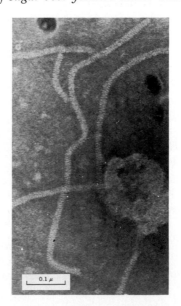

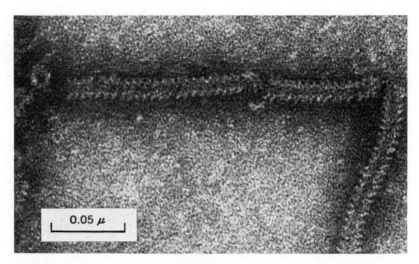

FIGURE 3-8. Helical capsid released from virion of Newcastle disease virus. From de-Thé and O'Connor, Virology **28**, 713 (1966). Courtesy Dr. G. de-Thé.

they have identical linear dimensions along orthogonal axes. A theory developed by Caspar and Klug (1962) accounts for the general class of shapes of these polyhedra and also for the specific appearance of various types of polyhedral capsids in electron micrographs. The theory starts from crystallographic considerations, which prescribe that only systems with cubic symmetry can generate isometric shells by locating identical units in equivalent manner on the surface of a sphere. Icosahedral symmetry satisfies this requisite: 60 identical subunits equally joined to each other and disposed over a spherical surface generating a regular icosahedron, a solid with 12 vertices, 20 triangular faces, and 20 edges (Figure 3-9, 3-10a). Some of the smallest viruses do have capsids consisting of 60 protein molecules arranged in icosahedral shape.

FIGURE 3-9. In this diagram 60 identical polygonal units are arranged according to icosahedral symmetry. Note that each unit makes identical contacts with its neighbors ("equivalence"). From Caspar and Klug, Cold Spring Harbor Symp. **27**, 11 (1962).

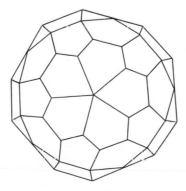

FIGURE 3-10. Deltahedra with icosahedral symmetry ("icosadeltahedra"). Each deltahedron has 20 T equilateral triangles on its surface, T being the triangulation number (see Table 3-1). Only a *and* b *are true icosahedra. Note the vertices with five-fold symmetry and the sites with six-fold symmetry. From Caspar and Klug,* Cold Spring Harbor Symp. **27,** *11 (1962).*

Given the small size of protein molecules as structural units, there is a limit to the size of icosahedra that can be constructed in this way. To build larger capsids more units must be used, but this creates the problem that the units can no longer be equivalent to each other in bonding. As soon as the faces grow some of the units are located away from the vertices. At the vertices the units are grouped by five (*pentamers*) while on the rest of the surface they are arranged in clusters of six (*hexamers*) as they would be on a flat plane (Figure 3-10b). The various subunits are still bound to each other in a quasi-equivalent way, the bonds being distorted to some extent from one part of the shell to the other.

Some of the larger capsids, like those of adenoviruses (Figure 3-2), are still icosahedra, but each of the 20 edges and faces contains additional subunits. Two major deviations from the icosahedral shape are often encountered. First, the symmetry may change while the faces remain triangular, generating solids called *icosadeltahedra*, still with 12 pentameric vertices but with faces less flat than the icosahedron. The possible classes predicted by theory are shown in Table 3-1 (and some are shown in Figure 3-10c–h). Second, and more important, the protein molecules that are the structure units of the shell may cluster together to form what is seen in electron micrographs as morphological units or *capsomeres* (Figure 3-11). It is often possible to recognize different

TABLE 3-1 CLASSES OF ICOSADELTAHEDRA AND REPRESENTATIVE VIRUS EXAMPLES

Triangulation No. $T = Pf^2$
$P = h^2 + hk + k^2$, where h and k are integers with no common factor
$f = 1, 2, 3, 4$

Classes
$H = 1; K = 0; P = 1$ $T = 1, 4, 9, 16, 25 \ldots$
$H = 1; k = 1; P = 3$ $T = 3, 12, 27 \ldots$

No. of triangular facets $= 20T$
No. of structural units $= 60T$
No. of morphological units (capsomeres) $= 10T + 2$
[$10(T - 1)$ hexamers $+ 12$ pentamers]

Examples:
Turnip yellow mosaic virus: $P = 3; T = 3$; 32 capsomeres (20 hexamers $+$ 12 pentamers); 180 structural units.
Adenovirus type 5: $P = 1; T = 25$; 252 capsomeres (240 hexamers $+$ 12 pentamers); (1500 structural units?)

Modified from Caspar and Klug, *Cold Spring Harbor Symp.* **27:**1 (1962).

morphological units as *pentons* and *hexons* in micrographs analyzed by special techniques (Figure 3-12). Caspar and Klug (1962) suggest that binding of subunits into capsomeres may help maximize the contacts between structure units at the outside surface, avoiding gaps between them.

An interesting consequence of the relative deformability of icosadeltahedra is that relatively minor influences can generate abnormal forms such as elongated instead of isometric capsids.

Giant virions and even long protein tubes or "polyheads" are produced by certain phage mutants (Showe and Kellenberger, 1975) and also in the presence of amino acid analogues (Cummings and Bolin, 1976). Presumably in these cases the protein subunits are structurally altered and their assembly is modified by the change in subunit shape (Figure 3-13).

Arrangement of Nucleic Acid in Isometric Capsids. DNA, which is fairly rigid, must presumably be folded in an orderly fashion in order to fit into a capsid. Analysis with polarized light and X-rays on oriented phage fibers has revealed some pattern of internal organization (North and Rich, 1961). Whether in viruses with single-stranded RNA or DNA the nucleic acid fibers are present as random coils or neatly folded is not known, but at least for turnip yellow mosaic virus it is believed from X-ray data that the RNA forms knobs located in symmetrical relation to the capsomeres of the shell (Klug et al., 1966; Figure 3-11).

Complex Capsids. Evidence of complexity of the capsids comes both from antigenic and from morphological studies. Careful microscopic observation often reveals presence of appendages or spikes, generally located at the 12 vertices of the icosadeltahedra. This is true for very large viruses such as adenovirus (Valentine and Pereira, 1965; see Figure 3-2) and also for small

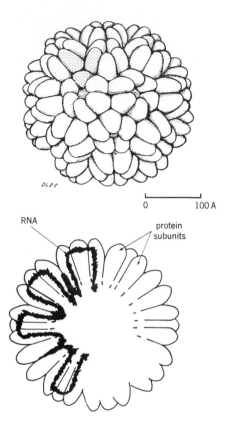

DLPC

0   100 A

RNA

protein
subunits

FIGURE 3-11. Top: *A drawing of the outer surface of the particle of turnip yellow mosaic virus (Finch and Klug,* J. Mol. Biol. **15,** *344 (1966)). The capsid consists of 180 subunits, each group of three subunits making up an asymmetric crystallo-graphic unit. Morphological units, pentons and hexons, are produced by clustering of subunits. Bottom: Section through a particle showing schematically the relation of the single RNA chain to the arrangement of subunits. From Klug et al.,* J. Mol. Biol. **15,** *315 (1966).*

viruses like phage ϕX-174. The spikes play a critical role in the initiation of infection. One phage with "hairy" virions, that is, with protein filaments attached on its head, has been described.

Figures 3-12 to 3-17 illustrate the morphology of a variety of bacterio-phages. The largest phages have tails that serve as organs of attachment to the host bacteria (see Chapter 9). There are few biological objects more remarkable than the T-even phages, in whose virions more than 50 proteins are assembled in a highly organized structural pattern of amazing complexity and regularity (Figures 3-14 to 3-18). The collar and the tail-plate structures of the T-even phages have hexagonal symmetry. The head shell of these phages is a deformed icosadeltahedron with an extra row of subunits that makes one dimension longer (Moody, 1965; Branton and Klug, 1975). The hexagonal tail structure must somehow be attached to a vertex of the head with five-fold

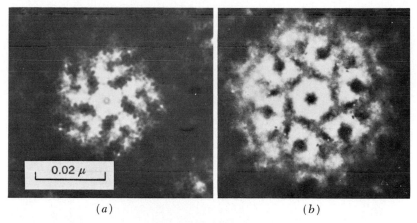

(a)   (b)

FIGURE 3-12. Virions of cucumber mosaic virus showing (a) a five-sided capsomere (penton) and (b) a six-sided capsomere (hexon). From Murant, Virology **26,** 542 (1965). Courtesy Dr. A. F. Murant.

FIGURE 3-13. Polyheads of mutant phage T4 am50 from a lysate of nonpermissive bacteria. A normal T4 virion is also shown. Note the molecular structure of the polyhead tubes and their unequal width. This is explained as due to formation of multiple concentric polyhead tubes around a central protein core in nonpermissive bacteria infected with this phage mutant (see insert). Courtesy Drs. E. Boy de la Tour and E. Kellenberger.

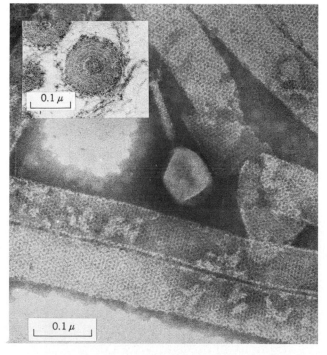

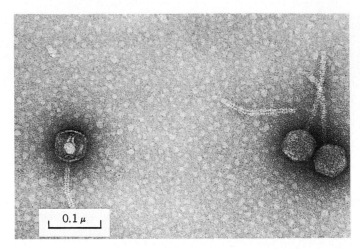

FIGURE 3-14. *Particles of bacteriophage* λ. *Note the striation on the tail and the inner core in the hollow capsid on the left. From Kaiser,* J. Gen. Physiol. **49** *(part 2), 171 (1966). Courtesy Dr. A. D. Kaiser.*

symmetry. Accidental formation of T4 particles with two tails instead of one is observed (Figure 3-15).

In some T4 mutants a change in the main structural protein of the shell produces a "petite," isometric form of capsid (Eiserling et al., 1970) which is also found in a very small number of the virions produced by the normal phage. The petite phage virions, observed also with other phages, contain DNA molecules shorter than those of normal size virions (Mosig et al., 1972). Giant virions, on the other hand, contain longer DNA molecules.

ENVELOPES.   Many animal viruses, some plant viruses, and at least one class of bacteriophage have external envelopes surrounding their capsids

FIGURE 3-15. *Virions of* subtilis *phage* φ29. *Insert: distal view of detached tail showing 12 appendages symmetrically arranged around the central tail. From Anderson et al.,* J. Bacteriol. **91,** *2081 (1966). Courtesy Drs. D. L. Anderson, D. D. Hickman, and B. E. Reilly.*

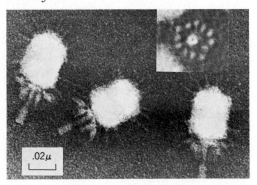

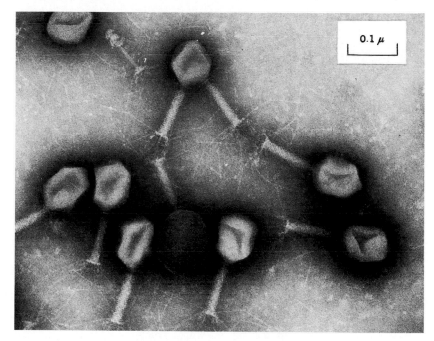

FIGURE 3-16. *Virions of bacteriophage T4; one with two tails, one with exploded head. Courtesy Dr. E. Boy de la Tour.*

FIGURE 3-17. *Details of the structure of bacteriophage T2L in negative staining. Note the absence of a collar, such as is present in phage T4. Upper right corner: a tail plate, showing hexagonal symmetry with spoke-like appendages. One empty virion with normal tail, one broken tail core, one core attached to an empty capsid. Another empty virion has a contracted sheath. Courtesy Dr. E. Boy de la Tour.*

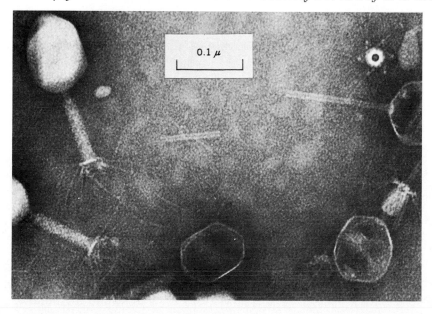

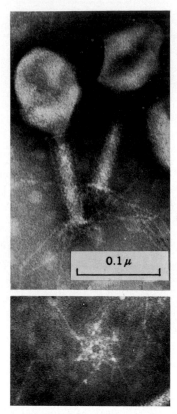

FIGURE 3-18. *Two virions of phage T4, one with some tail fibers attached to the whiskers to form a "jacket," and a detached base plate with fibers. From Kellenberger et al.,* Virology **26,** *419 (1965). Courtesy Dr. E. Boy de la Tour and E. Kellenberger.*

(Figure 3-19). The essential structure of the envelopes is that of all biological membranes, a phospholipid bilayer in which are embedded specific proteins. When the phospholipid bilayer is accessible at the virion surface, the enveloped virions are readily disrupted and inactivated by ether or other lipid solvents.

The phospholipids of the envelope may be similar or identical to those of the host cells, as in most enveloped animal viruses, or can deviate rather strongly, as in phage PM2 (Tsukagoshi et al., 1975). The envelopes of this phage differ in origin from those of animal viruses, being formed within the host bacteria without apparent relation with the bacterial membrane. The lipid bilayer of PM2 phage, like those of many animal viruses (Figure 3-20) is layered between two concentric shells. In PM2 the outer shell consists of two proteins, the inner one of a third protein. A fourth protein is associated with the DNA in the virus capsid (Harrison et al., 1971).

The envelopes of animal viruses are formed either at the cytoplasmic membrane or at the nuclear membrane. In electron micrographs of infected cells, viral proteins appear in patches in the cytoplasmic membrane at sites

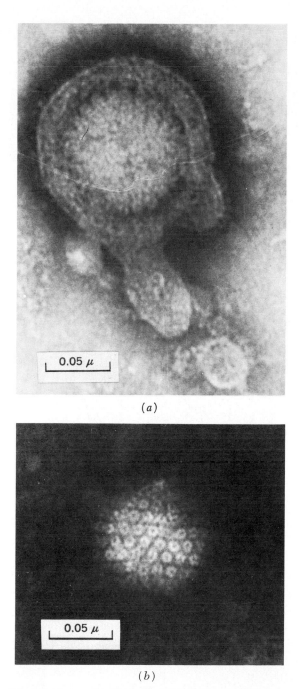

FIGURE 3-19. *Herpesvirus.* (a) *Enveloped virion, partially distorted;* (b) *capsid showing hollow prismatic capsomeres.* From Wildy and Watson, Cold Spring Harbor Symp. **27,** 25 (1962). *Courtesy Dr. D. H. Watson.*

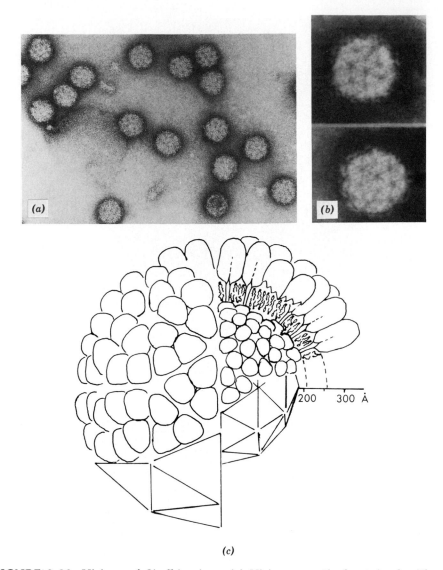

FIGURE 3-20. Virions of Sindbis virus. (a) Virions negatively stained with uranyl acetate. (b) Two virions stained with potassium phototungstate, showing the five-fold and six-fold arrangements of morphological units on the virion surface. (c) A "best-guess" summary of the structure of Sindbis virions. Note that both inner capsid and outer layer have icosahedral symmetry. Photographs by Dr. C. H. von Bonsdorff; courtesy Dr. S. C. Harrison. See von Bonsdorff and Harrison, J. Virology 16, 141 (1975).

where the viral capsids migrate and finally bud out (Morgan et al., 1962). In the envelopes of viruses such as influenza (Figure 3-21) the virus-specific proteins form characteristic projections.

The external protein layer of the envelope has, in some cases, icosahedral symmetry typical of viral capsids (Figure 3-20). Not all animal viruses have

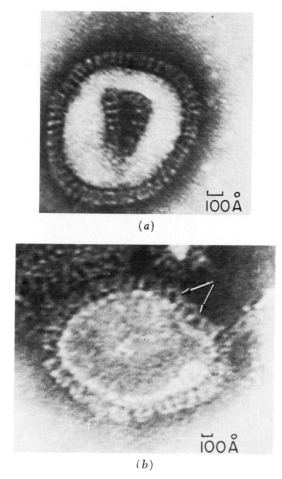

(a)

(b)

FIGURE 3-21. Interaction of influenza virions with homologous antibody. (a) Influenza virus alone; (b) antibody bridges between adjacent surface projections; (c) agglutination of virions by antibody. From: Lafferty and Ortelis, Virology **21,** 91 (1963). Courtesy Dr. K. J. Lafferty.

quasi-spherical virions, however. The virions of rhabdoviruses are thimble-shaped, with an envelope that originates by budding from the cell membrane (Figure 3-22). Other animal viruses, such as the pox viruses, have virions with complex envelopes that are formed entirely inside the cytoplasm. These virions are not ether sensitive, do not cross-react with host-cell antigens, and probably consist exclusively of virus-specific materials (Figure 3-23).

Some insect-pathogenic viruses give rise in the infected cells to remarkable structures (see Chapter 18). In one class, the *granulosis* viruses, the virions consist of rod-shaped capsids surrounded by envelopes, which in the cytoplasm of infected cells are enclosed in large oval protein granules or *capsules* (Figure 3-24). Another group of insect viruses causes the formation of intranuclear or intracytoplasmic *polyhedra*, consisting of virus-specific protein. The rod-shaped virions of these viruses resemble those of the granuloses but

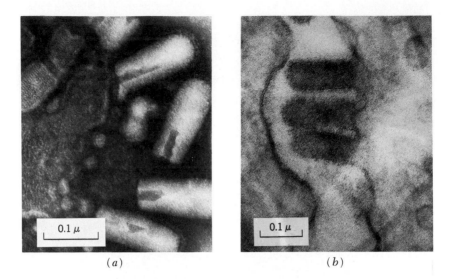

(a)   (b)

FIGURE 3-22. Particles of vesicular stomatitis virus. (a) Free virions; (b) virions being formed on a cellular surface in an intercellular space. From Howatson and Whitmore, Virology **16,** 466 (1962). Courtesy Dr. A. F. Howatson.

have no capsules and are found both within and outside the polyhedra. The polyhedral and capsular proteins are serologically unrelated to the virion proteins.

INCOMPLETE VIRIONS.   All crude preparations of viruses come from infected cells and, along with the complete virions, always contain other products of viral activity. Such virus-associated materials may be directly related, morphologically or chemically, to the components of the virion. For example, plants infected with TMV also contain TMV protein in various states of aggregation. Cells infected by viruses with isometric capsids yield, besides

FIGURE 3-23. Virions of vaccinia virus, showing an inner "triple element," probably a capsid, in different orientations. From Peters and Müller, Virology **12,** 266 (1963). Courtesy Dr. D. Peters.

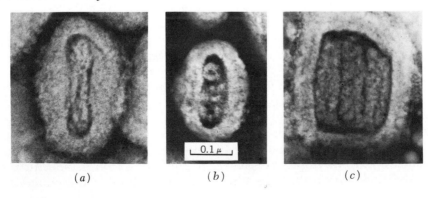

(a)   (b)   (c)

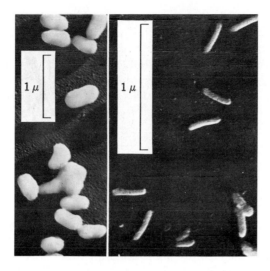

FIGURE 3-24. Capsules (left) and virions (right) from granulosis of Choristoncura fumiferana, a lepidopteron. Courtesy Dr. G. H. Bergold.

the complete virions, a certain amount of empty capsids. Negative staining with neutral phosphotungstate fills the empty capsids and reveals in micrographs the thickness and organization of the shell. The empty shells, which may co-crystallize with the full capsids, are lighter than virions and can be separated by centrifugation (Markham and Smith, 1949). They may contain some internal proteins that are normal components of virions (Figures 3-14

FIGURE 3-25. Empty capsids of human wart virus stained with uranyl acetate. Note the inner cores. From Klug et al., CIBA Foundation Symp., Principles of Biomolecule Organization, London: Churchill, p. 158 (1966). Courtesy Dr. J. T. Finch.

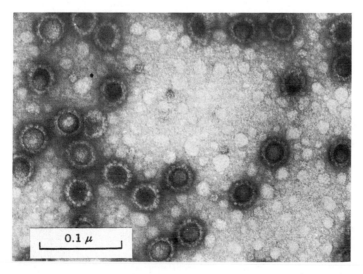

and 3-25). The empty shells are often precursors of the isometric capsids. That shells can be formed without a nucleic acid core means that some mechanism must exist to provide a substrate for assembly. Neither the protein subunits nor groups of them can form a shell as an arrangement of minimum free energy except upon surface (Caspar and Klug, 1962). In at least one case, that of the P22 group of phages, the matrix for assembly of the capsid is provided by a protein which is then removed from the capsid as DNA enters (Casjens and King, 1975; see Chapter 6).

ELECTRON MICROGRAPHIC COUNTS

Virions can be counted with the electron microscope in a variety of ways. In the *proportional count method* (Williams and Backus, 1949) a virus preparation is mixed with a measured amount of a polystyrene latex preparation consisting of uniform spheres of known diameter and density. The mixture is sprayed in a fine mist onto a specimen holder and counts are made on photographs of single droplets (Figures 3-26 and 3-27). From the ratio of latex spheres to virions the concentration of virions can be calculated.

In the *centrifuge method* (Sharp, 1965) virions sedimenting in a centrifuge tube impinge on a specimen holder and become stuck there so that subsequent drying of the specimen does not change their distribution. In the

FIGURE 3-26. *A droplet pattern from a mixture of tobacco mosaic virus and polystyrene latex particles. The concentration of the latex particles in the mixture is known, and the number of virus particles can be estimated from the ratio between the counts of the two types of particles. The enlarged sector shows the virus particles more clearly. Note that the rod-shaped virus particles orient themselves parallel to the edge of the droplet because of the changes in surface tension during drying. From* Research Today **8,** 28 (1952). *Courtesy Dr. R. C. Williams.*

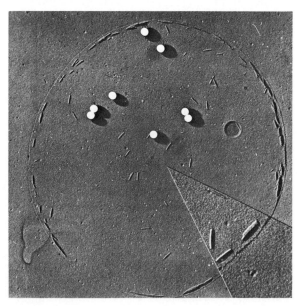

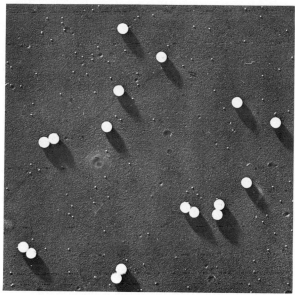

FIGURE 3-27. *A droplet pattern from a mixture of poliovirus and polystyrene latex particles. See Figure 3-26 for explanation. Courtesy Dr. R. C. Williams.*

filtration method (Kellenberger and Arber, 1957) droplets of a virus suspension are sprayed on a specimen holder membrane lying on top of a layer of agar. The diluent liquid passes rapidly into the agar leaving on the membrane an undistorted distribution of particles.

Physical-Chemical Methods in Virus Research

A large amount of information about the size and shape of virions was obtained by physical-chemical methods before electron microscopy became available and practical. Even today these methods provide the needed data about virion properties such as density, homogeneity, and stability, as well as constitute the basis for most virion purification schemes (see Chapter 4). They also contribute information about the properties of subviral fractions such as empty capsids or envelopes, subassemblies of capsid subunits, and nucleic acid free or in combination with protein. The following paragraphs summarize the essential elements of the relevant theories and their applications.

SEDIMENTATION OF SPHERICAL PARTICLES

A sphere of radius r in a liquid of density d_0 sediments in a gravitational or centrifugal field if its density d is greater than d_0. In a gravitational field the motor force P equals the gravity acceleration g multiplied by the difference between the mass m of the particle and the mass of a corresponding volume of the medium: $P_g = (4/3)\pi r^3(d - d_0)g$. In a centrifugal field, the gravitational acceleration g is replaced by the centrifugal acceleration $c = (2\pi\omega)^2 x$, where ω is the number of revolutions per unit time and x is the distance of the particle from the axis of rotation: $P_c = (4/3)\pi r^3(d - d_0)c$.

As a particle is accelerated, its velocity v increases. This, in turn, causes an increase in friction. The frictional force is $\phi = fv$, where f is the friction coefficient. The value of f for spherical particles is $6\pi\eta r$, where η is the viscosity of the medium. As the velocity increases a point is reached where the frictional force ϕ balances the centrifugal force P_c. The particle then continues to move at a uniform velocity. Therefore when $\phi = P_c$,

$$\tfrac{4}{3}\pi r^3(d - d_0)c = 6\pi\eta rv; \quad v = 2cr^2\left(\frac{d - d_0}{9\eta}\right); \quad r = \sqrt{9v\eta/2c(d - d_0)} \qquad [3\text{-}1]$$

These are Stokes' equations, which describe the motion of spherical particles in a fluid under ideal conditions (that is, with no interaction among the particles; particles large as compared with the solvent molecules; no disturbance due to convection).

From Equation 3-1 one derives the value of $S = v/c$, the *sedimentation constant*, characteristic for a given particle in a given medium at a given temperature.[2] Equation 3-2 can then be written

$$S = \frac{\tfrac{4}{3}\pi r^3 d - \tfrac{4}{3}\pi r^3 d_0}{6\pi\eta r} = \frac{m[1 - (d_0/d)]}{f} \qquad [3\text{-}2]$$

Since the centrifugal force, the density of the medium, and its viscosity can be measured, Equations 3-1 and 3-2 serve to calculate the radius and mass of a spherical particle provided we can measure its density and its sedimentation velocity in a centrifuge. Ultracentrifuges reach speeds of 60,000 rpm or higher. For a particle located 6 cm from the rotational axis, 60,000 rpm correspond to a centrifugal force 250,000 times gravity. The rotor is held in a high vacuum to reduce friction and control temperature. While analytical centrifuges are in motion, measurements can be made by photographing moving boundaries of sedimenting molecules.

The density of virions can also be measured by centrifugation in a series of solutions of graded density. The sedimentation rates are measured and the density of sedimentation rate zero is obtained by extrapolation (Figure 3-28). Because the material in solution is hydrated, the density measured in this way is generally lower than if measured by weighing dried preparations. The amount of hydration water in virions is between 50 and 100% of the dry weight, higher than is generally found with proteins, due to osmotic effects and to the high hydration of nucleic acids.

Density gradient centrifugation refers to centrifugation in a liquid column whose density is distributed in a continuous or discontinuous gradient. Particles collect in zones whose densities correspond to their solvated density. In *rate zonal centrifugation* the material to be analyzed is layered on top of a liquid column in which a gradient of the desired properties has been preestablished by layering appropriate sucrose solutions. The gradient prevents convection and stabilizes localized sedimentation boundaries of particles. The zones are sampled by orderly collection of fractions, generally starting from the bottom of the centrifuge tube.

In the *isopycnic gradient centrifugation* or equilibrium density gradient centrifugation the material is suspended in a solution of a high-density salt such as cesium chloride, cesium sulfate, or rubidium chloride. Prolonged centrifugation creates within the solution a stable density gradient due to a balance between sedimentation of dense solutes and their back diffusion. Particles collect and form a band at the place where the density of the medium equals their own. At equilibrium the force that keeps the

---

[2] Using CGS units, r and x are expressed in cm; g and c in cm sec^{-2}; ω in revolutions per sec; d in g cm^{-3}; and the viscosity in g cm^{-1} sec^{-1}. The sedimentation constant is obtained in seconds. The sedimentation constant is often expressed as S_{20} in multiples of the Svedberg unit, the rate of sedimentation in water at 20°C under one unit of centrifugal force.

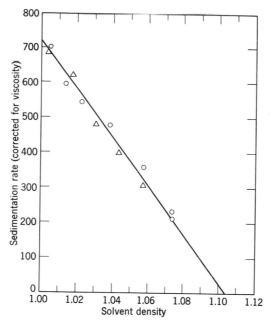

FIGURE 3-28. *The sedimentation rate of influenza virus A as a function of the solvent density. The density was varied by adding bovine serum albumin, 0 to 25%. The ordinate gives the sedimentation rate. The extrapolation to rate 0 (no sedimentation) gives a density of 1.104. Data from two experiments. From Sharp et al.,* J. Biol. Chem. **165,** *259 (1946).*

particles to their proper density level is exactly balanced by diffusion, which tends to make them move away from the region of highest concentration (see next section). The width of the band (that is, the standard deviation of the Gaussian distribution of concentrations about the midline of a band) depends on the mass of the particles. Mass and density obtained in this way represent properties of the particles in that particular medium with respect to hydration or salt formation. (The density of the Cs^+ salt of DNA is very different from that of the Na^+ or Mg^{++} salts.)

If the density of the nucleic acid in a virion is S_N, the density of its protein is S_P, and that of the virion is S_v, the fractional volume V_N of nucleic acid in the virion is

$$V_N = \frac{S_v - S_P}{S_N - S_P}$$

and the fractional mass M_N of nucleic acid is: $V_N S_N / S_v$. Virions with densities lower than 1.3 usually contain large amounts of lipids in their envelopes (Crawford, 1960). A useful variation is band-centrifugation (Vinograd et al., 1963), which can separate particles by size classes more efficiently than ordinary boundary sedimentation.

DIFFUSION

If a layer of a solution and a layer of the medium alone are in contact along a boundary plane, the random motion of the particles causes some of them to cross the boundary, leading to a progressive equalization of concentrations. If the molecules or particles of the solute are large compared with the molecules of the medium and do not interact with one another their kinetic energy depends only on the temperature. The resistance of the medium depends on the velocity of the particles and on their friction coefficient. The change in concentration that takes place in one second across 1 cm² of the boundary between solutions differing by 1 unit of concentration is the *diffusion constant D*.

$$D = \frac{RT}{Nf} \text{ cm}^2 \text{ sec}^{-1} \tag{3-3}$$

where T is the absolute temperature, R is the gas constant (8×10^{-7} erg °C^{-1} mol^{-1}), N is the Avogadro number (6.06×10^{23}) and f is the friction coefficient (which for spherical particles is $6\pi\eta r$). The diffusion constant D is obtained by measuring the concentration of particles at different times at known distances from the plane of the boundary. For spherical particles the radius is given by

$$r = \frac{RT}{ND6\pi\eta} \tag{3-4}$$

NONSPHERICAL PARTICLES

The situation for nonspherical particles is complicated by the fact that the frictional force depends on a complex function of the linear dimensions of the particles. One way of getting around this difficulty is to measure both sedimentation and diffusion constants. The ratio of the two constants is independent of the friction coefficient and depends only on their mass:

$$\frac{S}{D} = \frac{m[1 - (d_0/d)]N}{RT} \text{ sec}^2 \text{ cm}^{-2}$$

$$m = \frac{S}{D} \frac{RT}{N[1 - (d_0/d)]}$$

If all units are CGS, m is obtained in grams. An essentially equivalent method is to measure the so-called *sedimentation equilibrium* reached when sedimentation and diffusion balance one another. The mass of the particles can be determined directly from their concentration at various positions in a centrifuge tube at equilibrium.

Once mass and density of an asymmetrical particle are known it is easy to calculate the volume, radius, and friction coefficient f_0 that it would have if it were a sphere. The ratio of the actual value of f to the value of f_0 is the *asymmetry coefficient*, a measure of the extent of deviation from spherical shape.

A method of great usefulness in the study of virions and their components is *electrophoresis*. The migration of electrically charged particles in an electric field applied to a solution depends on the size and on the overall charge of the particles, which in turn depends on the dissociation of their acidic and basic groups. Among many varieties of electrophoretic techniques, acrylamide gel electrophoresis provides a convenient method for separation of a mixture of molecular species into separate bands. For proteins, if electrophoresis is carried out in the presence of a denaturing detergent such as sodium dodecylsulfate, the rate of migration is proportional only to the molecular weight of the molecules. Acrylamide gel electrophoresis is also applicable to nucleic acids, separating various classes of molecules by size.

Serological Properties of Virions

The possession of specific proteins gives to virions antigenic determinants that stimulate production of specific antibodies. The proteins of the viral capsid are virus specific. Cross-reactions with host-cell proteins are either accidental, such as may occur between any protein antigens, or, when an envelope is present, may reflect the incorporation of host cell proteins into the envelope.

Antivirion antibodies may be detected by any of the standard antigen-antibody reactions. Agglutination and precipitation of virions are easily demonstrated and can be used to quantitate either the amount of viral material or the amount of specific precipitating antibody. Neutralization—the abolition of infectivity—has been discussed in Chapter 1. For tailed phages, neutralizing antibodies are mainly those directed against the tip of the tail fibers that serve for specific attachment to host cells.

Even intact virions often have on their surface several different antigens recognized by one or another serological test and corresponding to different proteins (rather than to different antigenic determinants on the same protein). A clear example is adenovirus, whose component antigens can be separated and shown to correspond to different morphological sites on the virions (Figures 3-2, 3-29, and 3-30). Usually, some of the antigenic proteins of a virus are not accessible on the virion's surface and can be detected only after fractionation. Gel-diffusion precipitation is the most sensitive and rapid way to distinguish different antigens corresponding to antibodies present in a serum prepared against a virus.

Certain types of antigen-antibody reactions are especially useful in detecting and identifying virions and other virus-specific antigens within cells or tissues. In *immunofluorescence* a substance like fluorescine is covalently coupled to a purified antibody, which is then used directly or indirectly to trace the corresponding antigen microscopically (Coons, 1957). For electron microscopy an antibody is made electron-opaque by coupling with a protein such as ferritin, which is easily visible in micrographs.

Antibody reactions with virions may be reversible or irreversible depending on the types of antibody and on the nature of the antigen. An interesting case is that of influenza virus, whose virions are neutralized by complete,

FIGURE 3-29. *The morphology of antigenic components of adenovirus capsids.* (a) *The hexon unit, which in the capsid is surrounded by six adjacent capsomeres;* (b) *the penton unit, surrounded by five hexon units;* (c) *the fiber antigen, detached from the penton unit. See model of Figure 3-2. From Valentine and Pereira, J. Mol. Biol.* **13,** *13 (1965). Courtesy Dr. R. C. Valentine.*

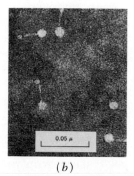
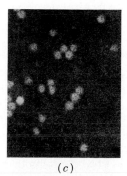

(a)   (b)   (c)

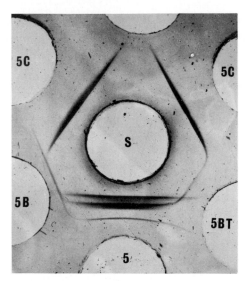

FIGURE 3-30. *Gel-diffusion precipitation plate showing the variety of antigens from adenovirus type 5 particles. The central well (S) contained rabbit antiserum against the virus; 5 was unfractionated antigen extract; 5B was purified B antigen (penton antigen); 5BT was trypsin-digested penton antigen; 5C was purified fiber antigen. Note the reaction of identity between the fiber antigen and the trypsin-digested penton antigen, and the partial identity (spur formation) between the complete penton antigen and the fiber antigen. From Valentine and Pereira,* J. Mol. Biol. **13,** *13 (1965). Courtesy Dr. R. C. Valentine.*

divalent antibody but not by univalent antibody fragments (Lafferty and Ortelis, 1963). It is suggested that the formation of antibody bridges between adjacent projections on a virion's surface may cause in the virion's envelope a reorganization like that produced by lectins on cellular membranes (Nicolson, 1974). Such reorganization may be the cause of loss of infectivity.

Serological cross-reactions among virions are recognized as the most reliable criteria of relatedness and, therefore, as criteria for classification. Often several viruses have in common certain antigenic determinants that allow classifying them in the same group or species. These can then be further subdivided into serotypes on the basis of finer differences detected with absorbed sera, that is, sera from which the antibodies reactive with a given virus or viral fraction have been removed by extensive precipitation. The residual antibodies serve to identify specific serotypes. Serological classification of virions agrees generally with morphological classification, but is sensitive to finer differences. Thus a series of morphologically indistinguishable phages or myxoviruses or rhabdoviruses or enteroviruses can be distinguished and typed serologically.

Most of the proteins of related but distinct viruses turn out to be serologically distinguishable, even though they may cross-react strongly. Presumably the evolution of these viruses has involved the accumulation of mutations, some selected for, others possibly neutral, which have caused most of the viral proteins to diverge. This suggests that related viral strains found in nature may be adaptive forms that have emerged from relatively long evolutionary processes rather than mutants recently derived from a common ancestor. In the case of viruses responsible for repeated epidemics, such as influenza there is evidence, however, that new mutant strains become established almost every year by selection in human populations that have acquired immunity to earlier strains (Francis, 1960).

As will be discussed in later chapters, in the course of their multiplication within host cells most viruses lead to the production of other proteins besides those that become part of virions. These nonvirion proteins are generally coded by viral genes and may be detected as antigens in infected cells or in their extracts. Moreover, virus-infected cells may acquire or lose antigenic determinants because of changes in glycoproteins and other cellular constituents (see Chapter 14). Particularly interesting instances are observed with tumor viruses, which transform normal cells into tumor cells. The transformation brings about the appearance of transformation antigens that are specific for each given virus irrespective of the species or tissue from which the cells originate. The relation of these virus-specific antigens to the mechanism of transformation will be discussed in Chapter 16.

The
Biochemistry
4 # of Viruses

The biochemistry of viruses is, in essence, the biochemistry of nucleic acids and proteins. Because viruses are one of the best sources of single molecular species of proteins and nucleic acids, some of the most significant advances in the understanding of the structure and function of macromolecules have come through the study of viral components.

It would not do justice to the present level of understanding of virus proteins and nucleic acids simply to describe their chemical characteristics. Functioning virus nucleic acids and proteins are *intracellular* constituents active in the dynamics of cell metabolism, and much discussion in later chapters centers on virus gene expression and virus proteins involved in intracellular steps of virion replication. Hence this chapter presents first the major ideas of gene structure, function, and regulation. An outline of recent advances in nucleic acid biochemistry is then provided, noting particularly the importance of the new techniques of sequencing, electron microscopic visualization, and molecular hybridization (strand reassociation) of nucleic acids, all of which allow determinations of the complexity and sequence arrangement within, and the relatedness between, nucleic acid samples. No separate section on protein biochemistry is given since most biochemistry textbooks concentrate on this area. A systematic treatment of nucleic acids and proteins occurring in virions concludes the chapter.

A Summary of Gene Expression

This section is not intended as a substitute for a thorough treatment of the well-established intricate details of gene expression, rather it is to remind the reader of those concepts, experimental approaches, and basic conclusions of molecular biology that are assumed to be known in the presentation of material elsewhere in the book.

63

Progress in the modern phase of the molecular biology of gene expression after the discovery of DNA structure by Watson and Crick in 1953, resulted from a marriage between genetic studies—isolation, mapping, and physiological analysis of mutants of bacteria and bacteriophages—and biochemical studies—the analysis of nucleic acid and protein biosynthesis in cells and cell extracts. Five central subject areas were explored with success, most complete in bacteria, but with some progress in eukaryotic cells as well:

1. Colinearity between genes and proteins
2. Messenger RNA (mRNA), the intermediate carrier of genetic messages
3. The genetic code
4. The protein-synthesizing apparatus and punctuation in the genetic code
5. Regulation in the manufacture of mRNA

COLINEARITY BETWEEN GENES AND PROTEINS

The basic tenet of molecular biology is that a linear sequence of bases in the DNA of a gene (or RNA if an RNA virus is under consideration) specifies the linear sequence of the amino acids in the polypeptide encoded by that gene. The first step in establishing a colinear relationship was to establish that a mutation caused an amino acid change in a protein. A mutant form of human hemoglobin provided the first evidence for this concept. Ingram (1958) demonstrated that sickle cell hemoglobin differed from normal hemoglobin by a single amino acid change (a "point" mutation eventually shown to be an exchange of valine for glutamic acid in position 6 of the β chain, Motulsky, 1975).

A great deal of additional combined biochemical and genetic work on the structural head protein of T4 bacteriophage (Sarabhai et al., 1964; see Chapter 7), as well as sequence analysis of the enzymes responsible for tryptophan synthesis in *E. coli* (Yanofsky et al., 1967) have clearly demonstrated that mutations that map in a linear fashion within a gene cause amino acid replacements or defects that are distributed in a corresponding linear manner in the polypeptide. The final proof of colinearity requires nucleic acid sequence as well as protein sequence data and consists of the demonstration that a change in amino acid sequence is accompanied by a change in nucleic acid sequence. This requirement has been fulfilled for the capsid protein of bacteriophage MS2. Not only has the RNA that encodes the capsid protein been sequenced, but so has the entire genome, 3569 nucleotides long (Fiers et al., 1976). The capsid protein has also been sequenced and the viral RNA coding for the capsid protein exactly matches the sequence expected from the genetic code. In addition, the RNA of virus capsid protein mutants exhibits the expected variations in sequence corresponding to mutant protein sites (Min-Jou et al., 1972).

MESSENGER RNA (mRNA): THE INTERMEDIATE CARRIER OF GENETIC MESSAGES

The mechanism by which genetic instructions from the DNA is "transcribed" into a messenger RNA which is then "translated" into protein became clear several years after molecular biologists had concluded that DNA sequences in genes were directly responsible for amino acid sequences in protein. A key reason for believing that RNA could be involved as a genetic intermediate rested on the discovery that certain plant and animal viruses contained RNA as their genetic material and that the viral RNA alone was infectious (Gierer and Schramm, 1956; Fraenkel-Conrat et al. 1957; Colter et al., 1957). When Jacob and Monod (1961) predicted the existence of a short-lived, unstable intermediate between the genes and the protein-synthesizing apparatus, the search for an RNA molecule with such characteristics was under way. Early hints of a phage-related RNA in bacteria infected with a DNA phage (Volkin and Astrachan, 1956) were followed by the convincing demonstration of an unstable RNA, newly synthesized after phage infection, that was associated with the preexisting bacterial ribosomes (Brenner et al., 1961). Definitive proof of the role of a messenger RNA in directing polypeptide assembly was provided by cell-free protein synthesizing systems. Extracts of normal *Escherichia coli* cells could be programmed by the RNA from the bacteriophage f2 to synthesize f2 specific proteins (Nathans et al., 1962).

mRNA has since been identified and studied in both bacterial and animal cells (Darnell, 1975). Many specific mRNA molecules, either viral or nonviral, have been shown to direct specific protein synthesis in a variety of cell extracts, confirming that specificity for protein synthesis lies with the mRNA not with the protein synthesizing system. In all cells the first step in the expression of genes is "transcription" of DNA to produce an mRNA.

THE GENETIC CODE

The evolution of the mRNA concept led Nirenberg to construct an artificial protein synthesizing system consisting of synthetic ribopolynucleotides as the mRNA and extracts of *E. coli* that would translate the synthetic polynucleotides (Nirenberg and Matthei, 1961). He found that poly(U) stimulated the synthesis of polyphenylanine, poly(C) the synthesis of polyproline, and poly(A) the synthesis of polylysine. The *code* contained within mRNA, and thus within the genes, that specified particular amino acids apparently could be discerned using synthetic RNA molecules to program synthetic polypeptide synthesis.

Frame-shift mutations, presumably the result of single-base deletions or insertions, within the rII genes of the T4 bacteriophages suggested the units that specified single amino acids, the *codons*, were three (or a multiple of three) nucleotides long (Crick et al., 1961). When three base-deletion mutations were grouped within the appropriate region of the rII locus the mutation was corrected; one, two, or four (single base-deletions) did not correct the lesion (see Chapter 7).

The chemical synthesis of ordered trinucleotides (Nishimura et al., 1965) plus the finding (Leder and Nirenberg, 1964) that each trinucleotide could direct the binding to ribosomes of individual aminoacyl tRNAs led to codon assignments for a considerable number of the 64, (4^3), possible trinucleotides. The chemical synthesis of alternating copolymers, followed by the translation of these copolymers, has allowed unambiguous assignment as amino acid codons of 61 out of the 64 possible triplets (Figure 4-1). To illustrate how

First position (5' end)	Second position				Third position (3' end)
	U	**C**	**A**	**G**	
U	Phe	Ser	Tyr	Cys	**U**
	Phe	Ser	Tyr	Cys	**C**
	Leu	Ser	TERM(OCH)	TERM	**A**
	Leu	Ser	TERM(AMB)	Trp	**G**
C	Leu	Pro	His	Arg	**U**
	Leu	Pro	His	Arg	**C**
	Leu	Pro	Gln	Arg	**A**
	Leu	Pro	Gln	Arg	**G**
A	Ilu	Thr	Asn	Ser	**U**
	Ilu	Thr	Asn	Ser	**C**
	Ilu	Thr	Lys	Arg	**A**
	Met	Thr	Lys	Arg	**G**
G	Val	Ala	Asp	Gly	**U**
	Val	Ala	Asp	Gly	**C**
	Val	Ala	Glu	Gly	**A**
	Val(Met)	Ala	Glu	Gly	**G**

FIGURE 4-1. The three-letter symbols for amino acids are standard (e.g. see A. L. Lehninger, "Biochemistry" 67, (1970), Worth, New York). TERM stands for termination, TERM (OCH) for the "ochre" termination triplet and TERM (AMB) for the amber termination triplet. GUG internal is for valine; can serve as initiator at beginning of mRNA.

definite codon assignments were made, consider the polymer CUCUCUCU, which presents only two codons, CUC and UCU, regardless of where translation begins; the polypeptide stimulated by this polymer contained alternating leucine and serine. A copolymer containing twice as much U as C, UUCUUCUUC, however, presents three possible codons, UUC, UCU, and CUU, depending on where the translation begins. Moreover, once a starting point is chosen the *same* codon repeats. This polymer directed the production of three pure products, polyphenylalanine, polyserine, and polyleucine. Each of the three codons was specific for one amino acid. From the examples given, it can be deduced that the one codon common to the two synthetic RNAs was UCU and the one amino acid present in both polypeptide products was serine, thus UCU must code for serine.

The presently accepted codon assignments given in Figure 4-1 are probably correct for all living systems on this planet, emphasizing the likely common evolutionary origin of all organisms from one source (Crick, 1968). A change in the meaning of a codon, that is, a change in the translation apparatus, would be a mutation so profound as to be lethal because every protein in the organism would be involved. A different dictionary of codons would probably demand an independent origin of life.

A number of general points should be noted about the code. Every amino acid except methionine and tryptophan has more than one codon, therefore the code is said to be *degenerate*. For a given amino acid, the multiple codons (in every case except serine, leucine, and arginine) have identical first and second positions but variation is allowed in the third position where C is frequently equivalent to U and G equivalent to A. Crick (1966) proposed that this variation in the 3' position of the codon was allowed because nonstandard base pairing with the 5' position of the anticodon in tRNA, together with standard pairing in the first two codon positions, was probably sufficient to allow one transfer RNA (tRNA) to recognize several codons in mRNA. This "wobble" hypothesis has been borne out particularly well, for example, by the findings with arginine and leucine, each of which has six codons. For both these amino acids there are two sets of codons that differ between sets in their second codon as well as within sets in the "wobble" position. The different codon sets are recognized by different tRNA molecules which discriminate between sets for the second position change, but which do not discriminate within sets for the change in the third position, the "wobble" position. The great importance of multiple tRNA molecules for individual amino acids in allowing "suppressor" mutations will become clear in the next section.

THE PROTEIN SYNTHESIZING APPARATUS AND "PUNCTUATION" IN THE GENETIC CODE

A list of the major structures involved in protein synthesis according to the scheme outlined in Figure 4-2 include ribosomes, tRNA, messenger RNA, and the protein factors for initiation, elongation, and termination of protein synthesis. All of these elements participate in protein synthesis in bacterial as well as eukaryotic cell-free systems (Zamecnik, 1960; 1965; 1969; Lengyel, 1974; Haselkorn and Rothman-Denes, 1973).

tRNAs AND RIBOSOMES.   Progress in this vast area of research began with the first cell-free incorporation of amino acids into peptide linkage by extracts of rat liver. (Zamecnik, 1969). The advantage of a cell-free system in uncovering each step in protein synthesis was illustrated when both a supernatant and particulate fraction were found necessary for amino acid incorporation. Crick's postulate (1958) of the need for an adaptor molecule that "translated" the language of nucleic acid sequence to the language of protein sequence was borne out by the demonstration that a key "supernatant" fraction in peptide synthesis consisted of small "soluble" RNA molecules, now known as transfer RNA, tRNA. The discovery of tRNA and the "activation" of amino acids by amino acylation of the common 3' adenosyl terminus of the tRNAs (Berg and Ofengard, 1958) was soon accompanied by the demonstration that specific amino acyl tRNA synthetases existed as did specific tRNAs for accepting each amino acid. In fact, for many amino acids, there are several different tRNA molecules capable of amino acid transfer.

The importance of the ribosome as the site for incorporation of amino acids into growing peptide chains was suggested in the earliest studies of cell-free protein synthesis in rat liver extracts. In addition, electron micrographic studies showed protein secreting tissues to be rich in granules which proved

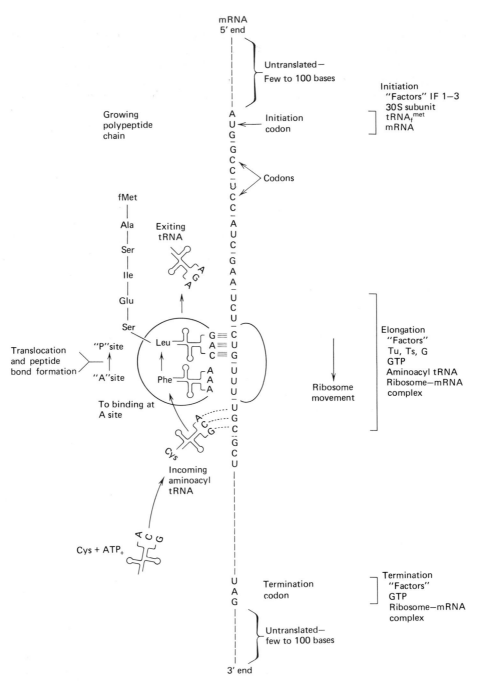

FIGURE 4-2. Bacterial protein synthesis. Most details are similar for eukaryotic systems. Three stages, initiation, elongation, termination and specific components involved in each are given on right. (Modified from A. L. Lehninger "Biochemistry" 652, (1970))

to be ribosomes (Jamieson and Palade, 1968). A conclusive demonstration that the ribosome was the site of active amino acid polymerization came after work on ribosomes in *E. coli* had established the uniformity in size and RNA content of ribosomes (Tissieres et al., 1959). When *E. coli* cells were pulse labeled with $^{35}SO_4^{-3}$ a precursor of methionine and cysteine, a large fraction of the newly made, "nascent" protein was found to cosediment with the ribosomes (McQuillen et al., 1959).

PROTEIN SYNTHESIS INITIATION AND TERMINATION: PUNCTUATION IN THE GENETIC CODE. With the establishment of tRNA as the amino acid carrier and the ribosome as the active site of peptide synthesis, considerations turned to the specific steps of starting and stopping protein synthesis. The first proof of the unidirectional nature of polypeptide chain growth, that is, for each chain there exists a fixed beginning and end, came from studies on hemoglobin synthesis. Dintzis (1961) found that ^{14}C leucine appeared in newly completed hemoglobin chains in a specific order. The trypsin-produced peptide fragments from the COOH-terminus became labeled before the NH_2-terminal fragments. Thus synthesis began at the NH_2-terminal end and proceeded to the COOH-terminal end. That the initiation step in protein synthesis might be similar for all proteins was suggested when about 30% of the proteins in *E. coli* were discovered to have methionine as their N-terminal residue (Waller, 1963). Soon thereafter, two methionyl tRNA molecules were identified in *E. coli* one of which carried N-formyl-methionine, an amino acid that could only be used to start a chain at the amino terminus. $tRNA_f^{met}$, as it was termed, recognizes the codon AUG as the specific start signal on the mRNA so that ribosomes can begin polymerization in the correct "reading frame" (Webster et al., 1966; Capecchi, 1966). Chain initiation in eukaryotic cells also involves a specific methionyl-tRNA but the methionine is not formylated (Lengyel, 1974).

The solution of polypeptide chain termination was much more complex than the problem of initiation. It involved partial protein sequence and genetic analysis of bacteriophage and bacterial (*E. coli*) mutants interpreted in light of knowledge of the genetic code (Garen, 1968). Mutations that cause insertion of one amino acid for another are termed *missense* mutations. Polypeptides bearing such mutant amino acids are often ineffective, for example, in carrying out enzymatic activity, but detectable as immunologically cross-reacting material (CRM, pronounced "crim"). Mutations that did not produce any CRM suggested a barrier in transcribing the gene or in translating the mRNA past the mutant sites; these mutations were termed *nonsense* mutations. The protein products were analyzed from a group of nonsense mutations in the head protein gene of T4 bacteriophage that mapped genetically in a linear order (Sarabhai et al., 1964). The mutant phages produced fragments of head protein containing the N-terminal peptides but terminating prematurely at various distances along the polypeptide chain, which allowed an ordering of the mutations on biochemical grounds. Most important, the genetic order corresponded to polypeptide length order. A similar group of nonsense mutants was isolated for the bacterial enzyme, alkaline phosphatase, and the corresponding protein products also lacked the –COOH terminal regions of the enzyme protein (Garen, 1968). The cause of the premature termination in the alkaline phosphatase nonsense mutants was revealed through the analysis of yet another type of bacterial mutation. Bacterial strains exist that correct or "suppress" certain of the nonsense mutations of both bacteriophage and alkaline phosphatase, although the structural gene defect still exists when the original nonsense mutation is transferred back to nonsuppressing strains.

The different nonsense mutations were grouped according to which *E. coli* strains could suppress each of the mutations. Two of the most important groups of mutants defined in this way were *amber* mutants and *ochre* mutants. When a set of *amber* mutations of alkaline phosphatase were each introduced into different amber suppressor strains, the amino acid inserted at the mutant site was determined to be always serine in one strain, glutamine in another strain, or tyrosine in a third strain. In addition, a series of alkaline phosphatase proteins purified from independent revertants at one particular *amber* site was found to contain a limited set of amino acids inserted at the reverted site. The data on amino acid substitution in suppressed strains plus the data from the amber revertants, together with knowledge of the genetic code, allowed the deduction that the amber codon was UAG (Figure 4-3). Since this codon had not been assigned to encode any amino acid, it was concluded that UAG encoded chain termination. The *amber* mutations then were abnormal introductions of the normal signal for polypeptide chain termination. Similar conclusions were also reached for UAA as the premature chain terminator in *ochre* mutants.

How did suppression operate? Since the chain terminations were ex-

FIGURE 4-3. *Identification of an* amber *nonsense triplet as UAG. The diagram summarizes the pattern of amino acid substitutions in the enzyme alkaline phosphatase produced by revertants of a phosphatase* amber *mutant. Each substitution occurred at the same position, which is occupied by a tryptophan residue in the standard phosphatase molecule. All of the codons for the substituted amino acids are listed around the outside of the circle; the underlined codons are those related to UAG by a single base change. UAG is the only triplet that can in one base change become at least one of the codons for each of the seven substituted amino acids. From A. Garen* Science **160**, *149–159 (1968))*

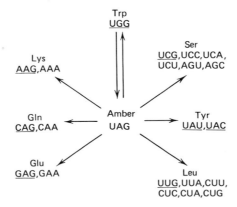

pressed during translation, correction was most probably caused by another mutation in some part of the translation machinery. Because multiple sets of tRNA molecules for individual amino acids were known, it was guessed that mutant tRNA molecules, probably in tRNA species present in low amounts, might be responsible. The proof of this idea came from *in vitro* protein synthesis; for example, RNA from the bacteriophage F2, carrying amber mutations in the capsid protein, (Notani et al., 1968) could be translated by *in vitro* extracts of wild type *E. coli* if tRNA from suppressor strains was added.

The use of suppressor strains and chain termination mutations have been extraordinarily powerful in bacteriophage genetics as discussed in Chapters 7 to 10. The experiments that revealed the nature of nonsense mutations also represent one of the most elegant applications of a unified genetic and biochemical approach to a problem of gene expression. With the conclusion of these experiments the genetic code was considered complete.

BIOCHEMICAL DETAILS OF PROTEIN SYNTHESIS.   More recent progress in understanding protein synthesis has featured isolation and characterization of proteins that play a role in each of the many steps of protein synthesis (Lengyel, 1974; Haselkorn and Rothman-Denes, 1973). There are *initiation factors* (at least three in *E. coli* and four in mammalian systems) which mediate proper, ordered initiation of peptide synthesis by $tRNA_r^{met}$-ribosome-mRNA complexes (Figure 4-2). One of the initiation factors in *E. coli* (F-3) and a separate factor in mammalian cells causes whole ribosomes to dissociate into subunits so that a new round of translation can begin. *Elongation factors* include several major proteins on the surface of both bacterial and mammalian ribosomes which are responsible for proper tRNA and amino-acyl-tRNA binding. These factors also cleave GTP, which releases energy for the translocation of the growing peptidyl-tRNA chain from one binding site (A) to another (P) site. Coincident with this translocation the mRNA and ribosome move with respect to each other a distance of one codon. The bacterial elongation factors, which are termed EF-Tu, EF-Ts, and EF-G, can exist as a single complex and are present in the cell in high concentrations. Tu and Ts have the interesting property of being part of the RNA replicase that synthesizes the RNA in RNA bacteriophage synthesis (Kamen, 1975). Finally, there are *termination factors*, proteins responsible for the cleavage of the peptidyl tRNA and release of the completed polypeptides. The binding of these factors to the ribosome is promoted by the termination triplets UAA and UAG. GTP is consumed during the termination process.

In addition to the many "factors" necessary for protein synthesis, the total protein composition of bacterial ribosomes is known (Wittman, 1974), and disassembly and reassembly of functional ribosomes has been achieved, making possible the assay for function of specific ribosomal proteins (Nomura and Held, 1974). Ribosomal protein separation from eukaryotic ribosomes has been greatly improved since the introduction of two-dimensional gel electrophoresis separations (O'Farrell, 1975) and reveals a somewhat more complex structure than with bacterial ribosomes. Ribosomes, however, appear to lie within the range of complexity that can be solved with present technology and work proceeds toward an eventual three-dimensional picture of ribosomal function during protein synthesis.

Almost all, if not all, of the machinery of protein synthesis used by viruses during their replication is borrowed or stolen from the cell. Only minor changes have thus far been detected, such as new tRNAs in bacteriophage infection (see section on RNA sequencing in this chapter). A widespread belief exists that subtle alterations in protein synthesis initiation play a role in the antiviral state induced in mammalian cells by interferon as well as in the inhibition by animal viruses of host-cell protein synthesis, but specific changes have not yet been documented (see Chapter 15).

THE REGULATION OF mRNA MANUFACTURE

The final area of molecular biology with which the student of virology should be familiar is the connection between the mechanical facts of gene expression—transcription and translation—and the regulation of gene expression—the call to action of a particular gene. The basic precepts of the regulation of genes are thus far understood only in bacteria. Hence the framework for considering gene regulation in all cells must at the moment derive from knowledge of bacterial regulation.

The steps in understanding gene regulation in bacteria were: (1) the discovery that certain genes varied in the level of production of their protein products under varying experimental conditions and (2) the isolation of mutants that retained normal structural information for the protein product but had lost the regulated response to external conditions. This work defined *regulatory genes* as separate from *structural genes* (Jacob and Monod, 1961). The first studies of regulatory mutations concerned genes that affected enzymes of lactose metabolism, histidine biosynthesis, and the regulation of growth of the temperate bacteriophages particularly phage λ (see Chapter 9). More recent work has involved, as important variants of the basic examples, genes concerned with arabinose metabolism (Wilcox et al., 1971; Goldblatt and Schleif, 1971), tryptophan metabolism (Bertrand et al., 1975), and bacteriophage P22 growth (Botstein and Suskind, 1974).

The central conclusion to come from all these genetic and biochemical studies is that sites exist in DNA, known now in at least two instances to consist of specific, unusual base sequences, which serve as recognition signals for the binding of specific proteins. Two of these regulatory proteins, one that represses the lactose operon and one that normally prevents the lytic cycle of bacteriophage λ, have been purified; their interaction with binding sites in the DNA termed *operators* has been shown to be site specific (Gilbert et al., 1974; Maniatis et al., 1975; see Chapter 9).

The regulatory proteins are activated in many instances by binding to small molecules, *effectors,* in combination with which the regulatory protein binds to the operator. Transcription of all the genes in a cluster, an *operon,* is then affected so that RNA polymerase is either prevented from initiating synthesis, *negative regulation,* (e.g. in β galactosidase formation) or caused to initiate more frequently, *positive regulation.* (e.g. as in arabinose metabolism) The specific site on the DNA to which the RNA polymerase binds and begins transcription is called the *promoter.* The promoter and operator are probably always close in bacteria and, in fact, in the lactose operon of *E. coli* are overlapping. All these regulatory elements—the promoter, the operator, the

repressor, and activator proteins—were discovered by characterizing mutations that rendered the element nonfunctional. By virtue of this rigid transcriptional control plus a very short turnover time for mRNA, bacteria very
efficiently regulate their protein synthesis.

For many bacterial genes, in addition to a major regulatory element which
exerts an almost all or nothing control, more subtle regulatory elements also
exist that affect the efficiency of already permitted transcription. For example,
growth of *E. coli* on glucose, simultaneously with lactose, causes β-galactosid-
ase to be only sluggishly induced. This "catabolite repression" is mediated by
a positive regulatory protein which does not bind to DNA in the absence of
cyclic (3',5') AMP (cAMP) (Zubay et al., 1970). When the cell's supply of
glucose is depleted, cAMP rises and the cAMP-binding protein plus the cAMP
now can bind to the lactose operon and promote transcription of more mRNA
from the region. Other newly discovered features of regulation include an
element in the tryptophan genes of *E. coli* termed an *attenuator* (Bertrand et al.,
1975). This site lies 166 nucleotides from the promoter. Normally about 80–
90% of the RNA polymerase molecules that start to transcribe the tryptophan
operon terminate transcription at the attenuator site. Severe tryptophan
deprivation calls on the cells to make enzymes for tryptophan synthesis, and
about half the polymerase molecules then do transcribe into the structural
gene regions for the tryptophan enzymes.

Perhaps the most complicated sets of interrelated regulatory proteins thus
far successfully studied are those concerned with temperate bacteriophage
regulation (see Chapters 9 and 10).

In eukaryotic cells a number of conditions are known in which the mRNA
content of cells changes, for example, during differentiation and after hormone
treatment. How many regulatory genes are involved and how the presumed
cycles operate is unknown at present.

Nucleic Acid Biochemistry

Many of the conclusions of the previous section of this chapter as well as
conclusions to be drawn in succeeding chapters are based on recent developments in nucleic acid research. The development of many of the newer
techniques, such as electron microscopic visualization of RNA and DNA,
nucleic acid sequencing, and quantitative hybridization, relied on viral
nucleic acids for appropriate material and greatly benefited virology in the
process. This next section is devoted to recent highlights in the rapidly
expanding field of nucleic acid research.

NUCLEIC ACID ENZYMES

Relatively little is yet known at the intracellular level about how the
function of nucleic acid enzymes is integrated with cell needs; on the other
hand, the science of enzyme biochemistry with crude cell-free systems and
purified enzymes is quite advanced and much of the progress in nucleic acid
sequencing and in the studies of DNA organization has depended on the
sophisticated use of various specific enzymes.

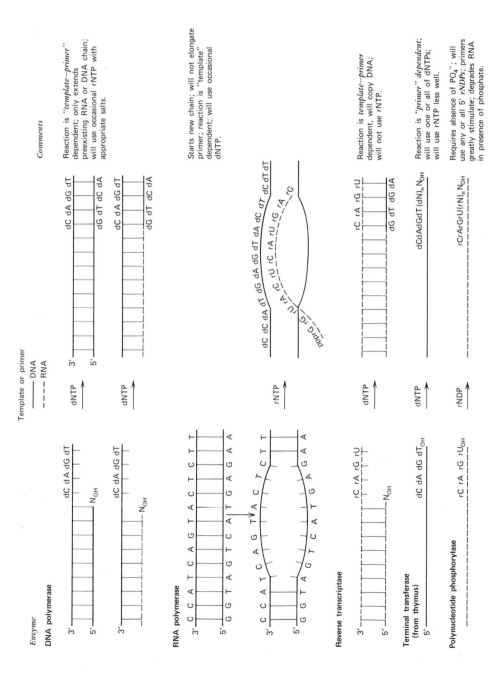

FIGURE 4-4. *Addition of nucleotides by polymerases and transferases.*

Enzymes involved in nucleic acid metabolism perform three basic reactions: *polymerization, modification,* and *degradation.*

1. Polymerization is carried out by *DNA polymerases* or *RNA polymerases* which add a monophosphate residue to the 3'OH end of a growing chain using ribo- and deoxyribonucleoside triphosphates as substrates. The order of nucleotide addition is determined by faithful copying of a *template* strand of RNA or DNA according to the rules of Watson-Crick base pairing. RNA polymerases can start a new chain; DNA polymerases require a *template-primer*; that is, a hybridized short "primer" segment of RNA or DNA which is elongated by copying the long "template" strand.

2. Modification of already synthesized nucleic acid chains occurs in three main ways: chemical changes in single nucleotides, a decrease or an increase in transcribed chain length. Enzymes exist that simply add or change chemical groups on individual nucleotides, for example, methyl group addition, rearrangement of ribose linkage to change uridine to pseudouridine, isopentenyl group addition, and glycosylation. Then, enzymes exist for cutting and trimming already formed chains in a controlled manner. For example, RNA processing enzymes exist that cleave chains at specific sites and remove unused parts of RNA molecules (Dunn, 1974; Smith, 1976). Other modifying enzymes increase the length of already existing chains. *DNA ligases* join either growing pieces of DNA or pieces of DNA that have suffered controlled endonucleolytic cuts. *Terminal addition enzymes* (ribo- or deoxyribonucleotidyl transferases) add onto either RNA or DNA *primer* chains, without need for a template, a few (for example, -CCA on the 3' terminus of tRNA) or many nucleotides (for example, poly(A) which is 75 to 200 nucleotides long on the 3' terminus of eukaryotic mRNA).

3. Degradation of nucleic acids is carried out both intra- and extracellularly, both endo- and exonucleolytically, by myriad RNAses and DNAses. The extracellular enzymes result in a recycling of the purine and pyrimidine bases as building blocks. No function is known for many of the intracellular nucleases. Some of the intracellular degradative enzymes in mammalian cells probably also serve a simple recycling function because they are contained in lysosomes (De Duve, 1959), organelles thought to digest incoming particles, but which may also play a role in intracellular turnover. Other intracellular nucleases may act only in a limited, highly specific manner; for instance, the restriction endonucleases in bacteria (Arber, 1974) which introduce a limited number of scissions in incoming foreign nucleic acids leading eventually to complete destruction.

DNA POLYMERASES. In bacteria such as *E. coli,* three DNA polymerases have been purified (Kornberg, 1974; Gefter, 1975). Each of these three polymerases uses deoxynucleoside triphosphates and incorporates a deoxynucleoside monophosphate on the 3'OH end of growing DNA chains (Figure 4-4). None of the three enzymes can initiate DNA synthesis *de novo* and it is generally believed that a short ribonucleotide chain serves as the initiator for elongation (Sugino and Okazaki, 1973; Bouche et al., 1976). Chain growth then proceeds from 5' → 3' with faithful Watson-Crick base-pair copying providing the instructions for the sequence of the new DNA chain. Although the exact role of each polymerase in cellular or viral DNA replication and/or repair is not completely settled, polymerase III would appear indispensable for cell replication. Bacterial strains with defective polymerase I are more sensitive to ultraviolet light than wild-type, suggesting a definite role for this enzyme in repair synthesis. A part of the polymerase I protein has a 5' → 3' exonuclease which can locate "nicks" in DNA, remove a few nucleotides, and then add nucleotides to fill the gap (Lehman and Uyemura, 1976). This so-called "nick-translation" may normally contribute to chromosome replication by the removal from the 5' end of DNA fragments of an RNA primer, thus promoting the joining of discontinuously synthesized DNA fragments. No definite function for polymerase II has been assigned (Gefter, 1975). Much less is known of the function of DNA polymerases of higher cells, although purification of several enzymes has been achieved (Bollum, 1975).

Many viruses stimulate the formation of new enzymes which have DNA polymerase activity. In some cases, such as bacteriophage T4 infection, new enzymes which are clearly the product of viral genes are necessary for viral DNA synthesis (see Chapter 7). However, the exact role of virus-specific, host-specific proteins and primer RNA in viral initiation is just becoming known for even the simplest bacteriophages (Bouche et al., 1976).

As will be described in Chapter 13, a new type of DNA polymerizing activity has been discovered in the virions of RNA tumor viruses (Figure 4-4). This enzyme can copy the RNA of the virion into DNA (Temin and Baltimore, 1972), and, in the infected cell, the new viral-specific DNA becomes integrated into the cell DNA. Because this enzyme reverses the normal direction of information flow (that is, not DNA → RNA but RNA → DNA) it has been termed "reverse transcriptase." Like all other DNA polymerases, however, it does not initiate chain growth but elongates from the 3-OH terminus of a preexisting RNA chain.

RNA POLYMERASES. The chief cellular enzymes for RNA polymerization are those responsible for copying DNA into RNA, the DNA-dependent RNA polymerases (for review on bacterial polymerases see Losick, 1972; for eukaryotic polymerases see Chambon, 1975). Given ribonucleoside triphosphates, Mg^{++} and a DNA template, the bacterial RNA polymerases can bind to and initiate RNA chains within specific DNA sites, *promoters*, in both bacterial and bacteriophage DNA. These specific sites are known in several cases to be the same *in vivo* and *in vitro* (see Chapter 8). The first nucleotide in the new chain is almost always a purine that remains at the 5' end of the new chain as a pppGp- or pppAp-. Chain elongation proceeds 5' → 3' by sequential addition of nucleoside monophosphates in an order dictated by Watson-Crick base pairing with one of the two DNA strands (Figure 4-4).

The basic complex of four polypeptide subunits of the bacterial polymerase ($\alpha_2 \beta\beta'$) is helped to choose promoters properly by a fifth subunit called sigma, σ (Burgess et al., 1969). In many bacteria this single enzyme apparently suffices for all DNA-directed RNA synthesis, although in special situations, such as sporulation of *B. subtilis*, there is evidence of a changed enzyme due to a different type of sigma-factor (Pero et al., 1975). Some DNA-containing bacteriophages either encode their own RNA polymerases or encode proteins that modify the cellular enzyme (see Chapters 7 and 8). The proper termination of RNA polymerase action, perhaps equally important with proper initiation, is much less well understood. A protein factor, termed *rho*, appears to induce chain termination at particular sites along the DNA chain and the participation of this protein in the transcription of viral genes is detailed in Chapters 9 and 10.

Eukaryotic cells, from yeasts to mammalian cells, possess at least three distinguishable RNA polymerases which function separately to synthesize pre-rRNA (I), pre-tRNA (III) and 5S RNA (III) and the (hnRNA) nuclear RNA related to mRNA (II) (Chambon, 1975). It is likely that some animal viruses use the cell RNA polymerase II for their mRNA formation. The poxviruses possess their own, virion-associated, DNA-dependent RNA polymerases for the manufacture of viral mRNA (see Chapter 14).

In addition to the DNA-dependent RNA synthesis, virus-infected cells can contain virus-specific, RNA-directed RNA polymerases. The virus enzyme(s) responsible for RNA virus replication, however, has been purified only in the case of RNA bacteriophages of *E. coli* (Chapter 8). Cells not infected by viruses have never been found to possess RNA-dependent, RNA-synthesizing systems.

METHYLASES. Methylation is perhaps the most widespread nucleic acid modification (Davidson, 1972): 5-methyl cytosine represents as much as 5–6% of the total cytidylate residues in plant DNA and 1–1.5% in animal DNA; about 40 different methylated bases have been reported for tRNA; r-pre-RNA molecules are extensively methylated both on bases and on the 2' OH of the ribose and mRNA from eukaryotic cells contains methyl groups including an unusual 5' blocking "cap" structure (see section on viral RNA).

Many methylation enzymes have been detected and at least partially purified. Almost all of the methylation reactions involve a methyl group transfer to the nucleic acid from the methyl donor, S-adenosylmethionine. These enzymes must be highly site specific because the same bases and ribose moieties are modified in all rRNA molecules (Maden and Salim, 1974) and in each specific tRNA molecule (Smith, 1976). By and large the enzymes which modify pre-tRNA and pre-rRNA do not affect mRNA molecules or viral molecules and no specific evidence of virus-induced destruction of the methylation enzymes for ribosomal or tRNA has been detected. During bacteriophage T3 infection, a phage-specific enzyme is synthesized which destroys S-adenosyl methionine, preventing any further cellular methylations (Gold et al., 1964). The role of most RNA methylations remains unknown; however, some evidence supports a necessity for certain tRNA modifications to promote maximal intracellular function of the tRNA (Smith, 1976). The role of methylation in protecting DNA from restriction enzyme cleavage in bacteria is discussed in Chapter 6.

TERMINAL ADDITION ENZYMES (non-template-dependent nucleotidyl transferases). Many enzymes have been detected that will add single bases or repetitively add many bases to the 3' terminus of an RNA or DNA chain (Figure 4-4). The –CCA terminus, which, of course, is necessary for proper tRNA amino acid acceptor activity, is perhaps one of the better known addition products. Enzymes for both the removal and the step-wise addition of –CCA to tRNA have been described from both bacterial and animal sources (Deutscher, 1973).

In recent years another posttranscriptional terminal addition product has received great attention. A majority of eukaryotic cell mRNA molecules are terminated at the 3′ end with segments of polyadenylic acid which is added after transcription of nuclear RNA is completed (Darnell, 1975; Sawicki et al., 1977). After arrival in the cytoplasm the terminus of the mRNA molecule undergoes both shortening to an eventual size of less than 50 nucleotides as well as stepwise single nucleotide additions that can reach 10 to 15 in length (Brawerman, 1976). The function of all this metabolic activity is unknown. A number of enzymes capable of poly(A) synthesis have been described (Edmonds and Winter, 1976) but the details and purpose of poly(A) addition and turnover have yet to be elucidated. The overall process of nuclear poly(A) synthesis and cytoplasmic appearance of poly(A) containing mRNA is strongly inhibited by 3′-deoxyadenosine (3′dA), while nuclear RNA synthesis is not so inhibited. Thus a poly(A)-synthesizing enzyme that is sensitive to 3′dA must exist (Sawicki et al., 1977). Many viruses also have poly(A) termini on their mRNAs and some appear to use cellular channels for poly(A) synthesis (for example, adenovirus, herpesvirus, retroviruses). Other viruses carry their own poly(A) synthesizing enzymes (for example vaccinia) and still others have poly(A) transcribed during replication from poly(U) (polio) (see Chapter 13).

In addition to poly(A), terminal addition enzymes have been discovered which produce no known products inside cells. For example, the widely used deoxynucleotidyl terminal transferase (Chang and Bollum, 1971), which can extend the 3′ end of single-stranded DNA chains with any single base, has proved very useful for *in vitro* polynucleotide synthesis (Figure 4-4).

DNA LIGASE.   Several situations in the normal nucleic acid economy of a cell require one strand of a DNA double helix to be specifically broken and

FIGURE 4-5. Enzymatic attacks on DNA and RNA: List of exo- and endonucleases

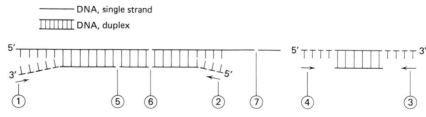

1. Exonuclease III of *E. coli*, duplex specific
2. DNA polymerase I of *E. coli*, duplex specific; λ exonuclease, duplex or single —strand
3. DNA polymerase I of *E. coli*, single—strand specific; snake venom diesterase, single—strand specific
4. Neurospora exonuclease and spleen phosphodiesterase, single—strand specific
5. DNAse I (pancreas)
6. DNAse II (spleen)
7. S1 (Aspergillus oryzae), single—strand specific; DNAse I and II

the pieces eventually rejoined, for instance, excision and repair of irradiation damage or breakage and reunion associated with recombination (Grossman, et al., 1975). In addition a joining activity is required during DNA replication (Kornberg, 1974; Gefter, 1975; see page 84). The enzyme responsible for these tasks is *DNA ligase* which joins a 5' phosphate terminus with a 3' OH terminus. *E. coli* ligase uses DPN as a cofactor while animal cell ligases and those induced during bacteriophage T4 and T7 infection use ATP as cofactor. It is not known whether animal cells infected with viruses produce virus-specific ligases. (Lehman, 1974)

ENDONUCLEASES—GENERAL. Among the earliest enzymes to be purified were DNAses and RNAses from mammalian pancreas where they are formed to serve as digestive enzymes in the gut (Davidson, 1972). These endonucleases, which digest DNA or RNA to small oligonucleotide fragments, have been widely used as laboratory reagents and the mode of action, especially for RNAse, well established. (Figure 4-5)

DNAses. DNAse I makes single-stranded breaks without a preference to neighboring nucleotide, leaving a 3' OH end. When these breaks are numerous enough, double-stranded DNA collapses. DNAse II, an enzyme purified from spleen and thymus, makes mainly double-stranded scissions leaving 3' PO_4 ends, again with no site preference. In recent years, a single-strand specific nuclease (for both DNA and RNA) termed S1 (from *Aspergillus oryzae*) has been widely used in experiments where duplex structures were to be retained and single strands removed (Sutton, 1971).

RNAses. RNAse A (pancreatic) and RNAse T1 are especially widely used. They are examples of a type of enzyme, present in virtually all kinds of cells, that cleaves *single-stranded* RNA endonucleolytically through the forma-

that will perform nucleolytic functions indicated.

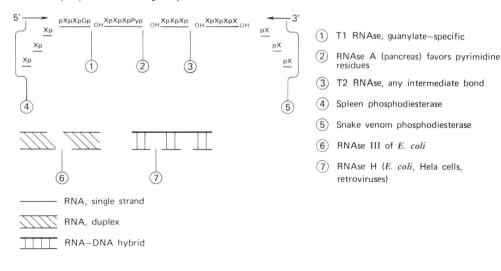

(1) T1 RNAse, guanylate—specific

(2) RNAse A (pancreas) favors pyrimidine residues

(3) T2 RNAse, any intermediate bond

(4) Spleen phosphodiesterase

(5) Snake venom phosphodiesterase

(6) RNAse III of *E. coli*

(7) RNAse H (*E. coli*, Hela cells, retroviruses)

———— RNA, single strand

RNA, duplex

RNA—DNA hybrid

tion of cyclic phosphate intermediates between the 2', 3' OH groups of ribose (Davidson, 1972). This cyclization is blocked by methyl group addition to the 2' OH of ribose, a frequent modification in cellular RNA molecules. RNAse A preferentially cleaves on the 3' side of a pyrimidine, RNAse T1 preferentially on the 3' side of guanylate, and RNAse T2 will cleave at any internucleotide bond. The property of cutting RNA at a single nucleotide has made RNAse T1 indispensible in RNA sequence studies (Fiers et al., 1976; Brownlee et al., 1968). Another endoribonuclease that has been recently used in sequence studies is RNAse U2 (from *Aspergillus ustilago*) which preferentially cleaves after purines, and thus if T1-created oligonucleotides (terminated by -Gp) are U2-treated they are cleaved after adenylate residues.

The *resistance of double-stranded* RNA to endonucleases presumably rests on the unavailability of bases as recognition sites when hydrogen bonds exist between the bases. This property of RNAse resistance of double-stranded RNA is widely used in studying viruses with double-stranded RNA as well as the replicative forms of single-stranded viruses.

A novel enzyme, termed ribonuclease H, has recently been discovered in cells (Hausen and Stein, 1970) and in the virions of retroviruses (Molling et al., 1971; Baltimore and Smoler, 1972). This enzyme hydrolyzes RNA that is hybridized to DNA and in uninfected cells may be involved in removing the RNA primer for DNA synthesis. In retrovirus-infected cells the enzyme may play a role in constructing the DNA to be integrated into the cell DNA. This enzyme provides a reagent that selectively destroys the RNA portion of hybridized RNA:DNA complexes (Figure 4-5).

EXONUCLEASES. Exonucleases for DNA probably exist with almost any conceivable type of specificity; for example, $3' \rightarrow 5'$, $5' \rightarrow 3'$, single- and double-strand specific enzymes are known (see Figure 4-5). Exoribonucleases, which are $5' \rightarrow 3'$ and $3' \rightarrow 5'$, are also known. The physiologic role for none of these enzymes is known with certainty but the exodeoxyribonucleases may participate in repair processes or error correction on damaged DNA (Gefter, 1975). As pure laboratory reagents, this group of enzymes has enormous value in tailoring DNA chains to the demands of the biochemist. For example, in addition to partial digestion to create molecules with partially single-stranded termini, these enzymes also afford a means of liberating 3' or 5' monophosphates where the phosphate is either attached to the nucleoside with which it entered the chain or transferred to its 5' neighbor. This enzymatic transfer (carried out with spleen $5' \rightarrow 3'$ exonuclease) on enzymatically synthesized DNA or RNA labeled with a single nucleoside triphosphate forms the basis for the so-called "nearest neighbor" analysis (Josse et al., 1961).

NUCLEASES POSSIBLY ASSOCIATED WITH DNA FUNCTION. DNAses that act endonucleolytically at specific sites are widely believed to have specific physiologic roles. For example, the thymine-thymine dimers created by ultraviolet light are enzymatically excised in order to begin the repair of damaged DNA (Grossman et al., 1975). Other endonucleases may act at highly specific sites but methods for detecting such specificity have yet to be developed. Recombination requires breakage and reunion of two double-stranded molecules and is not limited to specific sites. Bacterial strains mutant in the enzymes of recombination are known (see Chapters 6 to 10), but as yet no definite assignments of function to specific endonucleases have been made (Grossman et al., 1975). It may be that cell DNA replication in mammalian cells, where DNA synthesis occurs at many independent sites or *replicons,* requires a group of such specific enzymes. Which, if any, of the presently known nucleases subserve such functions is unknown.

A limited endonucleolytic activity exhibited by the so-called "nicking-closing" enzyme is a particularly good candidate to function in the growth of small covalently closed circular DNA molecules. This type of enzyme, first discovered in *E. coli* (Wang, 1971), was detected by its ability to relieve the supercoils that exist in closed circular duplex DNA molecules by "nicking" and then rejoining at the same site. As discussed later, SV40 is replicated bidirectionally from one point as a closed circular duplex DNA (Sebring et al., 1971; Salzman et al., 1973, Chapter 14). A nicking-closing activity would seem to be required to allow progression of the growing points. Such an enzyme could operate whenever a "swivel function" is required in DNA synthesis (Pulleyblank et al., 1975), and a large amount of this type of enzyme is found in mammalian cell nuclei. Nicking-closing enzymes might also function during RNA transcription when a topological problem results because the ever-lengthening RNA chain must rotate around the DNA helix as the RNA polymerase moves along the DNA. The necessity for RNA rotation might be relieved by a nicking-closing activity, which allows the DNA to rotate freely. Such an activity has recently been detected in vaccinia virions which have the capacity to synthesize and extrude RNA from DNA that is covered with protein (Bauer et al, 1977).

Many DNA-containing bacteriophages and animal viruses encode DNAses and nicking-closing enzymes which may play important but as yet undefined roles in synthesizing viral DNA and viral-specific RNA.

RESTRICTION ENDONUCLEASES. One group of bacterial DNAses clearly seems to serve a protective function against intruding foreign DNA (Arber, 1974; Nathans and Smith, 1975). This group of enzymes, termed *restriction enzymes*, function together with a paired set of *modification enzymes* (usually methylation enzymes) to recognize and preserve homologous DNA and recognize and destroy foreign DNA (see Chapter 6 and Arber, 1974). The recognition is thought to be accomplished by the presence of a methyl group at or near an otherwise "forbidden" DNA sequence. Without methylation the site is cleaved; with the methylation it is resistant. Although host restriction-modification systems were first studied in *E. coli* (Meselson and Yuan, 1968), the first site-specific restriction enzyme was isolated from *Hemophilus influenza* by Smith and his colleagues (Smith and Wilcox, 1970; Kelly and Smith, 1970). This enzyme, now referred to as Hin II, not only makes a limited number of breaks in a foreign bacteriophage DNA but all the termini are pGpTpPy,3' and 5'pPupApCp—. Thus, the double-stranded sequence from which these termini derive is

$$5'. . . .pGpTpPy \vdots pPupApCp—3'$$
$$3'. . . .pCpApPu \vdots pPypTpGp—5'$$

which possesses two-fold rotational symmetry. By now a battery of site-

TABLE 4-1 SITE-SPECIFIC RESTRICTION ENDONUCLEASES

Strain	Enzyme	Sequence (5' → 3')	Number of cleavage sites			Isoschizomeric enzymes
			λ	Ad2	SV40	
Escherichia coli (end I⁻, R⁺, RI)	EcoRI	G↓AATTC	5	5	1	
Escherichia coli (end I⁻, R⁺, RII)	EcoRII	↓CCTGG	>35	>35	16	
Bacillus amyloliquefaciens H	BamHI	G↓GATCC	5	3	1	
Haemophilus aegyptius (ATCC 11116)	HaeIII	GG↓CC	>50	>50	17	BsuX5, HhgI
Haemophilus influenza, serotype d	HindII	GTPy↓PuAC	34	>20	7	HincII
	HindIII	A↓AGCTT	6	11	6	HinbIII, HsuI, BbrI
Haemophilus parainfluenza	HpaI	GTT↓AAC	11	7	4	ApoI
	HpaII	C↓CGG	>50	>50	1	HapII, Mno1
Haemophilus haemolyticus (ATCC 10014)	HhaI	GCG↓C	>50	>50	2	
Serratia marcescens S$_b$	Sma I	CCC↓GGG	3	12	0	
Streptomyces albus	Sal I	—	2	3	0	

Additional, less well-characterized enzymes and the strains from which they are derived are as follows: Ava II, Anabaena variabilis; AluI, Arthobacter luteus, ATCC 21606, HaeII, Haemophilus aegyptus, ATCC 11116; HgaI, Haemophilus gallinarium, ATCC 14385; HinH-1, Haemophilus influenza H-1, HphI, Haemophilus parahaemolyticus; MboI, MboII, Moraxella bovis, ATCC 10900; SacI, SacII, Streptomyces achromogenes ATCC 12767; XamI, Xanthomonas amaranthicola ATCC 11645, HapI, Haemophilus aphrophilus, ATCC 19415.

Isoschizomeric enzymes possess same cleavage specificity; the isoschizomeric enzymes are derived from the following strains: ApoI, Arthrobacter polychromogenes, ATCC 15216; BsuX5, Bacillus subtilis X5 (a transformable strain): BbrI. Bordetella bronchioseptica, ATCC 19395: HapII, Haemophilus aphrophilus, ATCC 19415; Hhgl, Haemophilus hermoglobinophilus, ATCC 19416: HinbIII, Haemophilus influenzae, serotype b; HincII, Haemophilus influenzae, serotype c: Hsul, Haemophilus suis, ATCC 19417; Mnol. Moranella nonliquefaciens, ATCC 19775 (from Nathans and Smith, 1975).

specific restriction enzymes have been isolated (Table 4-1) and the great majority share the property of recognizing a 4 to 6 nucleotide sequence with two-fold rotational symmetry. Some of the enzymes such as *EcoR1* (from *E. coli*) make "staggered" breaks,

$$5'. . .pGp\vdots ApApTpTpCp. . .3'$$
$$3'. . .pCpTpTpApAp\vdots Gp. . .5'$$

that leave complementary "cohesive" ends.

When a purified species of DNA is digested by restriction enzymes, a regular set of fragments, *restriction fragments*, is produced which can be separated by gel electrophoresis. Through the use of two or more such restriction enzymes and secondary digestion of the products of a first digestion, an unambiguous map of the restriction fragments can be made (Figure 4-6). Virtually every virus DNA discussed in succeeding chapters has been subjected to extensive analysis using this method to obtain specific DNA fragments and physical maps of the viral genome.

FIGURE 4-6. Physical mapping of DNA with restriction enzymes. S_1, S_2 are sites for cleavage; R1, R2 are restriction enzymes for those sites.

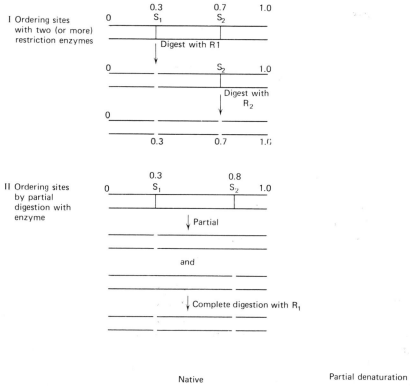

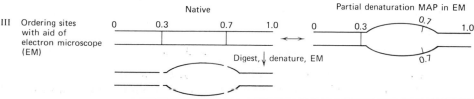

RNA PROCESSING ENZYMES.   Large RNA molecules (for example, pre-rRNA and pre-tRNA) that contain sequences of smaller molecules, and appear to serve as precursors to the smaller molecules, were detected in mammalian cells as early as 1963 (see Darnell, 1968 and 1975 for review). Bacterial cells also appear to make tRNA and rRNA through a similar pathway (Dunn, 1974). Recently a precursor to bacteriophage T7 mRNA was found (Dunn and Studier, 1973). In mammalian cells the enzymes responsible for the RNA precursor cleavages have not been identified whereas the specific enzymes responsible for the bacterial RNA cleavages have been characterized in a number of instances. Bacterial pre-tRNA has extra sequences on both the 5' and 3' side of the finished tRNA molecule and cleavage involves at least two enzymes termed P1 and P2 (Smith, 1976) (see sequence of pre-tRNA from bacteriophage T4 infected cells Figure 4-9).

Ribonuclease III, an endonuclease that cleaves double-stranded RNA but may also cleave specific sequences in single-strands, is very likely the enzyme responsible for processing bacterial pre-rRNA and for producing the mRNA of bacteriophage T7. A strain of *E. coli* with greatly depressed RNAse III activity makes ribosomes slowly, allowing the isolation of a large transcript containing ribosomal RNA before it is cleaved. T7 infection of these mutant cells also produces very low virus yields. In addition, the T7 precursor molecule, which can be synthesized *in vitro* by *E. coli* RNA polymerase from T7 DNA, is cleaved *in vitro* by RNAse III to give the same termini as exist for mRNA molecules inside the cell (Dunn et al., 1976; Figure 4-7).

Doubtless many more RNA processing enzymes will be discovered in coming years and virus nucleic acids will play a key role in their characterization.

SUMMARY OF ENZYMATIC SYNTHESIS OF DNA

Virion DNA replication is a theme that recurs repeatedly throughout this book. Certain problems of DNA replication are solved in a similar manner in all systems, both cell and virus, but other features present unique problems to individual viruses. Therefore, we shall summarize here the general solutions to general problems so that the detailed differences in the modes of DNA replication for individual viruses can be presented with greater economy later in the book.

The need for precision during the replication and repair of DNA, the genetic material, is obvious. The replication of a duplex helical structure with strands that are intertwined by one turn each 10 base pairs requires a most extraordinarily complex enzymatic system, for example, in *E. coli* at least seven genes play a role in DNA synthesis (Gefter, 1975). Two problems common to all DNA synthesis arise because of the biochemical capabilities of the DNA polymerases, while other general problems are related to the topology of the DNA molecule itself (Kornberg, 1974). First, the two DNA strands are of opposite polarity $\begin{smallmatrix} 5' \rightarrow 3' \\ 3' \leftarrow 5' \end{smallmatrix}$ and all known DNA polymerases elongate chains from $5' \rightarrow 3'$. A given replication point or "fork," however, moves along the double helix in one direction. Therefore one chain might be

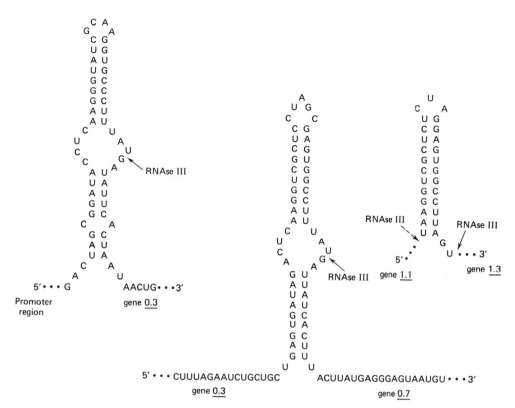

FIGURE 4-7. *RNAse III cleavage sites in precursor to T7 mRNA. Courtesy of Dr. John Dunn, quoted in* Prog. Nuc. Acid Res **19,** 273; *sequence for left-hand segment by Dunn and colleagues; for center segment by Rosenberg and Kramer, in press; and for right segment by Robertson, Dickson, and Dunn, in press.*

synthesized continuously but the other must be synthesized discontinuously and then the pieces linked together. This problem is solved by the existence of *DNA ligase* without which cells cannot grow.

A second common problem in all DNA synthesis is that no DNA polymerases can begin a new DNA chain. This problem is apparently solved by the existence of RNA primers which are elongated at their 3′ ends by DNA polymerases. Evidence for RNA involvement in virus DNA synthesis is reasonably strong both for DNA viruses and retroviruses (Bouché et al., 1976; Reichard et al., 1975; see Chapters 8 and 14). The earliest evidence for RNA involvement came from finding covalently linked RNA-DNA molecules in bacteria (see Sugino and Okazaki, 1973), and such complex molecules have now been found in animal cells (Tseng and Goulian, 1974). Because of the extremely rapid removal of the RNA from the DNA-growing point, less work on *in vivo* RNA-DNA complexes has been possible than *in vitro* systems, so that absolute proof is lacking that the mechanism detected *in vitro* operates *in vivo*. The recent discovery that the product of dnaG gene product in *E. coli* is a novel type of RNA polymerase, however, lends weight to the conclusion that DNA synthesis of all types is probably begun with an RNA-primer.

Most of the additional general problems of DNA synthesis are associated with the topology of the DNA molecule. In order to separate and copy the strands of the intertwined double helix a "swivel point" is necessary. The nicking-closing enzymes, which are widely distributed in cells, may furnish the answer to this general problem. The relief of topological restraint is a demonstrated feature of these enzymes (Pulleyblank et al., 1975). Clearly, not all DNA molecules are closed covalent circles, however, so this problem of DNA synthesis is presumably not universal. Other topological problems of DNA replication are specific to individual viruses and concern special sequences at the ends of molecules as possible starting points or special recognition sequences in the middle of molecules where replication is begun. These special problems are met with three general topological patterns of DNA replication: a single origin with one or two growing forks; multiple origins, probably used rarely in viruses but the rule in eukaryotic genomes; and rolling circle replication. These general modes of DNA chain growth were recognized mainly through electron micrographic studies, and the distinction between them is presented in that section. (pp 93–96)

RECOMBINANT DNA. DNA synthesis and repair may represent the apex of integrated intracellular enzyme functioning. Perhaps the culmination of the integration of laboratory studies in nucleic acid enzymology is exemplified by the recombination with purified enzymes of DNA molecules from any two sources, and the development of means to specifically replicate this recombinant DNA (for review see Nathans and Smith, 1975). The power of these new techniques is so great that in a discussion of nucleic acid enzymology they deserve special mention. Restriction enzymes cleave DNA only at specific sites and thus allow the physical separation of specific genes or gene fragments (Figure 4-8). In bacteria, there exist transmissible, independently replicating, circular DNA molecules termed "plasmids." If plasmid DNA is cut once by the same restriction enzyme as was used to isolate a DNA sequence of interest, the cut ends of the plasmid and desired DNA sequence can be complementary if a restriction enzyme has been used which leaves "cohesive ends." Further it is possible to assure hybridization of the plasmid and the desired DNA fragment by elongation with terminal transferase of both cut, 3' termini of one DNA sample (say the plasmid) with poly dT and the 3' ends of the other sample with poly dA. Upon mixing the two samples, rapid hybridization of the complementary ends ensues and the broken strands can be joined with DNA polymerase and DNA ligase. A new gene or genes has been "inserted" in the plasmid which upon reintroduction into a bacterium produces a cell capable of unlimited growth and amplification of the purified DNA segment. Because the plasmids contain antibiotic resistance genes, the rare cell containing the plasmid and the newly introduced DNA segment can be selected. This technique has been termed "cloning" of DNA segments and promises to revolutionize the approach to many experiments in virology as well as cell biology in general.

RNA AND DNA SEQUENCING

There are three general requirements for nucleic acid sequencing: (1) controlled means of fragmenting the nucleic acids at reproducible sites,

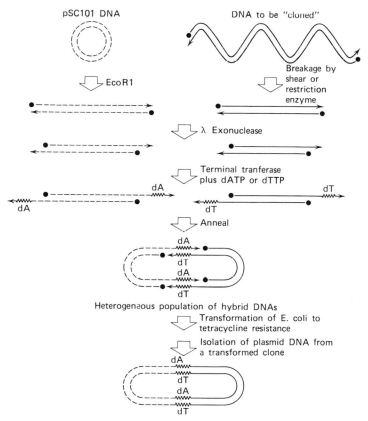

FIGURE 4-8. *Construction and "cloning" of Recombinant DNA.*

The DNA to be "cloned" is broken either by shear or by the action of bacterial restriction endonucleases. The DNA of a bacterial plasmid (exemplified in the figure by pSC101) can also be cleaved by restriction enzymes. Arrowheads and dots represent, respectively, the 3' and 5' termini of the DNA strands. If the same restriction enzyme has produced the cleavages both in the DNA to be cloned and in the plasmid DNA, spontaneous hybrid formation can occur due to cohesive ends. Hybrid formation is assured however by removing 5' terminal nucleotides with an exonuclease purified from E. coli infected with λ bacteriophage and elongating the resulting single-stranded 3' termini with the enzyme terminal transferase. Elongation of the plasmid DNA with deoxyadenylic residues and the DNA to be cloned with deoxythymidylic residues ensures hybrid DNA formation upon mixing. The DNA can then be closed with ligase or transferred into bacteria directly where ligation occurs. Selection of E. coli clones bearing the newly inserted sequences is achieved by selection for tetracycline resistance, a property transferred to the recipient bacterium by the psC101 plasmid DNA. (Redrawn from P. C. Wensink et al., Cell 3, 315–325. (1974))

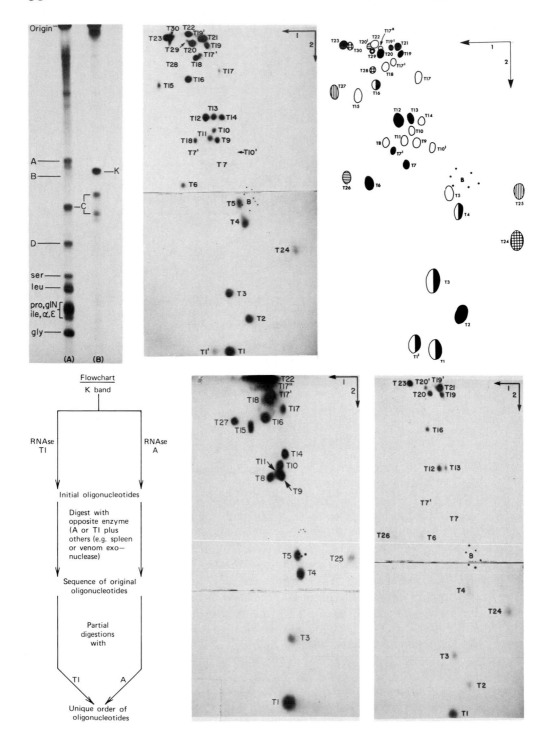

Component K: Gln—Leu tRNA dimeric precursor

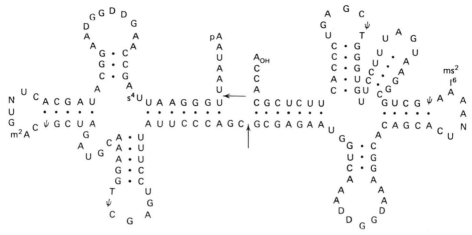

FIGURE 4-9. The sequencing of a bacteriophage tRNA precursor. Leucine and glutamine tRNA precursor encoded by bacteriophage T4 DNA was purified by gel electrophoresis and identified by radioautography (top left, band K); oligonucleotides from the precursor and the two final products (glutamine tRNA, right and leucine tRNA left) were compared revealing that oligonucleotides T28, 29, and 30 were not present in finished product. In addition the precursor was found to have distinctive oligonucleotides from each final product, e.g. T12, 13 (dark spots in diagram) from glutamine tRNA and T8, 9, 10, 11 and 14 (light spots in diagram) from leu tRNA are all present in precursor. The flow chart (on left) indicates the steps required for sequencing and ordering oligonucleotides and the final sequence and cleavage sites (arrows) are shown above. (Photographs courtesy Dr. C. Guthrie)

preferably at least two such methods that act at different sites so that overlap between fragments allows a linear order to be established; (2) techniques for separating the specific fragments from each other; (3) techniques for sequencing small fragments, which generally involve a controlled but eventually total degradation of a sample into constituent nucleotides. All these requirements are now met fairly well for both RNA and DNA.

RNA SEQUENCING. The success in sequencing RNA began with the discovery that T1 RNAse cleaved on the 3' side of guanylate residues, thus providing a reproducible catalog of fragments terminated with Gp, except for the original 3' terminus which may bear guanine at its 3' end. Sanger and his colleagues (Brownlee et al., 1968) used this enzyme coupled with an elegant two-dimensional separation (electrophoresis and "homochromatography") to obtain, for any specific RNA, a unique distribution of oligonucleotides called a "fingerprint." (The same term was applied earlier to two-dimensional separation of oligopeptides from proteins; Ingram, 1958.) The separation techniques are applicable to digests of RNA and to acid digested (depurinated) DNA where polypyrimidine tracts remain intact (Salser, 1973). Because sufficiently single-base-specific enzymes for RNA are lacking, the fingerprinting technique has been used more widely for RNA.

The basic method of detection is radioautography of ^{32}P labeled oligonucleotides, most frequently using *in vivo* labeled RNA. This obviously demands that the RNA be highly labeled inside cells. If the RNA or the DNA from which the RNA is transcribed can be highly purified, the RNA can be labeled *in vitro* chemically or enzymatically. Iodination of cytidylate with ^{131}I is the choice for *in vitro* chemical labeling, while ^{32}P triphosphates allow *in vitro* enzymatic synthesis of labeled RNA by RNA polymerase.

The T1 oligonucleotide RNA sequencing technique has been very widely used especially for small RNA molecules. A flowchart of the procedure for one application, the sequencing of a bacteriophage T4-specific tRNA precursor, is given in Figure 4-9.

The purified tRNA precursor molecule, highly labeled with ^{32}P, is cleaved with T1 to yield a number of major (longer) and minor (smaller) oligonucleotides. These are separated by electrophoresis at pH 3.5 followed by displacement chromatography (homochromatography) in a second dimension. The ordering of specific oligonucleotides can be achieved by digesting under less vigorous conditions and thus yielding products in which linkage of neighboring fragments is preserved. Specific oligonucleotides, either RNA or DNA, can be sequenced in a number of ways. One simple, rapid method is illustrated for a small deoxyoligonucleotide derived from φχ174. Venom exonuclease (an exonuclease from the venom of *Crotalus adamanteus* that is capable of digesting either RNA or DNA) removes one nucleotide at a time from the 3' OH end of

FIGURE 4-10. *Direct sequence determination of a deoxynucleotide. Snake venom phosphodiesterase (3' exonuclease) removed one nucleotide at a time to create a series of radioactive oligonucleotides detected by radioautography on left, sequence given at right. Photograph courtesy of Dr. E. Ziff.*

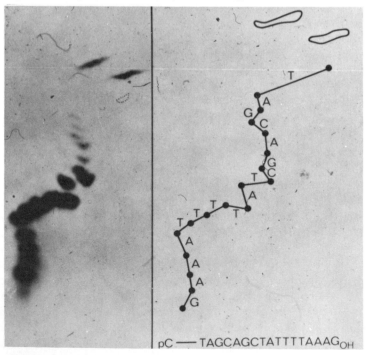

the oligonucleotide. The resulting series of oligonucleotides, differing in length by a single nucleotide, reveals a trail of radioactive species which separate as discrete steps due to removal of each nucleotide and to the left, right, right-center or left-center depending on whether a C, T (U for RNA), G, or A has been removed (Figure 4-10). Thus, the sequence of the oligonucleotides can be read directly from the fingerprint pattern of the partial exonuclease digest (Galibert et al., 1974).

DNA SEQUENCING. Rapid methods in DNA sequencing (Sanger and Coulson, 1975), which allow direct sequence reading from deoxyoligonucleotides differing in length by one base each, have been developed. One such technique (Figure 4-11) makes ingenious use of restriction fragments of $\phi\chi174$

FIGURE 4-11. *Rapid DNA sequencing technique of Sanger and Coulson (see text for details). The gel electropherograms at right allow the deduction of the sequence from position 50 to 140 as indicated from the diagram (left side) which points out T terminated and A terminated fragments. Photograph from Barrell, Air, and Hutchinson, in press.*

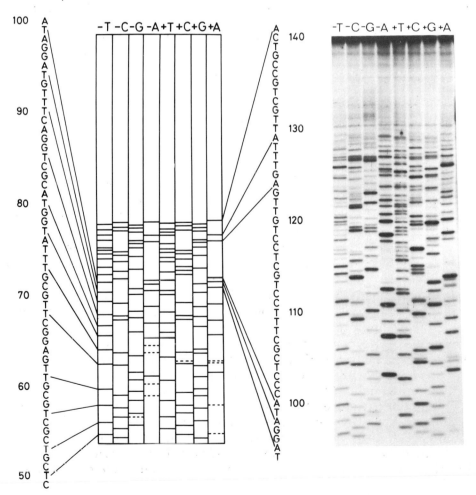

bacteriophage DNA plus DNA polymerase I. A restriction fragment is annealed to a single-stranded DNA chain and DNA polymerase I is used to extend the chain to varying random length with ^{32}P labeled deoxyribonucleotides. The reaction can be halted, the four triphosphates removed, and only a single nonradioactive triphosphate dA, dC, dG, dT added to each of four tubes. With only one triphosphate available, the DNA polymerase will either add this single next nucleotide or "chew back" due to the $3' \rightarrow 5'$ exonuclease activity of polymerase I; find the site where that nucleotide occurred previously, add only one nucleotide and stop. The varying length chains now terminate specifically with a known nucleotide and can be separated by gel electrophoresis. By inspection of the four specifically terminated parallel tracks the sequence can be read directly. A similar set of discrete bands can be achieved by supplying in the second part of such a reaction three of the four triphosphates; in this case the termination is just before the missing nucleotide. With the use of this technique Sanger and colleagues (1977) have sequenced the entire ϕX174 genome. A second rapid DNA sequencing technique has been developed (Maxam and Gilbert, 1977) which utilizes chemical treatment to specifically cleave at guanine, adenine, cytosine alone cytosine, plus thymine. This technique promises to be general because double-stranded DNA fragments can be terminally labelled; after random breakage at each nucleotide of the fragment, direct reading of the sequence can be made from acrylamide gels that separate each breakdown product.

These few examples of the sophisticated use of combinations of biochemical techniques should make it clear that any nucleic acid can now be sequenced. Many virus molecules are in fact currently being sequenced with a view to defining their coding regions, and from the size and primary structure of control regions gaining insight into the mechanisms of gene regulation.

ELECTRON MICROSCOPE (EM) TECHNIQUES

The limit of resolution of modern electron microscopes is in the range of 0.3 nm so that nucleic acid strands, which are about 2.0 nm in diameter, should be easily within resolving power of the electron microscopy (see Agar et al., 1974 for review).

In the late 1950s Kleinschmidt recognized that the need in visualizing nucleic acids was not higher resolution but a technique to spread the very long molecules on a uniform substrate so that their linear contours could be followed. His ingenious solution to this problem was to mix the nucleic acid molecules with a basic globular protein, cytochrome C. The nucleic acid and the protein form a complex that will spread on the surface of water and can attach to an electron microscope grid. After drying, the nucleic acid is embedded within a coat of protein and thus remains extended.

Shadowing with heavy metal reveals clearly the contour of the nucleic acid (Kleinschmidt, 1968). With more experience and adaptations of the Kleinschmidt technique, it became possible to distinguish single and double strands of both DNA and RNA (reviewed by Younghusband and Inman, 1974). Miller then discovered that the contents of cell nuclei and whole bacteria could be spread with nucleic acids still in association with cellular protein, allowing the observation of intracellular transcription complexes and functioning polyribosomes (Hamkalo and Miller, 1973; Miller and Beatty, 1969).

VISUALIZATION OF DNA REPLICATION AND TRANSCRIP-
TION. Figure 4-12 illustrates how microscopic techniques are used to study DNA replication. A central question in DNA synthesis is whether a single or multiple origin of DNA replication exists and whether the replication proceeds unidirectionally. Any replication site not at the end of a DNA molecule produces a bubble, and two bubbles within one DNA chain indicate multiple origins. Such multiple origins are characteristic of eukaryotic DNA replication, but most viruses and bacterial cells have single origins (Figure 4-13). Infected cells, of course, contain both viral and cellular DNA, but the viral DNA can be distinguished by its shorter, uniform size.

To determine whether replication begins at a fixed point, and whether replication is unidirectional or bidirectional from that point, some visible

FIGURE 4-12. *Use of Electron micrographs demonstrate modes of DNA synthesis.*

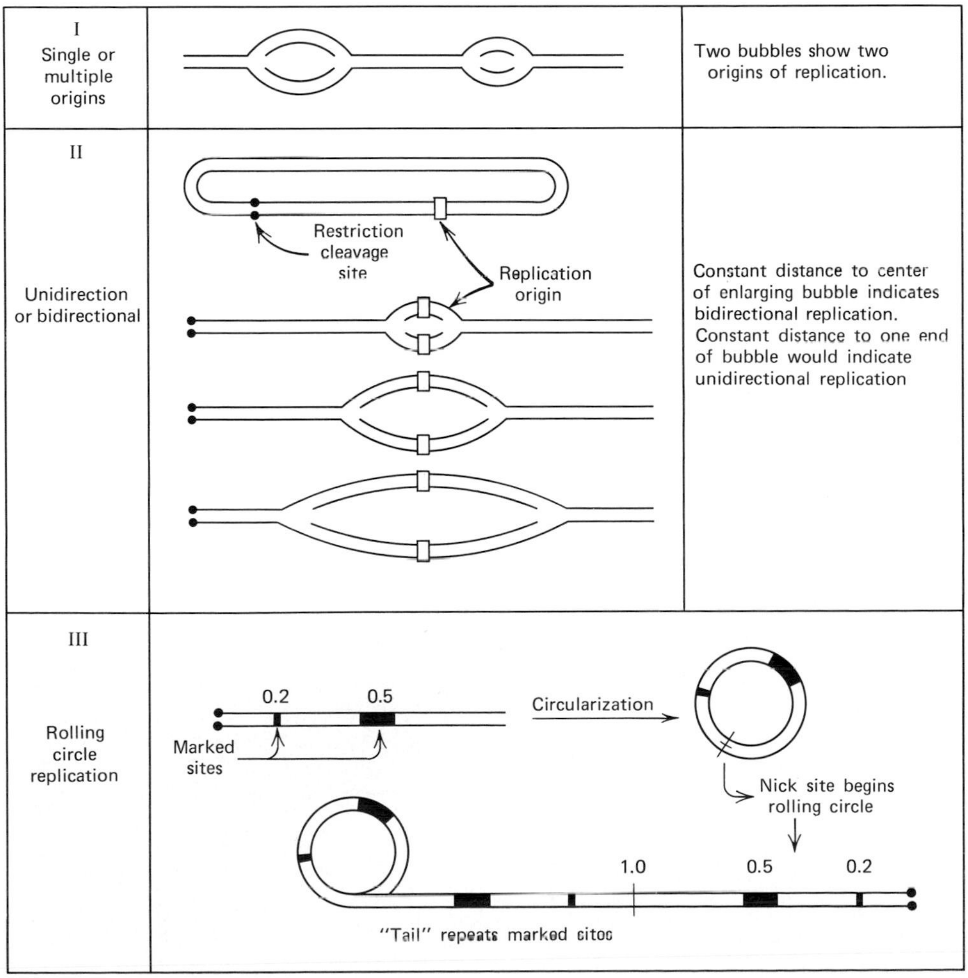

landmark on the replicating molecules must be available. Restriction enzymes provide one easy means of marking specific DNA loci; regions where local denaturation is easily induced are another important means of marking specific sites (Younghusband and Inman, 1974). Thus if a replicating bubble always has one end at a fixed distance from a marked site regardless of how big the bubble is, *unidirectional replication* is indicated. If, however, the centers of replicating bubbles are a fixed distance from the marked site, *bidirectional replication* is indicated.

A third mode of replication, termed *rolling circle* replication (Gilbert and Dressler, 1968) features a constant-sized circular region of DNA with a tail of varying length. In this case, replication begins at one site, but the replicating fork moves around a circle whose length equals the DNA being replicated. Emerging from such replicating circles are tails of varying lengths. Any landmarks on the circle then will also occur along the linear tail at equal distances from the point of emergence of the tail. All of the hypothetical types of replication mentioned above have in fact been observed and examples are discussed in later chapters.

Transcription of particular regions of DNA has been explored less frequently by the microscopic approach because distinction between various

FIGURE 4-13. *Electron micrographs of replicating DNA molecules.* (a) *Multiple sites of origin in chromosomal DNA of cultured* Drosophila *cells.* (b) *Bidirectional replication of SV40 DNA. Circular molecules cut once with Eco R1 restriction enzyme; center of bubble remains fixed distance from ends.* (c) *Rolling circle replication of* λ *bacteriophage DNA. At this magnification entire circle can be seen plus more than one genome length in "tail." Denaturation mapping at higher resolution confirms identical location of partially denatured regions in circle and tail. (Photographs courtesy of Drs. D. Hogness (a); N. P. Salzman (b), and D. Bastia (c))*

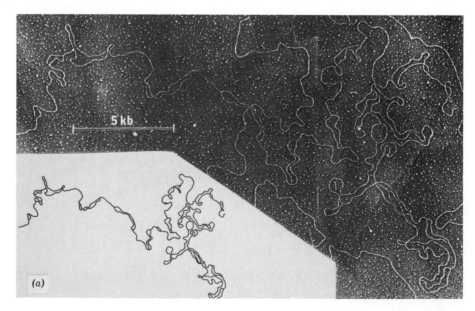

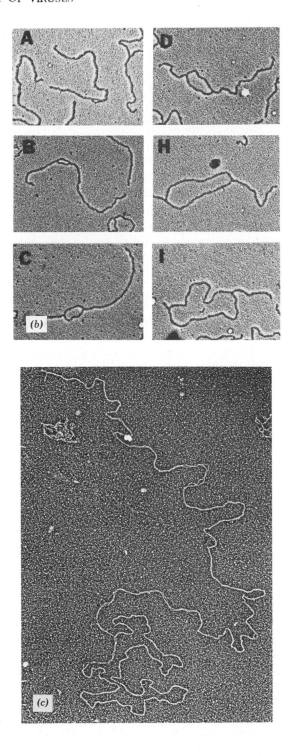

active regions is harder to make. However, certain situations where periodic transcription occurs, such as in the production of pre-rRNA from clustered ribosomal genes (Miller and Beatty, 1969), offer unusually informative pictures. From such pictures the length of the continuously transcribed and nontranscribed "spacer" regions can be discovered. The transcription patterns of many virus DNAs are not yet clear and it is obvious that adaptations of this technique will be useful in the future.

A final point about the power of the electron microscope for revealing details of events occurring within cells is well illustrated by the studies of Broker (1973) on recombination in bacteriophage T4 DNA (see Chapters 8 to 10). Recombination is a singular event which may occur in any position in the phage DNA at any one instant in time. Even studies on the products of single cells average the outcome of multiple recombinational events in the cell. The electron microscopic images of individual molecules isolate each recombinational event. It is this ability to visualize single molecules and their function that lends such power to electron microscope studies.

VISUALIZATION OF NUCLEIC ACID *IN VITRO*.

Heteroduplexes. Perhaps the most conspicuous use of the EM visualization of nucleic acids outside cells is in the determination of the length and topology of virus DNA molecules and of the arrangement of identifiable segments within these molecules.

Discontinuities between two related but distinct DNA samples can be located by mixing the two DNAs, denaturing, and allowing strand reassociation. So-called heteroduplexes are formed which diagrammatically reveal the areas of similarity and difference in the base sequence between the samples (Davis and Davidson, 1968). For example, if the samples are identical except for a deletion in one, then figures will form containing one strand of each molecular type in a DNA duplex except in the region of the deletion where a "deletion loop" will form. The location of the loop can then be measured exactly with respect to either end or any other identifiable site on the genome.

The technique of heteroduplex formation has been widely used in conjunction with bacteriophage genetic work and in recombinant DNA viruses such as adeno-SV40 hybrid viruses.

Denaturation Mapping. The distribution of bases within long stretches of DNA is not so random that physical differences within long segments cannot be detected. For example, the shear products ("halves") of bacteriophage λ (see Chapter 9) could be separated because of a different average GC content. When λ bacteriophage DNA was subjected to conditions of partial denaturation, a reproducible "melting" pattern was observed (Inman, 1967). This type of "denaturation mapping," which reveals regions of DNA that are relatively AT-rich (easily denatured) or relatively GC-rich (not easily denatured), has been successful for a wide variety of DNA molecules (Younghusband and Inman, 1974). Wellauer and David (1973) have made brilliant use of this technique in examining RNA structure (see Chapter 11, Figure 11-11). Specific regions within otherwise single-stranded RNA molecules can form stable double-stranded interactions. When the RNA molecule is of uniform size, as with rRNA, the double-stranded regions produce loops of a characteristic size at characteristic locations along the molecule. Physical maps of nucleic acid molecules may be expected in the future to form an important basic means of description.

Localization of DNA Sites Complementary to RNA. A very important question in the functioning of DNA is the determination of initiation and termination sites for RNA transcription, another problem whose solution is aided by electron microscopy. The site of origin of particular RNA molecules can be viewed by hybridizing an excess of RNA to a small amount of the specific denatured DNA and determining which region of the DNA is made double-stranded by the RNA. This technique has not yet been widely used, but with the advent of restriction fragments of DNA, it probably will be used more in the future.

A particularly novel technique for determining sites of RNA transcription is the so-called "R-loop" mapping (Thomas et al., 1976). If a region of local denaturation can be induced in DNA, any RNA complementary to that region will form an RNA-DNA hybrid, which is slightly more stable than the DNA-DNA hybrid. Since denaturation maps of DNA are relatively easy to achieve, mapping of RNA molecules on an already determined DNA structure should make localization of transcription sites easier.

The almost unlimited power of EM techniques in exploring nucleic acid topology is demonstrated by a group of new techniques that modifies nucleic acids to produce "decoration" for visualization of specific regions (Wu and Davidson, 1973). For example, tRNA molecules, which themselves are too small to be distinctly resolved in electron micrographs, can be coupled with ferritin, a protein containing electron-dense iron atoms (Figure 4-14). When hybridized to specific DNA such tRNA molecules are easily visible and locate the tRNA genes. This "positively" stained tRNA localization technique has recently been joined by a new technique that allows "negative" staining (Wu and Davidson, 1975). In this work the hybridized tRNA was located with the aid of a bacteriophage T4 protein, gene 32 protein, which binds to single-stranded DNA. All the DNA except that hybridized to tRNA is visualized as a thick rod while the tRNA:DNA hybrid stands out a discontinuity in the thick rod.

One of the newest but potentially most important applications of visualization of nucleic acids is the study of nucleic acid-protein interactions. It is clearly established now that some proteins can recognize specific regions of DNA or RNA, presumably on the basis of nucleotide sequence. Therefore, specific protein binding sites should be possible to locate either by using proteins that are large enough for direct visualization (for example, RNA-polymerase) or by marking the protein with an electron dense atom (Hg^{++} or ferritin conjugated antibody to the protein, for example). This technique was used to demonstrate that SV40T-antigen can bind at or near the origin of replication on the viral DNA molecule. (Reed et al., 1975).

MOLECULAR HYBRIDIZATION

One of the early proofs of DNA structure was the demonstration that DNA solutions heated to a temperature characteristic for each DNA suddenly

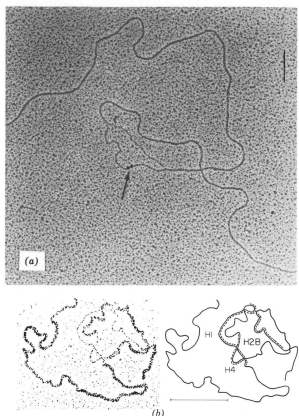

FIGURE 4-14. Mapping gene position by electron microscopy. (a) Ferritin-complexed $tRNA_{tyr}$ hybridized to a region of bacterial DNA contained in a derivative of bacteriophage $\phi 80$ ($\phi 80_p SU^+_3$). The molecule shown is a heteroduplex between $\phi 80$ DNA and $\phi 80_p Su^+_3$. The black dots, the ferritin, are separated by 140 ± 20 base pairs. The marker (——) is 1000 base pairs. (b) Purified histone DNA hybridized to histone mRNA. High-contrast photograph and tracing, which shows single-stranded DNA as thick line (covered with bacteriophage T4 gene 32 product, a protein that binds single-stranded DNA). Areas of DNA hybridized to histone mRNA then show as a thin line. Photographs courtesy of Dr. Norman Davidson.

lost viscosity and gained absorbancy at 260 nm (Marmur et al., 1963). This now familiar "melting" of the DNA double helix was also shown to reduce what, at the time, was the only easily measurable biological function of DNA—the ability of certain bacterial DNA, such as that from the pneumococcus, to transform cells of one serotype to another serotype. If the separated strands of pneumococcal DNA were allowed to interact at a temperature below their T_m (temperature where $\frac{1}{2}$ the "melting" reaction was complete) a reassociation or "reannealing" occurred with a restoration of transforming activity and a return to normal UV absorbance. Reassociation also occurred with DNA from genetically closely related strains but not with DNA from distantly related strains. The reassociation was most efficient if carried out at temperatures about 25° below the T_m of the DNA at a pH between 5 and 9, and was promoted by increasing monovalent cation concentrations at least through the range of 0.15M to 0.9M.

This first example of DNA:DNA reassociation was soon followed by the demonstration of RNA:DNA chain association. Hall and Spiegelman (1961) reasoned that if mRNA was transcribed from bacteriophage DNA by faithful base-pair copying, some of the RNA from cells infected with a DNA-containing bacteriophage should associate with the DNA of the infecting bacteriophage. Volkin and Astrachan (1956) had earlier described an RNA from phage-infected bacteria with a base composition like the phage DNA. RNA from cells infected with T2 did associate with T2 DNA but not with T5 DNA. This interaction was termed "molecular hybridization," and, through common usage in molecular biology, "hybridization" has come to signify molecular hybridization more often than cell hybridization or the classical artificial construction of genetically mixed individuals by grafting or inter-breeding species.

As we have stressed in previous sections of this chapter, the unique property of any segment of the DNA of an organism, or an mRNA derived from the DNA, is its primary sequence. The development of molecular hybridization represents an important milestone in nucleic acid studies because the technique provided a means of quickly measuring a specific primary sequence in a complex mixture of nucleic acids.

TECHNIQUES OF HYBRIDIZATION.   The earliest hybridization experiments allowed the bacteriophage mRNA and denatured phage DNA to interact in solution; separation of the "hybrid" molecules was then achieved by virtue of the distinct buoyant density of RNA:DNA hybrid in CsCl (Hall and Spiegelman, 1961). Because of the resistance of hybridized RNA to nucleases that attack only single-stranded RNA, a treatment of the reannealed mixture with RNAse A (pancreatic RNAse) either before or after CsCl equilibrium centrifugation allows an even clearer measure of the hybridized RNA molecules.

Various methods to facilitate separation of hybridized and unhybridized molecules have been used. The attachment of DNA to solid substrates, most notably to nitrocellulose filters (Gillespie and Spiegelman, 1965) provides a simple, rapid assay for the presence and quantitation of specific RNA sequences. Labeled RNA or DNA is exposed to filters bearing DNA, and after appropriate incubation the unhybridized molecules are washed away. RNAse digestion further removes unpaired RNA segments; perhaps this is the most widely used hybridization assay.

Many experiments require measurements of the rate of interaction not just a final yield, and interaction in solution is preferable for such measurements. (Britten et al., 1974). Hybrid molecules formed in solution can be conveniently measured by the binding of reassociated DNA to hydroxylapatite columns under conditions where single-stranded DNA does not bind or by attacking unpaired DNA or RNA with single strand-specific nucleases—S1 (from Aspergillus oryzae) for either RNA or DNA or pancreatic and T1 ribonucleases for RNA.

WHAT CAN BE MEASURED BY HYBRIDIZATION?

Final yield Measurements.   A useful measurement which allows the determination of the fraction of DNA complementary to a given RNA is the so-called *saturation* measurement, which is an *RNA excess experiment* (Figure 4-15). For example, 1.0 microgram of bacterial DNA binds 0.002 micrograms of 23S rRNA and 0.001 micrograms of 16s rRNA (Yankofsky and Spiegelman, 1963). Knowledge of the molecular weight of the rRNA and the DNA content per bacterium enables the calculation that bacterial DNA contains about 10 sites from which rRNA is transcribed.

A second type of final yield (that is, nonkinetic) measurement employs the technique of *exhaustion hybridization* which is a *DNA excess experiment*. Here a fixed input of highly radioactive RNA is exposed to increasing amounts of DNA until the maximum amount of RNA has hybridized. With mixtures of RNA, the exhaustion of a particular RNA sequence can be established by showing the residual unhybridized fraction is, upon exposure to DNA, totally depleted of hybridizable molecules. This technique is widely used in experiments with virus nucleic acids to measure the fraction of a given sample of RNA complementary to all or a part of a viral DNA molecule (Figure 4-15). One of the classic uses of this technique showed that when genes of the lactose operon are induced, *E. coli* cells contain greatly elevated levels of RNA (presumably lactose operon mRNA) hybridizable to phage DNA bearing the lactose operon (Hayashi et al., 1963).

Competition hybridization is another variation of a final yield measurement that can assess the overlap in sequence content of two related nucleic acid preparations, usually two samples of RNA. This valuable technique was first used to demonstrate different classes of phage-specific RNA during T2 bacteriophage replication. Increasing amounts of unlabeled RNA from cells early in infection "competed" with labeled "early" RNA for sites on phage DNA, reducing the amount of labeled RNA bound to DNA. So, also, could unlabeled RNA from late in the infection compete with "late" labeled RNA. Early RNA, however, could not compete with late RNA, a demonstration that the phage mRNA species were different at different times after infection (Hall et al., 1964). This technique has achieved wide application in the study of the time dependent expression of viral genes during infection with many different types of bacterial and animal viruses. *In competition hybridization experiments the labeled RNA alone need not be in excess of DNA sites, but the unlabeled RNA must be in excess of both the labeled RNA and in excess of DNA sites* (Figure 4-15).

Kinetic Measurements; Theory of Nucleic Acid Reassociation.   Britten and

FIGURE 4-15. Ideal hybridization experiments.

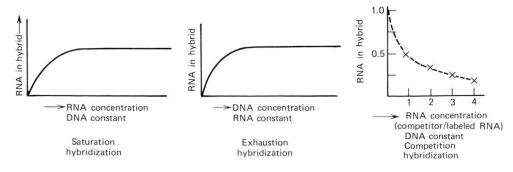

Kohne (1968) realized that the reassociation rates of DNA in solution should follow one main rule—the simpler the genome of an organism the more efficiently should the denatured DNA single strands find a correct partner during reassociation. The reassociation rate of any specific DNA should therefore be inversely proportional to the complexity of the genome (amount of unrepeated nucleic acid sequence) and should be second order because two strands must find each other. They amply demonstrated the validity of these ideas by showing (Figure 4-16) that complementary homopolymers, bacteriophage DNA, *E. coli* DNA, and a fraction of calf thymus DNA had reassociation rates inversely proportional to their complexity. By now a vast number of additional DNA samples have been similarly analyzed (Britten and Davidson, 1973). The equation for DNA reassociation is: $C/C_o = 1/(1 + KC_o t)$ where C_o and C are concentrations of single-stranded DNA in molecules per liter at time 0 and time t, respectively, and K is the rate constant for a given DNA (proportional to nucleotides in the genome). The convention in DNA reassociation experiments is to plot fraction of DNA reassociated as a function of $C_o t$ (DNA concentration × time). The $C_o t_{1/2}$, the DNA concentration × time necessary for $\frac{1}{2}$ of the DNA to reassociate, is a convenient means of describing the complexity of various DNA samples.

Wetmur and Davidson (1968), who also recognized the importance of complexity in determining the overall rate of the hybridization reaction, emphasized an additional important point—the hybridization reaction can be considered as consisting of (1) random collision, (2) the first effective pairing or "nucleation," and (3) spontaneous progression to long hybrids or "zippering." Not every collision is effective in causing a hybridization and therefore the limiting step in hybridization is the nucleation step. This causes the length of the interacting molecules to affect the rate of the reaction; for a pure

FIGURE 4-16. *Reassociation of double-stranded nucleic acids from various sources over a range of 10^9. The reassociation rate is proportional to the C_0t for half reactions. From R. J. Britten and D. E. Kohne,* Science **161,** *529–540 (1968).*

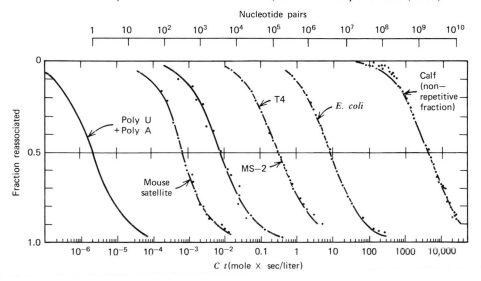

reassociating DNA, the longer the fragments the faster the reassociation, because each effective nucleation is followed by the zippering of more DNA into a helix. Therefore, only when all conditions are equal, for example, salt, temperature, or size of molecules, can two samples be accurately compared with respect to reassociation rate. These factors are taken into account by shearing the DNA to a uniform small size and carefully controlling salt and temperature (Britten et al., 1974).

Kinetic hybridization experiments have been used for two main purposes: (1) to measure the concentration of a specific sequence in a mixture of other sequences and (2) to measure the complexity of a nucleic acid by its rate of reassociation. The first type of measurement has been more frequently employed to study problems in virology.

Gelb, Kohne, and Martin (1971) recognized that the content of specific sequences existing within a mass of other sequences could be assayed by kinetic hybridization. If a highly radioactive, pure species of DNA (for example, a virus DNA) is denatured at high dilution its spontaneous reassociation is slow. The rapidity of reassociation of the labeled DNA can be increased by adding unlabeled viral DNA. A sample of cellular DNA that contains integrated viral DNA can also be used to promote the increased hybridization of the labeled viral DNA. By comparing the increase in reassociation rate produced by adding a known amount of pure viral DNA with that produced by a known amount of cell DNA, the content of viral DNA integrated into the cell DNA can be measured (Figure 4-17). This technique of promoting reassociation rates of a "probe"—a *DNA-driven* reaction—has been of great value in measuring concentrations of specific sequences within mixtures, usually total cell genomes. The data in such experiments is often plotted as 1/fraction of probe DNA remaining single-stranded vs. $C_o t$, a presentation that easily allows a determination of the number of integrated copies of the viral genome (Sharp et al., 1974).

An important variation of this type of experiment has developed since the discovery that a complementary DNA copy (cDNA) of RNA tumor virus DNA or a "cDNA" copy of an mRNA can be synthesized by the enzyme reverse transcriptase (Temin and Baltimore, 1972). Such "probes" can be made at very high specific activity and the rate at which they are driven into double strands by either cell DNA or RNA offers a measure of the content of that sequence in the cell DNA or cell RNA sample. Experiments of this sort where RNA excess is used are *RNA-driven reactions* and follow a pseudo-first-order equation: $C/C_o = e^{-kR_o t}$ where C and C_o are single stranded DNA concentrations at time t and t_o respectively, and R_o is the initial RNA concentration in moles per liter, a value that does not change appreciably during the reaction (see Hood and Wood, 1974). In analogy with the DNA reassociation experiments mentioned earlier, data from RNA-driven reactions is frequently plotted as 1/fraction of probe remaining single-stranded vs. $\log(R_o t)$, in order to estimate the concentration of RNA complementary to the probe.

The second major use of kinetic reassociation studies is to determine whether all the DNA of an organism reanneals with the same kinetics. As discussed in Chapter 11, this experiment has proved that eukaryotic cells contain repeated DNA fractions. The detection of the repeated fractions,

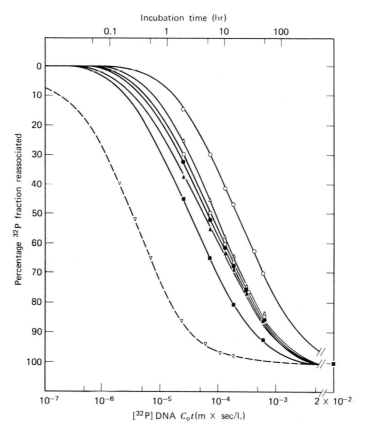

FIGURE 4-17. *The increase in SV40 reassociation by transformed cell DNA.* ^{32}P *labeled SV40 DNA was incubated for various C_0t's by itself (-o-), with DNA samples from transformed cells and with authentic SV40 DNA (-▽-) equivalent to 35 genomes per cell DNA equivalent. The SV40 DNA content of the transformed cells was calculated to be (in SV40 equivalents per cell), 1.04 -△-; 1.42 -□-; 1.56 - ●-; 2.08 -▲-; 3.56 -■-. From L. D. Gelb et al.* J. Mol. Biol., **57,** 129 (1971).

however, is based on their relatively high proportion of the total DNA (5 up to 50%) and their very high degree of repetition. In virology, the question often arises, whether a gene is repeated more than once per genome, or whether a genome is totally repeated once. The kinetic technique is less sensitive than other ways of answering these questions and, consequently, has not been widely used. In fact, kinetic measurements of DNA reassociation rates failed to locate repeated sequences in herpes DNA where approximately 10% of the genome consists of sequences that appear twice (Hayward et al., 1975) and these were not apparent in the reassociation curve of herpes DNA (Fraenkel and Roizman, 1971). Use of only the restriction fragments containing the repeated regions of the genome would be necessary to measure accurately partial repetitiveness of a portion of a genome repeated only a few times.

Virus Nucleic Acids and Proteins

VIRION PURIFICATION: INTRODUCTION

The key to the study of the chemical or physical properties of any substance is its availability in pure form. As discussed in Chapter 3 the surface architecture of virions is highly repetitive. However, since viral structure varies greatly from one virus to another, for example, from enveloped large animal viruses such as herpesvirus to small, naked icosahedral bacteriophages such as ϕX174, different techniques must be employed for each type of virion.

In addition to considering the general chemical nature and stability of various virions, the investigator, in planning a strategy for virus purification, must also consider the relative amount of virus versus contaminating cell debris in the starting material. This ratio is dependent on such experimental variables as the relative yield of virions in the cells or tissues used as host; whether cellular particles of similar size and chemical composition (for example, ribosomes or cell membranes) are likely to be bothersome contaminants; and, finally, whether the virions are efficiently released from the host cells or require disintegration of the infected cell for release.

Individual steps in the purification of virions are all based on one or more of the following characteristics of virions: size, density, and surface properties. How various techniques make use of these properties to allow virion purification is the subject of this section.

VIRION PURIFICATION: SIZE, DENSITY, AND CENTRIFUGATION

By far the single most widely used method of virion purification is centrifugation, which is based on the difference in either size or density or both between virus and contaminating material (see Chapter 3 for discussion of sedimentation theory). The earliest rather crude methods of centrifugation simply involved several cycles at high and low speeds which, respectively, did and did not sediment virions into a pellet. Sufficient purification of bacteriophages, vaccinia, and tobacco mosaic virus was achieved in this manner by 1940–45 so that the nucleic acid content of these particles was established.

Present-day use of ultracentrifugation makes available very discriminating separations on the basis of density as well as size.

Rate zonal sedimentation is usually carried out using gradients of sucrose or glycerol to stabilize the contents of the centrifuge tube against convection currents (Brakke, 1967). The density of the suspending medium is substantially less than the virion; therefore the size and shape of the virion are the major determinants in its rate of sedimentation (see Chapter 3 for discussion of sedimentation rates).

Although this technique is frequently referred to in the literature as "density gradient centrifugation," this is an unfortunate use of language because the density of the sedimenting particle plays a much smaller role in rate zonal sedimentation than does size and shape. *Equilibrium density or isopycnic sedimentation* is carried out by allowing a particle to sediment or float to a position in a density gradient equal to its own density (Vinograd, 1963). This technique most frequently employs Cs salts (chloride, sulfate, or formate) for naked virions where densities of 1.3–1.4 are the rule, or sucrose in H_2O or sucrose in D_2O for enveloped virions, almost all of which have densities of 1.10–1.20.

Several other techniques based on virion size have been used to achieve partial purification (Philipson, 1967). Filtration through membranes of various sizes, historically the means of proving the submicroscopic nature of virus infectivity, can also be used to retain particles larger than viruses. With newer technology, nitrocellulose membranes have been produced which will retain viruses and pass proteins and nucleic acids in solution thus affording a means of virion purification. In addition a whole range of substrates bearing pores of varying sizes allow so-called *molecular sieve chromatography*. Cross-linked substrates, such as polyacrylamide and polysaccharide beads (agarose, sepharose, etc.), contain channels that admit solubilized proteins and nucleic acids but not virus particles; thus the virus passes through such columns more quickly than smaller contaminants in the solution.

SURFACE CHEMICAL PROPERTIES. Because virion surfaces are largely proteins, the techniques of general protein chemistry have been applied to virion purification.

In the preparation of the first crystalline (actually paracrystalline) sample of tobacco mosaic virus, Stanley (1935) borrowed from the protein chemist, the so-called "salting out" technique, based on the fact that at a given pH different classes of proteins are rendered insoluble, one after the other, by gradual increases in salt concentration. Another physico-chemical property of proteins, insolubility in alcohols, can be used to concentrate viruses that withstand alcohol treatment. Alcohol precipitation has served as a first step in the purification of poliovirus (Charney et al., 1962) and bacteriophages (Sinsheimer, 1959b). Acid precipitation has been used (Herriot and Barlow, 1952) to concentrate some phages.

Another general method useful in partially purifying nonenveloped viruses is denaturation of cellular protein with the halogenated fluorocarbons, especially trichlorotrifluoroethane (Philipson, 1967b). These nonpolar lipid solvents tend to denature cellular proteins at an aqueous interface whereas the virion is unharmed. For viruses that are not released well from cells, such as adenoviruses, this technique is invaluable.

A complete list of general methods of virus concentration includes two-phase liquid polymer systems, which partition viruses and cellular material differently. For example, the polyethylene glycol-dextran sulfate system has been used to purify a wide variety of viruses including some with envelopes (Albertson, 1967).

Higher levels of virion purity can be achieved with a variety of additional methods based on surface chemical properties. *Column chromatography* can be accomplished by virion binding to a variety of substrates (treated cellulose columns; for example, DEAE, phosphocellulose, and ecteola) followed by elution with salt or pH gradients (Philipson, 1967a). As with other proteins, the virion surface charge, which changes as a function of pH and ion concentration, is the property used to advantage in purification. Another technique that detects extremely minor charge differences in virion surfaces is electrophoresis through the columns of sucrose. Single charge differences in capsid proteins are possible to detect and the method may distinguish charge differences when chromatography fails (Korant and Lonberg-Holm, 1974).

A final means of using the chemical surface of a virus for purification involves binding to a biological substrate. The myxoviruses, as will be detailed later, bind specifically to red blood cells, and then spontaneously elute due to digestion by a virion enzyme of the receptor on the red blood cell. Thus virion purification can be achieved by low-speed centrifugation of the virus bound to red blood cells followed by elution and resedimentation of the cells. Several hundred-fold purification of influenza virus has been achieved in this fashion (Laver, 1973).

ASSESSMENT OF CHEMICAL PURITY.   A proper appreciation of the molecular nature of likely contaminants of a virion preparation is essential in guarding against contamination. The special requirements of the experiments in which a purified preparation is to be used dictate the limits and types of allowable contamination. For example, in a preparation that will serve to produce antiserum specific for the virion capsid protein, contamination by small amounts of cellular nucleic acid would be of little consequence, whereas the presence of contaminating cellular protein could affect the specificity of the resulting antiserum. In studies on virus-cell interaction using isotopically labeled amino acids incorporated into the viral protein, it is necessary to remove from the preparation all free, labeled amino acid which, if present, can enter the cells much more rapidly and effectively than does the viral protein. For experiments with labeled virions the preparation must be adequately free both of extraneous macromolecules and small molecules that may carry the relevant label. In general, adequately pure virion preparations require a combination of at least two of the more powerful techniques of purification: zonal sedimentation, isopycnic banding, or column chromatography.

ISOLATION OF NUCLEIC ACIDS AND PROTEINS FROM VI-RIONS.   Early studies on the component parts of virions were hampered by lack of techniques for separating the nucleic acids and proteins in undamaged form. Gierer and Schramm (1956) made a major contribution to this problem, as well as to all studies on nucleic acids, by introducing the technique of phenol extraction of aqueous suspensions. Virion protein is denatured at the phenol:H_2O interface leaving intact virion RNA or DNA in solution. Even in cases where the virion DNA is over 10^8 daltons or the virion RNA threatened with traces of contaminating RNAse, whole molecules could be obtained. Additional techniques have been added to the basic phenol extraction. For example, some virions release nucleic acid when simply treated with detergents such as sodium dodecylsulfate and this compound is now frequently added during phenol extraction (Noll and Stutz, 1968). Powerful proteases, such as pronase and proteinase K, which resist denaturation by detergents can be used along with detergents to remove the capsid of almost any virus (Doerfler, 1969). Thus a variety of techniques are now available for nucleic acid release from virions.

Many of the same techniques are used in releasing cellular nucleic acids from normal and virus-infected cells, and, in fact, it was the discovery that whole infectious viral RNA of >2 × 10^6 daltons could be released from cells

(Colter et al., 1957; Wecker, 1959) that gave impetus to characterization of high molecular weight RNA from mammalian cells. (Darnell, 1968; 1975)

Virion protein release presents special problems. Clearly the virion is designed for maximum stability and effectiveness when its protein or proteins are integrated into a capsid structure. In fact, in the cell the polypeptides which make up the virion in many cases never exist as free individual polypeptide chains; that is, they enter membranes before becoming part of virion or they enter provirion structures as larger polypeptides and are cleaved. Therefore, attempts to solubilize virion proteins in simple buffered aqueous solutions are frequently unsuccessful. Resort to solubilization with protein denaturing reagents such as urea, guanidinium chloride, and sodium-dodecylsulfate is common (Fraenkel-Conrat and Rueckert, 1968). Nevertheless, purified soluble virion proteins plus RNA can reconstitute both helical and isometric plant viruses and RNA bacteriophages. Also, purified proteins with enzymatic or hemagglutinating capacities have been isolated from animal viruses (see Chapters 8 and 13).

VIRAL DNA

The major structural consideration for most viral DNA molecules as for DNA in general is, of course, the existence of two antiparallel base-paired chains. The DNA genome of viruses, however, is small enough so that questions can be raised not about the somewhat monotonous, virtually unending "midsection" of a double helix, but about the ends of the helix, and, in fact, about the overall shape of the DNA molecule. Fascinating answers have resulted—viral DNA molecules can be linear *or* circular, they can be double-stranded *or* single-stranded throughout their entire length *or* single-stranded for short stretches on the ends of otherwise double-stranded molecules. In addition, while most of the sequence arrangements in a viral genome occur once, the ends of the genome may contain repeated or redundant regions.

The details of these variations make up a considerable part of the subject matter of later chapters, but a brief experimental definition of some of the above unusual features of viral DNA is given here.

SINGLE AND DOUBLE STRANDEDNESS: LINEARITY AND CIRCU-LARITY.
Upon extraction from virions, most viral DNA imparts to solutions all the characteristics that the biochemist or physical chemist expects of the rigid, double-stranded DNA molecule—base composition where A = T and G = C, high viscosity, hyperchromic shift upon heating, retarded sedimentation and extended radius of gyration because of rigid form, and so forth.

Sinsheimer and his colleagues were therefore puzzled in the late 1950s when they observed the aberrant behavior of the DNA of the small bacteriophage ϕX174. This DNA did not give a distinct "melting profile," its light-scattering properties resembled denatured DNA and its bases reacted with formaldehyde, a property of denatured not double-stranded DNA (Sinsheimer, 1959). Also despite repeated analyses the base ratios were A, 1.0; T, 1.3; G, 1.06; C, 0.82;—distinctly not base-paired. This extensive DNA-detective work led to the very firm conclusion that the $\phi\chi174$ DNA was a single-stranded molecule.

Furthermore, they discovered that the isolated $\phi\chi174$ DNA could not be digested by an exonuclease until after a chain scission by an endonuclease; that is, in its native state the molecule had no ends. In addition, two sedimentation forms of ϕX174 DNA were observed, S1 (faster sedimenting) and S2 (slower sedimenting). Careful endonuclease digestion studies revealed that the first scission in S1 produced S2 without decreasing the molecular weight and the first scission in S2 decreased the molecular weight (Fiers and Sinsheimer, 1962). Thus ϕX174 was not only single-stranded but a covalently closed circle. When electron microscopy of nucleic acid molecules was developed these elegant biophysical and biochemical studies received a pictorial confirmation (Figure 4-18).

In the wake of the work on ϕX174, a second round of reports ensued on the aberrant physical behavior of double-stranded DNA of polyoma virus (Vinograd et al., 1965). In this case, the DNA exhibited base-pairing and an appropriate hyperchromic shift but two distinct strands were not released after melting. Sedimentation of virion DNA frequently revealed three components. Proof of the polyoma structure involved isolation of virion DNA in a

FIGURE 4-18. *Closed circular and supercoiled DNA molecules.* $\phi\chi174$ *molecules* (a) *single-stranded virion DNA.* (b) *Double-stranded circular replicative form.* (c) *Supercoiled replicative form. Photographs courtesy of Dr. M-T. Hsu.*

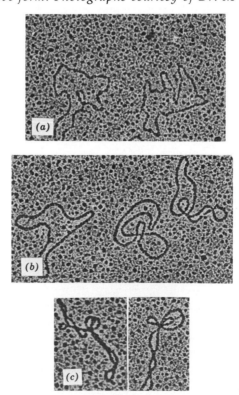

native Form I, 20S, which by sequential nuclease digestion was shown to give rise to Form II, 16S, and finally Form III, 14S. Denaturation was performed on each form revealing that Form III gave two linear single-stranded chains, sedimenting at 16S, Form II a single-stranded chain sedimenting at 16S and a closed single-stranded chain of the same length which sedimented at 18S (reminiscent of ϕX174). Thus Form I, which could not be denatured, was a covalently closed, double-stranded circle. Again electron microscopy confirmed these studies (see Chapter 14).

Further physical studies with SV40 DNA, another closed circular molecule, led to an appreciation of higher order structural features imposed by the covalently closed double-helical form. Such a molecule is under rigid topologic constraint and any change in the helix winding or unwinding must be reflected in an overall conformational change. This results in "superhelical" twists in closed circular DNA of variable number and were originally thought due to the covalent closure of the double helix prior to the formation of the last Watson-Crick base pairs; the last Watson-Crick helical turns might therefore induce superhelical twists in the opposite direction (Pulleyblank et al., 1975). Recently however, Gellert and his colleagues (1976) have identified and purified an enzyme from *E. coli* which can take "relaxed" (nonsupercoiled) closure circular DNA (plus ATP and Mg^{++}) and convert the DNA to closed, covalent, supercoiled DNA. They have named this enzyme *DNA gyrase*.

The third major surprise in the early studies of the topology of viral DNA was the circularization of linear virion molecules demonstrated with bacteriophage λ DNA. If the normal linear double-stranded helices were allowed to interact under conditions favoring DNA annealing, end to end aggregates as well as unstable circles formed, suggesting cohesive, single-stranded ends that allowed base-pairing over short regions (Hershey, et al., 1963). After entering cells, λ DNA becomes closed by the DNA ligase of the cell and a covalently closed duplex DNA can be detected early in infection (Young and Sinsheimer, 1964). As emphasized in Chapters 9 and 10 much of the interest in the circular forms of DNA molecules derived from genetic studies implying that circularization preceded integration of λ DNA into the bacterial cell genome. Thus the physical proof of circularity was received with wide interest.

Figure 4-19 presents a summary of forms of various viral DNAs. As studies continue, both bacteriophage and animal and plant viruses may well be found to present examples of all permutations of strandedness and circularity. One particularly interesting and unusual situation is the defective parvovirions which each contain a single DNA strand; the DNA from these virions easily reanneals, after extraction, however, demonstrating that strands of opposite polarity are separately encapsidated (Rose, 1975).

After the work on "cohesive ends" in λ, interest in the termini of other viruses increased. Genetic evidence in bacteriophage T4 suggested redundancy of phage DNA which might occur at termini (Streisinger et al., 1964 and see Chapter 8) with the aid of electron microscopy and physical chemistry the demonstration of several types of redundancy followed (Thomas, 1966). T-even bacteriophages do not have "sticky ends" but the DNA of each particle contains about 5% of the total molecule repeated at its terminus, so-called *terminal redundancy*. In each particle, however, a different set of sequences is repeated at the ends. This is referred to as *circular permutation*. Such a

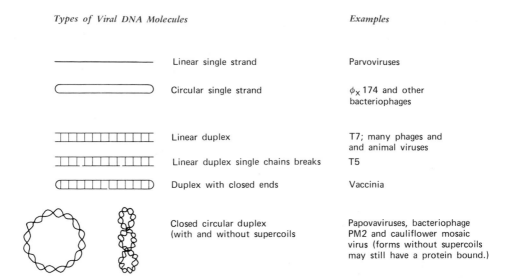

FIGURE 4-19. Forms of DNA found in virions.

situation arises from randomly encapsidating an amount of DNA about 2% longer than the total genome from still longer DNA molecules (see Chapter 7). In distinction to the T-even phages, the T-odd bacteriophage DNA molecules share the same repeated terminal sequences and this is called simply *terminal redundancy*.

A different type of terminal structure has been found in adenovirus (Wolfson and Dressler, 1972) and in the defective parvoviruses (Koczot et al., 1973). Here the two ends of each strand are self-complementary (Figure 4-19), which implies that they are inverted repeated sequences. This self-complementarity was recognized in the electron microscope when single strands were seen to circularize. In addition to this terminal self-complementary feature, adenovirus, type 2, DNA also has another short inverted repeat sequence (possible "hairpin") of about 50 base pairs approximately 180 nucleotides in from each end of the duplex (Padmanbhan et al., 1976). Although the importance of the variations in the repetitive structures at the termini of DNA molecules is not understood it seems highly likely that these regions play a role in either transcription initiation or DNA replication or both (see Chapter 14).

Herpes DNA represents one of the most complicated viral DNA arrangements so far deciphered. The genome appears to be built of two large joined sections, each of which has repeated terminal sequences (see Chapter 14). There are four possible end to end arrangements of the two major DNA segments and all are apparently found in every virion preparation (Hayward et al., 1975).

Vaccinia virus, the largest known virus, has a DNA genome of about 1.5 $\times$ 10^8 daltons. DNA prepared from fresh virions appears to be *cross-linked* because it cannot be denatured into two strands. One possible model for such a molecule involves closing the ends of a linear duplex to make one giant circle which could not be denatured (Holowcyzk, 1976).

RANGE OF MOLECULAR WEIGHTS OF VIRAL DNA. In addition to the very interesting variations in shape and terminal constructions in viral DNA, there is also a wide range of sizes of viral genomes (Table 4-2). The smallest complete viruses, that is, those capable of multiplication within a host cell, are exemplified by $\phi\chi174$, the parvoviruses and the papovaviruses, polyoma, and SV40. The single-stranded $\phi\chi174$ has a DNA of 1.7×10^6 and encodes only nine proteins. Parvovirus DNA, also single-stranded, is about the same size or slightly smaller. A number of the parvoviruses are defective, that is, they must be grown in association with a "helper" virus, suggesting that they approach the lower size limit below which independent replication fails. Polyoma and SV40 both have double-stranded DNA of about 3×10^6 daltons which therefore has a coding capacity equivalent to the small, single-stranded viruses. At the other end of the spectrum are the large bacteriophages such as the T-even phages of *E. coli,* the SP series of *Bacillus subtilis* phages and animal viruses, such as herpesviruses and vaccinia, whose genomes are between 1 and 1.5×10^8 daltons, sufficient to encode more than 100 proteins. In fact, more than 100 genes have been identified in bacteriophage T4 (see Chapter 7).

UNUSUAL COMPONENTS OF VIRION DNA. In 1953 Wyatt and Cohen made the surprising and experimentally useful discovery that the DNA of T-even bacteriophages contain 5-hydroxy methyl cytosine rather than cytosine. This discovery allowed phage DNA to be studied independent of host DNA and led to the discovery of bacteriophage encoded enzymes that perverted the DNA metabolism of the infected cell to the virus' purposes (see Chapter 8). An additional biochemical variation in bacteriophage DNA, the addition of glucose residues to hydroxymethyl cytosine, apparently serves the purpose of preventing phage DNA digestion by certain host enzymes (Kornberg, 1974).

These early studies presaged the discovery of a wide variety of chemical modifications of DNA of other bacteriophages such as the methylation-host restriction system in *E. coli* (discussed in Chapter 6) and the extensive

TABLE 4-2   MOLECULAR WEIGHTS OF DNA OF COMMONLY STUDIED VIRIONS

	Virus	Host	Type of DNA	Mol. wt × 10⁻⁶
Phages	φX174	E. coli	ss, circ.	5,375.7
	fd	E. coli	ss, circ.	1.7–2.0
	M13	E. coli	ss, circ.	2.0
	T1	E. coli	ds	30
	3	E. coli	ds	24
	5	E. coli	ds	75
	7	E. coli	ds	38
	λ	E. coli	ds	30
	T2	E. coli	ds	120
	4	E. coli	ds	110
	6	E. coli	ds	120
	Mu1(Mu)	E. coli	ds	25
	HP1	Hemophilus	ds	20
	P1	Salmonella, E. coli	ds	60
	P2	Shigella, E. coli	ds	22
	P22	Salmonella	ds	26
	PBSX	B. subtilis	ds	9–12 (only host DNA)
	PBS1,2	B. subtilis	ds	200
	SP01	B. subtilis	ds	100
	SP02	B. subtilis	ds	25
	PM2	Pseudomonas	ds, circ.	6
	N1	Blue-green algae	ds	43

Plant viruses

Cauliflower mosaic virus		ds	4.7
Dahlia mosaic virus		ds	4.7
Carnation etched ring virus		ds	4.7

Animal viruses

Parvoviruses		ss	1.2–1.8
Papovaviruses			
Polyoma		ds, circ.	3.0
SV40		ds, circ.	3.2
Adenoviruses			
Human Type 2,4,5,6 etc		ds	23.
Human Type 7,12		ds	21.
Avian, CELO virus		ds	30
Herpes			
Herpes simplex		ds	100
Cytomegalovirus		ds	100
Epstein-Barr		ds	95
Marek's disease virus		ds	120
Iridoviruses			
Frog virus 2 and 3		ds	130
Tipula iridescent virus of insects		ds	126–140
Poxviruses			
Vaccinia		ds	150–160
Fowlpox		ds	200

Data entered in table taken from Fraenkel-Conrat (1976); Fenner et al., 1974, and general literature; ss and ds indicate single- and double-stranded, Circ. is covalently closed circular DNA.

modifications found in phages of *B. subtilis* (see for example Marmur et al., 1972).

In contrast the DNA of animal viruses is remarkably free of modifications. For example, although the DNA of host cells is extensively methylated (mostly, if not exclusively as 5-methyl cytosine) virus DNA contains at most a few methyl groups per genome (Gunthert et al., 1976). Since the majority of virus deoxynucleotides are clearly unmodified, the interest of any definitely established modifications will be enhanced.

VIRAL RNA

Studies on virus RNA represent some of the most important contributions of virology to molecular biology. That plant viruses containing only RNA could be replicating genetic systems established the possibility that RNA could serve to store genetic information. The infectivity of TMV-RNA (Gierer and Schramm, 1956; Fraenkel-Conrat et al., 1957) and the demonstration that the whole molecule was necessary for infection (see Gierer, 1960) were landmark discoveries that established the necessity of the integrity of high molecular weight RNA to its function. By no means last in importance on the list of achievements of the early TMV work was the establishment of techniques for extracting and characterizing high molecular weight molecules; these techniques have formed the basis of studies on the diverse kinds of RNA found in other viruses.

A description of the chemical and physical properties of the various forms of viral RNA and an indication of how the various forms were discovered is not possible without some reference to biological functions of the RNA. Most of the newly recognized virus RNA structures are from animal viruses (Figure 4-20), and greater detail of virion RNA function is presented in Chapter 13.

INFECTIOUS, SINGLE-STRANDED RNA. Soon after the initial TMV work, single-stranded, infectious RNA was reported for a variety of animal viruses (picornaviruses by Colter, 1959; togaviruses by Wecker, 1959) as well as icosahedral plant viruses (Kaper and Steere, 1959). In 1960 RNA bacteriophages were discovered (see Zinder, 1975) and found to contain single-stranded infectious nucleic acid in common with the then most frequently studied animal and plant viruses. In keeping with the presumed single-stranded nature of these viral RNAs there was no sharp melting profile as seen with double-helical DNA and the base composition did not show base pairing. From the RNA content of virions, the molecular weight of isolated RNA molecules, and the single hit UV-killing curves, it was concluded that all the viruses possessing single-stranded infectious nucleic acid contained one RNA molecule per particle (see Gierer, 1960 and Zinder, 1975).

Failure to extract infectious RNA from a number of different types of animal viruses known to contain RNA and the extensive genetic mixing demonstrable with influenza virus (Hirst, 1962) led to speculation that perhaps some of the RNA viruses did not contain simply one single-stranded RNA molecule.

DOUBLE-STRANDED RNA. A series of discoveries since 1962 has borne

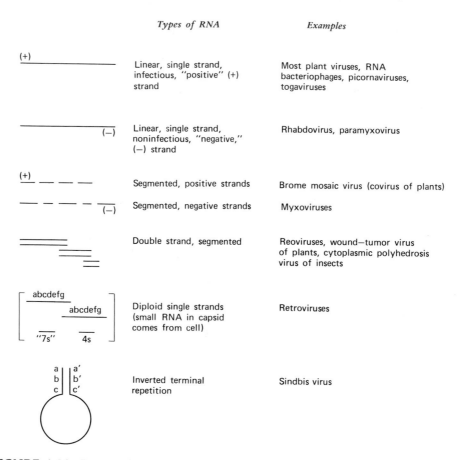

Types of RNA	Examples
Linear, single strand, infectious, "positive" (+) strand	Most plant viruses, RNA bacteriophages, picornaviruses, togaviruses
Linear, single strand, noninfectious, "negative," (−) strand	Rhabdovirus, paramyxovirus
Segmented, positive strands	Brome mosaic virus (covirus of plants)
Segmented, negative strands	Myxoviruses
Double strand, segmented	Reoviruses, wound—tumor virus of plants, cytoplasmic polyhedrosis virus of insects
Diploid single strands (small RNA in capsid comes from cell)	Retroviruses
Inverted terminal repetition	Sindbis virus

FIGURE 4-20. *Forms of RNA found in virions.*

out this suspicion (Figure 4-20). Reoviruses contain double-stranded RNA, which is not all in one piece in the capsid but rather exists in 10 discrete units (Gomatos and Tamm, 1963; Shatkin, 1971). In addition to reoviruses, a number of animal and plant virions are now known to contain segmented double-stranded RNA.

NEGATIVE-STRAND VIRUSES.   The failure to find infectious RNA from paramyxoviruses such as NDV, even though extremely high molecular weight RNA could be isolated from such viruses, was finally explained by experiments of Robinson and Kingsbury (reviewed in Kingsbury, 1972). The polyribosomal fraction of cells productively infected with NDV contained an RNA that would hybridize to the genome RNA. Thus these viruses did not contain the "sense" strand of their genetic information. They have been termed *negative strand* viruses (see Chapter 13) and they contain within their virions an enzyme that produces the genetically effective (*positive strand*) RNA. Thus far only animal viruses of this type have been discovered.

SEGMENTED GENOMES.   The lack of infectious RNA from certain virions might have reflected technical inability to obtain whole RNA molecules. Although the size of any one of the RNA molecules released from, say, reovirus or influenza was not equal to the particle content, the fragments of RNA were very reproducible in size, and if a particle contained one of each fragment, the total molecular weight does equal the particle content. It appears, therefore, that a number of animal viruses possess segmented genomes (Shatkin, 1971). Influenza viruses (and presumably all myxoviruses) contain at least six to eight single-stranded segments, which are negative strands that must be copied by a virion polymerase.

An unexpected type of segmented genome is exemplified by brome mosaic virus (a plant virus). In this case four distinct virion RNA species are recognized. These RNAs are contained in three types of virions of different density that have the same capsid proteins. The three largest RNA molecules are all necessary for infection. The fourth smallest RNA apparently codes only for the capsid protein (Shih et al., 1972) and its sequences are contained within the third largest RNA. A number of plant viruses share this property of divided genomes (Table 4-3).

SIZE RANGE OF RNA.   The size range of *virions* of RNA viruses is quite great from picornaviruses, $\sim 7 \times 10^6$, to retroviruses, over 2×10^8; however, the size and thus the informational content of the RNA fall within narrower limits (Table 4-3; Fraenkel-Conrat, 1974). Picornavirus RNA, perhaps the smallest, is about 7500 nucleotides, and paramyxovirus RNA, perhaps the largest, is less than twice as large. Presumably all the independently replicating RNA viruses require the minimum information for a replication system and a capsid protein, but the elaborate additional information which the large DNA viruses can possess is not found among RNA viruses. However, defective RNA viruses with still simpler genomes do exist—satellite necrosis virus of plants contains RNA ($\sim 4 \times 10^5$ daltons) encoding only a capsid protein and presumably binding sites for replication; however, it must rely for survival and growth on simultaneous infection with a "helper" virus, tobacco necrosis virus, that possesses a compatible replication system (Reichmann, 1964; Rees et al., 1970). No defective animal viruses of this type have been found but among the RNA tumor viruses (retroviruses) gradations of defectiveness and "helper" functions exist depending on the host cell infected (see Chapter 16). Table 4-2 gives the size ranges of a variety of RNA viruses.

UNUSUAL COMPONENTS OF VIRAL RNA.   One of the most prominent "unusual" components of nucleic acids are methylated nucleotides. Because methylated polyribonucleotides are poorly translated, mRNA was thought probably to lack methylated bases until Perry and Kelley (1974) found an average of two to three methylations per mRNA molecule in mouse L cell mRNA. Soon thereafter, methylation was found in the mRNA produced by virions of vaccinia (Wei and Moss, 1974), reovirus (Shatkin, 1974), and cytoplasmic polyhedrosis virus (CPV) of insects (Furuichi, 1974). Simultaneous with the work on methylation, attempts to add labeled phosphate to the presumed 5' pXp. . . or 5' pppXp. . . end of the viral mRNA and virion RNA had failed.

Three groups of workers, puzzled by this blockage of the 5' terminus of

TABLE 4-3 PROPERTIES OF VIRION RNA FROM COMMONLY STUDIED VIRUSES

	Molecular Weight $\times 10^{-6}$
Phages	
f^2	1.2
R17	1.2
MS2	3569 bases, totally sequenced
Qβ	1.2
Plant Viruses	
Potato spindle tuber viroid (?virus)	0.05
Potato virus	2.1
Pea enation virus (2 virions, some infectious RNA)	1.68 (infectious) 2.2 and 0.36 (noninfectious, from larger virion)
Coviruses (multiple component viruses)	
Cowpea mosaic virus (2 virions)	1.4, 2.5
Alfalfa mosaic (4 virions)	1.3, 1.0, 0.7, 0.34
Brome mosaic (3 virions)	1.05, 0.95, 0.71, 0.30
Tomato bushy stunt	1.5
Tobacco mosaic	2
Tobacco necrosis	1.3
Tobacco necrosis satellite virus	0.4
Turnip yellow mosaic	2.1
Wound tumor virus (double-stranded RNA)	12 species, 1.2×10^7 daltons total
Rice dwarf virus (double-stranded RNA)	12 species, 1.2×10^7 daltons
Lettuce necrotic virus (?negative strand)	
Animal Viruses	
Positive strand	
Picornaviruses	2.7
Togaviruses	4.0
Negative strand	
Paramyxoviruses	5.5–7.5
Myxoviruses	Total 3.9 to 4.9, for 8 segments of 0.4–1.0
Rhabdoviruses	
Infectious virions	3.8–4.5
Defective virions	1.2–1.3
Double-stranded, segmented	
Reo	Total 15; 10 segments of 0.6–2.8
Blue tongue	Total 12.1; 10 segments of 0.3–2.7
Cytoplasmic polyhedrosis virus of insects	Total of 14.6; 10 segments 0.35–2.8
Retroviruses	6–7; complex of 2 segments $\sim 3 \times 10^6$ each plus 2 cell RNAs

Data from Fraenkel-Conrat (1974) and general literature.

the RNA molecules and suspecting a connection with methyl incorporation, solved the structure of the 5′ end of virus mRNA and in so doing learned about a structure of great general importance in biology of higher cells (reviewed by Shatkin, 1976). The 5′ terminus of mRNA produced by vaccinia, CPV and reovirions was found to be blocked in a 5′-5′ linkage with a 7-methyl guanylate residuc (Figure 4-21). In addition, the first nucleotide in the mRNA chain is methylated on the 2′ OH position of the ribose, thus protecting the structure from endonucleolytic or alkali cleavage. This entire methylated terminal oligonucleotide has been termed a "cap" structure. It is present not only in the three viral mRNAs (produced *in vivo* and *in vitro*) but also in most, if not all, cell mRNAs from yeasts through human cells. Viral mRNA of DNA viruses produced in the nucleus (for example, SV40 and adenovirus), viral mRNA of negative strand viruses such as VSV (both *in vitro* and *in vivo* produced), virion RNA of retroviruses produced by transcription of integrated DNA, the plus strands of double-stranded virion RNA of reovirus and CPV, and the virion RNA and mRNA of togaviruses, probably among others all contain the cap structure (see Shatkin, 1976). Picornaviruses, so far alone among RNAs that serve the messenger function, definitely do not possess either a blocked terminus or any methylations in virion RNA, replicating RNA or in viral polyribosomal mRNA from infected cells (see Chapters 13 and 14).

All four genome segments of brome mosaic virus, a plant virus, have the cap structure. The RNA bacteriophages as well as bacterial mRNA lack the cap; the RNA phages contain pppGp at their 5′ ends.

FIGURE 4-21. Structure of 5′ blocked, methylated "cap" of eukaryotic mRNA. Features of note are presence of 5′-5′ linkage of methyl guanylate N_1; N_2, and N_3 can be any of four nucleic acid bases. 2′O-methyl group on ribose of N_1 nucleotides is always present, methyl group on N_2 present in some cases only.

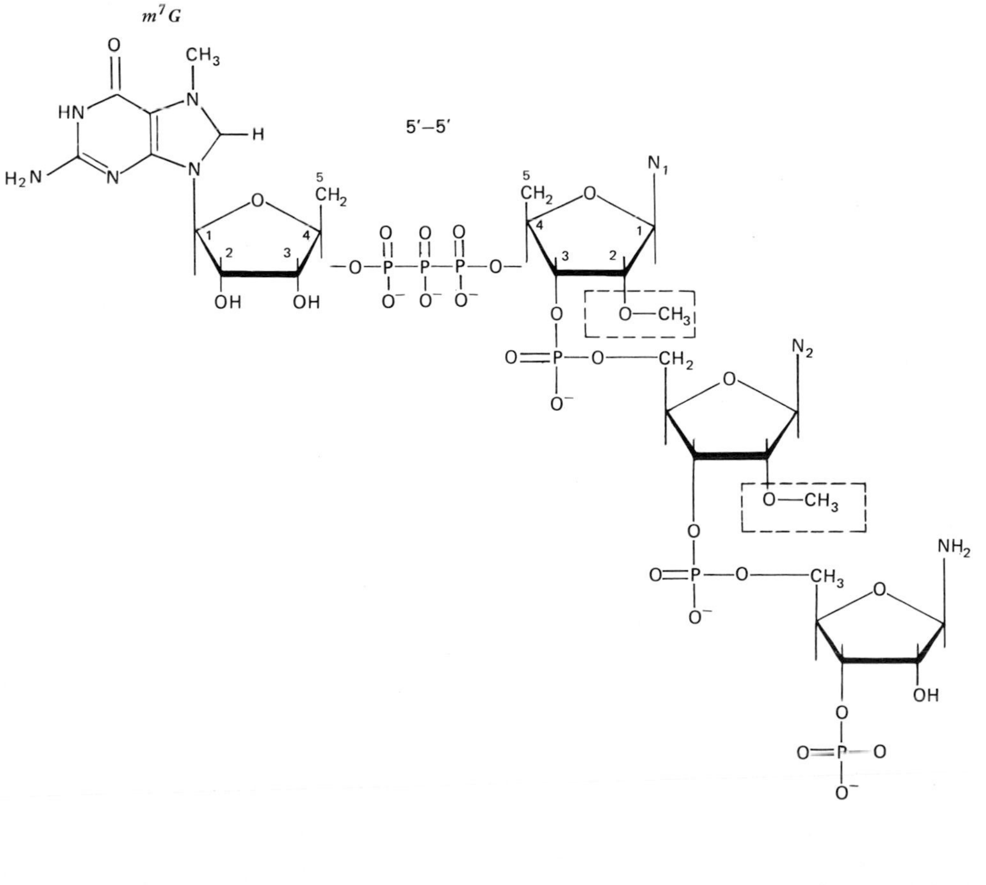

The function of the cap structure is to enhance initiation of translation of eukaryotic mRNA (Shatkin, 1976) but it is not recognized by bacterial ribosomes.

The biosynthesis of the cap structure proceeds either of two ways. In vaccinia and reovirus mRNA synthesis the pppG, which becomes m^7G in the cap, contributes only one phosphate to the cap structure. The first nucleotide of the RNA chain laid down by polymerase contributes two phosphates. The reaction thus proceeds:

$$pppXpXp. . .$$

$$\downarrow \text{Phosphohydrolase of virion}$$

$$ppXpXp$$

$$\downarrow pppG + \text{S-adenosylmethionine}$$

$$m^7GpppX_mpXp$$

In vesicular stomatitis virus "cap" synthesis, the pppG forming the m^7G portion of the cap contributes *two* phosphates allowing for the possibility of "capping" an internal cleavage site where a new pXp. . . terminus has been created, thus:

$$. . .pXpX \,|\, pXpXp. . .$$
$$\downarrow$$
$$. . .pXpX_{OH} + pXpXp. . .$$
$$\downarrow + pppG + \text{S-adenosylmethionine}$$
$$m^7GpppXmp. . .$$

In both cases at least two types of enzymatic activities are necessary: a guanylyl transferase activity and methylases for m^7G and 2'0 methylation of the ribose of the first nucleotide. Methylation of the ribose of the second nucleotide frequently occurs in the cytoplasm (Shatkin, 1976).

The concentration of interest in methylation brought about by the work on "caps" has led to the discovery of other methyl groups in the genome RNA of certain viruses and in the mRNA of DNA viruses which are transcribed in the nucleus. SV40 and adenovirus mRNA contain both a cap and from 2 to 4 methyl adenylic acid (m^6A) residues per molecule. The retrovirus virion RNA, which is transcribed in the nucleus from integrated viral DNA, has as many as 8 m^6A residues. No RNA viruses that replicate in the cytoplasm nor the mRNA from vaccinia, a DNA virus that replicates in the cytoplasm, contain this methylated base. Sindbis virus RNA, which functions as an mRNA, has a "cap" in addition to several 5-methyl cytidylate residues (Dubin and Stollar, 1975). Doubtless other specific methylations will soon be reported. The role of the base methylations in these mRNAs is unknown.

A final point about unusual characteristics of viral genomes refers to 3' ends of animal and plant viruses. All the infectious single-stranded animal viruses contain a polyadenylic segment at their 3' termini, which in general, is shorter than the 3' poly(A) found in cellular mRNA or virus mRNA derived from DNA viruses (see Shatkin, 1974 for review). Some functional viral mRNAs, however, definitely lack poly(A), for example, reovirus mRNA. The purpose of the poly(A) is as yet unknown and experiments dealing with poly(A) in picornavirus and adenovirus replication are discussed in Chapter 13.

The 3' ends of a number of plant viruses have been studied and do not contain poly(A) but do possess an unusual structure. The RNAs of TYMV, brome mosaic virus, and CCMV (cowpea chlorotic mosaic virus), among others, can accept amino acids as if they were tRNA molecules and the 3' end sequence can be arranged in a folded form like a tRNA molecule. The meaning of this structure is unknown; no experiments demonstrating amino acylation of the viral RNA within cells have appeared, and the "charged" viral RNA does not function well as an amino acid donor in cell-free protein synthesis (Lane, 1974).

VIRAL PROTEINS

The first viruses to be purified were among the simplest, for example, TMV where a single RNA molecule is coated by a single type of polypeptide chain. As information grew it became clear that viruses of more complex capsid structure existed and so too did multiple capsid proteins. The earliest recognized cases of complex capsid structures were, of course, the bacteriophages, with their auxiliary structures ("tails") for attachment. However, all of the proteins present in virions were generally thought of as "structural" and enzymatic activities in virions were thought of as cellular contaminants. Again bacteriophage tails, which contracted with the splitting of ATP, were regarded as a special case as was the neuraminidase in the envelope of influenza virus.

In the past 10 years, however, beginning with the discovery (Kates and McAuslan, 1967a and b; Munyan et al., 1967) that vaccinia virions contained a DNA-dependent RNA polymerase, a whole range of virion-specific enzymes have been found in animal viruses. Most, but not all, of these activities are confined to enveloped viruses. For example, as discussed in the previous section, double-stranded RNA viruses contain enzymes for synthesizing viral mRNA, including the addition of a "cap." No such enzymatic activities have been detected in bacterial or plant viruses thus far except for the double-stranded RNA viruses of plants.

Thus the idea of capsid proteins as inert coverings for viral nucleic acid has had to be modified. Hypotheses of capsid protein arrangements must include not only structural components, whose primary function probably still can be properly viewed as constructing a shell to harbor the nucleic acid, but must also include either a few molecules of highly specific enzymes or the structural proteins themselves must be enzymatically active.

In addition to capsid proteins that enclose the nucleic acid, other types of virion proteins are encountered in the enveloped viruses, examples of which are found among animal, insect, plant, and bacterial viruses. Here in addition to nucleoprotein cores, virions may contain virus-specific proteins that are inserted into the plasma membranes of the infected cell and enclose the entire particle as it exits or "buds" from the surface. In addition, in some enveloped viruses a submembranous matrix protein exists between the envelope and the nucleocapsid (see Chapter 14).

A second major division of virus-specific proteins are the *noncapsid* virus proteins. These proteins are largely concerned with the manufacture of virion nucleic acid and are therefore properly dealt with in discussing the replication cycle of various viruses.

STRUCTURAL (NUCLEOCAPSID) PROTEINS. No isolated virion proteins have yet been crystallized so that a complete X-ray diffraction analysis of three-dimensional structure at the atomic level is not available. However, the structural properties of TMV protein have been studied with other techniques (see Chapter 3). Without RNA present, monomers, oligomers, disks, and stacked disk and helical ("lock-washer") arrangements of TMV proteins exist, depending on various conditions of salt and pH. The assembly of TMV particles when both protein and RNA are present is probably similar to the acid-induced helical configuration of protein alone (Richards and Williams, 1976). As indicated from the discussion of virus morphology in Chapter 3, in icosahedral virions, the proteins must interact at specific sites to generate the two-, three-, and five-fold axes of symmetry. For viruses with only a single capsid polypeptide, this requires that there be precise, self-complementary segments of the protein. To satisfy the requirements of quasi-equivalence—the principle that allows a single polypeptide complex to assume configurations producing either six-fold or five-fold symmetry—the polypeptide must also have flexibility. In one case this flexibility comes from a hinge that attaches separate domains of the polypeptide (S. Harrison, personal communication). In addition to protein-protein binding sites, substantial evidence exists for protein-nucleic acid binding sites. For example, the reconstitution of F2 bacteriophage, a spherical RNA virus, is strictly dependent on the presence of viral RNA as an "organizing" component (Hohn and Hohn, 1970). Since complicated capsid symmetries are now recognized in which all polypeptides are not symmetrically arranged with all other peptides, for example, reoviruses and herpesviruses contain a double-shelled capsid, it is not surprising that many virions contain multiple polypeptide chains. It is, of course, no surprise either that the highly developed morphology of bacteriophage particles is accomplished with a large variety of virion polypeptides. Some 30 different polypeptide chains have been identified in T4 (see Chapter 7).

The chief technique for examining the structural proteins of virus capsids is polyacrylamide gel electrophoresis (Maizel, 1971). This technique is very widely used in virus protein work and the earliest experiments using gel separations led to the discovery of an important general feature of protein synthesis of many capsid proteins; that is, cleavage from a higher molecular weight precursor protein.

When purified labeled virion preparations are denatured in hot SDS (sodium dodecyl sulfate) solutions the polypeptides separate and bind SDS in proportion to their length. If the chains are then moved through polyacrylamide gels by electrophoresis, the sieving properties of polyacrylamide allows separation of polypeptide-SDS complexes on the basis of size. Proteins with a difference in molecular weight of only a few percent can be separated. One caveat is that a few proteins, for example, glycoproteins of enveloped viruses, may not bind SDS equivalently to their molecular weight so that while they give a reproducible gel banding pattern, their molecular weight cannot be determined with this technique.

When the technique of gel separation of polypeptides was first applied to briefly labeled HeLa cells infected with poliovirus, large virus-specific polypeptides not found in virions were detected (Fig. 4-22; Summers et al., 1965). These large peptides were not so prominently labeled after longer exposure to labeled amino acids. This large size and apparent metabolic instability, coupled with the finding that one larger polypeptide was found in place of two smaller ones in a precursor particle (Jacobson and Baltimore, 1968), led to the suggestion and ultimate proof that peptide cleavage was important in poliovirion maturation (see Chapter 13 for details).

Cleavage of proteins in capsid assembly has now been found to be a general property of many polypeptides in animal viruses (Korant, 1975), both for structural nucleocapsid proteins and envelope proteins. A number of bacteriophage proteins also derive from cleavage of larger precursors (Laemmli, 1970; King and Mykolajewycz, 1973). It would appear that in some cases binding sites that are finally used for constructing complicated capsid structures are allowed to form only after the peptide has first been brought into position as a larger polypeptide; cleavage occurs and the final structure then takes shape. The older idea of virion proteins simply achieving structures totally by self-assembly is probably valid only for the simplest viruses for

FIGURE 4-22. *Virus-specific peptides formed in poliovirus-infected cells. ^{3}H amino acid-labeled extract of infected cells (prepared when only virus-specific synthesis was occurring) was mixed with ^{14}C-labeled polio virions. The proteins were disrupted to polypeptides and analyzed by gel electrophoresis (from D. F. Summers et al., Proc. Nat. Acad. Sci. U.S.A.* **54**, *505–513 (1965)).*

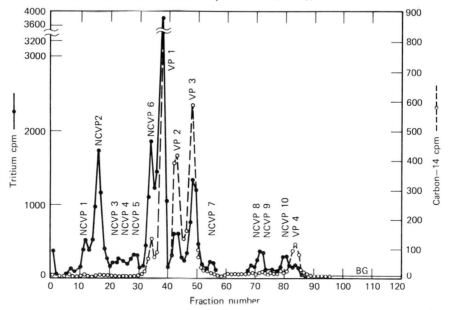

example, TMV and perhaps RNA bacteriophages and some spherical plant viruses (Bancroft, 1970).

ENVELOPE PROTEINS.   Envelope proteins are distinctive proteins from several points of view. Although they are synthesized in the cytoplasm, they locate in and penetrate a lipid bilayer; in some cases they may span the bilayer leaving a portion of their chain inside the cell, a portion outside the cell, and a portion in the lipid bilayer (Garoff and Simons, 1974). Thus they probably have *domains* that are hydrophobic and domains that are hydrophilic. For example, proteolytic digestion of SFV (Semliki forest virus) particles leaves behind a lipid-embedded portion of the envelope protein which is rich in hydrophobic amino acids (Uterman and Simons, 1974).

The envelope proteins are special in a second way also. In common with cellular membrane proteins, they are frequently glycosylated. The sugar residues are probably determined by cellular enzymes since the same virus grown in different hosts possesses different polysaccharide moieties (Burge and Strauss, 1970). The role of the polysaccharides is not yet firmly established but the glycoprotein surface structure probably plays an important role in cell attachment.

VIRION COMPONENTS OTHER THAN NUCLEIC ACID AND PROTEINS

The most prominent virion constituent that is not a nucleic acid or protein has already been mentioned in discussing enveloped viruses—it is the bilayer of lipid that makes up the mass of the envelope. It is generally believed that the lipid of the envelope is simply pirated from the host cell plasma membrane, thus strictly speaking it might not be considered "virus-specific." In fact, paramyxoviruses grown in different host cells may contain different lipids reflecting the cell of origin. The specificity of the viral envelope thus resides in the virus glycoproteins projecting from the envelope surface (Choppin and Compans, 1974).

A number of small molecular weight components are known to be present in highly purified virions and a function for some of these can be inferred. About 140 molecules of ATP and 140 Ca^{++} ions are bound by the contractile sheath in the phage tail and the ATP is cleaved during the contraction process that injects phage DNA into the host cell, thus presumably providing the energy for this process. In addition, a folic acid derivative, dihydropteroyl polyglutamate is found in the tail assembly of T-even bacteriophages where it may play a role in initiating phage infection by changing the structure of the base plate of the tail (Kozloff et al., 1970).

Polyamines are found in bacteriophages, plant viruses, and animal viruses but no physiologic role other than neutralization of the charge on nucleic acids has been suggested for these. For example, herpes virus contains enough spermine to neutralize half the virus DNA and spermidine is supposedly found in the viral envelope (Gibson and Roizman, 1971). A number of plant viruses (turnip crinkle, broad bean mottle, and tobacco mosaic viruses) contain a polyamine, bis(3-aminopropyl)amine which presumably functions, as the polyamines in phage, to neutralize charges on the viral RNA (Johnson and Markham, 1962), but may be a new product of infected cells since it is not found in healthy leaves.

General Features of Virus-Host
5 Interaction

Problems and Methodology

In the preceding chapters the properties of virions and related materials have been described. Virions represent the static, inert form of a virus. No reproduction, no metabolism take place when virions are in their extracellular state. All dynamic events—biosynthesis of viral materials, damage to the host organism—take place when viruses interact with host cells. Even in multicellular hosts the critical events take place at the cellular level; the spread of virus occurs by repeated cycles of virus-cell interaction accompanied by dispersal of virions through the extracellular fluids. What we know about the various components of virions leads us to anticipate that within host cells the viral materials will generally not be organized as in the virions. In virus-infected cells there occurs a profound reorganization of viral materials and often also of cellular materials. A novel entity is generated, the *virus-cell complex*, whose functional organization results from the interplay between viral functions and cellular functions. The machinery of the complex differs from the machinery of the host cell.

The events that take place in virus-infected cells are not the same for all virus-host systems. The outcomes can also be different, ranging from multiplication of the virus (*productive infection*), with or without destruction of the host cell, to more-or-less permanent association between virus and cell. Infection may be abortive and the infecting virus may be eliminated either with concomitant cell death or without recognizable cell damage.

The study of virus-cell interaction is the exploration of the events that take place in the virus-cell complex, considered either as a new entity generated by the entry of a virus into a cell or as the expression of viral materials present in the cells of a virus-carrying cell line. The goal is to interpret the functional organization of the complex in terms of specific viral functions, of specific

125

cellular functions, and of their interactions. These functions must be approached in the light of the known principles of molecular biology, a discipline that the study of viruses has itself helped to create:

1. Each unit function is the result of the specific activity of one or more genes.

2. Each gene provides, through transcription and translation processes, a polypeptide that becomes either a protein or a subunit of a protein.

3. Some nucleotide sequences in the genetic material are the sites of regulatory processes, specifying signals for attachment or release of transcription enzymes or of regulatory proteins.

4. Gene function may be regulated at the level of transcription (RNA synthesis) or at the level of translation (protein synthesis). Transcriptional controls are well understood for procaryotic cells, much less so for eucaryotic cells; translational controls are rare or absent in procaryotes. Suggestive but indecisive evidence exists for their occurrence in eucaryotes.

5. Replication of genetic material requires not only a set of enzymes and cofactors but also the existence of specific "starting" sequences in the nucleic acid template. A genetic element possessing such a sequence is called a *replicon*.

6. The bonds among polypeptides in a protein and among specifically associated proteins, as in the capsids of viruses, are not covalent; they are more or less stable patterns of weak bonds. Assembly of large protein structures often occurs through intermediate subassembly steps and results from specific protein-protein interactions as well as from protein-nucleic acid interactions.

7. Polypeptides are not necessarily used in enzymes or structural assemblies in the same form in which they are first synthesized, but may first be "processed" by exopeptidase or endopeptidase action.

8. Some proteins may serve as structural intermediates by providing transient frameworks that facilitate specific assembly of other proteins or of proteins and nucleic acids.

Once the problems of virus-cell interactions are stated in terms of molecular biology, these interactions come to be viewed as *developmental sequences,* that is, interactions between genes, gene functions, and gene products specified in space and time. In the same way as the goal of cellular biology is to resolve the processes of cellular life cycles into specific sequences, and the goal of organismic physiology is to interpret complexity and differentiation in comparable terms, so the goal of virology is to unravel the processes by which viruses generate in their host cells novel sequences of events that yield virus multiplication or other outcomes. Because viruses are exogenous elements added to cells they do in fact represent valuable probes to explore the functional organization of living cells.

BIOCHEMICAL AND GENETIC APPROACHES

The above formulation of the problem also implies that the relevant methodologies are those of molecular biology: biochemical and genetic approaches employed jointly with observational approaches described in the preceding chapters. Each step of the interaction can be analyzed in these terms. The specificity of virus-cell attachment and the nature of the elements that take part in the attachment, determining its specificity and thereby defining the host-range of the virus, are examples of a set of phenomenological observables to be analyzed genetically and biochemically. The entry of viral materials into the host cells, their organization, and their amounts and reproduction within the cells are studied by analysis of cellular extracts at various times. Serological and enzymological assays are used in the identification of the proteins involved in various phases of the interaction. Nucleic acids of viral or cellular origin are monitored by chemical and radiochemical analysis, nucleotide sequencing or, more commonly, by measurements of specific density, using either natural differences between viral and host nucleic acids or differences created by feeding density labels. In some instances chemical analysis permits identification of unique components of viral nucleic acids such as 5-hydroxymethyl cytosine in T-even phage DNA (Wyatt and Cohen, 1953). DNA:DNA hybridization (or DNA:RNA or RNA:RNA) serves to identify and quantitate viral vs. cellular nucleic acids. Radioisotopes are of course the prime tools to label specific components of specific macromolecular species.

The biochemical tests as a whole reveal that the interactions between virus and cell are not like those between a bacterium and its culture medium. They involve a parasitism both at the biochemical level—provision by the host cell of enzymes and enzyme products for viral biosynthesis—and at the genetic level, that is, the destruction or incapacitation of the host's genome or its alteration by insertion of viral genes.

Genetic approaches add new dimensions to the biochemical ones. With or without mutagenesis one generally can select out mutants of a virus differing from the "wild-type" strain in almost any one of the observable characteristics: host range, morphology of characteristic local lesions or overall course of a disease, antigenic specificities, stability of virions in different osmotic or chemical environments, sensitivity to radiation, to mention only a few. The availability of mutants makes it possible to analyze specific phases of host virus interactions such as attachment and penetration or the modalities of host-cell damages. Most informative are *conditional lethal mutants* of viruses, especially mutants that grow in a given host at one temperature but not at another (temperature-sensitive mutants; *hs* or heat-sensitive, *cs* or cold-sensitive) (Epstein et al., 1963). These temperature mutants have undergone genetic changes that lead to amino acid replacements in specific proteins such that the protein is nonfunctional at the nonpermissive temperature but functional at the permissive one.

Another important class is that of *suppressor sensitive* (*sus*) mutants, which can grow only on host cells that possess specific suppressor genes, which cause the insertion of incorrect amino acids in response to certain genetic codons. These suppressor genes usually are genes for mutant tRNAs, which accept their normal amino acids but have a mutant anticodon and therefore pair with the wrong codon and insert the wrong amino acid. More rarely, the suppressor gene codes for a mutant activating enzyme that puts the wrong amino acid onto its tRNA (see Chapter 4). A viral mutant whose mutation has altered a nucleotide in an essential gene can be lethal in a host without suppressors;

but in a host that has a suppressor capable of placing an acceptable amino acid in place of the unacceptable one the mutant virus may grow (suppression of unacceptable missense in the genetic code). More important still, a suppressor mechanism may place a usable amino acid at the site where a mutant codon spells chain termination (suppression of nonsense mutation).

Conditional mutants, either temperature-sensitive or suppressor-sensitive, have been powerful tools in the analysis of intracellular events. The permissive condition allows propagation of the virus. The stage at which viral development is arrested in the nonpermissive condition is generally the stage at which the product of the mutant gene is required. In this way various gene functions can be assigned to specific stages of the process. With *hs* or *cs* mutants one can learn by temperature shifts the actual order of functioning of various genes (Jarvik and Botstein, 1973): for example, until the product of gene X has been made in response to a shift to permissive temperature, the protein of genes Y,Z. . . may fail to be made. Conditional mutants have been used extensively in analyzing phage development and are proving useful in animal virus studies as well. Similar methods, in the phage-bacterium system at least, can be extended to the host genome: conditional mutants of bacteria, which fail to support growth of some phage, reveal the involvement of specific host genes in specific stages of phage development.

Most viruses have more than one gene. Recombination, known to occur in all DNA-containing phages and in many animal viruses, is analyzed by mixed infection of cells with two related viruses differing by two or more genetic traits. This mixed infection can be compared with a genetic cross between two mutant fruit flies or two mutant strains of maize, although the situation is more complex (see Chapter 7). Such viral crosses make it possible not only to generate a variety of combinations of mutants in the same viral genome, but also to locate the genes of viruses onto linear genetic maps analogous to those of any other organisms. The genetic maps of viruses, in accord with the amounts of nucleic acid, range from fewer than 10 genes for some small viruses to 100 or more for some of the largest viruses such as phage T4 or vaccinia. Deletions and duplications of genetic materials do occur and sometimes also insertions of new nucleic acid, which may be derived from the host cells.

There is another important aspect of genetic recombination. At least for bacteriophages and probably also for other viruses, a mechanism for genetic recombination—that is, physical exchanges among or within nucleic acid molecules—is required at some stage of viral development (see Broker and Doermann, 1976). Circles can be generated by recombination from linear molecules of DNA. *Concatemers*—multiple lengths of viral genomes—are often intermediate forms in nucleic acid replication. Circular forms of viral DNA mediate the integration of viral genes into the host-cell genome. Examples of these mechanisms will be illustrated in subsequent chapters.

A powerful genetic test is that for dominance and complementation. One tests for the occurrence of a specific viral function or product in host cells infected with two virus mutants affected by mutation in the same phenotypic trait. The *cis-trans* test (Figure 5-1) illustrates the power of complementation tests to assess whether two mutations have affected the same gene. The tests also reveal which, if either, of the two alleles of the gene is *trans*-dominant

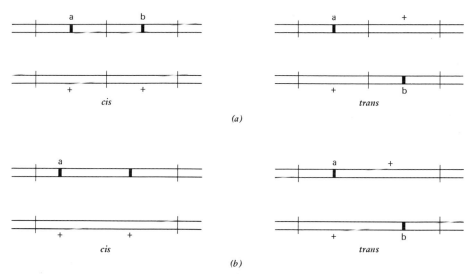

FIGURE 5-1. The cis-trans *test in diploid organisms or in mixed infection of a cell with two mutants of a virus.* (a) *Mutations a and b in different cistrons; the function is present whether the mutations are in* cis *or* trans. (b) *Mutations a and b in the same cistron; the function is absent when mutations are in* trans.

(and therefore, presumably, functionally active). The other allele is then recessive to the dominant one. *Cis*-dominance is also encountered, generally for mutations that affect regulatory circuits between adjacent gene groups (see Hayes, 1968).

Developmental Phases: Eclipse, Replication, Maturation

The various approaches, taken all together, reveal a variety of situations, which have in common, however, the property that for each virus the interaction with the host can be interpreted as a specific developmental sequence. Each virus is an organism with its own ontogeny and morphogenesis as well as phylogeny. In their broadest lines, the developmental cycles of viruses share certain essential features.

After a virion attaches to a susceptible cell there occurs a series of reactions that leads ultimately to release of the genetic material of the virus within the cell. The infecting virion ceases to exist as an organized unit. Since the infectivity of free viral nucleic acid is generally much lower than that of the complete virion, the release of the viral genome into the host cell is accompanied by decrease or disappearance of infective units detectable in cell extracts. This is called the *eclipse* phenomenon. The release of viral nucleic acid into the newly infected cell may take various forms. It may take place directly at the cell surface, as with phages that inject their DNA in an oriented fashion through the bacterial envelope (Hershey and Chase, 1952). Some phages attach specifically to bacterial pili or flagella; they may inject their genomes through these organelles or use them to approach the cell surface. Enveloped virions may fuse with the cell membrane and release into the cellular cytoplasm the inner capsid, which in turn releases the nucleic acid genome.

Once the viral genome is released from the virion it becomes available as a source of information both for replication and for transcription, serving as template for the biosynthesis of specific products. Multiplication of viral genomes occurs by replication of the genetic material, DNA or RNA. The essential events of DNA replication are biochemically on the same level as the replication of cellular genetic material. Replication of a DNA viral genome in a host cell is possible provided the genome is a replicon (Jacob et al., 1963) recognized by replication mechanisms of cellular or viral origin. Both cellular and viral enzymes may cooperate in the replication process. Before replication begins a number of steps may be required and special conditions may have to be fulfilled. The enzymatic apparatus of the cell may be expanded by virus infection, sometimes by enzymes brought into the cell by the virion (vaccinia, vesicular stomatitis, influenza and retroviruses), sometimes by enzymes newly made as products of viral genes. This is particularly evident for certain phages that require unique DNA components and contain the information to make the corresponding synthetic enzymes (Cohen, 1968). Viruses may also determine the synthesis of enzymes that catalyze reactions already fulfilled by cellular enzymes (see Mathews, 1971).

Most RNA viruses multiply by the production of RNA replicas without intervention of an intermediate DNA template: their replication can occur normally in cells in which DNA synthesis is inhibited. These viruses code for their own RNA *replicases* (Baltimore and Franklin, 1963; Spiegelman and Haruna, 1966). The host cells have no such RNA-replicating enzymes. Some groups of RNA viruses replicate their RNA by the intermediate of a DNA complementary replica of the RNA present in the virions. This DNA replica is synthesized by a reverse transcriptase (Baltimore, 1970; Temin and Mizutani, 1970), which is brought into the host cell by the virion along with the RNA. Such introduction of preformed viral enzymes into the host cells is not a rare occurrence. Various examples will be discussed in Chapter 13.

There is an intrinsic limitation to the amount of synthetic machinery a viral genome can code for: the size of its nucleic acid. The smallest viruses contain of the order of 10^6 daltons of DNA or RNA or about 3500 nucleotides. Because the molecular weight ratio of coding nucleic acid to coded protein is about 9 to 1 for RNA or single-stranded DNA and 18 to 1 for double-stranded DNA, these viruses have only enough information for a few proteins and these usually include structural proteins of the virions. Thus all viruses are dependent to a large extent on the enzymatic machinery of their host cells. Some have also become curiously dependent on help from other viruses. For example, a "satellite" virus of tobacco necrosis has only 1200 ribonucleotides while the protein subunit of the capsid for which it codes consists of 400 amino acids (Reichmann, 1964). There is evidently no room in the RNA for any other information. In fact, this virus reproduces only in plant cells that are also infected with tobacco necrosis virus, which provides the required replicase. There are other examples of viruses that persist in nature (or in the laboratory) only by the cooperation of "helper" viruses infecting the same cells.

During its replication the viral nucleic acid is not associated with the characteristic proteins that are present in mature virions. In fact, under certain conditions replication of viral nucleic acid can occur even when protein synthesis is chemically inhibited. In viral developments that lead to production and release of a progeny crop of virions, the synthesis of virion proteins—the beginning of maturation—usually starts after nucleic acid replication is already under way. There is not a one-to-one association of newly formed virion proteins with viral genomes: the proteins accumulate, forming "precursor pools" from which individual molecules or groups of molecules are finally withdrawn for capsid assembly (Casjens and King, 1975). This is a complex process and an irreversible one: nucleic acid or structural virion proteins, once they have entered a virion assembly or subassembly, are not again released within the same cell. Capsid assembly removes viral genomes from the replicating population and capsid proteins from the pools. Outer envelopes, if any, are then added either within the cell cytoplasm or upon interaction with cellular membranes. Such an assembly process, going through stages of pools of precursor molecules, accounts for the occurrence of *phenotypic mixing* (Novick and Szilard, 1951), which is the production, in cells infected with two different but compatible viruses, of virions whose capsids are mixtures of protein subunits derived from different viral genomes.

Newly formed virions are finally released into the external milieu (often together with related immature forms) either by lysis of the host cells brought about by viral enzymes, as in phage infected bacteria, or by extrusion of portions of cytoplasm, or by release of individual particles or groups of particles. Some animal viruses are ineffectually released by cells in cultures: in the entire animal, viral release and spread may be aided by phagocytic digestion of cells damaged by infection. Plant viruses are normally not released by cell lysis but can be transported from cell to cell through intercellular connections.

REGULATION AND VIRAL CYCLES

Like all developmental sequences, the life cycles of viruses require not only specific machinery and specific components but a set of specific regulatory controls. These controls are presumably similar to those that regulate gene functions in cells, but this is a rather weak proposition. Only for bacteria we have a reasonable understanding of gene regulation—the operon theory and its subsequent refinements (Jacob and Monod, 1961; Hayes, 1968)—and even then a goodly part of our knowledge stems from the study of bacteriophages. We know almost nothing of regulation in eucaryotic cells: for example, why a brain cell, a liver cell, and a muscle cell with the same genetic heredity produce different sets of proteins. Viruses will probably be a fruitful source of information for years to come. Control mechanisms in viral development include, for example, the presence in a viral genome of promoter sequences that are recognized only by transcribing enzymes generated or altered by the action of other viral genes. This mechanism generates a specific sequential pattern, which requires transcription and translation of a subset A of genes before subset B can be read, and so on. Other regulatory events take place at operator sequences. The establishment of phage lysogeny, for example, requires circularization of the viral genome and the blocking and unblocking of certain phage genome operators by one or more repressors and antiblocking factors made by phage genes (Botstein and Susskind, 1974). The repressors inhibit the sequence of events that would lead to virus maturation and cell death and make possible for the viral genome to become integrated into the host cell chromosome.

To these types of genetic regulation there are added other interactions between virus and host. Several phages and animal viruses specifically inhibit replication or transcription of host cell genes. Alternatively, a virus persisting in a cell line may alter the developmental properties of the cells. The most glaring example is that of tumor viruses, which cause mature, nonproliferating cells to acquire the proliferative capacity of their embryonal ancestors and to become cancer cells (see Chapter 16).

Certain events in the viral maturation sequence, such as the processing of capsid proteins by viral or host cell peptidases, may also be considered as regulatory because they contribute to the precise sequentiality of the developmental cycle (Showe and Kellenberger, 1975). The unprocessed proteins would block the cycle. Some viral proteins, in the course of virion maturation, undergo a transient but indispensible association with other proteins of the viral capsid (Casjens and King, 1975). Their role is not truly regulatory: it rather calls to mind the temporary scaffoldings used by builders in intermediate stages of construction.

Even the precise processes of assembly that produce the admirably regular structures of the virions are subject to what might be called molecular regulation. In the discussion of bacteriophage maturation (Chapters 7 and 8) there are examples of virions that, by maturation accidents or because of mutations, incorporate nucleic acid segments shorter or longer than the modal length into well-formed but dwarf or giant capsids. These phenomena reveal, on the one hand, some role of the nucleic acid in capsid assembly and, on the other hand, a plasticity of protein-protein interaction in assembly. Plasticity reflects a certain range of deformability in the patterns of weak interactions that hold the capsid proteins together.

Some plant pathogens never produce virions but only infectious nucleic acid; these are the so-called *viroids* (Diener, 1972). It is not clear whether these agents may be considered as maturation defective viruses or as a completely different class of pathogens.

Interesting complexities are encountered in the multiparticulate plant viruses, discussed in Chapter 17. A single replicase apparently replicates several RNA fragments, which are packaged separately but are all needed for infection.

Other viruses present a different complexity—a multipartite genome that must be incorporated into a single virion. Altogether, the mechanisms involved in the precise packaging of viral genomes in their capsids and envelopes constitute a fascinating aspect of molecular biology.

Model Systems in the Study of Virus-Cell Interactions

Knowledge about the life cycles of many viruses has grown at such rate in the last 30 years that it would be futile to attempt to review it all in the following chapters. Rather, the developmental processes of the best known viruses or groups of viruses will be used as model systems to illustrate both general and specific situations. Even for the best known viruses knowledge is incomplete, just as comparable knowledge about the most thoroughly studied bacterium, *Escherichia coli*, is still incomplete. The materials presented in the following Chapters aim at exemplifying:

1. sequences of events, called *productive infections*, that lead to production of more virions through viral genome replication and virion synthesis and assembly;

2. sequences that lead to the integration of viral genome into the host-cell genome and the functioning of viral genes in the integrated state;

3. sequences leading to abortive viral development attributable to specific viral or cellular genetic properties;

4. the relation between developmental sequences and structural properties of viral genomes, such as different sequences reflecting the nature (DNA or RNA) of the nucleic acid, the one strandedness versus two strandedness of the viral genome, the presence of one or more nucleic acid elements per virion, and the topology of genetic sequences in these elements (circularity, redundant ends, "sticky" ends, etc.);

5. the extent of dependence of viral development on host cell machinery for unspecific, general metabolic functions or products and for specific steps in the viral cycle;

6. the mechanisms that determine the specific host range of viruses and the suitability of a given cell to serve as a host, including specific virus receptors, rejection mechanisms against foreign nucleic acids, and other classes of restrictions.

Certain well-studied groups of bacteriophages are considered first. Many of the generalizations of virology were founded on the study of these viruses. Similarity of developmental processes among certain groups of phages reflects most closely the criteria of relatedness based on morphology of viruses, type and amount of nucleic acid, and sequence homology as revealed by formation of hybrid nucleic acid duplex. Knowledge of the development of animal viruses is less advanced than for phages, but it includes information of unique nature, particularly on the relation of the developmental sequences of viruses to the topography of the host cells (cytoplasm, nucleus, organelles).

Plant viruses are the least well-known in terms of the sequences of events during development. The reasons for this are mainly the intrinsic difficulties of generating cell systems suitable for the analysis of cell-virus interaction on a one-to-one basis. Plant tissue cultures have not yet proven manageable. More useful are cultures of cells from the tissues of insects that are vectors for the viruses and whose tissues support the multiplication of some plant viruses. Expanding on this possibility, it is conceivable that useful information will be obtained by the study of "cross-kingdom" interactions: the artificial introduction of the genome of viruses into cells cultured from unnatural, utterly unrelated hosts: plant viruses in bacteria or in mammalian cells and the converse.

In learning about the various kinds of interactions, the readers may do well to remember the existence of other kinds of extrachromosomal genetic elements such as plasmids (Lederberg, 1952) and episomes (Jacob and Wollman, 1958) as well as of less well-understood extrachromosomal elements in eucaryotic cells (Sager, 1972). The possible evolutionary relation of viral genomes to these genetic elements, which share with them the property of autonomous replication, will be discussed in Chapter 19. Information about the properties and reproduction of these elements may contribute to an understanding of viral life cycles and vice versa.

Phage-Bacterium Interaction: General Features

6

Bacteriophages have been found to exist as parasites of some members of almost every group of procaryotic organisms, from the miniscule *Bdellovibrios*, themselves parasites of other bacteria, to some of the large blue-green algae. The common characteristics of these phages reflect the common properties of their hosts. The rigid cell wall of most procaryotes requires special mechanisms for penetration or release of virus. Since procaryotic organisms usually do not differentiate into stem cells and specialized cells, but consist of a population of more-or-less equivalent cells, which continue to multiply as long as food is available, phage-host interactions can be repeated cyclically within a bacterial culture under reasonably steady-state conditions, the relevant parameters being the numbers of host cells and viral elements and their rates of reproduction. Singularities appear when bacteria undergo differentiation, as in sporulation (Losick and Sonenshein, 1969) or the alternation of states of *Caulobacter* (Shapiro et al., 1971).

Attachment and Penetration

Attachment of phage virions to bacterial cells is a first-order reaction and occurs usually at the cellular surface, which may have different structures in different bacterial groups (Figure 6-1). Some phages, however, attach specifically to certain appendages, called F or I pili, which are involved also in bacterial mating (Hayes, 1968; Figures 6-2 and 6-3). The virions of one group of phages, χ, first attach themselves reversibly to bacterial flagella and then crawl down the flagella to the bacterial surface, propelled presumably by flagellar motion (since nonmotile mutants do not serve as hosts for these phages).

On the bacterial surface one assumes the presence of specific phage

135

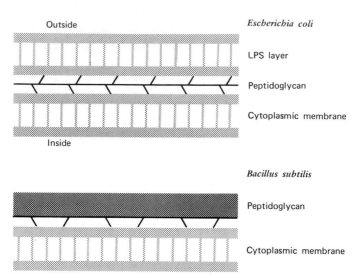

FIGURE 6-1. *The envelopes of* E. coli *(gram-negative) and* B. subtilis *(gram-positive). Both the lipopolysaccharide (LPS) layer of* E. coli *and the thick peptidoglycan of* B. subtilis *contain a variety of proteins. The peptidoglycan layer is shown as connected with the other layers.*

TABLE 6-1   NATURE OF THE PUTATIVE RECEPTORS FOR SOME BACTERIOPHAGES

Phage	Receptor substance
C21	LPS* incomplete at site Ⓐ
P1	LPS core of *E. coli* complete beyond site Ⓐ
T3, T4, T7	LPS of *E. coli* without 0 antigen
ε15[†]	LPS, 0 antigen of *Salmonella* group E with $n > 1$ and Ⓑ linkage in α configuration
ε34	Same as ε34, but Ⓑ linkage in β configuration
P22[†]	LPS, 0 antigen of *Salmonella* groups A, B, D1
T2, T6	Protein of *E. coli* outer membrane
T5	LPS-lipoprotein complex of *E. coli*
SP-50	Glucosylated teichoic acid of *B. subtilis*
Vi φII[‡]	Vi antigen of *Salmonella*
χ	Motile flagella of *Salmonella*
F1, F2 (male-specific)	Male pili of *E. coli*

* Schematic representation of the complete lipopolysaccharide (=LPS) of a *Salmonella* (for further details see Lindberg, *Ann. Rev. Microbiol.* 27: 205 (1973).

$$(Man \rightarrow Rha \overset{Ⓒ}{\rightarrow} Gal)_n \overset{Ⓑ}{\rightarrow} Glc \rightarrow Gal \overset{Ⓐ}{\rightarrow} Glc \rightarrow Hep \rightarrow \left| \begin{array}{c} KDO \\ Etn \\ P \end{array} \right| \rightarrow Lipid\ A$$

$$\begin{array}{ccc} \uparrow & \uparrow & \uparrow \\ GlcNAc & Gal & Hep \end{array}$$

| 0 antigen | outer core | inner core |

† An enzyme in the phage virion splits LPS at linkage Ⓒ
‡ An enzyme in the virion deacetylates the Vi antigen.

receptors. Knowledge about these receptors is scanty. That a bacterial mutant fails to adsorb a given phage need not mean that it has lost the chemical groups that act as receptors for that phage: they may have been masked by other constituents of the cell envelope. Receptors may be nonessential cell components: for example, they may be missing in bacteria grown at certain temperatures (La Montagne and McDonald, 1972). A specific substance extracted from the envelope of a phage-susceptible bacterial strain may inactivate a phage and may be either the receptor itself or only part of the true receptor structure on the surface of the bacteria. Receptors may provide only a first-stage, reversible adsorption. They may also be involved in other interactions such as iron ion transport (Hantke and Brown, 1975).

Table 6-1 lists some of the instances in which the full or partial nature of phage receptors has been fairly well characterized. The reactions between receptor and virion have not yet been characterized chemically in terms of protein-protein or protein-polysaccharide interactions, so they do not provide

FIGURE 6-2. Escherichia coli cells with F pili showing attached particles of MS2 (icosahedral) and M13 (filamentous) phages. Courtesy Dr. C. C. Brinton.

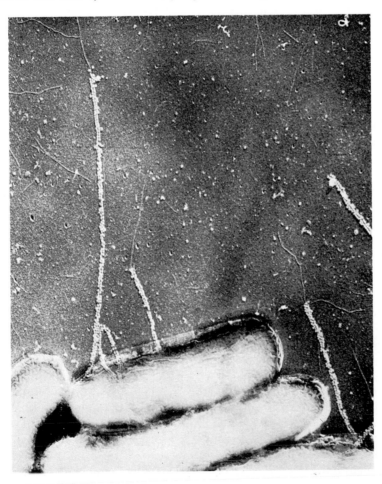

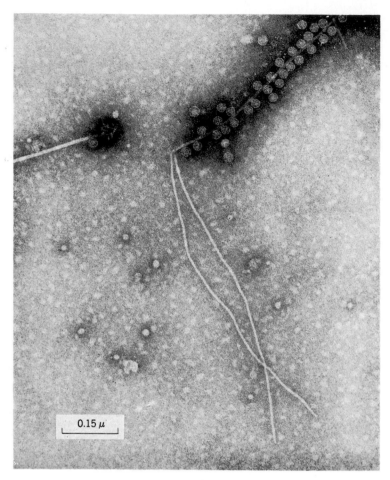

FIGURE 6-3. *Two filamentous particles of phage f1 attached to the tip of an F-pilus of a male* E. coli *cell. Many virions of spherical phage f2 are also attached to the same pilus. Courtesy Dr. L. Caro.*

an insight into relevant mechanisms leading to reorganization or entry of phage material. The closest approach is that of phages whose virions can enzymatically attack specific links in the lipopolysaccharide (LPS) molecules of the outer membrane of gram-negative bacteria (Choy et al., 1975; see Table 6-1).

Following phage attachment susceptible bacteria remain morphologically more or less intact for the duration of the latent period even when the infection ultimately leads to cell lysis, which finally occurs as a sudden event. The entry of the phage genome is a physical separation of the nucleic acid from most of the capsid proteins, which remain on the outside of the cell envelope. This was first revealed by a celebrated experiment performed by Hershey and Chase (1952): most of the protein of infecting T2 phage virions, as assyed by ^{35}S label, could be removed from the cells by a shearing force without removing the nucleic acid and without interfering with the course of

productive infection (Figure 6-4). The phage nucleic acid could be found by its
^{32}P label to have moved into the bacterial cell. In a complementary experiment,
virions of the same phage equilibrated with a 4 N NaCl solution and plunged
into distilled water (*osmotic shock*) could be separated into empty protein
capsids plus free DNA. These experiments, which incidentally illustrate one of
the many uses of radioactive labeling in virus research, were done with
phages with slender, fragile tails, which in fact act as syringes to inject the
nucleic acid. With tailless phages, capsid-genome separation on the cell
surface is more difficult to demonstrate experimentally.

As stated in an earlier chapter, each phage virion contains a nucleic acid
element, DNA or RNA, linear or circular, single-stranded or double-stranded.
Genetic experiments suggest that the injection of linear nucleic acid elements
with recognizable ends is unidirectional. Only one Pseudomonas phage, $\phi 6$,
appears to have a genome composed of several segments of double-stranded
RNA (Semancik et al., 1973).

Small amounts of proteins and other materials, including oligopeptides
and polyamines, may be injected by phage virions into the infected bacterium
(Hershey, 1957). No function has yet been assigned to these injected substances in the subsequent processes of phage development; some of them are
residues of proteolysis of capsid proteins during morphogenesis of the
virions.

If bacterial cells are "competent" to accept free DNA from the medium,
phage genomes may enter as free DNA molecules. This process is called
transfection. Bacterial competence to accept DNA molecules may arise either
from physiological events at some stages of bacterial growth, as in *B. subtilis*
(Spizizen et al., 1971) or from artificial manipulations as in *E. coli* (Mandel and
Higa, 1970; Sabelnikov et al., 1975). The events following transfection are not
essentially different from those in normal phage infection, except that resist-

FIGURE 6-4. *Diagram of the DNA-injection process. 1. Phage and bacterium. 2.
Adsorption. 3. Separation of the DNA core (P, stippled) from the protein shell (S).
4. Removal of shell by stirring in a Waring Blender. 5. Production of new phage
without participation of the shell of the infecting phage.*

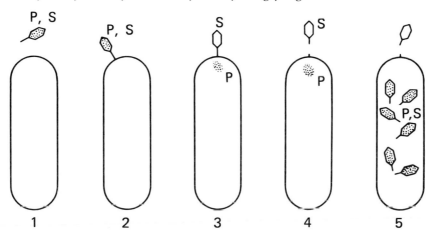

TABLE 6-2 CHARACTERISTICS OF SOME PHAGES OF *ESCHERICHIA COLI*

Phage	Virion morphology		Nucleic acid, type and amount (daltons)	Latent period (minutes)	Average yield per cell	Growth cycle at 37°	
	Head	Tail				Lysogeny	Peculiarities
T1	Icosahedral, 500 Å	100 × 1500 Å	DNA, 2.5 × 10^7	13	150	−	Resistant to drying
T2, T4, T6	Prolated icosahedral, 650 × 950 Å	250 × 1100 Å	DNA, 1.2 × 10^8	21–25	150–400	−	Contain glucosylated HMC
T3, T7	Icosahedral, 470 Å	100 × 150 Å	DNA, 2.4 × 10^7	13	300	−	Give semitemperate mutants
T5	Icosahedral, 650 Å	100 × 1700 Å	DNA, 7.5 × 10^7	40	200	−	DNA injection in two steps
λ, φ80	Icosahedral, 540 Å	100 × 1400 Å	DNA, 3.3 × 10^7	35	100	+	DNA circularizes *in vitro* or *in vivo*
P1	Icosahedral, 650 Å	120 × 1500 Å	DNA, 6 × 10^7	45	80	+	General transduction
P2	Icosahedral, 500 Å	100 × 1500 Å	DNA, 2.2 × 10^7	30	120	+	Multiple chromosomal locations
φX174, S13	Icosahedral, 300 Å	None	DNA, 1 strand, 1.7 × 10^6	13	180	−	Circular DNA
f2, MS2	Icosahedral, 240 Å	None	RNA, 9 × 10^5	22	20,000	−	Male specific, attach to F pili
f1, fd	None	60 × 8000 Å	DNA, 1 strand, 1.3 × 10^6	30	100–200 (continuous release)	−	Male specific, circular DNA
χ*	Icosahedral, 675 Å	125 × 2300 Å	DNA	60	200	−	Attaches to motile flagella

* Not a coliphage; grows on many strains of *Salmonella*.

ance to phage caused by the absence of receptors or other cell envelope properties may be bypassed.

Entry of the phage genome into a susceptible bacterium can give rise either to a *lysogenic* or to a *lytic* (or *productive*) infection, depending on the nature of the phage (and sometimes also of the bacterium) and on environmental conditions such as temperature. In lysogeny the phage genome is carried and replicated in the bacterial cells from generation to generation without lysis or production of virions. Virions are only rarely synthesized from lysogenic bacteria. When this happens, cell lysis follows. The lysogenic bacteria are *immune* to the phage whose genome they contain: if reinfected by the few virions present in their cultures they do not lyse (see Chapter 9) so that lysogenic bacterial cultures grow quite normally. The presence of virions in such a culture can be detected by their action on susceptible *indicator* bacteria. Phages that can establish lysogeny are called *temperate* in contrast to virulent. One must realize, however, that even temperate phages, when they first infect sensitive bacterial cells, give rise to productive infection in many or even most of the cells. Establishment of lysogeny, and the prevention of virion maturation and cellular lysis, require a series of developmental events that is not successfully accomplished in every infected bacterium. The relative probabilities of lysogeny vs. productive infection vary from phage to phage and with culture conditions.

Lysogeny and the pertinent observations are discussed in Chapter 9. The present chapter outlines the main features of productive phage infection. Chapters 7 and 8 illustrate them and the relevant experimental procedures for a few well-studied phage-host systems. Table 6-2 lists the properties of several such systems including a set of phages, T1 to T7, that played an important part in the development of phage research.

Productive Phage-Host Interaction

THE ONE-STEP GROWTH EXPERIMENT

The study of productive phage-bacterium interaction was made possible by the translation of this process into a quantitative procedure called the *one-step growth experiment* (Ellis and Delbrück, 1939). Its essential features are as follows: Concentrated sensitive bacteria under standard cultural conditions are mixed with phage. Phage fixation or *adsorption* is allowed to proceed for a brief period, shorter than the minimum *latent period* between adsorption and lysis. Then the mixture is diluted to an extent that reduces drastically the chances of phage-bacterium collisions, so that little or no further adsorption of phage takes place, and samples are taken at intervals and plated for plaque counts. Typical results are illustrated in Figure 6-5. The low plateau represents the latent period: the plaque count is constant because each infected bacterium, whatever its intracellular phage content was at the moment of plating, produces only one plaque on the solid medium (virions and bacteria do not swim around on solid media). The rise in plaque count corresponds to lysis and the final plateau to the completion of lysis. The newly liberated virions, failing to meet new bacteria because of the high dilution, remain free until

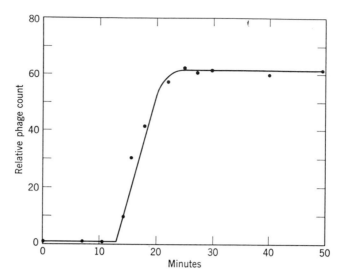

FIGURE 6-5. *A one-step growth experiment with bacteriophage T1 on* E. coli *strain B in nutrient broth at 37°C. Phage and bacteria, at a ratio of 1:10, were mixed at time 0. The mixture was diluted after 4 minutes, when 45% of the phage had been adsorbed. Assays were made at intervals after dilution. Average phage yield per infected cell = 62 × 100/45 = 138.*

plated. This procedure allows observation of a single cycle of bacteriophage growth.

The average yield of phage per bacterium, called *burst size*, is given by the ratio: (total liberated phage)/(initially infected bacteria). If not all the initial phage inoculum has been adsorbed, a correction must be made by using the ratio $(F - U)/(I - U)$ where F = final titer; U = inoculum phage that remained unadsorbed; I = plaque count during the latent period (= infected bacteria + unadsorbed inoculum). Typical yields for a series of coliphages are given in Table 6-2. The yields vary with temperature, bacterial nutrition, and other metabolically relevant parameters.

If no dilution were made from the initial mixture the phage growth curve would exhibit, instead of a single step, a series of growth cycles of phage adsorption and liberation blurring into one another; this is how mass phage lysates are produced. High titer lysates often contain 10^{11} infectious units per milliliter or more, corresponding to about as many virions (Luria et al., 1951).

Innumerable modifications of the one-step growth experiment have been devised to study specific variables such as single, multiple, mixed, and synchronized infection. An important modification is the *single-burst experiment* (Burnet, 1929; Figure 6-6). After phage is mixed with bacteria, the mixture is diluted until it contains a very small number n of infected bacteria (say, two per milliliter). Before lysis begins (that is, within the time span of the latent period) many small aliquots, for example, 0.2 ml each, are distributed into a series of containers. Most of these aliquots will contain no infected bacterium (in the foregoing example, a fraction $e^{-n} = e^{-0.4} = 0.67$), some will contain one infected bacterium (a fraction $ne^{-n} = 0.4e^{-0.4} = 0.268$), and very

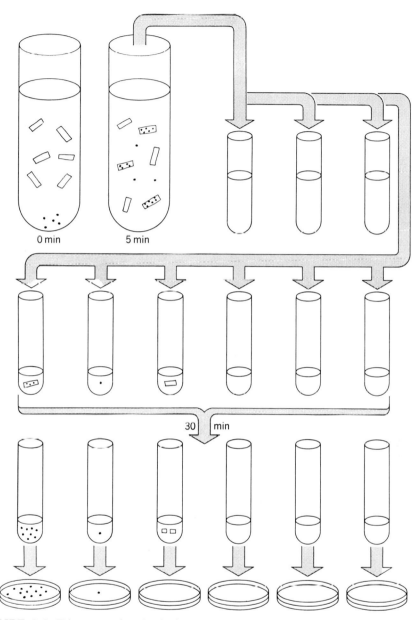

FIGURE 6-6. *Diagram of a single-burst experiment with phage. In the diagram, one out of six tubes (second row) contains an infected bacterium, one tube contains an unadsorbed phage, and one tube contains an uninfected bacterium. After 30 minutes the infected bacteria have lysed. The contents of each tube are plated on individual plates for plaque count. The first plate (left, bottom row) shows the yield of phage from the infected bacterium. Modified from Delbrück,* Harvey Lecture Series **41,** *161 (1946).*

few contain more than one (a fraction $1 - (1 + 0.4)e^{-0.4} = 0.062$). These frequencies are calculated from Equation 2-1 (page 23), under the reasonable assumption of a random distribution of infected bacteria. After incubation and until lysis is completed the entire contents of each container are plated for phage assay using a separate plate for each tube. Most plates show either no plaque or possibly one or a few plaques, representing residual unadsorbed phage. The other plates will have numbers of phage plaques representing the yields from single cells (rarely from two or more cells). The burst size distribution is broad (Table 6-3), broader than expected, for example, from the size distribution of cells in a culture. The fluctuations must reflect wide variations in the initiation and rate of virion production within individual bacteria. Clearly there is no specified tie between amount of intracellular virions and lysis.

In one-step or single-burst growth experiments the unadsorbed phage can be eliminated before lysis begins, either by centrifugation or more conveniently by adding some antiphage serum, which inactivates the free phage without affecting the adsorbed phage (Delbrück, 1945). The action of serum is then stopped by further dilution to avoid inactivating the newly liberated phage.

In the same experimental conditions as used for a one-step experiment one can measure the kinetics of phage attachment, which generally turns out to be first order with respect to the concentrations of both phage and bacteria:

$$\frac{dP_{ft}}{dt} = kP_{ft}B; \qquad P_{ft} = P_0 e^{-kBt} \qquad [6\text{-}1]$$

where P_{ft} is the concentration of free phage infective units per cm^3 at time t, P_0 the initial phage concentration, B is the bacterial concentration in cells per cm^3, t is the time in minutes, and k is the adsorption rate constant (cm^3 $min^{-1}cell^{-1}$). Since Equation 6-1 is often valid for mixtures with phage in excess up to ratios P_0/B over 100, many virions can evidently be adsorbed by one cell before the receptor sites are saturated. If the adsorbed phage at time t is P_{at}, then $P_{at} = P_0 - P_{ft}$. The ratio P_{at}/P_0 is the average number of infective units adsorbed per bacterium or *multiplicity of infection* (m.o.i.). Infection of bacteria with very large numbers of virions may cause *lysis from without*; that is, lysis by massive damage to the cytoplasmic membrane before phage reproduction has even started.

The fraction B_n/B of bacteria that adsorbs n particles follows a Poisson distribution

$$\frac{B_n}{B} = \frac{(P_{at}/B)^n}{n!} e^{-(P_{at}/B)} \qquad [6\text{-}2]$$

The number B_u of uninfected bacteria will then be

$$B_u = B \, e^{-(P_{at}/B)} \qquad [6\text{-}3]$$

Colony counts of surviving bacteria in mixtures with varying amounts of phage follow rather closely Equation 6-3. This indicates that a single virion of a virulent phage can kill a bacterium by initiating a lytic sequence of phage development.

TABLE 6-3 DISTRIBUTION OF YIELDS FROM INDIVIDUAL *ESCHERICHIA COLI* BACTERIA INFECTED WITH PHAGE T2 (SINGLE-BURST DISTRIBUTION)

(a)	Average number of infected bacteria per plate, calculated from phage input	0.73
(b)	Total number of plates	96
(c)	Expected number of plates without bursts	$96 \times e^{-0.73} = 46$
(d)	Number of plates found without bursts	39
(e)	Average number of infected bacteria per plate, calculated from (d)	$0.9 = -\ln(39/96)$
(f)	Calculated number of plates with 1 burst	$96 \times 0.9 \times e^{-0.9} = 35$
(g)	Calculated number of plates with 2 bursts	$96(0.9^2/2)e^{-0.9} = 16$
(h)	Calculated number of plates with 3 or more bursts	$96[1 - (1 + 0.9 + 0.9^2/2)]e^{-0.9} = 6$

Plaque count distribution

9	67	92	114	161	260
31	67	92	117	163	278
41	70	99	118	165	291
45	75	101	124	192	298
48	76	102	130	201	407
49	81	103	130	206	413
55	86	105	132	209	477
56	89	106	135	215	
58	90	110	136	216	
65	91	110	151	230	

Total count	7938
Average yield, calculated from (e)	$7938/(96 \times 0.9) = 92$

Note that many virions can attack and infect the same cell. Neither the duration of the latent period nor the burst size are significantly changed by such multiple infection. Yet it is easy to show that several infecting phage genomes can participate in productive infection in a given cell provided they belong to the same phage type (mutants or related phages). Detection and counting of different phages in mixed infection is done by plating on *mixed indicators* (Figure 6-7); that is, a suitable mixture of two compatible bacteria, each of which is sensitive to one of the two phages and not to the other. Most useful indicators are bacterial mutants isolated as resistant to specific phages. Such mutants are often designated by the symbol X/Y, where X relates to the bacterial symbol and Y to the phage symbol: for example, B/4 signifies *E. coli* B resistant to phage T4. In bacterial genetics symbols like *ton* for a gene controlling resistance to phage T1 are usually preferred (Hayes, 1968).

A bacterium infected with two genetically related phages generally gives rise to a mixed burst of virions of the two types and often also to new types that arise by genetic recombination. As many as 20 or more infecting phages can contribute to the phage yield from one bacterium. Phages unrelated by genetic, biochemical, and serological criteria usually give *mutual exclusion*; that is, they block each other's productive infection in a given cell, probably because of incompatible biochemical steps in their developmental sequences.

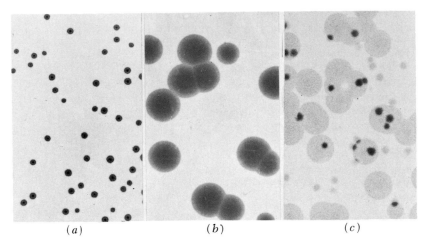

(a)   (b)   (c)

FIGURE 6-7. Platings of a mixture of bacteriophages T1 and T2 on various indicator bacteria. (a) Indicator B/1, sensitive to phage T2 only. (b) Indicator B/2, sensitive to phage T1 only. (c) Mixed indicators B/1 and B/2; both phages form turbid plaques, with clear areas of lysis where the plaques overlap. Enlarged about $2\frac{1}{2}\times$.

An interesting situation is that of mixed infection of a bacterium with two mutants *a* and *b* of a given phage, each at a low m.o.i., for example, $P/B = 0.2$ for each. The fraction of bacteria that receive at least one phage of each kind is about $0.2 \times 0.2 = 0.04$, which represents about 10% of all infected bacteria. Out of 100 single bursts from such bacteria about 10 should be mixed bursts and almost all these will contain the progeny from *one* single *a* and *one* single *b* parent. This kind of experiment is useful in the quantitative analysis of mechanisms of genetic recombination in phage.

Adsorption of phage virions as defined by Equation 6-1 is irreversible adsorption, leading to injection of nucleic acid. In pure distilled water or at low concentrations of monovalent ions (10^{-4} M or less) most phages do not adsorb to bacteria. The optimal concentrations of cations permitting maximal adsorption rates are characteristic for each phage (Tolmach, 1957). Adsorption at low cation concentrations is generally reversible by dilution (Puck et al., 1951). At higher ionic concentrations (10^{-2} to 10^{-1} Na$^+$ or K$^+$, for example) the reversible adsorption is overshadowed by fast irreversible fixation. Even under optimal conditions the irreversible adsorption rate constant k (3×10^{-9} cm^3 min^{-1} per bacterium for phage T4) is still lower than the value calculated on the assumption that every collision between phage and bacterium results in fixation (Delbrück, 1940a). Thus the rate-limiting step in phage adsorption is not the frequency of collision: mutual orientation of the colliding elements may be critical, or the phage may have to reach specific sites by "browsing" on the bacterial surface (Benz and Goldberg, 1973).

The Latent Period of Phage Development

A phage virion attacks a bacterial cell; 19 or 20 or 60 minutes later the cell lyses and hundreds or thousands of phage virions emerge. The study of phage development is the analysis of the processes that relate the initial infecting event, and more specifically the various parts of the infecting phage virion, to the new crop of phage.

If infected bacteria are disrupted at intervals (by means of treatments that do not damage mature virions, such as decompression, lytic enzymes, or detergents) one finds at first a period, the eclipse period, in which no intact, infectious virions are present. Mature virions appear later. These are not, of course, the infecting virions, but newly synthesized ones, which accumulate in increasing numbers until sudden lysis occurs. Under the light microscope one can watch infected bacteria lyse and disappear suddenly one by one. Such bursting cells may also be caught in electron micrographs (Figure 6-8; Luria et al., 1943).

To unravel the finer events of the latent period one must bring to bear a whole rage of technologies, a good part of which was developed precisely to study phage development. Electron microscopy provides information about the presence of virions and related structures in lysates or in sections of bacteria and reveals the effects of infection on bacterial organization. Serological study with antisera against entire phage virions or their specific fractions or subunits serves to trace and quantitate viral materials in extracts from infected cells.

Electrophoresis, especially gel electrophoresis, detects specific proteins and evaluates their sizes and times of appearance. Electrophoresis coupled with labeling with radioactive amino acids in a variety of pulse-chase protocols provides information not only about the time course of synthesis of various phage proteins but also about changes in the patterns of synthesis and utilization of these proteins. An early protein, for example, labeled between 3 and 5 minutes after infection, may be found as such in gels of extracts up to 15 minutes and then disappear from the gel pattern because it becomes incorporated into virions (which are too large to migrate into the gel). A protein may change size and electrophoretic mobility as a result of processing, that is, by partial proteolysis. In early extracts from infected cells such a protein will behave differently from when it is incorporated in capsid precursors or in mature virions. Electrophoresis also detects phage-specific proteins that never enter the virions—enzymes and structural intermediates. This approach is especially fruitful in the study of those phages whose productive development is accompanied by the cessation of synthesis of host-specific proteins.

Messenger RNA of viral origin can be identified by appropriate labeling followed by gel electrophoresis or sucrose gradient fractionation. The *in vivo* instability of mRNA in procaryotes, with half-lives of only a few minutes (see Chapter 4), must be reckoned with in interpreting the appearance and disappearance of individual mRNA bands. Hybridization with viral and bacterial DNA identifies mRNA fractions according to their templates. In hybridization competition tests (Figure 4-15) mRNA collected, for example, 5 minutes after infection is competing for DNA sequences with mRNA labeled between 10 and 15 minutes. Such an experiment tests what proportion of mRNA molecules two samples have in common. As a further identification of mRNA classes it is possible to test mRNA fractions for their ability to direct the synthesis of specific phage (or bacterial) proteins in cell-free systems that contain all components of the protein synthetic machinery. Novel, phage-

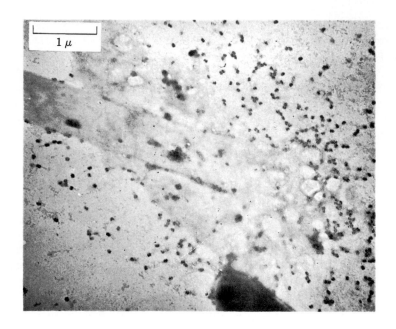

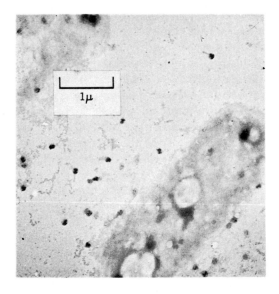

FIGURE 6-8. (a) *Lysis of a cell of* E. coli *infected with bacteriophage T2.* (b) *Cell wall of* E. coli *after lysis by bacteriophage T2. From Luria et al.,* J. Bacteriol. **46,** 57 *(1943).*

determined tRNA molecules differing from the bacterial are sometimes also found in phage-infected bacteria.

Various classes of novel enzyme activities may appear after phage infections. These enzymes are probably coded for by phage genes, although the evidence is solid in a few instances only. Only the largest and most complex phages produce enzymes of all three classes. One class consists of nucleases, whose function is either to attack the host cell DNA or to catalyze certain steps in the replication and maturation of the phage genome. A second class consists of enzymes that catalyze the biosynthesis of acid soluble, low molecular weight substances needed as precursors for the synthesis of phage nucleic acids. These are enzymes that take part ·in the interconversion of nucleotides and nucleoside di- and triphosphates. Such an enzymatic equipment is not limited to phages whose DNA contains unusual constituents such as hydroxymethylcytosine or uracil. The occurrence of such metabolic enzymes poses interesting problems concerning the origin and evolution of the phage genome. Instances of such situations are presented in the following chapters.

A third class of phage-initiated enzymes is specifically related to the replication and expression of the phage genome: DNA polymerases, RNA replicases, and RNA polymerases (transcriptases). Some enzymes of this class appear to represent the whole replicating machinery, capable at least *in vitro* of carrying out viral nucleic acid replication without cooperation by bacterial enzymes (for example, the replicases made by RNA phages). Often, however, the phage produces protein subunits that confer new specificities to a host enzyme (for example, T-even phages and several *B. subtilis* phages alter the host's RNA polymerase). The complex DNA polymerizing systems used by phage are often mixtures of viral and cellular enzymes and cofactors. Many of these are identified only by defects mapped in specific genes.

Interesting situations have come to light in which specific bacterial mutations make the mutant cells unable to support the growth of a given phage while still capable of adsorbing its virions. Rarely (Arber, 1974) the defect is in penetration. Most commonly the mutant bacteria are altered either in their RNA polymerase, which therefore fails to transcribe correctly some essential phage genes, or in bacterial proteases that participate in the processing of phage proteins in the course of virion assembly. The degree of specificity of these cellular proteases is unknown.

Whereas the assignment of nucleic acids from infected bacteria as either viral or host-coded can be done by molecular hybridization tests, as discussed in Chapter 4, for proteins this must be done by genetic tests. Loss of a function as a result of a phage mutation is *prima facie* evidence for assignment of that function to a phage gene. A function of a protein or polypeptide may be absent either because of a mutation in the corresponding gene or because of alterations in genes with regulatory activity. Definite assignments of a phage protein to a given phage requires, from the point of view of a geneticist, the identification in the genetic map of the phage of a coherent linear set of mutable sites colinear with a set of amino acid changes in a specific polypeptide. Such precise matching is theoretically feasible for phage genes: genetic recombination occurs in all phages with DNA genomes in the course of DNA replication and can be studied by mixed infection with appropriately "marked" phages, that is, mutants differing by two or more mutational changes (Hershey, 1946). Recombination analysis in phage can be pushed to extreme refinement thanks to the relative ease of isolation of mutants and to strong selective methods for isolation of recombinants (see Chapter 7 for T-even phages). In this way one can construct "fine-structure" maps of individual genes (Benzer, 1961) and compare them with amino acid substitution maps. Even without fine-structure mapping it is possible to attribute to a given gene the structural information for a given polypeptide when mutations in that gene cause an alteration in the properties of the polypeptide. It is customary to refer to the protein made by gene X as gp X (= gene protein X).

Mapping phage genes is not in itself essential to the study of the programmed function of these genes in phage reproduction. Genetic mapping plays many significant roles, however. It makes it possible to assign mutations to specific positions in the phage map. Tests of genetic complementation in mixed infected bacteria can then show whether mutations in a given genetic region are actually within the same gene or in neighboring genes controlling the same function—for example, genes coding different polypeptides of the same enzyme. Mapping also reveals the presence or absence of clustering of genes whose products are involved in a complex sequence of reaction. Clustering, when present, may indicate regulatory mechanisms such as the occurrence of jointly regulated operons. In temperate phages, for example, the genes that control the establishment of lysogeny are arranged into two or more mutually interacting operons (see Chapter 9). In other phages such as the T-evens the genes that determine the many proteins for phage enzymes, or phage head, or phage tail, or attachment fibers are more or less clustered. There is evidence of coordinate expression, that is, multigenic messengers for groups of adjacent genes (see Figure 7-1). Gene clustering might also reflect some unrecognized novel regulatory mechanisms, for example, an orderly exposure of certain parts of the phage genome to transcription enzymes at various times or in various sites of the host cell.

More directly relevant to the events in phage replication is the information that the genetic map of a phage provides about the topological relation between the phage genome and its nucleic acid molecules. As already mentioned and as discussed in detail in the following chapters, the infection, replication, and maturation of a phage involve changes in the physical structure of the nucleic acid molecules of viruses. These changes do not involve dispersion and reassembly: the sequence of phage genes is always maintained, but the topological relation may change. The genes of a phage can often be represented on a circular genetic map even if the nucleic acid molecules are linear. This is accounted for by the finding that circular structures or their topological equivalents are produced at some stages of development. Some examples are illustrated in Figure 6-9.

The genome of phage λ is present in the virions in the form of linear double-stranded DNA molecules, 50,000 nucleotide pairs in length, with single-stranded, mutually complementary "tails" of 12 nucleotides (Figure 6-9a). In the host cells these tails anneal and convert the molecule to a circle, which is then covalently sealed by polynucleotide ligase (see Figure 9-3). The nucleotide sequence is the same in all molecules. In the linear DNA from the virions of phage λ one can actually identify by electron microscopy a series of well-defined regions, recognizable in every molecule, by exposing the DNA to controlled partial heat denaturation: loops of single-stranded DNA appear at characteristic places in the molecules because of preferential melting of

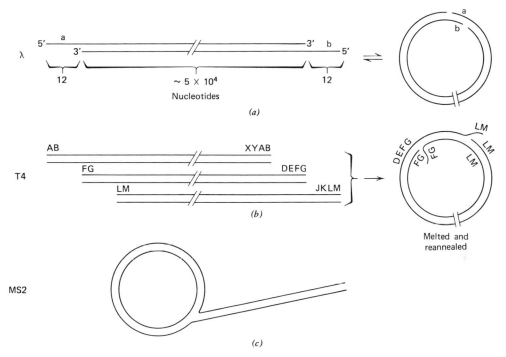

(a)

(b)

(c)

FIGURE 6-9. *Structures generated by DNA molecules of various phages. (a) A molecule of phage λ with complementary single-stranded ends can generate a noncovalently closed circle; the ends can also generate concatemers of two or more molecules. (b) Terminally redundant, circularly permuted molecules of phage T4. Such molecules, after heat-melting and reannealing, generate circles with single-stranded tails. (c) Structure generated in the course of replication of phage MS2 by the rolling circle mechanism.*

sequences rich in A-T pairs (Inman, 1966). Many complex problems of genetic rearrangements can be clarified by means of this technique.

The T-even phages on the other hand consist of DNA molecules that are circularly permuted and terminally redundant in their genetic map (Figure 6-9b). In the host cell pairing and recombination between the redundant ends of a phage DNA molecule can generate circles. Multiple infection as well as replication provide the opportunity for recombination between molecules, which can rise to multiple length circles and/or to multiple-length linear molecules. Circular permutation of the genetic map in the DNA from the virions implies a mechanism that "measures" and cuts a definite length of DNA for packaging into the virions (see Chapter 7). Genetic deletions can increase the length of terminal redundancy by an amount corresponding to the deleted length provided the measuring device remains the same. A direct way to visualize and determine the size of deletions is to prepare artificial hybrid DNA molecules by melting, reannealing, and examination by electron microscopy: in the hybrid molecules the nondeleted strand shows a single-strand loop at the site missing from the other strand (Figure 6-9b).

Other complexities arise from the mechanisms of replication. At least some phages (and other viruses and plasmids) replicate by the "rolling circle" mechanism (Gilbert and Dressler, 1968; see Chapter 4) which can generate long DNA filaments wherein the gene sequence is repeated many times (Figure 6-9c).

Repetitive structures, whether they arise by recombination, replication, or a combination of both, must in the course of maturation be cut into appropriate linear segments to be packaged into virions. Each group of phages has its own specific cutting device. Some phages package single-stranded circular DNA molecules in their virions (see Chapter 8).

The DNA of many phages, like that of bacteria, replicates by mechanisms that involve the participation of circular molecules. This specialization obviates certain problems arising from the intrinsic properties of the polymerizing enzymes. Replication of a $5' \rightarrow 3'$ DNA strand (by the laying down of short $5' \rightarrow 3'$ fragments in a backward direction and then joining them) would meet a snag at the end of the molecule. This is because the $3'$ end of the template strand, freed from its complement, would fail to provide an initiation site (see Chapter 4). One way out of this difficulty without recourse to circularization has been suggested (Watson, 1972; see page 200).

The details of DNA replication mechanisms are still uncertain, but their complexity and the fact that they involve not only replication per se, but also the removal of extra lengths of polynucleotides, the sealing of breaks in the polynucleotide chains, and probably also the cutting of chains at appropriate points, explain why many phages (and probably also other DNA viruses) code for enzymes such as DNA polymerases, ligases, endo- and exonucleases. One of the most remarkable features of phages is the variety of DNA-replicating mechanisms they display. Likewise, the molecular mechanisms of phage genetic recombination, by which strands of nucleic acid interact by exchanging parts, breaking, rejoining, and separating again, are catalyzed by enzyme mechanisms that are involved in DNA replication as well as in the repair of damaged DNA. Even though bacteriophages have played an important role in these studies, the molecular biology of recombination will not be dealt with systematically in this book (for an up-to-date discussion see Grell, 1975). Specific findings relevant both to recombination mechanisms and to phage replication are discussed in the following chapters.

MUTATION FREQUENCIES AND THE PROCESS OF MUTATION

The mutation process of bacteriophages has been studied extensively, not only for purposes of genetic analysis but also for the information it provides on the properties of phage itself. Frequencies of mutants in phage stocks vary from below 10^{-9} for some host range mutants to 10^{-3} or higher. The effects of high mutation rates are usually checked by some selective disadvantages. For example, a phage mutant may be displaced by revertants to wild type because of a higher phage yield by the wild type phage.

High spontaneous mutation frequencies are typical of those mutations that can occur at many different sites within a gene locus. If the normal character represents the functional form of the gene and the mutant character is due to a change anywhere in the gene locus, the forward mutations will be much more frequent than the reversions, since the latter must restore the normal condition. The revertants are sometimes pseudorevertants, either because of suppressor mutations in other genes or because of changes within the gene itself that generate a different but functional form of the gene product.

Spontaneous mutations do not occur with significant frequency in mature virions, but can be induced by mutagenic agents such as X-rays, ultraviolet light, nitrous acid, hydroxylamine, and alkylating agents (Freese, 1961). Nitrous acid acts by deaminating the bases in nucleotides; ethylmethanesulfonate, by ethylation of the bases. Hydroxylamine converts cytosine to uracil. When such altered phages infect host cells, mutations occur because of mistakes in the replication of chemically changed nucleic acid strands, so that a mixture of normal and mutant phage emerges from a single bacterium. Mutagenized phages with single-stranded DNA produce, of course, mutants in pure clones.

The kinetics of the mutation process during phage replication has interesting implications for phage development. Consider a bacterium in which a spontaneous phage mutation occurs. Such a bacterium will yield a mixture of normal and mutant phage. The proportion of mutants in individual phage bursts will depend on when during reproduction the mutation takes place and also on the mode of phage reproduction, because new phage genes must arise by replication of preexisting genes. If a mutation has a constant probability of occurring at each replication, the number of mutants to which a mutation gives rise depends on the sequence of replication. For example, if each gene replica were produced independently of the others, mutated replicas would be distributed at random among the bursts. If, instead, a gene replica can in turn reproduce and give rise to further replicas the mutated replicas will be found in groups or clones of mutated sibs. These and other possibilities are illustrated in Figure 6-10.

The experimental test was done by counting all mutant plaques (r or rapid lysis mutants of phage T2) in individual bursts from thousands of T2 infected bacteria T2 (Luria, 1951). The frequency distribution of clones containing 1, 2, 3, . . . mutants fit quite well the distribution calculated by assuming that

FIGURE 6-10. *Diagrammatic representation of possible sequences of production of viral genomes. (a) Increase by repeated reduplications. (b) Increase by successive replications of the initial element. (c) Increase by replication of the last element produced. Solid dots indicate mutants. Mutants are produced in clones of identical sibs in case (a), singly in case (b), and in series in case (c). Case (a) agrees with the experimental findings (see Figure 6-11).*

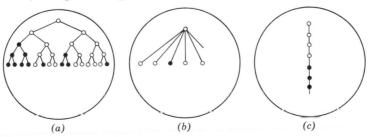

(a) (b) (c)

phage genes were replicated by a process of successive duplications, with a constant probability of mutation at each gene replication. The calculated frequency y_x of clones with $x = 2^n$ mutants is

$$y_x = \frac{mN}{x} = \frac{mN}{2^n}$$ [6-4]

where m is the probability of mutation, N is the final number of gene copies, and n corresponds to the number of generations of phage genes that have taken place. Since the replication of phage genomes is not synchronized, a convenient form of the distribution is in terms of the frequency Y_x of clones with x or more mutants:

$$Y_x = \frac{2mN}{x}$$ [6-5]

The experimental distribution (Figure 6-11) and the theoretical one fit adequately well, especially if one takes into account the effects of limited burst size and random maturation of phage from a pool of replicating elements (Steinberg and Stahl, 1961). This result is interpreted to indicate that the

FIGURE 6-11. *The experimental distribution of spontaneous r mutants of phage T2 in single bursts. Y_x, the proportion of clones with x or more mutants, is plotted on a log-log scale as a function of x. The solid line corresponds to the expectations from Equation 6-5. The broken lines are calculated for two different values of T, the ratio of total burst size to the vegetative pool size (Steinberg and Stahl, J. Theoret. Biol. **1**, 488 (1961).*

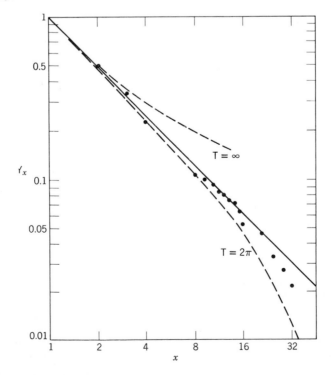

replication of phage genes not only fits the semiconservative replication of DNA, in which each double-stranded molecule gives rise to two molecules, but also that new replicas participate again in the process of replication, which entails a constant probability of mutational error.

A peculiar distribution of mutants was observed with phage T4 treated with the mutagen ethyl-methane sulfonate before infection (Green and Krieg, 1961). The frequencies of clones with 1, 2, 4, 8 mutants were about equal. This mutagen acts by ethylating guanine in DNA. Each ethyl-guanine group does not prevent phage DNA replication but has a constant probability of being removed from the DNA and replaced by some wrong base. The mutation, therefore, has equal chances of occurring at each phage generation and all the copies made after the event will be mutants. The experiment provides evidence that DNA strands from the initial phage particles are copied repeatedly within a single infected bacterium.

HOST-INDUCED MODIFICATION OF BACTERIOPHAGES

In addition to mutations, bacteriophages are subject to a class of nongenetic changes in which the host cell exerts the directing role. This phenomenon is called *host-induced modification* (Luria and Human, 1952; Arber, 1965 and 1974) and is important in molecular biology because it reveals that the cellular environment can impart to DNA certain chemical changes that serve as recognition mechanisms, identifying the specific cell lines where the DNA has been synthesized. These phenomena were discovered on phage but affect any DNA made in bacterial cells. There is suggestive evidence that similar phenomena occur in eucaryotic organisms (Sager, 1975a).

A typical instance is illustrated in Figure 6-12. Phage λ has two hosts, *E. coli* strains K and C. Phage λ grown on K (symbol λ·K) can infect and multiply in K cells but is effectively rejected by C cells (for example, the efficiency of plating or e.o.p. may be 10^{-4}). The few phages that grow on C cells (called λ·C)

FIGURE 6-12. *Host-controlled modification of phage* λ *in* E. coli *strains K12 and C. Single arrows represent infection; the numbers near the arrows give the growth probability of* λ·K *or* λ·C *on the two hosts. Double arrows indicate phage production.*

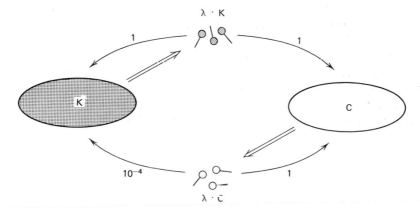

are now derestricted (e.o.p.$_K$ = e.o.p.$_C$ = 1). One cycle of growth on K cells restores the original restriction, that is, $\lambda \cdot K \cdot C \cdot K = \lambda K$. This is described by saying that growth on strain K *modifies* λ to a form that is *restricted* against strain C, and growth on C modifies λ so that it is no longer restricted in C. No mutation is involved; the modification usually affects practically all the phage produced in the modifying host. Specific situations are variously complicated, as shown in Table 6-4. There may or may not be a reciprocal restriction of a phage by two host strains.

A restricted phage is adsorbed normally by restricting cells and injects its DNA, but some of this DNA is rapidly degraded and no replication occurs. Degradation is initiated by specific endonucleases (*restriction* or R nucleases; Eskin and Linn, 1972) which recognize specific sequences of DNA and attack them unless the sequences have been "modified" by specific modification or M enzymes. Once the DNA is cut, degradation to single or oligonucleotides is carried out by exonucleases. A bacterial strain may have one or more R nucleases and corresponding M enzyme activities that protect its own DNA. A convenient nomenclature has been proposed (Smith and Nathans, 1973). There is evidence that for some R nucleases the recognition sequence is not identical to the site of the endonucleolytic attack (Horiuchi and Zinder, 1972) suggesting that the enzyme can migrate along the DNA before reaching the site of attack. The modifications consist of methylation of adenine or cytosine groups within specific DNA sequences, the methyl donor being S-adenosyl methionine. The recognized sequences are often palindromes recognized by the M and R enzymes, for example:

$$5' \ldots \text{pGpTpPypPupApCp} \ldots 3'$$
$$3' \ldots \text{pCpApPupPypTpGp} \ldots 5'$$

where Pu and Py stand for purine and pyrimidine, respectively. Methylation of even one strand suffices to protect the sequence against restriction. The M and R activities are generally exerted by a multiunit enzyme complex coded for by a three-gene system (Arber and Linn, 1968). One unit functions in sequence recognition.

The significance of host-induced modification is not yet understood. It can protect a bacterial strain from massive destruction by phages grown on different bacteria. More generally, it provides a device to prevent the entry and establishment of unacceptable foreign DNA into bacterial cells. A bacterium A that rejects a phage grown on strain B also rejects DNA of bacterium B if this is introduced by conjugation or transduction.

The restriction enzymes, because of their specific nuclease action, have become a valuable tool in the manipulation of DNA for purposes of so-called genetic engineering (Cohen et al., 1973). They make it possible to produce double-strand breaks at specific sites of a viral genome or plasmid, providing ends that can then be joined in precise manner with the ends of other genetic elements to produce new combinations of genes (See Chapter 4).

Productive Phage Cycle: The T-Even Coliphages

7

The T-even phages—T2, T4, T6—are prototypes of a vast group of virulent bacteriophages found in natural environments such as human stools or in sewage. They have played an important historical role in the development of phage science. T2 was used in the first studies of genetic resistance in bacteria (Luria and Delbrück, 1943), of mutations in phage itself (Luria, 1945), and of genetic recombination (Hershey, 1946). T4 has the most thoroughly mapped genome (Figure 7-1) and was used in key studies of gene structure and function (Epstein et al., 1963), of genetic recombination mechanisms (Benzer, 1957), and of the genetic code (Crick et al., 1961). A recent review (Wood and Revel, 1976) presents a comprehensive description of the T4 phage genome.

The injection of DNA from virion to host cell was discovered on phage T2 (see Figure 6-4). The complex morphology of the T-even virions prompted the first and most detailed analysis of capsid assembly including *in vitro* assembly (Edgar and Wood, 1966). Members of the T-even group undergo recombination in mixed infection, making possible to study the homology and evolution of genes and gene groups and to generate hybrid phages. The T-even phages have even evolved in the brief period of their service as experimental objects. T2 phage has given rise to at least two well-established lines, T2H and T2L, comparable to varieties of plant or animal species. Likewise, T4 has branched into T4B and T4D in two different laboratories as a result of unrecognized, unselected mutations. The T-even phages will serve here as paradigms for the description of virulent phage development.

Relation of Virion Structure to Initiation of Infection

Figure 7-2 illustrates schematically the attachment of a T4 virion to a sensitive bacterium and the penetration of the tail tube through the envelope as a result

157

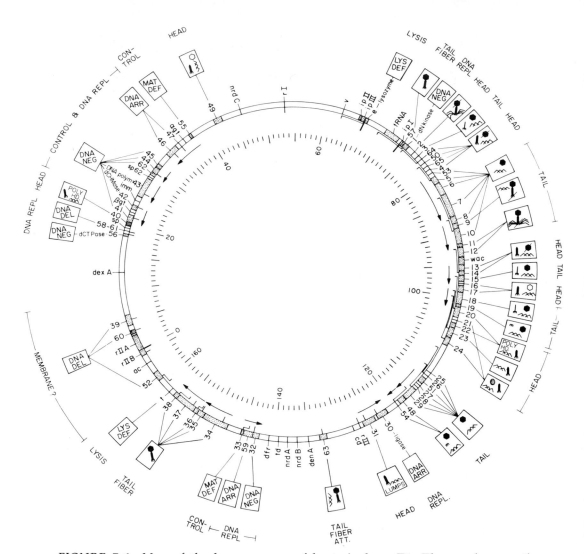

FIGURE 7-1. *Map of the known genes of bacteriophage T4   The numbers on the inner dial are map distances in recombination units (see page 179). The symbols in rectangles indicate the functions altered by mutations in the corresponding genes (NEG = negative; DEL = delayed; ARR = arrested; HD = head; LYS = lysis; MAT = maturation; DEF = defective). The arrows within the gene map indicate the direction of transcription of the known operons (groups of cotranscribed genes). Map after Wood,* Handbook of Genetics *(R. C. King, ed.),* **1;** *327 (1974).*

of a contraction of the tail sheath. This interpretation is based primarily on high-resolution electron micrographs (Figures 7-3 and 7-4) by Simon and Anderson (1967) and others. The long fibers are the sites of specific recognition for attachment to host cells. Mutations in the genes for the fiber proteins alter or abolish attachment. In fact, only antibodies specific to the distal part of the protein on the tip of the fibers interfere with phage attachment to cells (Beckendorf and Lielausis, 1973).

The fibers are kept folded around the tails with their midpoints held by "whiskers" that are attached to the head-tail joint and are probably the product of gene *wac* (Conley and Wood, 1975). Contact of a fiber tip with a receptor may trigger the unfolding. Phage T4 has the remarkable property, readily lost by mutation and selection, that release of tail fibers from the whiskers is dependent on L-tryptophan as a cofactor (Stent and Wollman, 1950). This is a rare instance of a reversible protein-protein interaction mediated by a low molecular weight specific cofactor. The interesting dependence of fiber release and phage attachment on tryptophan concentration suggests that attachment of some of the six tail fibers to a bacterium facilitates

FIGURE 7-2. Schematic representation of phage T4 virion and its component parts and of the mechanism of penetration of the phage core through the bacterial envelope. (a) Virion with extended tail fibers. (b) Virion with the tail sheath contracted and the spikes of the tail plate pointing against the bacterial cell wall. Modified from Simon and Anderson, Virology **32,** *279 (1967).*

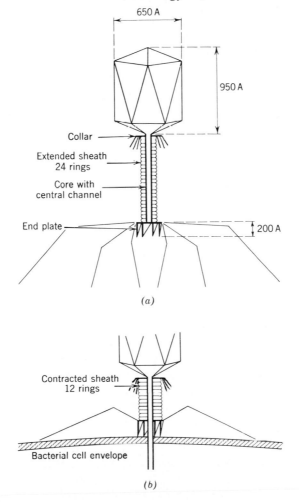

650 A

950 A

Collar

Extended sheath
24 rings

Core with
central channel

End plate

200 A

(a)

Contracted sheath
12 rings

Bacterial cell envelope

(b)

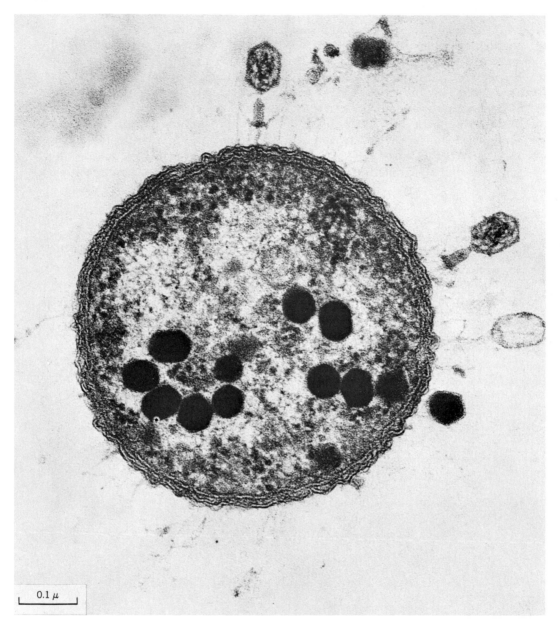

0.1 μ

FIGURE 7-3. A section of a cell of E. coli strain B from a culture infected with bacteriophage T2. Clearly visible are: the cell envelopes (wall and cytoplasmic membrane); the clear area of the phage DNA pool containing many condensed phage DNA cores; and, attached to the cell surface, three phages, one empty and two still partially filled. The phage at the top shows the long tail fibers and the spikes of the tail plate in contact with the cell wall. The tail sheath is contracted and the tail core has apparently reached the cell surface but has not penetrated it. Courtesy Dr. Lee D. Simon.

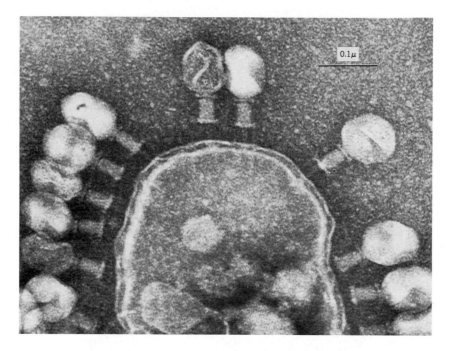

FIGURE 7-4. A partially lysed cell of E. coli B *with attached virions of phage T4. Note an empty capsid, the contracted tail sheaths, and the interval between tail plate and cell wall. The "collar" is also well evident. Courtesy Dr. L. D. Simon.*

the release of the others. All six fibers must probably be attached to cell receptors in order to position the tail plate for the next step of interaction.

The exact roles of the long and short fibers of the phage in attachment and entry are still uncertain and will be until the interactions with receptors are clarified. The receptors are the lipopolysaccharide molecules of the outer cell envelope (Wilson et al., 1970). The fiber attachment may allow some browsing to bring the phage to a suitable site for entry of DNA. A remarkable observation (Bayer, 1968) suggests that irreversible attachment and penetration occur only at relatively few (about 300) sites where cytoplasmic and outer membranes form stable contacts that resist mild osmotic shock (Figure 7-5). This may also be true of phages other than T-even. It would be important to establish the relation of these special sites to sites of synthesis of membrane components and of phage receptors.

The next step of interaction is sheath contraction, which drives the tail tube into the cell envelope. Contraction is triggered from the baseplate, which in turn gets a conformational stimulus from the fibers (Arscott and Goldberg, 1976). Sheath contraction involves a complex displacement of the 144 subunits of the sheath (Moody, 1973) to a form half the original length. A role of phage-associated ATP in this contraction has been suggested but not convincingly established. The distal end of the tail tube must be driven to the inner, cytoplasmic membrane, not necessarily through it. E. coli spheroplasts, in which outer membrane and rigid layer have been removed or loosened, can receive DNA from urea-treated phage particles whose sheath is precontracted

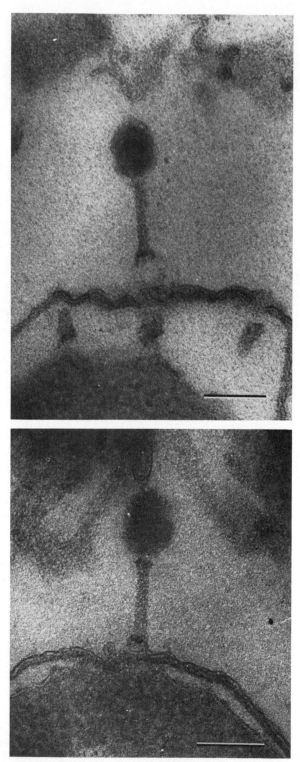

FIGURE 7-5. Attachments of T-even virions to points of adhesion between cytoplasmic membrane and outer membrane (cell wall) of E. coli. Bars correspond to 100 nm. From Bayer, J. Virology **2,** 346 (1968).

and whose tail tube is exposed. Such infection of spheroplasts (kept of course in hypertonic media) leads to production of normal phage progeny. Spheroplasts can also accept entire or fragmented phage DNA molecules, which may then replicate and participate in recombination. Infection of spheroplasts bypasses the normal receptors for attachment: urea-treated T4 can infect T4-resistant mutants of *E. coli* or even resistant bacteria of distant species (Wais and Goldberg, 1969). Attachment of urea-treated particles to spheroplasts can be blocked by phosphatidyl glycerol, which may be the membrane component that triggers DNA ejection (Benz and Goldberg, 1973).

A few minutes after bacteria are infected with a T-even phage they exclude more phage of the same kind from participating in replication (*superinfection exclusion*) and cause breakdown of the DNA of the superinfecting phage (*superinfection breakdown*) (Yutsudo and Okamoto, 1973). Exclusion and breakdown are expression of functions of phage genes within the host bacterium and can be abolished by mutations in these genes (Childs, 1973).

Sequence of Events in T4-Infected Bacteria

The phage enters accompanied by several minor peptides and proteins. At least one protein, the product of phage gene 2 (= gp2), probably enters attached to the lead part of the DNA and protects it against the bacterial exonuclease V. Mutants 2⁻ inject DNA but this is broken down (Silverstein and Goldberg, 1976a). Later the phage produces an inhibitor of exonuclease V (Silverstein and Goldberg, 1976b). Examination of T4-infected bacteria at various times after infection by electron microscopy and by a combination of analytical methods reveals the pattern illustrated in Figure 7-6. Phage DNA increases after a brief delay; virion-specific proteins begin to appear somewhat later and their appearance is soon followed by appearance of organized capsid precursors and then by the formation of mature infectious capsids. Some mRNA of viral specification is made promptly after infection and throughout the latent period. Bacterial mRNA and bacterial proteins stop being synthesized within a few minutes after the entry of phage DNA. Bacterial DNA is rapidly degraded to acid-soluble fragments and the "nuclear bodies" or DNA-containing areas of the bacterium become dispersed. It will be our task to explore this complex sequence of events, including the takeover of the cell machinery by phage, in its various aspects. We may point out that the pattern shown in Figure 7-6 is temperature dependent. The latent period for phage T4 is 24 minutes at 37° and 40 minutes at 28°C.

DNA SYNTHESIS

T-even phages have hydroxymethylcytosine (HMC) instead of cytosine in their DNA (see Cohen, 1968). In mature DNA, HMC has a pattern of glucosylation specific for each T-even phage (Table 7-1). HMC DNA is first synthesized without glucosylation: phage-coded glucosylating enzymes (see below) glucosylate HMC groups in DNA, not in deoxynucleoside mono-, di-, or triphosphates (Revel and Luria, 1970).

HMC is a good chemical marker for T-even DNA. Its glucosylated derivatives in phage DNA also provide antigenic specificities recognizable by

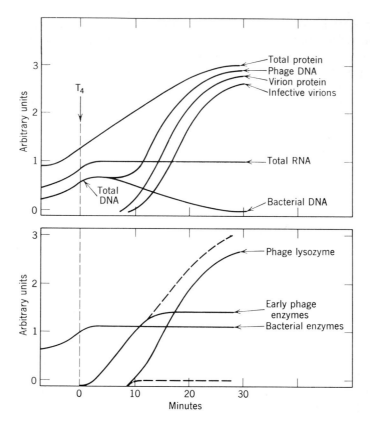

FIGURE 7-6. Schematic representation of the pattern of biosynthesis in E. coli bacteria infected with bacteriophage T4. The broken lines in the lower part of the figure represent the course of these enzyme syntheses in bacteria infected with conditional mutants that cannot initiate DNA replication.

specific antibodies (McNicol and Goldberg, 1973). By tracing DNA from the infecting phage it has been established that replication of T-even phage DNA is semiconservative (*à la* Watson and Crick) and dispersive, in the sense that segments of the infecting DNA strands are parcelled among several progeny phages (Kozinski, 1961). This dispersion is correlated with the occurrence of repeated molecular exchanges that are revealed by genetic recombination (see page 186).

The DNA molecules that enter from T-even virions into the bacterium are linear, circularly permuted, and have a terminal redundancy of 2 to 4% (depending on the phage) corresponding to 2–4 × 10^6 daltons of DNA (3000–6000 nucleotide pairs). If all the required phage genes are functional, replication begins and is bidirectional (Delius et al., 1971). The major initiation site is near gene 42 (see Figure 7-1). Additional initiation sites may exist, possibly functional only under special conditions such as in DNA fragments arising in the course of repair of molecular damage. Initiation presumably requires synthesis of an RNA primer (Morris et al., 1975) and proceeds by the addition of short 5′ → 3′ segments (Okazaki fragments, see Chapter 4), which are then

TABLE 7-1　GLUCOSYLATED HYDROXYMETHYLCYTOSINE
DEOXYRIBOTIDES IN THE DNA OF T-EVEN PHAGES

R = H
= α-glucosyl
= β-glucosyl
= β(1-6)-glucosyl-
α-glucosyl

deoxyribose-5'-phosphate

	HMC	α-glucosyl HMC	β-glucosyl HMC	β(1-6)-glucosyl-α-glucosyl HMC
T2	25%	70%	0	5%
T4	0	70%	30%	0
T6	25%	3%	0	72%

Data from Lehman and Pratt, *J. Biol. Chem.* 235: 3254 (1960).

joined to the growing strands by ligase action. About 20 copies of the DNA accumulate and then *concatemers* begin to appear, probably coincident with expression of late genes.

The formation of concatemers has been related to the 3'-end problem: replication of a linear DNA molecule cannot be complete at the 3'-OH end since there is no primer for the last Okazaki fragment to attach to. The single-stranded regions at the end of the T-even DNA molecules, being redundant, allow for concatemer formation between the two daughter molecules (Watson, 1972; see Chapter 8). Joining of the concatemers will generally require the action of a variety of enzymes—exonucleases, endonucleases, polymerase, plus DNA ligase—to form covalent joints. Concatemers of great length are actually found and explain among other things the presence of DNA molecules 10 times longer than normal (almost as long as those of an *E. coli* cell) in abnormal phage with giant heads (Uhlenopp et al., 1974). The actual organization of the intracellular T4 DNA is in fact more complex than a simple collection of concatemers or even of group of concatemers joined by regions of polynucleotide pairing (Worcel and Burgi, 1972).

It is reported that large DNA complexes are attached to some membrane sites (Siegel and Schaechter, 1973). Claims of membrane association for phage or bacterial DNA may be more or less solid depending on the methods used; adhesions may be established during extraction and centrifugation. For T-even phage DNA the claims are not fully convincing. It is likely, however, that *in vivo* the phage DNA is associated with RNA, proteins, and possibly membrane components in functionally meaningful complexes (Curtis and Alberts, 1976) resembling those present in uninfected bacteria (Worcel and Burgi, 1972; Bernstein and Bernstein, 1974).

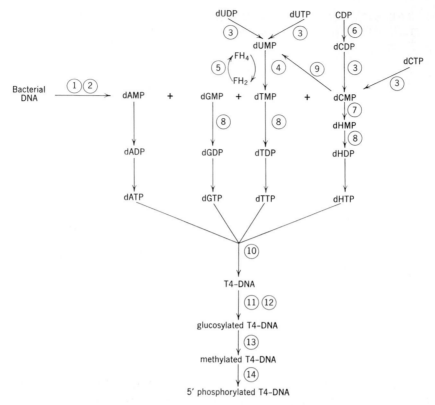

FIGURE 7-7. *Enzymatic biosynthesis of T4 phage DNA and concomitant reactions. The enzyme activities that appear or increase after infection follow. Modified from Cohen in:* Viral and Rickettsial Infections of Man *(Horsfall and Tamm, eds.), Lippincott; Philadelphia, page 203.*

1. *Endonuclease*
2. *Exonuclease*
3. *Deoxycytidine and deoxyuridine tri- and diphosphatase*
4. *Thymidylate synthetase*
5. *Dihydrofolate reductase*
6. *Cytidine diphosphate reductase*
7. *Deoxycytidylate hydroxymethylase*
8. *Deoxynucleotide kinase*
9. *Deoxycytidylate deaminase*
10. *DNA polymerase*
11. *α-Glucosyl transferase*
12. *β-Glucosyl transferase*
13. *DNA methylase*
14. *DNA kinase*

Joining of phage DNA molecules explains why successful infection can occur after infection with two (or more) "petite" or minivirions of T4 each of which has only two-thirds as much phage DNA (see below). Two such "two-third-DNA" molecules, being circularly permuted, can by recombination generate a full phage genome if they overlap by not more than one-third their map (Mosig, 1963). In single infections with petite T4 only some of the infecting DNA molecules undergo replication: the others probably lack a starting point or origin needed for replication (Mosig and Werner, 1974).

PHAGE-CODED ENZYMES

The early transcription of newly entered phage DNA is done by the bacterial RNA polymerase. A number of phage proteins are made from such transcripts. Then the RNA polymerase becomes modified by the action of early phage products, its specificity changes, transcription of host genes stops even before the host DNA is broken up and other phage genes are transcribed (see below under Program of Phage Gene Expression).

Among the phage-coded proteins are enzymes belonging to several classes: (1) enzymes that produce unique phage DNA constituents such as HMC deoxynucleotides, or glucosylate these nucleotides, or destroy precursors of cytosine deoxynucleotides (Figure 7-7); (2) enzymes that play specific roles in DNA replication and recombination; (3) enzymes that destroy host-cell DNA and make its nucleotides available for phage DNA synthesis; (4) enzymes that take part in the processing of virion proteins.

Phage mutants defective in a given phage enzyme present characteristic defects. For example, a defect in deoxycytidine triphosphate and/or dCMP deaminase causes some dCMP to find its way into phage DNA and makes it subject to degradation by a phage nuclease (Elliott et al., 1973). Appropriate selection in turn makes it possible to isolate phage mutants that lack the ability to degrade cytosine-containing DNA to free deoxyribonucleotides. Selection is done for phages that fail to grow in the presence of hydroxyurea, an inhibitor of ribonucleotide reductase. Under these circumstances the only source of deoxynucleotides is the degradation of host DNA, which therefore becomes a requirement for phage growth (Warner et al., 1975). Figure 7-8 illustrates the pathway for ribonucleotide reduction by T4-specific enzymes as an example of a particularly well analyzed pathway (see Berglund, 1975).

Of course, mutations in essential genes can be obtained only as conditionally lethal mutations. Most genes for phage enzymes are clustered in one or two regions of the T4 map (Figure 7-1). Note that phage T4 brings into the host cell the gene for its own DNA polymerase, the product of gene *43*, whose mutations can prevent phage DNA replication. Mutant forms of gp 43 can act as mutagens by causing insertion of wrong nucleotides into phage DNA or by affecting their excision (Bessman et al., 1974), confirming the essential role of this protein in DNA replication.

A critically important protein is gp 32, the product of gene *32*. This is a DNA unwinding protein (Alberts et al., 1970) that associates with single-stranded DNA stretches in stoichiometric amounts, one molecule per 10 nucleotides. This protein actually can promote some unwinding of double-stranded DNA to allow initiation of replication to occur at sites where single-strand nicks are present. A peculiarity of gp 32 is to regulate its own rate of synthesis by a feedback mechanism (Krisch et al., 1974): when more single-stranded regions are present in DNA, more gp 32 is synthesized. Proteins functionally equivalent to gp 32 are present in *E. coli* and probably in all cells.

Synthesis of several phage enzymes and proteins has been achieved in *in vitro* systems prepared from *E. coli* and directed either by added mRNA from

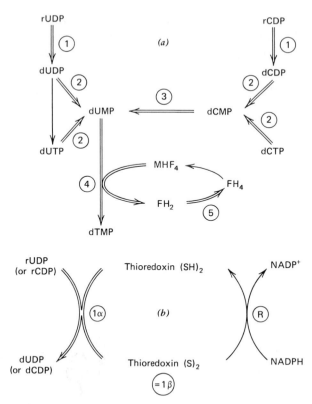

FIGURE 7-8. (a) Enzymes of thymidylate biosynthesis in T4-infected E. coli. Double arrows indicate reactions catalyzed by phage-directed enzymes. The corresponding genes in Figure 7-1 are: (1), genes nrdA, B, C; (2), gene 56; (3), gene cd; (4), gene td; (5), gene wh (near td). (b) Components of reaction (1): phage-directed enzyme (1) consists of a ribonucleotide diphosphoreductase (subunit 1α) and a thioredoxin (subunit 1β). Enzyme (R) is bacterial thioredoxin reductase.

T4-infected bacteria or directly by T4 DNA. Such systems make it possible to study directly the expression of different phage genes or groups of genes in the course of infection by testing the template function of mRNA present at various times (O'Farrell and Gold, 1973b; Wu et al., 1973).

In vitro synthesis of T4 DNA by a set of T4-specific proteins has been accomplished and has clarified some aspects of the in vivo process (Morris et al., 1975). The six proteins—gp 43 (polymerase), gp 32, gp 41, gp 44, gp 45, and gp 62 (Figure 7–9) have been purified. Together and in the presence of the four deoxynucleotide triphosphates, the four ribonucleotide triphosphates, and Mg^{++} ions these proteins carry out faithful replication of DNA at nearly the in vivo rate (about 800 nucleotides sec^{-1}) even though the origins of replication in vitro are presumably not the normal initiation sites. The gp 44 and gp 62 supposedly form a complex that has ATPase activity and is stimulated by gp 45, while gp 32 maintains single-stranded regions of DNA available to the other proteins (Fig. 7–9). Together these proteins would act on the polymerase by increasing the length of DNA that the enzyme copies

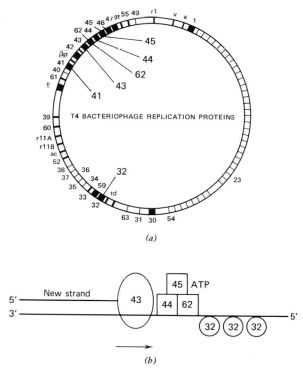

FIGURE 7-9. (a) *The six genes in the T4 map whose proteins can replicate T4 DNA in an* in vitro *system.* (b) *Proposed role of some of the proteins at the growing point of one of the new DNA molecules. The gp44-gp62 complex may be an ATPase. The gp41 may be an RNA polymerase needed to synthesize the RNA primers. Modified from Alberts et al.,* ICN-UCLA Symp. Molecular and Cellular Biology, **3,** 241 (1975).

before "falling-off" the template. This T4-derived *in vitro* system provides clear evidence for several features of DNA replication such as the occurrence of Okazaki fragments on both growing strands at a replication fork and the requirement for RNA priming at each fragment start. (The gp 41 may be an RNA polymerase that synthesizes the RNA primers). Extensive replication on single-stranded DNA templates generates typical rolling circle forms, with long, unbranched double-stranded tail (see page 151).

The six proteins needed for the DNA synthesis may form a multienzyme complex. Thus T-even phage produce in their host bacterium a complex of enzymes and subsidiary proteins that make their replication independent of the DNA synthesizing systems of the bacterium. This is not true of other phages, which, as will be discussed in subsequent chapters, are variously dependent on bacterial enzymes for replication of their genomes.

PROGRAM OF PHAGE GENE EXPRESSION

Based on the time of appearance of their messengers or gene products one distinguishes immediate early, delayed early, quasi-late, and late genes. The immediate early, whose mRNA is made even in the presence of an inhibitor of protein synthesis such as chloramphenicol, are transcribed from phage DNA by *E. coli* RNA polymerase either *in vivo* or *in vitro*. Another early class is that of delayed early genes, whose mRNA is initiated early but whose products are made only after a delay because they are coded by genes located distally from their promoters on the phage operons (O'Farrell and Gold, 1973a; Figure 7-10).

The quasi-late genes are transcribed by RNA polymerase that has been altered by the action of certain products of immediate early phage genes.

FIGURE 7-10. *A simplified diagram of the program of gene expression in T4 phage development. Bacterial RNA polymerase transcribes the "early" genes. The distal genes in long early operons produce "delayed early" proteins. Some of the early proteins alter or replace some of the polymerase subunits (reaction (a)), which are then shown in parentheses. The altered polymerase transcribes the genes for "quasi-late" proteins. Some of these (X, Y, Z) join the polymerase (reaction (b)), which can then transcribe "late" genes provided DNA replication has started.*

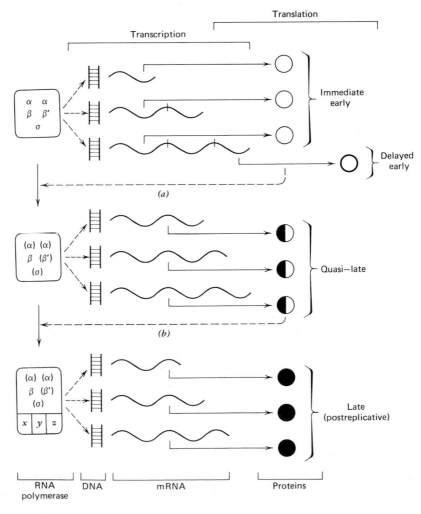

Particularly important are the loss or modifications of polypeptide σ and alterations in other subunits of the polymerase (Burgess, 1971), which thereby acquires new promoter-recognition specificities. Especially interesting is the alteration of the α polypeptides of the enzymes, which acquire an adenosine diphosphoribose residue donated by nicotinamide adenosine dinucleotide (Goff, 1974). Host promoters are no longer recognized, and no more host-specific RNA is transcribed. A second shift in RNA polymerase specificity occurs later (about 6 to 8 minutes in bacteria infected at 30°C) and is due to the association of *E. coli* polymerase with three polypeptides including gp 33 and gp 55 (Stevens, 1972; Horvitz, 1973). At this point in the program two events occur: a more or less complete shut-off of the synthesis of the phage enzymes under control of phage gene 62 (Wiberg et al., 1973; see Figure 7-6) and the start of synthesis of all the late gene products, including the virion constituent proteins and phage lysozyme.

The start of late gene functions requires not only the activity of genes *33* and *55* but some replication of phage DNA. Even *in vitro* the late genes are not transcribed unless the phage DNA is derived from infected bacteria late in the latent period. The explanation appears to be that only DNA with some single-strand breaks is a suitable template for late transcription (Riva et al., 1970). In fact, unreplicated DNA can serve as template for late gene transcription if some single-strand breaks are introduced into it. This suggests that DNA with discontinuous strands has a special conformation that makes the late promoters available to RNA polymerase as modified by gp 33 and gp 55. Single-strand breaks are produced all the time in the course of replicative recombination (see below) and are then sealed by a polynucleotide ligase, the product of T4 gene *30*.

The most convincing confirmation of the program as a series of feedback loops involving products of specific phage genes interacting with RNA polymerase is, of course, that the program can be reconstructed *in vitro*. Protein-synthesizing systems programmed by the addition of mRNA, either extracted from cells or synthesized *in vitro*, duplicate the situations deduced from *in vivo* studies. Early gene messengers are made early; late gene messengers are made only after the RNA polymerase has been altered; delayed immediate proteins are made distally from early messengers (O'Farrell and Gold, 1973b). These relations are still observed in *in vitro* systems in which DNA itself serves as template for mRNA, which in turn is translated into protein.

A simplified view of the major loops of the T4 program is given in Figure 7-10. The occurrence of phage-directed changes of specificity in RNA polymerases, through the action of specific phage gene products, has been convincingly demonstrated with other phages, particularly *Bacillus subtilis* phages (Fox and Pero, 1974). Such conversion can be produced *in vitro* (Duffy et al., 1975).

The switch from bacterial to phage program caused by changes in RNA polymerase has one important consequence. Since no more ribosomal RNA or proteins are made after T-even infection and since the phage does not code for new ribosomes, all phage protein synthesis must take place on bacterial ribosomes made before the first shift in program. This circumstance made possible one of the classical experiments that demonstrated the existence of

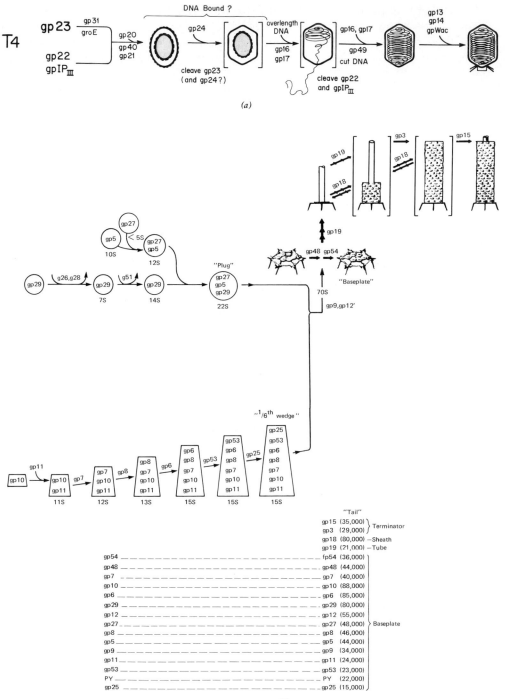

(a)

(b)

mRNA as distinct from ribosomal RNA (Brenner et al., 1961) confirming an earlier surmise also based on experiments with T-even phage (Volkin and Astrachan, 1956).

Not all the synthesizing apparatus for phage proteins is purely bacterial; several tRNAs different from the bacterial ones are coded by phage T4 (Wilson, 1973). Deletions of the phage DNA region that carries these genes are not lethal, although the burst size may be reduced. Here again, as with some enzymes, a T-even phage codes for dispensable genetic information, whose presence may be a remnant from earlier stages of phage evolution. At least some amino-acids activating enzymes of the bacterium become altered by combination with phage gene products (Müller and Marchin, 1975).

Assembly of Virions

In contrast to the early phases of phage development, the assembly of capsids and complete virions is not programmed in time by the successive expression of phage genes. All virion proteins and other late proteins such as phage lysozyme appear to be synthesized more or less simultaneously and accumulate in "precursor pools." From the pool they are then withdrawn by direct specific interactions with other protein molecules to construct subassemblies that are then assembled into complete virions. Although the overall assembly pattern could be fairly well deduced by *in vivo* studies with mutant phages

FIGURE 7-11. *Steps in the assembly of T4 phage virion, indicating the gene products involved in each step. (a) Assembly of the phage head. (b) Assembly of the phage baseplate and complete tail. (c) Assembly of the tail fibers (the tail fibers are assembled separately and added to the tail after head and tail have joined). After Casjens and King,* Ann. Rev. Biochem. **44,** *555 (1975) and Wood and Bishop,* Virus Research *(Fox and Robinson, eds.) page 303.*

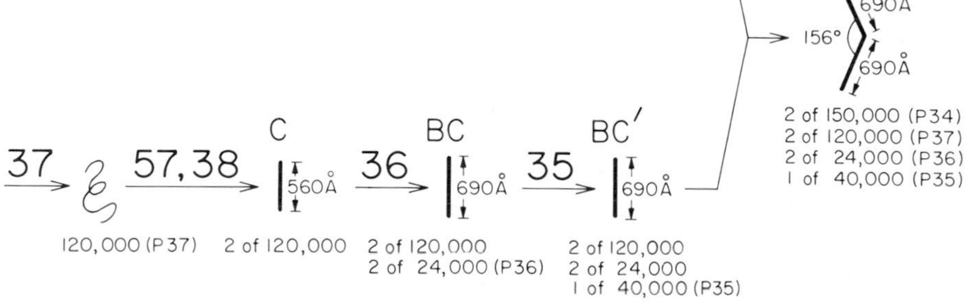

(c)

and examination of lysates, a powerful tool became available with the discovery of *in vitro* assembly of preformed phage precursors. For example, extracts of infected bacteria containing phage heads and no tails when mixed with extracts with tails and no heads gave rise to complete infective virions (Edgar and Wood, 1966). This *in vitro* complementation, reproduced since then for several other phage systems, is not limited to joining of subassemblies. An extract of bacteria infected with phage T4 X^-, mutated in late gene X and lacking gp X, provides a complementation test for the presence of free gp X in another extract. In this way the role of many genes in the structure and assembly of the T4 capsid has been established. Some of these genes code for capsid proteins, others for proteins that interact, enzymatically or otherwise, with capsid protein during assembly. An overall picture is given in Figure 7-11.

Essentially, virion assembly consists of four major subassembly processes interacting with each other only at some critical points (Casjens and King, 1975).

1. The baseplate of the phage tail consists of 15 gene products and its synthesis involves also several other genes. Surprisingly, the baseplate appears to contain some molecules of two phage-coded enzymes; dihydrofolate reductase and thymidylate synthetase, as well as some folic acid (Kozloff et al., 1975a,b,c).

2. The complete baseplate, specifically after the addition of gp 54 (Berger and Warner, 1975), provides a primer for the assembly of the tail tube (144 copies of gp 19). Around the tube the sheath assembles as a polymer 144 copies of gp 18. The products of two other genes stabilize this assembly. The constancy of tube length, without any guiding metric within, is a puzzling feature. Either there exists some other linear protein that serves as an external measure, or contact with the baseplate imparts to the tube subunits a specific conformation, which makes the precise length of the tube a state of minimum free energy. The implications of the latter hypothesis for the constructional role of protein assemblies will not escape the reader.

3. The shell of the phage head is the product of many genes and consists of more than 10 proteins, one of them, gp 23, representing the bulk of it. In complete heads gp 23 is present as a partially cleaved form due to removal of a 10,000 molecular weight piece from the original polypeptide. The cleavage is due to the proteolytic action of proteins, mainly gp 22 and possibly gp 21, which do not appear in the mature virions. Protein gp 22, in fact, is an internal protein that ultimately digests itself to small peptides, some of which remain within the phage heads (Laemmli and Quittner, 1974). Other internal proteins are also present and are processed by gp 22 to their final form.

4. Once the tail and the head of the phage are assembled separately they combine spontaneously *in vitro* as well as *in vivo*.

5. The tail fibers are made up of four gene products. Their assembly is independent of the rest of the virion, but they attach to the baseplate only after head and tail have been joined. The gp 63 plays a role in this reaction, which also involves an interaction with whiskers attached to the collar between head and tail (Dewey et al., 1974).

The specific shape of the phage head is determined by gp 23 and by other proteins, whose mutation may alter the head structure. The normal T4 head consists of a distorted icosadeltahedron with an inserted extra band of sub-units in the long axis with about 840 copies of gp 23; the gp 20 is located at the vertices (Branton and Klug, 1975; see Figure 7-12). This shape must reflect precise constraints resulting from protein-protein interactions. When these constraints are altered odd things can happen (Cummings and Bolin, 1975). It has already been mentioned that 1% or more of T4 virions are symmetrical, petite, and contain two-thirds lengths of DNA (Mosig, 1963). In some head protein mutants as many as 70% of the virions produced are petite. In 22⁻ infections, protein 23 is uncleaved and the transverse diameter of the head is variable. When gp 20 is missing the cells contain "polyheads" of normal

FIGURE 7-12. *The shape of T4 capsid head shown by a construction model. Note the five-fold symmetry vertices and the presence of one extra row of subunits, distorting the simple icosahedral shape. From Branton and Klug,* J. Mol. Biol. **92,** *559 (1975). Courtesy Dr. A. Klug.*

transverse diameter but exaggerated lengths, similar to the "lollipops" that are produced by bacteria infected with T4 phage in the presence of the amino acid analog canavanine (Uhlenhopp et al., 1974).

Amber mutations in gp 23, of course, prevent the assembly of phage head altogether. Peptide analysis of the fragments of gp 23 made by a set of amber mutants in gene 23, coupled with mapping of the *amber* mutations, provided one of the striking confirmations of the assumptions of the genetic code (Sarabhai et al., 1964).

Phenotypic mixing has already been mentioned: the incorporation into the phage virion of homologous proteins derived from two different alleles of a given phage gene. It is most easily demonstrated for proteins of the tail fibers since these embody the specificity of phage attachment to bacteria (Novick and Szilard, 1951). Thus a phage virion produced in bacteria mixed-infected with T4 and T2 may have a T4 genome but tail fibers of T2 specificity. It will infect a bacterium B/4 (lacking receptors for T4) but give progeny of T4 type.

PACKAGING THE DNA

In the T-even virions the phage DNA is tightly packed (500 μm in a cavity less than 0.1 μm in linear dimensions). The topography of packing—like a ball or a spindle of yarn—is unknown; there may be a central hollow (Klimenko et al., 1967). It is believed that the DNA collapses into a compact form within the capsid when the gp 22 protein inside the capsid is cleaved. How DNA is "sucked" into the capsid head is not clear; it may simply be brought in by a displacement reaction caused by exit of internal proteins (Laemmli et al., 1975). There is a definite if still somewhat unclear relation between capsidation of DNA and the cutting, out of long DNA fibers, of precise "headfuls" of DNA, each about 2% longer than a complete phage genome or gene set. This is the mechanism that generates the terminal genetic redundancy (Streisinger et al., 1964; Séchaud et al., 1965). Empty capsids are precursors of mature phage (Laemmli and Quittner, 1974). Which basic protein or proteins provide the lead for winding the DNA into the capsid is not clear. The energetics of winding are probably as unintuitively simple as are other kinetic phenomena involving the relatively rigid DNA molecules (Davison et al., 1961). By its structure DNA can rotate fairly freely around its axis, like a wire in a speedometer cable. The rolling up of DNA filaments may be facilitated by rotation.

The gp 49 is needed to fill the heads normally with DNA. In temperature-shift experiments with T4 temperature sensitive in gene 49, empty heads are formed at the nonpermissive temperature and become filled on lowering the temperature (Laemmli et al., 1974). Whether gp 49 is an endonuclease and whether the cutting occurs within or outside the phage head is not known. There are reports, however, that DNA segments of normal headful length can be cut off in bacteria infected with mutants that make no head capsid. The remarkable headful cutting phenomenon evidently requires further clarification.

It may be useful to point out here that many of the features of phage development have been deduced from experiments in which the materials extracted from bacteria abortively infected with phage mutants were tested either by biochemical tests or by *in vitro* complementation. It is conceivable that unrecognized complications are generated by this methodology, so that the interpretations may not be completely congruent with the features of normal infection. Yet the overall picture is self-consistent and represents one of the most remarkable achievements of molecular biology—the reconstruction of practically the entire life cycle of an organism in terms of specific gene functions and their regulation.

Lysis

One of the late gene products of T-even phages is a lysozyme (gp *e*) that cleaves bacterial peptidoglycans (as egg white lysozyme does) and whose amino acid sequence is known (Terzaghi et al., 1966). Lysozyme is made within infected cells well before the onset of lysis, that is, before the end of the latent period. Lysis takes place, however, only if the lysozyme gains access to the peptidoglycan layer. This requires a phage-initiated reaction (gene *t*) that damages the cytoplasmic membrane and arrests energy coupling within it (Joslin, 1971).

The well-known *rII* genes of T4, to be discussed below, apparently act to prevent or delay the activity of gp t, so that *rII* mutant phage lyses bacteria more rapidly than the *rII*⁺ normal phage (*r* stands for "rapid lysis"). More specifically, bacteria infected with *rII*⁺ exhibit inhibition: lysis can be delayed several hours provided a second infection of the same kind takes place within minutes (Doermann, 1948). Apparently this causes the membrane to become insensitive to gp t.

One other gene, *sp*, is involved in the lysis control program. Mutants *sp*⁻ lyse at the appropriate time even when the phage is without lysozyme (Emrich, 1968). This gene may normally act to maintain the integrity of the cytoplasmic membrane, making possible continuing aerobic metabolism and preventing the rigid layer from being attacked prematurely by bacterial enzymes.

Role of Host Cell Genes

It has already been mentioned that host-specific macromolecular syntheses stop after T-even infection because of arrest of transcription and breakdown of host DNA. The breakdown is caused by phage-coded endonucleases specific for cytosine-containing DNA. The microscopically recognizable disruption of bacterial DNA masses is under control of the phage gene *ndd* (Sunstad et al., 1974).

Most interesting are certain host gene mutations that interfere specifically with the performance of certain steps in the phage program. Some of these mutations, collectively called *gro*, alter the RNA polymerase subunits so that this enzyme does not respond to the modifying factors made by phage (Snyder and Montgomery, 1974). Other *gro* mutations may affect proteins needed to assemble or to process phage proteins (Coppo et al., 1974).

TABLE 7-2   THE KNOWN GENES OF BACTERIOPHAGE T4 ACCORDING TO
WOOD AND REVEL (1976)

At least 10 additional genes are known whose products have unknown metabolic
functions.

Conversely, certain phage mutations collectively called *ogr* can render the phage able to multiply on *gro* cells, either by compensating for the host defect or by bypassing it.

An interesting observation (Chao et al., 1974) is that the *pol A⁻* mutant of *E. coli,* defective in DNA polymerase I, produces many more "petite," symmetric-headed virions than the normal phage. A plausible suggestion is that the *pol A* defect may upset the regulating system that ties DNA replication to the expression of late genes and that this in turn may affect the head assembly process.

The Genetic System of T-even Phages

Table 7-2 shows the distribution of the known genes of phage T4 according to function. About 140 gene loci have been recognized in the genome of these phages, not all of which have been accounted for in terms of specific protein products. A genome with 1.2×10^8 daltons of DNA, consisting of 1.66×10^5 nucleotide pairs, can code for a maximum of 5 to 6×10^4 amino acids or about 300 average size proteins. The map, therefore, is already quite well filled although not nearly saturated. There is more compartmentalization in the map, in terms of genes coding for early and late functions, than is accounted for by operon structure (see Figure 7-1). Presumably the major groupings for tail fibers, baseplate, early enzymes, and so forth represent traces of the evolutionary history of these phages. The genes for tail tube and sheath are less clustered than the other groups and may represent a later evolutionary development (Wood and Revel, 1976).

A remarkable feature is that the RNA is generally transcribed "clockwise" for the late genes and "counterclockwise" for the early and the quasi-late genes. Since transcription always proceeds from the 5'- to the 3'-OH end of the messenger, the template strand of the DNA is transcribed in the 3'-5' direction, so that clockwise and counterclockwise transcriptions mean transcriptions of different strands. Usually only one strand of a given segment is transcribed, but exceptionally the two opposite strands of a segment may be transcribed (probably not translated) in different phases of the program (Bautz et al., 1966).

Figure 7-13 illustrates the relation between the physiological map, the physical map, and the recombinational map of phage T4. The total map length of phage T4 is about 2500 map units, one map unit corresponding to about 70 nucleotide pairs.

The physical genetic map is obtained by an elegant method that employs the petite virions of phage T4, whose DNA molecules are about two-thirds of the genome length (Mosig, 1968). The method scores the frequency with which two genes are present together in such a short genome. This frequency depends for obvious reasons on the physical distance between the two genes on the DNA. In general the physical map and the recombinational map coincide within the limits of experimental errors. An exception is the calculated length of the tail-fiber genes *34* and *35*, which appear to be much longer in the recombinational map than in the physical map of T4 but not of T2. The explanation is the existence in gene *34*, near gene *35*, of a "hot-spot" for

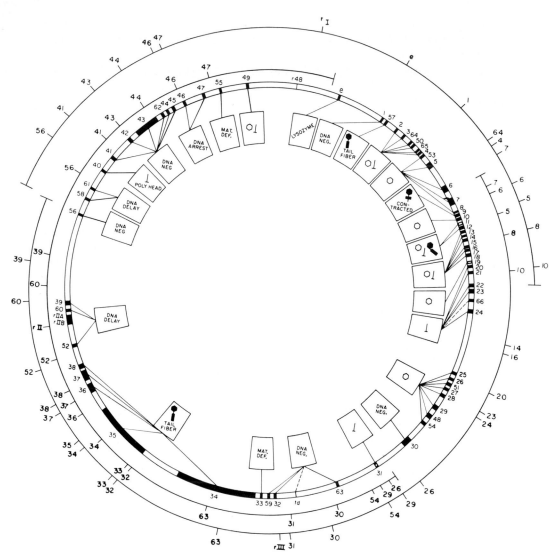

FIGURE 7-13. Distances between genetic markers in the T4 chromosome. The inner circle shows functions of genes and recombination distances between them (see Figure 7-1). The three outer arcs show physical distances obtained from the frequency of joint presence of genetic markers in petite T4 virions. From Mosig, Genetics **59**, 137 (1968).

recombination, that is, of a sequence that undergoes recombination about 10 times more frequently than its physical length would account for (Beckendorf and Wilson, 1972).

The recombinational map of phage T4 is derived from experimental recombination frequencies $X_{a,b}, X_{a,c}, A_{b,c}, \ldots$ in two-marker or three-marker crosses between T4 mutants

$$a \times b;\, a \times c;\, b \times c$$
$$ab \times c;\, ac \times b \ldots$$

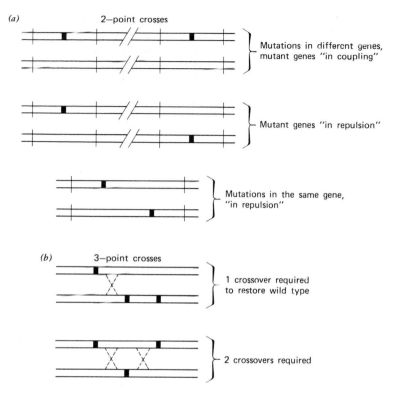

FIGURE 7-14. Examples of 2- and 3-point crosses used in genetic mapping.

Each gene mutation provides a genetic *marker,* distinguishing the gene that carries it from other alleles of that gene. The above crosses should in fact be represented as:

$$ab^+ \times a^+b; \; ac^+ \times a^+c; \; \text{etc.}$$

where a; a^+, b; b^+ represent alternative forms of different genes or of specific sites within a gene (for example, a cross $Ha \times Hb$ may be a cross between two mutants altered at sites a and b within gene H; see Figure 7-14). A classical set of recombination data (for phage T2 rather than T4) is presented in Table 7-3. From the experimental frequencies of recombination one derives a genetic map. For phage T4, as for all other phages, all genetic markers are linked and the map corresponds to a single chromosome. Moreover, the recombinational T4 map is circular (Streisinger et al., 1964), a direct consequence of the packaging mechanism that segregates terminally redundant headfuls of phage DNA into the virions (see Figure 6-9b).

There are interesting differences between recombination in a phage and in an eucaryotic organism such as *Drosophila.* For any three linked markers a, b, c on an eucaryotic chromosome one can write a mapping function

$$R_{a,c} = R_{a,b} + R_{b,c} - 2IR_{a,b}R_{b,c} \qquad [7\text{-}1]$$

where the R's are the observed frequencies of recombinants emerging from pairwise crosses. The third term on the right takes care of those coincident

TABLE 7-3 RECOMBINATION IN BACTERIOPHAGE T2

This table includes some of the data from the classical phage recombination experiments of Hershey and Rotman, *Genetics* **34**:44 (1949) and Levinthal and Visconti, *Genetics* **38**:500 (1953). All the data are in percent of total phage yield.

I. Frequency of various genotypes in the progeny from mixed infection experiments with mutants of phage T2.

Cross	Parents		Recombinants		Recombination frequency
a	*h*	*r13*	*hr13*	*++*	
	59	39	0.74	0.94	1.7
b	*hr13*	*++*	*h*	*r*	
	48	50	0.8	0.8	1.6
c	*h*	*r7*	*hr7*	*++*	
	56	32	6.9	6.4	13.3
d	*h*	*r1*	*hr1*	*++*	
	42	34	12	12	24

Conclusions: (1) Frequencies of recombinants are similar whether the mutant characters are in coupling (cross *b*) or in repulsion (cross *a*). (2) *h* is progressively less linked to *r13, r7, r1* (crosses *a, c, d*).

II. Frequencies of recombination in two- and three-point crosses.

Cross	Two-point crosses*	Recombinant r^+	Recombination frequency
e	*r1* × *r7*	15	30
f	*r1* × *r13*	15	30
g	*r7* × *r13*	7	14
h	*r4* × *r7*	3.5	7
i	*r4* × *r13*	7	14
	Three-point crosses		
j	*r4r13* × *r7*	1.8	
k	*r7r13* × *r4*	0.8	
l	*r4r7* × *r13*	5	

Conclusions: (1) The most probable order is *r7-r4-r13* since the frequency of r^+ recombinants is minimal in cross *k*, which for that order requires two crossing overs (*r7 + r13/+ r4 +*). (2). Assuming the foregoing order, the recombination frequencies in the two-point crosses are not additive (crosses *g, h, i*); there is less recombination between the outside markers than expected from the equation $z = x + y - 2xy$. This indicates negative interference ($z = x + y - 2Ixy$; $I > 1$). (3) In three-point crosses there is some excess of double recombinants r^+ over the frequency expected from the two-point crosses, confirming negative interference. (For example, in cross *k* the expected frequency of r^+ recombinants would be $0.07 \times 0.07 = 0.0049$ or 0.49% instead of 0.8%.)

---

* In the two-point crosses the recombination frequencies are taken as twice the frequencies of the r^+ recombinants because the *rr* recombinants cannot be distinguished from the *r* parental types.

TABLE 7-3   *(CONTINUED)*

III. Triparental crosses

Cross	Parents	Frequency of +++ recombinants
m	*r4* × *r7* × *r13*	7.3
n	*r4r7* × *r4r13* × *r7r13*	0.16

Conclusion: Multiparental recombination, probably by successive mating acts, can take place since in cross *n* the +++ recombinant cannot be formed in any one mating between two parental types.

IV. Increase in frequency of recombinants with increase in burst size.

Cross	Parents	Burst size	Recombination frequency
o	*h* × *r13*	120	2.6
		300	6.0
		450	9.5

Conclusions: The frequency of recombinants increases with increasing yield of phage per bacterium.
Note: The recombination frequencies are higher in cross *o* than in cross *a* with the same phages because cross *o* was done under conditions assuring simultaneous mixed infection of all bacteria, thereby increasing the frequency of recombinants. Under these conditions and for large phage yields, the recombination frequencies for the crosses shown in *d, e,* and *f* would be about 40%.

exchanges that bring the outside markers back together. The coefficient of coincidence *I* equals 1 when there is no interference; that is, when exchanges in one interval do not alter the frequency of exchanges in the next interval. The coefficient is less than unity when there is *positive interference*; that is, when one exchange reduces the chances of another exchange in the adjacent interval, as is normally the case for eucaryotic chromosomes.

The phage situation is more complex. A phage cross differs from a *Drosophila* cross because it provides, not one single opportunity for crossing over at meiotic prophase, but repeated opportunities for exchanges between multiple phage genomes, which usually are also being replicated. In addition, the nature of the act of genetic exchange is not known, not even whether each exchange yields one, two, or more recombinants. The first evidence that things are not simple is that for phage recombination the experimental frequencies $X_{a,b}$ $X_{a,c}$, $X_{b,c}$ reveal *negative interference*; that is, recombination in the *a, b* interval *increases* the chances of recombination in the *b, c* interval.

This situation has been interpreted in the following manner (Visconti and Delbrück, 1953). One assumes that between any two markers *a* and *b* there is a probability $p_{a,b}$ of recombination, comparable to a true map distance as in Equation 7-1. In order to relate the experimental values $X_{a,b}$ . . . to $p_{a,b}$ . . . one proceeds as follows. If *p* is the probability of a recombinant to originate in a mating act, then the total fraction of recombinants should be *mFp*, where *F* is the fraction of recombinants among the genomes that emerge from a recombinational event (for example, *F* = 1 for a breakage and exchange between two genomes); m is the "mating experience" or average number of mating acts that have occurred in the line of descent of the progeny particles emerging from a cross. The probability of a given progeny phage being nonrecombinant in the a, b interval is then $e^{-mFp_{ab}}$ (no exchanges in the a, b interval) and the probability of its being recombinant is

$$1 - e^{-mFp_{a,b}}$$

The relation between $X_{a,b}$ and $p_{a,b}$ can then be written

$$X_{ab} = k(1 - e^{-mFp_{ab}})$$

For small values of $mFp_{a,b}$ (and of $X_{a,b}$) this becomes

$$X_{a,b} \approx k(mFp_{a,b}) \qquad\qquad [7\text{-}2]$$

The near proportionality of $X_{a,b}$ to $p_{a,b}$ by Equation 7-2 justifies the use of experimental frequencies X of recombination for mapping so long as one uses only low X values, which are nearly additive, derived from experiments done under standard conditions.

The proportionality constant k includes a number of factors such as effects of unequal amounts of the two parent phages in the cross (which can be eliminated by using high equal multiplicities of the two parents). The most significant factor, which is introduced to account for negative interference, originated in what we may call "incomplete mixing." If not all the phages in a cross experiment have the same chance to engage in mating, then the ones that do so will undergo a greater than average numbers of matings and therefore of exchanges. Specifically, in a cross $abc \times a^+b^+c^+$ a progeny phage a^+b has a greater than expected chance of being a^+bc^+ for the very fact that it has participated in mating. (This is something like saying that for an individual who has had a child the probability of having more than one child is greater than for the average person in the population). Unequal distribution of mating events, therefore, can explain the observed negative interference.

Another source of negative interference, a tendency for repeated exchanges to occur within narrow regions in the course of a mating act, is called *high negative interference* and will be mentioned in a later section.

The factor m in Equation 7-2 must be estimated by experiment. Specifically, m, the average number of matings, can be evaluated from the dependence of the X values on the amount of phage produced. It turns out that the more phages are made in a mixed-infected bacterium, the greater is the proportion of recombinants, indicating greater mating experience per progeny phage (Levinthal and Visconti, 1953). The average value of m for T4 phage progeny in a normal burst has been estimated as about 3 under specific assumptions (Doermann, 1972). (The reader should keep in mind that in a phage cross matings and exchanges occur also among identical phage genomes, obviously without leading to genetic recombination.)

The conclusion that many recombinational events occur within a mixed-infected cell is directly confirmed by triparental recombination. Bacteria infected with three T4 mutants a, b, and c produce some progeny $a^+b^+c^+$ and abc. Recombination events actually leave direct evidence of their occurrence in the form of phages heterozygous for one or more genetic markers (Hershey and Chase, 1961). A phage *heterozygote* for markers a and a^+ produces in

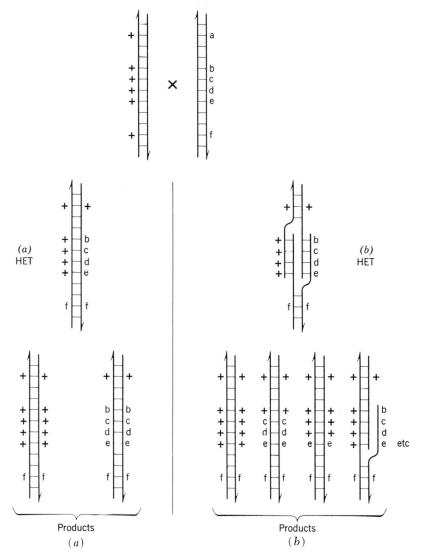

FIGURE 7-15. Schematic representation of two types of partially heterozygous (HET) phage genomes that can arise in a cross and of the progeny types they may generate. (a) heteroduplex type; (b) overlap type.

single infection a mixture of a and a^+ phages. There are two classes of heterozygotes: redundancy heterozygotes, with different alleles in the redundant ends, and true heterozygotes, with locally noncomplementary nucleotide sequences (Figure 7-15). All heterozygotes derive from recombinational events.

MOLECULAR ASPECTS OF DNA RECOMBINATION

Experiments with isotopically marked T-even phages have indicated that the replication of these phages actually involves physical fragmentation of the parental strands with exchanges of segments between phage DNA molecules (Kozinski, 1961). Any attempt to interpret genetic recombination more precisely in molecular terms meets the difficulty that DNA replication mechanisms are involved, which are themselves only partially known. The picture presented below is a condensed summary of conclusions reached in a recent review (Broker and Doermann, 1976).

Valuable information comes from the examination of DNA molecules from bacteria infected with damaged phage or with phage mutants which, under nonpermissive conditions, are variously impaired in DNA replication. Three classes of such mutants are recognized: DO or DNA negative (block in an early gene required for phage DNA synthesis); DD or delayed DNA synthesis; and DA or arrested DNA synthesis. Each class includes mutations in any one of several genes (see Figure 7-1) and each provides information about the effects of certain genes on DNA structure and on recombination.

The key features of the proposed sequence of events (Broker and Doermann, 1976; Figure 7-16) are the occurrence of single-strand "nicks" in the

FIGURE 7-16. *Idealized pathway for T4 recombination in the absence of DNA replication. From Broker and Doermann,* Ann. Rev. Genetics, **9,** *213 (1975).*

I.	*Intact circularly permuted T4 DNA.*
II.	*DNA with single-strand nicks.*
III.	*DNA with single-strand gaps and termini in strands of opposite polarities.*
IV.	*Branched DNA with a heteroduplex region formed by the pairing of complementary single strands.*
V.	*Branched DNA with a longer heteroduplex region and a displaced single strand ("branch migration").*
VI.	*Insertion heteroduplex.*
VI'.	*Overlap heteroduplex.*
VII, VIII.	*DNA with partially or completely repaired junctions.*

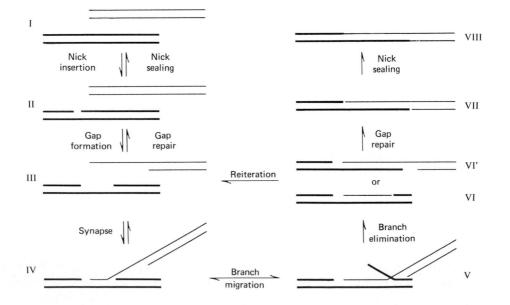

DNA and the special role of termini of DNA strands. Parental DNA molecules prevented from replicating by DO mutations undergo nicking and can therefore take part in the later steps of the process. These steps include: (1) gap formation by exonuclease action at the nicks; (2) pairing or "synapse" of a terminus complementary to the single-strands exposed in a gap; (3) strand displacement, which can extend the region of intermolecular pairing; (4) removal of unpaired ends by nuclease action; (5) formation of heterozygotes, which in successive replications generate recombinants. The DD and DO mutants affect recombination either by allowing for greater numbers of nicks and gaps or by increasing or decreasing their repair. This picture accounts for the presence of single- and double-strand branches in electron micrographs of replicating DNA. It also accounts for the molecular dispersion of parental DNA strands.

The special role of the termini in recombination and the elimination of some terminal segments explains instances of polarized loss of genetic markers located near the termini (Doermann and Boehner, 1963). Preferential pairing localized at termini and at gaps accounts for *high-negative interference* (Chase and Doermann, 1958) the tendency for multiple recombination events to occur within very small intervals in greater frequency than accounted for by the negative interference caused by incomplete mixing.

Finally, the model predicts that actual fragmentation will occur when replication is arrested at a damaged site. The partial replicas thus produced can participate in recombination and even reconstitute complete phage genomes. This is presumably the explanation of the phenomenon of *multiplicity reactivation*, that is, the reappearance of active phage in bacteria multiply-infected with irradiated, inactive phage (Luria and Dulbecco, 1949).

DELETION MAPPING AND GENETIC FINE STRUCTURE

Among the interesting features of phage genetics is the occurrence of genetic deletions, both spontaneous and induced by mutagenesis. The T-even headful mechanism that generates permuted, terminally redundant phage DNA segments makes it possible for phage to cope with any deletion that does not eliminate any essential gene function. The shorter genome will simply pack into a virion along with a longer terminal redundancy (Séchaud et al., 1965). A phage with a deletion in a nonessential gene can be viable if the deletion has no polar effects on other genes. In genetic crosses a deletion mutant fails to give recombinants with any mutant whose mutated site is comprised within the deletion (or overlaps it, if it also is a deletion mutant). The availability of a set of variously overlapping mutations within a gene makes it possible to map rapidly any point mutations located within the span encompassed by the deletions. Figure 7-17 illustrates deletion mapping for the *rIIA*, *rIIB* genes of T4 (Benzer, 1961). These famous genes owe their suitability for a variety of experiments to the circumstance illustrated in Table 7-4. New *rII* mutants are easily detectable by their large, rapid-lysis plaques and hundreds of them have been isolated and mapped. Any *rII*+ revertant or recombinant is easily detected as a plaque-former on λ-lysogenic bacteria (Benzer, 1957). The map of the T4 gene region *rIIA* and *rIIB* containing about 3000 nucleotide pairs is shown in Figure 7-18 and is the most detailed fine-

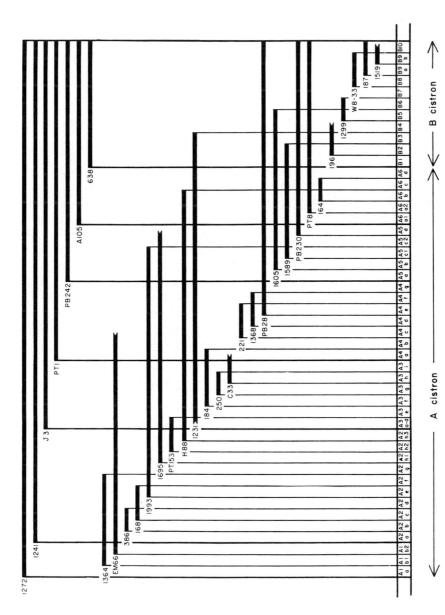

FIGURE 7-17. *Genetic deletions in the* rII *region of the genetic map of phage T4. Each deletion is indicated by a black line and the corresponding mutant fails to give* r+ *recombinants with any of the point mutants in the corresponding segment (see Figure 7-18). From Benzer, PNAS* **47**, *403 (1961).*

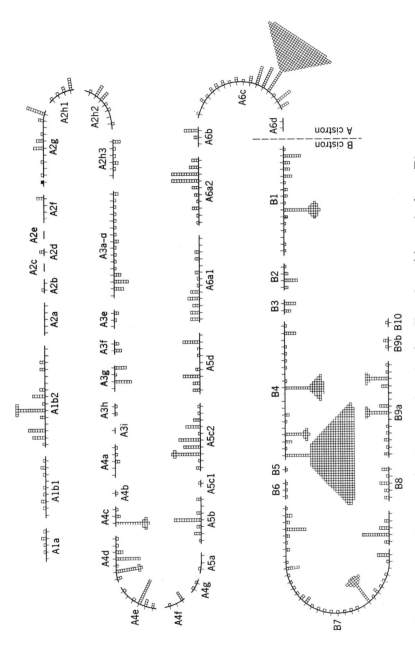

FIGURE 7-18. The map of point mutations in the rII region of bacteriophage T4. The segments of this genetic region correspond to those defined by the genetic deletions shown in Figure 7-17. From Benzer, PNAS **47**, 403 (1961).

TABLE 7-4   THE *rII* MUTANTS OF BACTERIOPHAGE T4

	Growth on host			Plaques on B
	B	K	K(λ)	
rIIA	+	+	−	large
rIIB	+	+	−	large
rII⁺	+	+	+	small
rIIA and *rIIB*	+	+	+	

structure map available for any one gene. The lowest recombination frequencies, about 2×10^{-5}, represent exchanges between two adjacent nucleotides.

Although the functions of the *rII* genes are not yet fully understood, they appear to relate to some of the membrane-associated phases of the phage program. The product of the *rIIB* gene is tentatively identified as an early phage protein associated with the membrane (Peterson et al., 1972).

It was on the basis of observations on the *rII* genes of T4 that Benzer (1957) defined the operational units of all genetic systems: the *muton* or unit of mutation (a nucleotide or a sequence of nucleotides in the case of deletions); the *recon* or unit of recombination (whose lowest limit is the interval between adjacent nucleotides); and the *cistron* or unit of function. The cistron is defined for two genes A and B by a *cis-trans* test in diploid state or in mixed infection: two function-abolishing mutations x^- and y^- are in the same cistron if the *cis* diploid x^+x^+/x^-y^- is functional and the *trans* diploid x^+y^-/x^-y^+ is nonfunctional (Figure 5-1). A classic application of the *rII* system was the demonstration that the genetic code is translated in commaless fashion starting from one fixed point for every gene (Crick et al., 1961). This experiment is illustrated in Figure 7-19. The mutations studied were a set of "frameshift" mutations induced by acridine dyes, which if present during DNA replication cause the addition or loss of single nucleotides or rarely two or more nucleotides. The set of mutations used was located in the "start" region of gene *rIIB*, a region that appears to be insensitive to most amino acid replacement mutations (probably it codes for a useless flap of protein at the N-terminal end). The critical finding is that within this dispensable region a frameshift mutation eliminates the *rIIB* function, but functionality may be restored by a second frameshift mutation within the same region. Also, three frameshift mutations that do not compensate for one another pairwise can sometimes restore the *rIIB* function when combined in a single phage.

The explanation (Figure 7-19) is that translation from *rIIB* mRNA (and generally from any mRNA molecule) always begins at a fixed point corresponding to the "start" of the gene, and then it proceeds so that each successive group of three nucleotides codes for an amino acid. If the initial segment of the protein product is unessential for the function of the protein, the requirements for successful function are that: (1) mutations in that segment do not give rise to nonsense codons that arrest translation; and (2) the mutations keep the "translation frame" in phase so as to avoid misreading of the rest of the gene. (Note that removal or addition of one or two or any

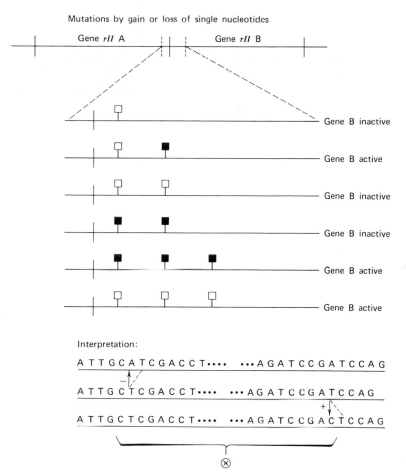

Mutations by gain or loss of single nucleotides

Interpretation:

A T T G C A T C G A C C T···· ···A G A T C C G A T C C A G

A T T G C T C G A C C T···· ···A G A T C C G A T C C A G

A T T G C T C G A C C T···· ···A G A T C C G A C T C C A G

FIGURE 7-19. *Scheme of experiments by Crick et al.,* Nature **192,** *1227 (1961) using frame-shift mutations to test the triplet nature of the genetic code.* □ = *addition (+) mutations;* ■ = *deletion (−) mutations. Certain pairs of + and − mutations compensate each other to give a functional gene; the requirement is that the "frame shift" between the two mutations does not generate any nonsense codon that stops protein synthesis. Some sets of three + or three − mutations also restore gene function.*

number of nucleotides other than $3n$ will generate "gibberish" all the way down.)

These experiments provided the first genetic support for the triplet nature of the genetic code. A direct confirmation of the theory was then provided by the T4 lysozyme gene (Streisinger et al., 1966). A lysozyme gene whose function had been abolished and then restored by two compensating frame-shift mutations, as in the *rII* experiments described above, produced an enzyme whose polypeptide chain differed from the normal one by a single sequence of five amino acids. The substitutions were exactly those predicted by the genetic code assuming two frameshift mutations of opposite sign 15 nucleotides apart.

Radiobiology of T-even phages

Radiation acts on cells and viruses primarily by damaging the nucleic acids. Ionizing radiation—X-rays, γ-rays and, indirectly, protons and other nucleons—causes single- and double-strand breaks. (It also causes production of free radicals in water, and these in turn can damage proteins or nucleic acid. In media with high contents of organic material this effect of radiation is usually negligible; Matheson and Thomas, 1960.) Ultraviolet radiation (UV) causes primarily formation of pyrimidine dimers, thymine-thymine or cytosine-thymine, from adjacent pyrimidines in the same DNA strand (Figure 7-20).

Bacterial cells, like most other cells, have repair systems that restore one-strand damages by ultraviolet or X-rays (two-strand scissions are not repaired). T-even phages and other large phages also have some genes for repair mechanisms. One repair mechanism is the enzymatic splitting of pyrimidine dimers. This enzyme is activated by visible light (photoreactivation; Kelner, 1949). Bacteriophages apparently make no photoreactivating enzymes of their own, but their DNA can be reactivated by the bacterial enzyme.

Another system is *excision repair*, which involves a variety of enzymes, some of which also play roles in genetic recombination. The mechanism can be represented in a very simplified way as in Figure 7-21. T-even phages have genes for at least some of the relevant enzymes. Such genes are identified by the effects of their mutations. Gene *v* codes for an endonuclease that attacks only DNA that has been irradiated (Friedberg and King, 1971). Since nuclease *v* does not free the dimers it must cut the DNA strand on one side of the dimers only. Genes 46 and 47 make an exonuclease that removes nucleotides from the dimer-containing strand. Other enzymes of known functions participate: DNA polymerase and polynucleotide ligase refill and seal the gap using as template the remaining strand. The process is certainly more complex still (Maynard-Smith and Symonds, 1973). It involves other proteins such as gp 32, which lines up along the single-strand gaps and protects them. Many other genes exist whose mutations increase the sensitivity to ultraviolet light, presumably by impairing the repair process (Hamlett and Berger, 1975).

The structures generated by the excision of damaged DNA resemble the

FIGURE 7-20. *Thymine dimers.*

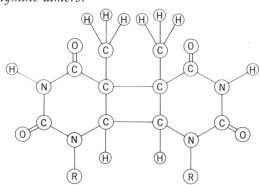

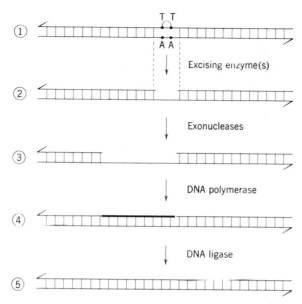

FIGURE 7-21. Scheme of repair of ultraviolet damage in DNA.

structures generated in the course of genetic recombination; both processes involve excision of redundant strands as well as rebuilding and sealing of new strands (Broker and Doermann, 1976). The excision repair system recognizes and corrects other local distortions in DNA strands besides the thymine dimers. For example, the repair system can carry out "loop removal," that is, excision of a single strand DNA loop opposite a deletion in a phage heterozygote (Benz and Berger, 1973). The nuclease gene v^+ is required as well as other repair genes. Repair of damaged DNA can also occur in T4 without excision, by the elimination of damaged segments in the course of recombination. Any damage to DNA molecules tends to increase the frequencies of recombination, by increasing the numbers and persistence of recombination structures.

RADIOACTIVE DECAY

The polynucleotide strands of DNA are held together by phosphodiester bonds. If the phosphate bond contains ^{32}P atoms, nuclear disintegration to give ^{32}S nuclei can result in a rupture of the polynucleotide strand. This "suicide" effect, first studied in phage T2 (Hershey et al., 1951; Figure 7-22), has since been observed in bacteria and other cells and viruses. Disintegration of incorporated ^3H atoms is much less efficient in damaging DNA, but can serve as a means to select mutants.

Inactivation of phage by ^{32}P radioactive decay is a one-hit phenomenon. The efficiency of inactivation for phages with double-stranded DNA is of the order of 10% for phage kept at 0° and 4% at −196°. (For RNA phages or for phages with single-stranded DNA the efficiency is unity, that is, every disintegration destroys infectivity.) The low efficiency of decay in double-stranded DNA is explained (Harriman and Stent, 1964) by postulating that ^{32}P

decay can produce three types of damages: (1) two-strand breaks, with inactivating efficiency near 1; (2) single-strand breaks, which can generally be repaired and are lethal only if they occur in the messenger-producing DNA strand of those genes that determine essential functions required for initiation of phage replication and repair; and (3) damages due to free radicals, protectable by radical-quenching reagents (Matheson and Thomas, 1960).

Decay of ^{32}P-labeled phage DNA after infection can be studied by deep-freezing bacteria infected with suicidal phage, allowing ^{32}P to decay, and after thawing assaying for phage-producing ability. This provides information on the role that the initial phage DNA is still playing at various times after infection (Stent and Fuerst, 1960). Integrity of the parental DNA appears to be needed only before replication of DNA has occurred.

Host-Induced Modification

The T-even phages, probably because of the HMC in their DNA, are not subject to the host-strain specific restriction phenomena described in Chapter 6. They do, however, exhibit a restriction system of their own.

It has already been mentioned that T-even DNA is variously glucosylated. If a T-even phage multiplies in a host bacterium unable to synthesize uridine-diphosphoglucose, the glucosyl donor for phage DNA, the phage DNA receives no glucose (Table 7-5). Phage virions are produced, but when they infect new host cells, they are usually rejected: the DNA enters and is broken

TABLE 7-5 HOST-INDUCED MODIFICATION AND RESTRICTION OF BACTERIOPHAGE T2

$$UTP + Glucose\text{-}1\text{-}P \xrightarrow{P} UDPG + iPP$$

$$UDPG + HMC\text{-}DNA \xrightarrow{G} Glu\text{-}HMC\text{-}DNA + UDP$$

$$Entering\ HMC\text{-}DNA \xrightarrow{R} fragmented\ HMC\text{-}DNA$$

| | Glucose on DNA | Plating efficiency on | | | | |
| | | E. coli | | | S. dysenteriae | |
Phage		B	B_{P-}	B_{R-}	S	S_{P-}
T2·B	+	1	1	1	1	1
T2·B$_{P-}$	−	10^{-5}	10^{-5}	1	1	1
T2·B$_{R-}$	+	1	1	1	1	1
T2G$^-$·B$_{R-}$	−	10^{-5}	10^{-5}	1	1	1
T2·S	+	1	1	1	1	1
T2·S$_{P-}$	−	10^{-5}	10^{-5}	1	1	1

Genetic blocks either in the bacterial enzyme P (UDPG-pyrophosphorylase) or in the phage enzyme(s) G (HMC-DNA glucosyl transferase) prevent glucosylation of phage DNA and expose it, upon entering a host cell, to degradation by restriction enzymes R, resulting in a 10^5-fold reduction in plating efficiency. P$^-$, G$^-$, R$^-$ indicate mutational losses of the respective enzyme activities. T2·B, T2·S, etc., stand for T2 grown on cells of strain B, strain S, etc.

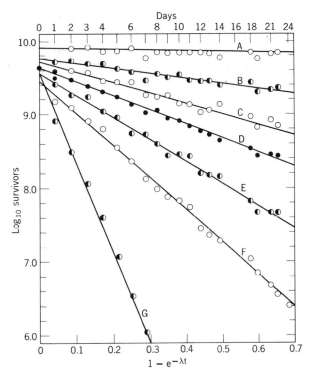

FIGURE 7-22. The "suicide" inactivation of bacteriophage T2 by decay of incorporated ^{32}P atoms. Lines A to G refer to preparations labeled in media with the following concentrations of P^{32} (in millicuries per mg phosphorus): A—0; B—10.5; C—19; D—30; E—45; F—58; G—>93. The origins of the lines have been adjusted to clarify the graph. The upper values on the abscissa correspond to the time allowed for decay; the lower values correspond to the proportion of radioactive atoms that have decayed. λ = radioactive decay constant for ^{32}P atoms = 0.0485 per day. From Hershey et al., J. Gen. Physiol. **34**, 305 (1961).

down by E. coli nucleases (Hattman and Fukasawa, 1963). Rejection is due to the products of two bacterial genes, r_6 and $r_{2,4}$ (Revel and Luria, 1970).

Some bacterial strains, however, are permissive for nonglucosylated phage, either naturally or because of mutations in the r_6 and $r_{2,4}$ genes. These bacteria make it possible to isolate and propagate phage mutants that have lost the ability to glucosylate their hydroxymethylcytosine (Figure 7-23). It is apparently the presence of nonglucosylated hydroxymethylcytosine at certain positions in the phage DNA that makes it vulnerable to attack by E. coli r_6 and $r_{2,4}$ enzymes (Revel and Luria, 1970).

These enzymes have not yet been fully characterized; at least one of them, r_6, is membrane bound (Fleishman and Richardson, 1971). Since T-even DNA when newly synthesized within its host cell is not glucosylated, it must somehow be protected from any cytoplasmic enzyme. It appears that normal T4 causes the production of an inhibitory protein that prevents breakdown of T4 DNA by the $r_{2,4}$ enzyme (Dharmalingam and Goldberg, 1976).

The interplay of phage-coded mechanisms with bacterial mechanisms revealed by this restriction—hydroxymethylation, glucosylation, inhibition of restriction on the side of the phage; two restriction enzymes on the side of the bacterium—suggests an interesting evolutionary history, including selection for mutual defense mechanisms.

Phage-Host Interaction Characteristics of Selected Phage Groups

8

Among virulent phages one encounters a whole range of developmental sequences differing to various extents from that of the T-even phages described in Chapter 7. The greatest differences, of course, lie in the size or nature of the phage genome and also the peculiarities of the phage capsids. In this chapter a few groups will be described whose features contribute to the achievement of a more complete picture of phage-host interactions. The T-odd phage groups to be discussed below have recently been the object of a comprehensive review (McCorquodale, 1975) and the genetic aspects have been reviewed by Hausman (1973).

Bacteriophage T5 Group

Phage T5, infective for coliform bacteria, with about 8×10^7 mol wt DNA with the ordinary nucleotides, has several features in common with T-even phages. Like them, T5 codes for many enzymes related to DNA biosynthesis; it too matures by a complex assembly process that includes proteolytic processing of some of the proteins and shows dependence on *gro* genes of the host bacteria (Zweig and Cummings, 1973). The DNA in T5 virions is peculiar in that it has several "nicks," that is, broken phosphodiester bonds, which can be sealed by polynucleotide ligase (Bujard, 1969). The nicks are all in the same DNA strand. There is no hint as to the functional role of this peculiarity.

The T5 genome can tolerate deletion of a substantial amount of DNA up to 10% from its central portion. Such long deletions are detected as phage virions resistant to heat in citrate buffer (Scheible and Rhoades, 1975). Evidently the region of DNA that can be deleted has no essential function in phage development, at least in normal *E. coli* cells as hosts.

The most unusual aspect of T5 phage is a two-step process of phage

197

injection from virion to host (Lanni, 1968). If the medium is deficient in Ca^{2+} ions or if protein synthesis is inhibited by chloramphenicol, the phage injects only a segment of DNA about 8% of the total length. The phage DNA is not circularly permuted, so that the injected part is always the same. Following this first-step transfer (FST) of DNA, if protein synthesis is allowed the genes of the FST segment of DNA are transcribed. There are at least three known genes in this segment; the unknown products of one or more of these genes causes the rest of the phage DNA to enter the cell. After FST and before the second step of injection the capsids can be sheared off the bacteria taking with them the noninjected DNA. Alternatively, one can denature the capsids after phage attachment and then DNA filaments remain attached to the cells and can even be made to enter following resumption of protein synthesis (Labedan et al., 1973). The most likely explanation is that for the second injection event some specific process must take place at the membrane and that the gene product(s) from the FST segment activates this process, which also has a requirement for Ca^{2+} ions.

It was believed at first that T5 provided the most direct evidence that injection of DNA from phage to cell was unidirectional, that is, always from the same end. It now seems well established, however, that T5 DNA has a terminal redundancy of about 9% of the genome, including the genes known to be in the FST (Rhoades and Rhoades, 1972). Hence the evidence for unidirectional injection is less persuasive.

One interesting aspect of the FST phenomenon is its possible similarities to events that accompany the transfer of DNA molecules among bacterial cells during mating. There too one observes an apparent, if not well understood, requirement for the expression of certain genes of the donor and the recipient cells (Hayes, 1968). Although there is no definite analogy in terms of specific gene functions, it is attractive to speculate that the similarities reflect some evolutionary relationship. What we see in the FST phenomenon of T5 may be a derivation of a DNA transfer mechanism that either evolved into a phage mechanism or was introduced into a phage genome by recombination.

The T7 Phage Group

These phages, including T7, T3, and many less well studied ones, have about 24×10^6 mol wt DNA and are of interest for a number of reasons. Their genetic program (Table 8-1) is expressed in a linear sequence from one end of the DNA genome to the other end (Studier, 1972). The first essential gene on the conventionally designated "left" end of the DNA, gene *1*, codes for an RNA polymerase whose specific presence is required to make mature phage. Conditional mutants in gene *1* do not grow under nonpermissive conditions, even though low levels of messenger and proteins of many of the later phage genes are made by host polymerase. The active T7 DNA polymerase consists of gp 5 and a host-coded subunit identified as *E. coli* thioredoxin (Mark and Richardson, 1976).

Gene *1* and the adjacent ones (Class I) are transcribed as early messengers by host polymerase, but gp 1 is required for full orderly transcription of genes in Class II and III. Class II has genes for DNA replication, Class III genes for

TABLE 8-1 PROTEINS SPECIFIED BY GENES OF T7 BACTERIOPHAGE

Gene	M.W.	Function
0.3	8,700	Nonessential
0.5	40,000	Nonessential
0.7	42,000	Host shut-off
1	100,000	RNA polymerase
1.3	40,000	Ligase
1.7	17,000	Nonessential
2		Reduced DNA synthesis
3	13,500	Endonuclease
3.5	13,000	Lysozyme
4	67,000	Reduced DNA synthesis
5	81,000	DNA polymerase
6	31,000	Exonuclease
7	14,700	Found in phage particle
8	62,000	Head protein
9	40,000	Head assembly protein
10	38,000	Major head protein
11	21,000	Tail protein
12	86,000	Tail protein
13	14,000	Found in phage particle
14	18,000	Head protein
15	83,000	Head protein
16	150,000	Head protein
17	76,000	Tail protein
18		DNA maturation
19	73,000	DNA maturation

From Studier, *Science* 176: 367, 1972.
The gene numbers correspond to the gene order in the T7 DNA molecules. The order of genes 0.3, 0.5, 0.7 is uncertain. M.W. = approximate molecular weights estimated by electrophoresis in SDS-polyacrylamide gels.

virion proteins. There is a "stop" signal between Classes I and II. If a genetic deletion removes this signal, early messenger RNA proceeds to be made along the entire genome by host polymerase.

The products of transcription by T7 specific RNA polymerase consist of 12 or 13 RNA segments, which are apparently produced by the cutting action of an mRNAse (RNAse III of *E. coli*) on long mRNA strands. Thus the genetic system of T7 appears to have only a few promoters, determining the starting points for long mRNA molecules. The correct programming depends on the stop signal that prevents early transcription by host polymerase.

The replication of T7 DNA in a newly infected cell starts at a specific site in the DNA molecules, which are not permuted and are about 0.6% terminally redundant (Ritchie et al., 1967). Replication proceeds bidirectionally to the ends (Wolfson et al., 1972). Later, concatemers appear, presumably formed by the mechanism proposed by Watson (1972; see Figure 8-1). Cutting of the concatemers into virion-length DNA segments for packaging must be done by

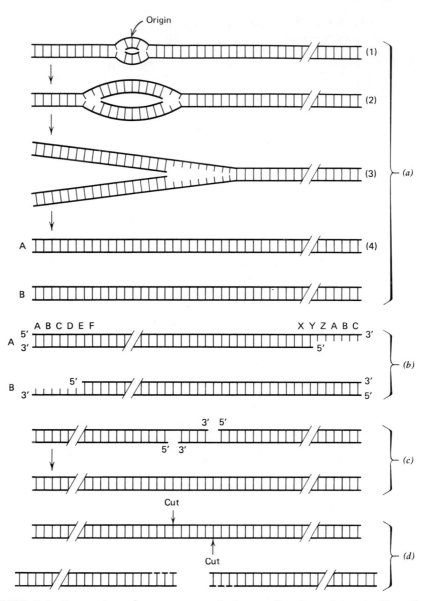

FIGURE 8-1. *Formation of concatemers during replication of terminally redundant, nonpermuted DNA molecules. (a) The first cycle of replication by bidirectional synthesis from one origin. The ends of the two daughter molecules A and B are illustrated in detail in (b); the single-stranded tails at the 3'—OH termini are generated by the failure of DNA polymerase to initiate synthesis at the 5' terminus without a primer (see Chapter 4). (c) How two molecules come together to form a dimer, which can then be sealed by the action of DNA polymerase and DNA ligase. (Excision of 3'—OH termini may be required if the single-stranded regions are shorter than the redundant section.) (d) The dimers and longer concatemers are finally converted to monomers with redundant ends by specific endonucleolytic action and resynthesis during phage maturation. After Watson,* Nature New Biol. **239**, *197 (1962).*

specific nucleases. In this case, however, the cutting mechanism must be different from the one that cuts headfuls off the T4 concatemers. Since all molecules have (approximately) the same ends and the same sequence, the cutting must be made by a mechanism that recognizes sequence rather than length. The cutting must probably be accompanied by polymerase action to restore the terminal regions to the full length.

T7 capsids contain a cylindrical protein core, apparently oriented specifically within the capsid and possibly involved in DNA cutting and/or folding within the virion.

Phage T7 (but not other phages of the same serologically related groups such as T3) has the peculiarity of being "female specific." It does not grow on *E. coli* strains that have the F mating factor (Blumberg and Malamy, 1974). In such bacteria, synthesis of phage components ceases about halfway in the latent period and a variety of low molecular weight cellular constituents leak out (Britton and Haselkorn, 1975). Mutations either in the F factor or in the phage eliminate the effect and allow phage to grow. A reasonable interpretation is that a gene of the F factor alters the cytoplasmic membrane in a way that makes it vulnerable to the product of a gene of T7. The T7 function might be direct or might be an interference with a bacterial function that maintains the integrity of the membrane in F-carrying cells. Other interpretations are possible, however. Phages related to T7 but not excluded by F lack the T7 gene.

Phage T3, a close relative of T7 (see below), has a peculiarity of its own: it induces in infected bacteria an S-adenosyl methionine cleaving enzyme or SAMase (Gefter et al., 1966). Because of the absence of SAM the T3-infected bacteria become nonrestricting for entering DNA from certain other bacterial strains since some restriction enzymes require SAM (Krueger et al., 1975). For T3 itself the only function of the SAMase appears to be allowing the phage to establish, in slowly growing cells, a so-called pseudolysogenic condition: the infected cells survive and can even divide several times before lysing (Fraser, 1957). Mutants of T3 without SAMase activity lyse promptly, like T7. Presumably, methylation either of DNA or possibly of a tRNA prevents the appearance of some phage function needed for prompt lysis.

The relation between member phages of the T7 group illustrates the use of DNA-DNA hybridization in the study of virus evolution. Hybridization, which detects full or partial homologies between DNA from two different sources, shows that T7 and T3 have very extensive regions of partial homology, as if they had diverged by many mutations from a common ancestor. Phage ϕ2, another member of the group, shows a mixture of regions fully homologous with T7 and regions without any significant homology (Hyman et al., 1973). This suggests that ϕ2 originated from T7 relatively recently by nonhomologous recombination with some other phage or possibly with bacterial or plasmid DNA.

The Phages with Single-Stranded Circular DNA

This group includes two main subgroups: phages with symmetrical icosahedral virions, prototype ϕX174, and male specific, filamentous phages, prototype f1. In both groups the virions have single-stranded, covalently closed circles of DNA, 1.5 to 2×10^6 daltons (5000 to 7000 nucleotides). At least some of these DNA molecules have one methylcytosine nucleotide at a fixed position: function (if any) unknown.

Entry of DNA from the symmetrical capsids occurs after one of the spikes, located at the 12 vertices of the icosahedron, interacts with the lipopolysaccharide (LPS) of the cell surface. Phage ϕX174 requires the entire LPS core, terminating in N-acetyl glucosamine, for specific recognition (Jazwinski et al., 1975a).

For the filamentous phages entry of DNA requires fixation of the virion to a male pilus (see Figure 6-2). Whether the capsid introduces its DNA into the pilus (both are protein tubes with a 2 nm hole) or the virion is pulled into the cell by the pilus is still uncertain.

The subsequent events are similar, at least in their broad lines, for both symmetrical and filamentous phages of this group. One of the proteins from the virion capsid, referred to as pilot protein, accompanies the DNA into the cell, anchoring it at first to some membrane site (Jazwinski et al., 1975b). Then the entering single strand, identified as the "plus" strand, serves as template for the synthesis of a "minus" strand to generate a double-stranded *replicative form* or RF. This replication involves only bacterial mechanisms since no transcription of the plus single strand has occurred. The RF synthesis has been studied in great detail (Figure 8-2) and duplicated *in vitro* using well-defined bacterial proteins (Schekman et al., 1974). Surprisingly, the specific requirements are different even for phages of the same group, probably reflecting differences in the specificity of the "initiation sequences" that serve as recognition sites for DNA polymerase, for RNA polymerases that make the RNA primers, and possibly for other required factors. It is even possible to carry out *in vitro* the conversion of single-stranded DNA in intact ϕX174 virions into RF in the presence of the various enzymes plus the LPS receptor plus a cell membrane fraction (Jazwinski and Kornberg, 1975).

Once the first RF is made, it replicates to yield many copies. Replication requires that one of the strands be nicked. The nicking occurs at a specific site in the viral strand of the RF by the endonucleolytic action of the pilot protein (Henry and Knippers, 1974).

The circular DNA molecules of these phages replicate by the rolling circle mechanism (see Chapter 4), which is compatible with either unidirectional or bidirectional replication. Rolling-circle DNA molecules are actually found in the infected bacteria. By a variety of configurations, about which there is still much uncertainty, rolling-circle structures generate new RF circles, long filaments that can again produce RF circles, and finally single plus strands for the progeny virions. Single strand production requires a special mechanism whose appearance is correlated in time and function with capsid assembly, either of the icosahedral or the filamentous types. The exact sequence of *in vivo* reactions is still uncertain. A simple overall scheme of RF and plus strand production has been proposed (Horiuchi and Zinder, 1976).

Figure 8-3 shows a recombination map of the ϕX174 genes and their function. Note that most genes relate to capsid components. In the icosahedral phage ϕX174 the capsid shell consists mainly of 60 molecules of one protein, called F. Since an icosahedron must have 12 pentons, that is, 12 groups of 5

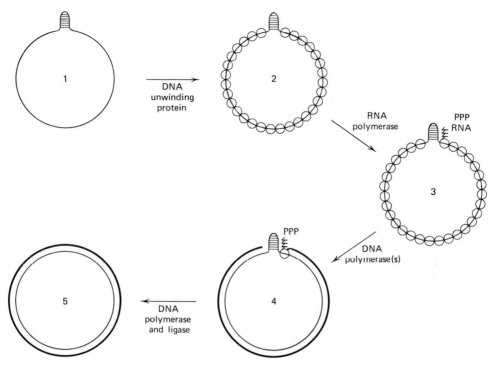

FIGURE 8-2. Scheme for conversion of single-stranded DNA of phage M13 to RF1. The hairpin structure includes the initiation site. Modified from Scheckman et al., Science 186, 987 (1974).

FIGURE 8-3. Genetic map of bacteriophage ϕX174 according to Sinsheimer, Handbook of Genetics 1, 323 (1974).

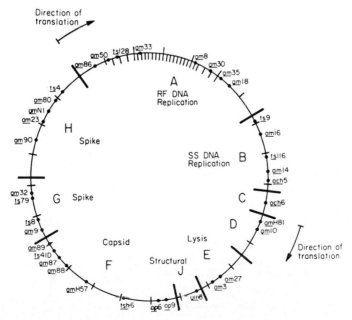

subunits arranged around a vertex, each penton must consist of five F molecules. Each spike supposedly consists of five molecules of G protein and five of J protein, plus also one molecule of pilot protein (Siden and Hayashi, 1974). Protein of gene *E* is needed for lysis.

A surprising and remarkable conclusion has recently been drawn from a study of DNA and amino acid sequences in φX174 genes and gene products (Barrell et al., 1976). The E polypeptide appears to be coded for by a portion of gene D. (The map of Figure 8-3 will require revision.) Translation of E starts

FIGURE 8-4. *A schematic illustration of some features of the life cycle of a filamentous male-specific phage. Successive stages* a *to* f *are drawn counterclockwise around the bacterium.* a: *The phage attaches to the sex pilus by means of the A protein;* b: *the phage DNA and A protein enter the bacterium;* c: *the phage DNA is converted to a duplex form, which replicates to give several hundred duplexes;* d: *a progeny duplex spins off a single-stranded tail that becomes coated with dimers of the protein of gene 5;* e: *the DNA is closed to give a circular DNA molecule with A protein attached;* f: *as the DNA passes out of the plasma membrane, the gene 5 protein is displaced by coat protein previously deposited on the plasma membrane. The phage is released without lysis or cell death. Modified from Marvin and Wachtel,* Nature **253,** *19 (1975).*

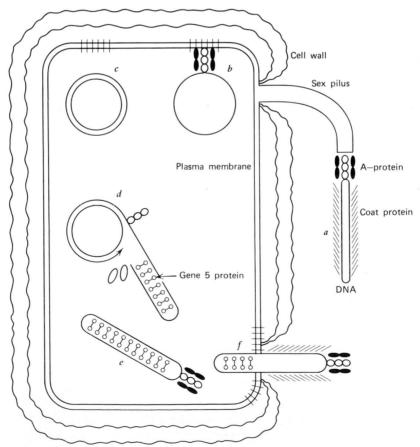

from an internal signal, which is frame-shifted by one nucleotide from the translation frame of gene D. The possible implications of such a situation for protein evolution are still unclear but may turn out to be far-reaching.

The filamentous phages of this group do not lyse their host cells, so that cell growth and phage production can continue for a long time. The newly made virions are excreted through the cell envelope. The filamentous capsid, about 1000 nm by 6 nm, consists of a tube of hydrophobic protein, 5000 daltons, almost 100% α-helical, plus a few molecules of the pilot protein. The hydrophobic protein is derived from a larger precursor polypeptide by removal of an N-terminal segment. In infected cells this protein accumulates in the cytoplasmic membrane, where it receives the single-stranded DNA (led by pilot protein). The coat protein there replaces a small protein that inside the cell protects the single-stranded DNA (acting like gp 32 of T4) and assembles around the DNA into a left-handed helix (Marvin and Wachtel, 1975). This sequence of events is illustrated in Figure 8-4.

Bacteria infected with filamentous male-specific phages are almost normal but, not surprisingly, exhibit altered membrane properties. At least six proteins are synthesized by an *in vitro* system from an f1 RF-DNA template. These include the two capsid proteins as well as others whose function has not yet been identified with that of specific genes (Model and Zinder, 1974).

The RNA Bacteriophages

RNA bacteriophages are important for a variety of reasons: (1) the replication of their RNA genome involves unique enzyme mechanisms; (2) since RNA genome is a template for both replication and translation, functioning as messenger RNA, special features are present in the developmental program; (3) the simplicity of the genome makes it possible to obtain a complete picture of the relation between genetic structure and function; (4) the expression of the RNA phage genome *in vitro* contributes powerfully to the analysis of mechanisms of protein synthesis; and, (5) these phages serve as models for many RNA viruses of practical importance. A detailed description of these phages is available in monograph form (Zinder, 1975).

All RNA phages are specific for bacteria with male pili (F pili in *E. coli* or their analogs in other organisms; see Fig. 6-2). All but one have in their virions a single-stranded RNA with messenger properties. (The exception is $\phi6$, a *Pseudomonas* phage with two-stranded RNA and, also an exception, a lipid envelope; Semancik et al., 1973.) The genome of most single-stranded RNA phages is 1.2×10^6 daltons, about 5000 nucleotides in an icosahedral virion about 4×10^6 daltons in weight and 25 nm in diameter. The virion consists of 180 molecules of the coat protein plus one molecule of A protein or "maturation" protein. Of the coat protein, 60 molecules form the 12 pentons at the vertices of the icosahedron, the other 120 form the hexons corresponding to the 20 faces. Coat protein monomers *in vitro* can form subassemblies that, with the RNA, can reconstitute a virion. Addition of maturation protein confers infectivity to the assemblies (Roberts and Steitz, 1967). Such direct assembly is possible because the single-stranded RNA provides a compact

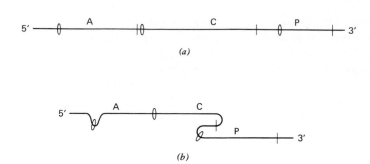

FIGURE 8-5. *Gene sequence in RNA phages. A = maturation protein; C = coat protein; X = not translated or product deleted; P = polymerase (= replicase). Ovals represent initiation sites. Secondary structure that partially restricts initiation at A and at P is indicated in* (b).

nucleating center for the coat protein shell. The host specificity of the various phages resides in their serologically distinct protein coats.

In the presence of a male bacterium the virion attaches itself to a pilus and delivers its RNA in a way that is not yet understood. Apparently the RNA enters the cell from its 3' end (Wong and Paranchych, 1976), together with its associated molecule of A protein. The next step, the expression of the phage RNA as messenger, is necessary for replication since the bacterium has no RNA replicase. Expression is not uncomplicated (Lodish, 1975). Of the three genes of the phage (Figure 8-5) only the coat protein initiation sequence is directly available to ribosomal attachment. Experiments with partially denatured RNA show that the initiation sequence for the replicase protein is blocked by secondary structure. It becomes available only after ribosomes, in translating the coat protein gene, open up the blocked sequence. The initiation sequence for the A protein is probably also partially blocked by secondary structure. Between coat protein gene and replicase gene there is a silent, usually noncoding region. In one phage called Qβ, however, this region is translated, although inefficiently, to produce a peptide that is incorporated into the virions but has no known function (Weiner and Weber, 1971).

By introducing whole phage RNA or fragments of it into a mixture of ribosomes plus all other factors needed for initiation of translation, but not the factors needed for extension of polypeptide synthesis, the initiation regions become caught in the ribosomes and protected against nuclease action (Steitz, 1975). In this way it has been possible to determine the nucleotide sequences of the three initiation sequences of several phages (Figure 8-6). In the newly entered RNA some of these sequences are blocked. Figure 8-7 presents a best-guess "flower" structure of the central part of phage MS2 RNA. A complete sequence of the entire MS2 RNA has also been published (Fiers et al., 1976) with a "bouquet" proposal of tridimensional structure.

The replicase protein fulfills multiple functions. It combines specifically, surprisingly, with two host proteins, the translation factors Tu and Ts, in a complex that associates itself with the initiation sequence of the coat protein gene, which is not at the 3' end of the phage RNA (Weber et al., 1964). Only then it begins to replicate the RNA, recognizing the 3' end of the RNA strand from the virion (plus strand) and synthesizing in the 5' → 3' direction a "minus" replica strand. The replicase can then copy either strand, the "plus" or the "minus." The 3' end sequence, at which replication must begin, is identical in both strands: —C—C—C, to which before encapsidation an A nucleotide is added, not by template transcription but by a special enzyme. At first "minus" strands are produced by copying "plus" strands supposedly membrane bound. Later a number of minus-strand-directed *replicative intermediates* (Figure 8-8) generate many plus strands, which combine with coat protein to form virions.

The overall program is regulated not only by the conformation of the RNA but also by the remarkable properties of the phage-coded proteins: the replicase combines with the coat protein initiation sequence to inhibit its translation, and the coat protein inhibits translation of the replicase! The A protein, needed in 1:180 ratio to coat protein for the virions, is made in a somewhat higher ratio *in vivo*: either the regulation is less perfect or the A protein has some additional function in the program.

It is remarkable how much functional wisdom has accumulated in so little

FIGURE 8-6. *Initiator sequences of the three genes of RNA phage R17 and one of the possible secondary structures for each of these sequences. According to Steitz, in* RNA Phages, *Cold Spring Harbor Laboratory, p. 319 (1975).*

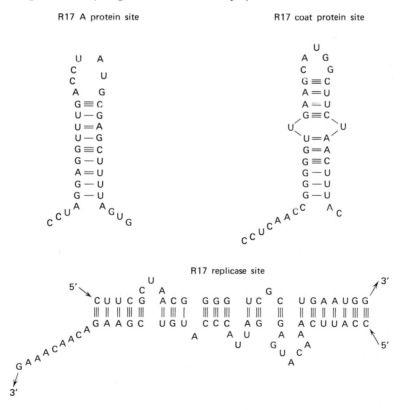

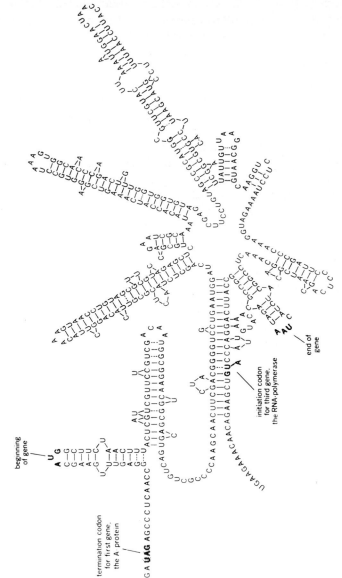

FIGURE 8-7. The "flower" model for the RNA of phage MS2. The nucleotide sequence of the last part of the A protein gene, the first intercistronic region, the coat gene, the second intercistronic region and the first part of the polymerase gene is given in the form of a secondary structure. The initiation codons and termination signals are shown boxed. Modified from Fiers, in RNA Phages (N. Zinder, ed.) Cold Spring Habor Laboratory, p. 353 (1975).

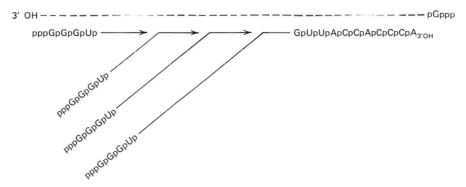

FIGURE 8-8. *Structure of the MS2 RNA replicative intermediate. The broken line stands for parent strand.*

genetic material. Even more remarkable is that the genome of an RNA phage can "evolve" *in vitro* into smaller elements (Kacian et al., 1972). By repeated selection of RNA molecules that replicated faster *in vitro* in the presence of a limiting amount of replicase enzyme, shorter and shorter deletion mutants were isolated, finally down to a few hundred nucleotides. These, of course, had no coding information left for useful proteins but must have retained the specific sequences that the replicase recognizes. This experiment has been called an example of test-tube evolution, the shorter strands having an advantage in the competition for replicase molecules. It is certainly an instructive model of how a genome may lose functionally dispensable segments under conditions of replicative stress.

9 Bacteriophage λ

Bacteriophage λ is a *temperate* phage; that is, it can either be transmitted from cell to cell by infection or passed from mother to daughter within a cell line. The latent phage genome in such a cell line is called a *prophage*. The cells harboring the prophage are *lysogenic*. The presence of the phage genome within a lysogenic culture can be detected by the spontaneous liberation of phage from a small fraction of the cell population in which viral development is spontaneously activated.

Because the genetics of its natural host, *E. coli* K12, is well developed, λ was chosen as the object of concentrated efforts to understand lysogeny. The original wild type K12 strain is lysogenic for λ. Phage λ does not form plaques on this lysogenic strain, which, like most lysogenic bacteria, is immune to infection by phage of the carried type. Phage λ is generally grown on K12 derivatives that have been cured of the prophage. Such cured derivatives are found, at low frequency, among the survivors of high doses of irradiation.

Every temperate phage must solve two basic problems if it is to produce stably lysogenic cell lines. First, the prophage must establish itself in such a manner that, at cell division, each daughter cell receives at least one prophage copy. Phage λ achieves this end by inserting its DNA into the continuity of the bacterial chromosome, so that the prophage DNA is passively replicated and segregated by host machinery. Second, those viral genes whose products are potentially disruptive to cellular integrity must be so regulated that the cell can survive, grow, and divide. This is accomplished by repression of gene transcription.

In a cell lysogenic for λ, none of the viral genes needed for productive infection is transcribed. Very little viral mRNA is detectable in lysogenic cultures. Most of the mRNA that is formed comes from one λ operon that includes two genes: *cI*, which determines a repressor, and *rex*, whose product prevents T4 *rII* mutants from plating on λ lysogens (for unknown reasons).

The rex product has no demonstrated function in the λ life cycle. The repressor coded by gene cI prevents transcription of the rest of the viral genome. Inactivation of repressor induces phage production in the entire cell population (reviewed in Hershey, 1971).

The simplest way to inactivate repressor is by heating a bacterial culture lysogenic for a λ mutant whose repressor is thermolabile. Such mutants have been isolated by screening for variants that form clear plaques at high temperature but turbid plaques at low temperature. Wild type λ forms turbid plaques at both temperatures, because, during the development of a plaque, some of the infected cells, rather than lysing and liberating phage, survive and grow as immune lysogens. A culture of bacteria lysogenic for a cIts mutant can be grown indefinitely at 30°. If the temperature is raised to 43° for a few minutes, repressor is inactivated, and the same temporal program of transcription is initiated as in an infected cell.

A more general method of destroying repressor (applicable to wild type λ and some other temperate phages) is to expose lysogenic cells either to a relatively small dose of ultraviolet light or to various chemical agents. The amount of UV used would kill only a small fraction of cells if applied to a nonlysogenic culture, but it induces viral development (and therefore death) in the majority of lysogenic cells. Rather little is known about the pathway whereby UV triggers repressor destruction. The wavelengths of ultraviolet light most effective in induction are those maximally absorbed by nucleic acid, suggesting that nucleic acid is the primary chromophore. Direct irradiation of the prophage is unnecessary; mating a λ lysogen with irradiated F⁺ bacteria, or infection by irradiated P1 phage, can induce λ development. Thus it seems likely that either irradiated DNA or cellular by-products thereof can activate repressor degradation. Many other inducing treatments (such as thymine starvation of a thymine auxotroph, or heating of lysogenic bacteria whose host carries a thermosensitive mutation in the gene for polynucleotide ligase) resemble UV in inhibiting DNA replication, thus some products of arrested DNA synthesis may be the actual inducers.

However induced, repressor degradation takes place by proteolytic cleavage generating a specific fragment of the repressor molecule (Roberts and Roberts, 1975). Some of this cleavage product is found even in untreated cells. The product of the host recombination gene recA is needed for cleavage; in recA mutants, neither spontaneous nor induced cleavage is observed. Certain mutations within the cI gene (called ind for "noninducible") render the repressor resistant to such cleavage. Whether the inducing agents cause the production of a specific protease or sensitize repressor to a preexisting one is not known. It is also not established for certain that proteolysis of repressor is the cause rather than the consequence of derepression.

Because repressor is present, lysogenic bacteria are specifically immune to superinfection by phage of the carried type. If a bacterium lysogenic for λ is infected by another λ particle, the entering DNA is repressed and cannot initiate a productive cycle of viral development. When repressor is inactivated, this immunity is lifted. Thus infection of a derepressed lysogen with a λ genetically distinguishable from the prophage leads to liberation of phage of both the superinfecting and the carried type.

Lysogeny is very common in nature. As with most parasites, the most stable host-virus associations probably represent balanced situations in which the virus does minimal damage to the host. Since the prophage has little visible effect on the growth of the lysogenic culture, its presence is frequently unnoticed. The K12 strain of *E. coli*, for example, had been the object of intensive experimentation for over 20 years before it was discovered to harbor the λ prophage. Some isolates of almost all bacterial species have proven, on close inspection, to be lysogenic for one or more phages. Many different temperate phages have thus come to light, some of which will be discussed in the next chapter.

All temperate phages face the same general problems of natural survival, although they vary in the strategies used to solve them. The nature of both the problems and the solutions is better appreciated by analyzing one particular phage in depth. For this purpose, detailed knowledge of the viral genome, and the functions and interactions of all its gene products, is essential.

As we shall see, such analysis of λ has revealed regulatory circuitry of considerable sophistication. The observed interactions can be rationalized after the fact, but their exact role in the natural life of the virus is still unknown. We may expect that other phages, when examined by equally sensitive methods, will prove as complex as λ is.

Genetic Methods

The analysis of many λ mutants (especially conditional lethals of the amber or thermosensitive types) has allowed the identification of 39 different genes and their accurate location along the DNA molecule (Figure 9-1). The physical positioning of genes in terms of molecular distances has been achieved mainly through the use of genetic deletions and substitutions.

For example, the *cI* gene of λ can be located by studying recombinants formed between λ and a related phage, 434. Phage 434 is one of several temperate phages independently isolated from nature that can recombine with λ. Phages λ and 434 differ in several ways, one of which is the specificity of repression by the *cI* product. Phage 434 forms plaques on bacteria lysogenic for λ, and vice versa. Likewise, purified λ repressor protein can bind *in vitro* to λ DNA, but not to 434 DNA. From a series of backcrosses between 434 and λ, a hybrid phage was isolated in which all of the DNA is derived from λ except for a small segment (called the immunity region) that includes the *cI* gene, the operator sites at which repressor acts, and some closely linked genes.

The DNA of this λ-434 hybrid phage (called λ*imm*434) has been separated into single strands and allowed to reform double helices with the complementary strands of λ. Such heteroduplexes (Figure 9-2) are perfectly paired except for one segment in which about 2400 λ nucleotides have been replaced with 1500 nucleotides from 434. This technique located the *cI* gene between 21% and 27% of the distance from one molecular end. Using the immunity marker as a reference point, other substitutions or deletions can be located along the DNA. The physical map emerging from such studies (Figure 9-1) is in general quite similar to the linkage map generated from tabulations of recombination frequencies.

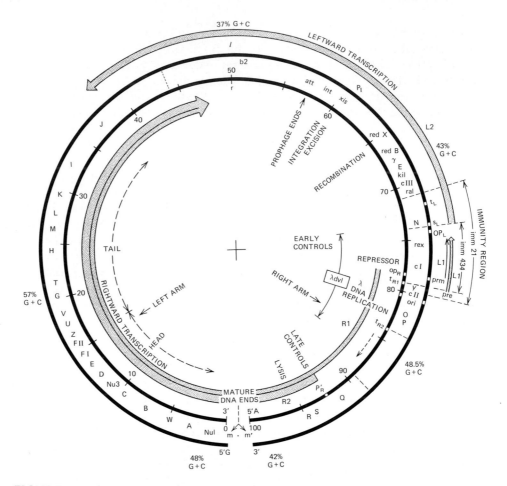

FIGURE 9-1. *Genetic map of λ (modified from Szybalski, Handbook of Genetics 1, 309, 1974), showing the locations on the phage DNA of genes (A, W, B, etc.) and target sites for interaction of macromolecules with DNA (op$_R$, att, etc.). Because the viral DNA is linear in the virion but circularizes upon infection, it is drawn here as an open circle with the left end (m) close to the right end (m'). A leftward direction along the linear map thus corresponds to a counterclockwise direction in the circular representation.*

In λ*imm*434, a DNA segment from λ has been replaced by a functionally equivalent portion of the 434 genome. Thus the hybrid phage, though different from λ, can form plaques and lysogenize as λ does. On the other hand, many λ variants in which DNA segments have been deleted or substituted by nonviral DNA are unable to complete their infectious cycle because genes essential for normal development are absent. Variants unable to form plaques are termed *defective* phages. Bacteria harboring such incomplete phage genomes are called *defective lysogens*.

In order to use deletions or substitutions for locating genes on the DNA molecule, one must not only visualize the deletion in DNA heteroduplexes, but also determine which λ genes have been removed. Experimentally, the

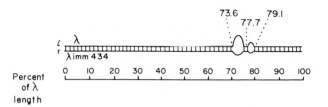

FIGURE 9-2. *Diagrammatic representation of a heteroduplex formed between the* l *strand of* λ *and the* r *strand of* λimm434. *The two phages are homologous throughout their length except for the DNA between 73.6% and 79.1% of the molecular length of* λ. *This DNA segment of* λ *contains the* rex, cI, *and* cro *genes and the operator-promoter sites* op_L *and* op_R. *The corresponding segment of* λimm434 *contains* cI^{434} *and* cro^{434} *genes, whose products specifically recognize the* op_L^{434} *and* op_R^{434} *sites but not* $op_L^{λ}$ *and* $op_R^{λ}$. *Phages 434 and* λimm434 *have no* rex *function, and their DNA is correspondingly shorter. The constriction at 77.7 is due to a short segment of homology, about 140 base pairs long, within the* cI *gene (Wilgus et al.,* Virology **56,** 46, 1973).

latter is ascertained by the ability of the variant phage to recombine with various point mutants and generate wild type progeny. (An example is given in Table 9-2.) For defective deletion phages superinfected with conditionally lethal phage mutants, the method is extremely sensitive, because under nonpermissive conditions only the wild type recombinant can grow. The approach is exactly analogous to the deletion mapping of the *r*II region of phage T4 described in Chapter 7.

Deletions and substitutions allow the physical localization not only of genes but of messenger RNAs as well. Thus messenger coding for the *cI* gene will be complementary to λ DNA, but not to DNA from λ*imm*434.

Preparations of the two complementary DNA strands can be obtained separate from each other, because of a small difference in base composition between them. The strands are designated *l* and *r*, respectively. The *l* strand runs in a 5' to 3' direction as we go from left to right (clockwise) on the map of Figure 9-1, and is thus transcribed leftward (counterclockwise). The *cI* message hybridizes to the *l* strand, and thus must be transcribed leftward.

INITIAL STAGES OF INFECTION

The λ virion comprises an icosahedral head (T = 7), 54 nm in diameter, enclosing a DNA molecule of 30.8×10^6 daltons; and a flexible tail about 180 nm long, that terminates in a single fiber (reviewed in Casjens and King, 1975). The DNA has a unique sequence and is double-stranded throughout its 46.5 kb length, except for the 12 bases at the 5' termini of the polynucleotide chains. These 12 bases are single-stranded, and the single strands at opposite ends are complementary to each other. Thus DNA extracted from virions can establish an equilibrium between linear molecules and doubly nicked rings (Figure 9-3). At high concentrations, end-to-end aggregates also can form.

Adsorption of λ virions to the cell surface requires specific interaction between the tail fiber protein (product of the λ *J* gene) and a cellular protein

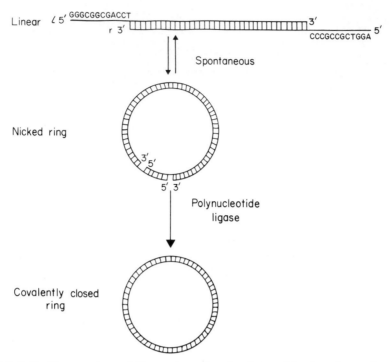

FIGURE 9-3. *Conversion of linear λ molecules into nicked and covalently closed rings. DNA extracted from virions can spontaneously equilibrate between linear molecules and nicked rings. Nicked rings formed following injection are converted enzymatically into closed rings.*

that also determines chemotactic response to the disaccharide sugar maltose. Some bacterial mutations to λ resistance alter this protein, whereas others inactivate the regulatory genes controlling maltose utilization.

The λ DNA is arranged within the virion with a specific end of the DNA molecule (the right end of the genetic map drawn in Figure 9-1) in close proximity to the head-tail junction. This end can be chemically cross-linked to the proximal tail proteins with formaldehyde (Saigo and Uchida, 1974), and is accessible to digestion by micrococcal nuclease in heads that have not yet combined with tails (Bode and Gillin, 1971). Where within the head the other DNA end is located is unknown. Whereas this regular arrangement might suggest that injection would always proceed right end first, X-ray inactivation studies indicate that injection may proceed from either end (Sharp et al., 1971). DNA injection is specifically reduced by certain host mutations of unknown biochemical nature (Scandella and Arber, 1974).

After injection, intracellular λ DNA is found as rings whose nicks have been covalently closed by the host enzyme polynucleotide ligase (Figure 9-3). When viral development is induced in a λ lysogen, the prophage seems to be excised from the chromosome as a ring identical to that found after infection. The productive cycle of viral development, whether initiated by infection or by induction, thus starts from the same DNA template and seems to follow the same programmed sequence of gene expression.

EARLY INTRACELLULAR DEVELOPMENT

In a lysogenic cell, the amount of λ-specific message is very small. Most of this message hybridizes to the *l*-strand of λ, and not to that of the λ*imm*434. This message comes from transcription of the *cI* and *rex* genes. A small amount of *l*-strand message from outside the immunity region comes from transcription of the *int* gene, whose product catalyzes the insertion of viral DNA into the host chromosome (Shimada and Campbell, 1974).

The rest of the viral genome is not transcribed because the repressor binds to two operator sites (op_L and op_R, Figure 9-1), within the immunity region, and prevents transcription by host RNA polymerase. The nucleotide sequences of these repressor-binding sites have been determined. To do this, advantage was taken of the fact that bound repressor protects the operators from degradation by added nucleases. Both op_L and op_R turn out to consist not of a single repressor-binding sequence, but rather a series of three such sequences, (Figure 9-4).

Following infection (or induction of a lysogen), host RNA polymerase binds to op_L and op_R DNA and begins transcription leftward from op_L and rightward from op_R. The polymerase binding sites, identified physically as DNA segments protected by polymerase from nuclease digestion, and genetically by promoter mutations that decrease the rate of transcription, lie within the operator regions upstream from the initiation point for transcription (Ptashne et. al., 1976).

This initial transcription from op_L and op_R sets in motion a chain of events that can lead to phage reproduction and eventual destruction of the cell. (What happens in those newly infected cells that choose the alternative course of becoming lysogenic will be discussed later.) The first genes to be transcribed (*N* and *cIII* in the leftward transcript and *cro* and *cII* in the rightward) produce regulatory proteins that influence the timing of events in the productive cycle and also, together with *cI*, constitute a switching mechanism that allows stable commitment to either the productive or the lysogenic cycle. The qualitative effects of these and other regulatory proteins are shown in Table 9-1.

The *N* protein alters the character of transcription initiated at op_L and op_R so that the RNA polymerase no longer responds to normal signals for transcription termination. In an N^- mutant, leftward transcription stops somewhere between *N* and *cIII*, and most rightward transcription terminates between *cro* and *cII*. When gpN is present, these transcripts do not terminate, but continue more or less indefinitely. The precise mechanism of antitermination is not known. It is specific for transcription initiated at op_L and op_R, and allows readthrough of sites that would cause termination if transcribed from other promoters (Franklin, 1974).

The λ transcripts formed by an N^- mutant *in vivo* are the same ones that are produced when λ DNA is transcribed *in vitro* by RNA polymerase in the presence of rho, a bacterial protein that is required for correct termination. The gpN must in some way render transcription resistant to rho. It cannot simply inactivate rho, however, because its effect is strictly confined to transcription initiated at op_L and op_R. Furthermore, various λ-related phages isolated from nature make gpN of different specificity. Phage 21, for example, will grow in mixed infection with λ, and the two phages can recombine with

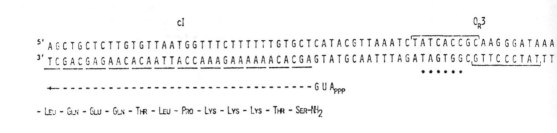

cI O_R3

5' AGCTGCTCTTGTGTTAATGGTTTCTTTTTTGTGCTCATACGTTAAATCTATCACCGCAAGGGATAAA
3' TCGACGAGAACACAATTACCAAAGAAAAAACACGAGTATGCAATTTAGATAGTGGCGTTCCCTATTT

- GUA$_{PPP}$

- LEU - GLN - GLU - GLN - THR - LEU - PRO - LYS - LYS - LYS - THR - SER-NH$_2$

O_L3

5' TAAAAAACATACAGATAACCATCTGCGGTGATA
3' ATTTTTTGTATGTCTATTGGTAGACGCCACTAT

FIGURE 9-4. *Nucleotide sequences of repressor binding sites in* λ. *Top line:* op$_R$ *and adjacent DNA. Bottom line:* op$_L$. *In order to align equivalent sections of the two operator regions, the orientation of* op$_L$ *is inverted with respect to the map order of Fig. 1. The promoter mutation* sex1 *lies 31 bases upstream from the origin of transcription of the* N *operon. (Modified from Ptashne* et al., *1976.)*

each other. However, the gpN of phage 21 will not extend transcription of an N^- λ phage in the same cell, or vice versa. Each gpN must therefore recognize some specific sequence in the op_L and op_R sites and somehow modify the nature of transcription initiated at those sites so that rho is unable to cause termination of it.

The *cro* gene product has not been completely characterized biochemically, so its role in λ development is inferred from the behavior of *cro*$^-$ mutants. Its main effect on the productive cycle is to inhibit transcription from the promoters op_L and op_R (Takeda et al., 1975). This inhibition increases with time, as though gpcro were accumulating in the cell. As the *cro* gene itself is transcribed from op_R, the expression of the rightward operon is self-limiting. It cannot sustain its maximum potential rate of transcription, which therefore decreases later in the cycle.

Once gpN appears, transcription continues leftward through genes *ral*, *cIII*, *kil*, *gam*, *redX*, *redB*, *xis*, and *int*. Most of these genes determine products that act in recombination or related processes. None of them is essential for viral development. Phage variants from which all these genes have been deleted or substituted still form plaques, although *red*$^-$ and *gam*$^-$ mutations decrease the burst size.

Transcription extension by gpN was the first case in which a regulatory protein was observed to exert its effect on a control point within a transcript,

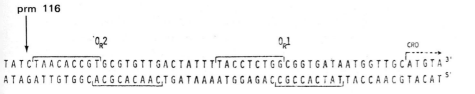

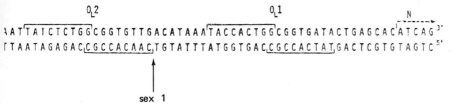

rather than at its origin. Subsequently, controlled attenuation of transcription has been discovered in bacterial operons (Bertrand et al., 1975) and may be a common regulatory device. Such systems enlarge the concept of the operon as a unit of coordinate expression. Depending on which regulatory genes are active, two genes separated by such a control point may behave sometimes as though they belonged to the same operon, and at other times as though they did not.

The novel properties of N control in fact came to light originally as an

TABLE 9-1   REGULATORY PROTEINS OF PHAGE λ

Regulatory gene	Site of DNA interaction*	Transcript affected*	Nature of effect
cI	prm	L1	Activation and repression[+]
	op_L	L2	Repression
	op_R	R1	Repression
cII and $cIII$	pre	L1'	Activation
N	op_L	L2	Extension beyond t_L
	op_R	R1	Extension beyond t_{R1} and t_{R2}
cro	prm	L1	Repression
	op_L	L2	Partial repression
	op_R	R1	Partial repression
Q	p_R'	R2	Activation

* Locations of genetic sites and transcripts are given in Figure 9-1. L1 and L1' are both leftward transcripts that include genes cI and rex, but have different origins. L2 includes genes N through int. R1 is a rightward transcript of genes cro through Q. R2 includes the "late" genes $S-J$.
+ The direction of the effect of cI repressor on prm depends on repressor concentration.

unexpected deviation from simple genetic expectations. According to the methodology devised by Thomas (described in Hershey, 1971), λ genes of the productive cycle can be classified into two groups. Genes of the first group are directly under repressor control, because they are transcribed from promoters op_L and op_R at which repressor acts. Repressor control of genes in the second group is indirect; these genes are unexpressed in lysogenic cells because their transcription is activated only by products of genes that are themselves under direct repressor control. The basic experimental test was to infect a bacterium lysogenic for λ, without induction, with a mixture of wild type λ and a λimm434 phage that carried a mutation inactivating the gene under examination (Figure 9-5). The hybrid phage, not being repressed, should carry out any steps prior to the mutant block. It still cannot turn on a gene like the N gene of

FIGURE 9-5. Experimental test for direct repressor control of λ genes. Cells lysogenic for λ, and thus containing λ repressor, are infected with a mixture of λX⁺ and λimm434X⁻ where X is the gene under test. In the upper panel (X = N), transcription of gene N is directly blocked by repressor. No gpN protein is formed, hence the infection is unproductive. In the lower panel (X = R), the λimm434R phage makes Q protein, which activates transcription from p_R' on chromosomes of both genotypes. Endolysin is thus formed from the R⁺ gene, and infection is productive. Because repression blocks replication directly, the λ chromosomes do not replicate. Therefore the yield consists predominantly of λimm434R⁻ phage.

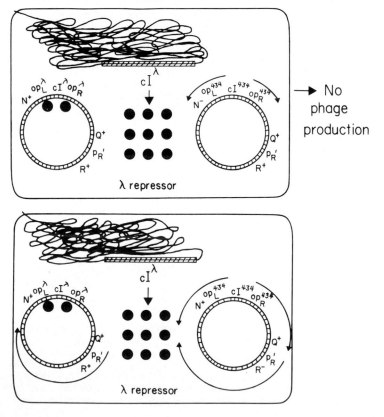

a coinfecting λ phage, because the N gene is transcribed from a promoter that is blocked by λ repressor. On the other hand, a gene like the endolysin gene R (which, we will see later, is controlled indirectly) can be transcribed from a phage that is still repressed at op_L and op_R. Infection of a bacterium lysogenic for λ with a mixture of $λR^+$ and $λimm434R^-$ thus results in endolysin synthesis and phage production.

When this test was applied to the redX gene by measuring its product, exonuclease, the result was that exonuclease was not made from a repressed template, indicating that it was under direct repressor control (Luzzati, 1970). On the other hand, it was equally clear that N^- mutants failed to make exonuclease, indicating that exonuclease synthesis required prior expression of the N gene. The conclusion was paradoxical in terms of the classical operon concept; the redX gene, like N, must be transcribed from op_L, yet redX transcription required prior formation of gpN.

Why does λ introduce this added complexity into early operon transcription, rather than simply evolving a leftward operon with no termination sites that must be overridden by gpN? One reason may be to stabilize the repressed condition. If a lysogenic cell is momentarily depleted of repressor, some early λ message may be formed. However, this transcription will not initiate viral development unless derepression is continued long enough so that gpN can be formed, and new transcripts can then be made under its influence. It thus provides a double-lock mechanism to stabilize the lysogenic cell against accidental vicissitudes in repressor synthesis.

Replication and Recombination

After gpN is formed, rightward transcription continues through genes cII, O, P, and Q. The products of genes O and P initiate DNA replication at a site (called "ori" for "origin") between cII and O. Replication initiated at ori is generally bidirectional. At early times, phage DNA can be extracted from the cell as replicating "theta forms." The positions of the replication forks with respect to ori in any given molecule can be deduced by subjecting the extracted molecules to partial denaturation. (Chapter 4.)

Replication of λ DNA depends on transcription from op_R in two distinct ways: first, because the products of genes O and P are needed; second, because transcription of the ori site itself promotes replication initiation there. This conclusion has been deduced from experiments in which the same cells were infected with a mixture of phages, only one of which was undergoing transcription and producing O and P products. For example, bacteria lysogenic for λ (and therefore containing λ repressor) have been infected with a mixture of λ and λimm434. The λimm434 replicates in this situation. The superinfecting λ does not. Even though gpO and gpP are present, replication is not initiated on the λ molecule, because its ori site is not being transcribed. The mechanism of transcriptional activation is unknown, but may be related to the general need for RNA primers in DNA synthesis (Shekman et al., 1972). Phage λ fails to replicate in certain host mutants that are blocked in chain elongation (for example, dnaB ts mutants at high temperature), indicating that it uses some of the replication machinery of the host in the synthesis initiated by the viral proteins gpO and gpP.

Later in infection, replication generates molecules longer than the λ genome, in which unit genomes are arranged end to end. These long DNA chains sometimes comprise tails attached to circular molecules, indicating a rolling circle mode of replication. As will be discussed later, these long chains probably comprise the direct precursor of progeny virion DNA, which is cut at specific sites as it is packaged.

Somewhere during intracellular DNA development, genetic recombination can take place. Neither the time of occurrence nor the biochemical mechanism is yet understood. Recombination can follow at least three distinct pathways, distinguishable by their genetic requirements. (1) The host genes (recA, recB, and recC) can allow recombination among phage mutants having no known recombination pathway of their own. (2) In rec⁻ hosts, λ can still recombine, due to the presence of genes redX and redB. (3) Inactivation of both rec and red pathways still allows recombination at the site at which λ DNA inserts into the host chromosome during lysogenization. This recombination, as we will see later, requires a specific viral gene, int.

Mutants with inactive redX or redB genes have lower burst sizes than wild type, and are deficient in their ability to generate rolling circle forms late in infection (Skalka and Enquist, 1974). This suggests that the principal function of these gene products is to resolve replication intermediates in single infection rather than to reshuffle genes within natural populations.

Most of the recombination observed in infection of rec⁺ E. coli by red⁺ λ probably occurs by the red pathway, because one component of the rec pathway, exonuclease V (coded by genes recB and recC) is specifically inactivated by the protein product of a λ gene, gam. Exonuclease V is believed to destroy some intermediate in λ replication, and gpgam prevents this destruction by neutralizing the exonuclease. In the presence of exonuclease V (for example, in infection of the recBC⁺ host with a gam⁻ mutant of λ), an active recombination pathway becomes essential for λ development, perhaps by providing a route that circumvents the sensitive intermediate. Thus a red⁻ gam⁻ double mutant fails to plate on a recA⁻ recBC⁺ host.

On a wild type host, λ red⁻ gam⁻ grows very poorly and forms tiny plaques. These conditions are strongly selective for additional mutations (chi mutations) that create "hot spots" for rec-promoted recombination. Chi mutations can occur at several specific sites on λ DNA. This selective system has allowed the detection of natural "hot spots" of the same type in bacterial DNA, which can become incorporated into λ transducing phages (to be discussed later). Natural chi sites may account for much of the recombination observed in bacterial crosses. They act as sites of initiation for recombinational events, although the actual crossover may take place some distance from the chi site that initiated it (Lam et al., 1974).

Whatever the steps in recombination, the genetic map of λ, recombining by either red or rec pathway, is linear; Recombination may take place between circular molecules, but the DNA cuts that occur during packaging effectively constitute a mandatory crossover at the molecular ends of virion DNA.

Phage λ was the first organism in which it was demonstrated directly that recombination can involve physical breakage of parental DNA molecules

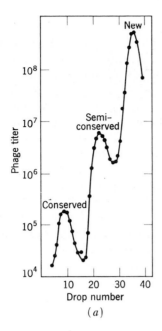

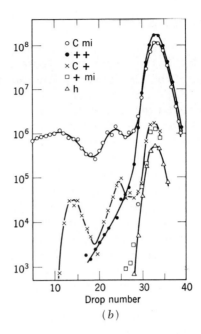

(a)   (b)

FIGURE 9-6. *The distribution of parental and recombinant phage particles in the progeny of a cross between heavy* λ *cI mi and light* λ ++. *On the left is a diagram showing the positions in a CsCl gradient of phage particles with conserved-heavy, semiconserved, and new (= light) phage from an infection with heavy phage alone in light bacteria and light medium. Note on the right the presence of both parental and recombinant phages with conserved or semiconserved DNA. The* λ h *was added as a density marker for light phage.* mi *is a plaque-morphology mutation lying near gene* R *on Figure 9-1. The host-range* (h) *mutation is in gene* J. *From Meselson and Weigle,* Proc. Natl. Acad. Sci. **47,** 857 1961.

(Figure 9-6). Mixed infections were made between parents of different isotopic composition, so that the physical origin of recombinant molecules could be identified. Normal infection entails extensive DNA synthesis, so that most progeny DNA of all genotypes is composed of atoms derived from the medium rather than from infecting particles. The methodology can thus tell us that some recombinants arise through DNA breakage, but is technically incapable of showing that they all do.

More recent investigations have examined recombinant production under conditions of restricted DNA synthesis (λ O^- or P^- mutants infecting a *dnaB$^-$* mutant of the host). Such studies have been complicated by the fact that λ monomer circles are not packaged into infectious particles, so that the small number of virions formed constitute a biased sample with respect to both replication and recombination. The results indicate that recombination by *red, rec,* and *int* pathways proceed by physical breakage of parental DNA, but that the *red* pathway differs from the others in that each recombinational event typically yields a single recombinant product, rather than a pair of reciprocal recombinants (Stahl and Stahl, 1974).

Virion Formation

At about the same time that rolling circles appear, the protein constituents of virions start to accumulate. These proteins are determined by genes *A–J* at the left end of the λ map. The product of gene *Q* is required to initiate transcription of these genes, as well as that of genes *S* and *R*, whose products are needed for cellular lysis. In *Q⁻* mutants these genes are all transcribed at about one-tenth the normal rate. The small amount of *Q*-independent transcription is not seen in *N⁻* mutants and presumably represents rightward transcription from op_R that has been made termination resistant by gpN (reviewed by Herskowitz, 1973).

Although *S* and *R* are at the opposite ends of the genetic map from genes *A–J*, they become contiguous within the cell, once the ends are joined to form a closed ring. All the genes from *S* through *J* constitute one operon, as judged by the behavior of lysogenic strains in which prophage DNA between *Q* and *S* has been lost by genetic deletion. If all DNA between *Q* and *S* has been deleted, none of the prophage genes *S–J* is expressed, even when the cell is infected with a phage that produces *Q* product.

Even in the absence of *Q* product, a small *r*-strand messenger RNA molecule complementary to DNA between *Q* and *S* is formed (Roberts, 1975). One possible mechanism for *Q* function is that, acting as gpN does, it extends this short transcript into a much longer one. Another possibility is that gpQ is a new RNA polymerase, or a subunit thereof. This hypothesis is not necessarily distinct from that of an antiterminator, because an antiterminator must interact with polymerase at some stage.

Nine phage genes (*A–F_{II}*), plus one host gene, *groE*, are needed for construction of normal heads, while tail assembly requires eleven genes (*Z–J*) (reviewed in Casjens and King, 1975). The protein products of most of these genes are known, and *in vitro* assembly studies define the roles of some of these proteins (Figure 9-7).

Five proteins (gpE, gpB, gpC, gpgroE, and gpNu3) interact to form empty protein shells (petit λ), which are found in all normal λ lysates, and which can interact *in vitro* with λ DNA, gpA, gpD, gpW, gpF_{II}, and tails to give complete infectious particles. The first five proteins are thus needed to form petit λ, and

FIGURE 9-7. Pathway of in vitro *assembly of infectious λ heads. From Kaiser et al., J. Mol. Biol.* **91,** *175, 1975.*

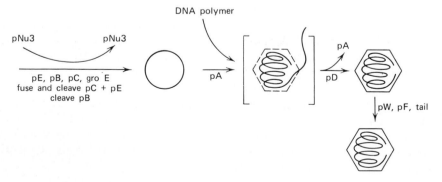

only for that purpose. One of them (gpNu3) is not a constituent of petit λ, but is required for its assembly. The major coat proteins, gpE and gpD, are present in about 420 copies per virion.

As in T4 assembly, packaging involves concomitant protein processing. Two kinds of processing are seen in λ: cleavage and splicing. Many examples of cleavage are known from other systems. In λ assembly two proteins coded by separate genes (*C* and *E*) are joined, apparently by covalent bonds, to generate one polypeptide chain.

Once protein shells have been formed, viral DNA is packaged inside them. The immediate precursors of virion DNA are long concatemers, probably tails of rolling circles, in which viral genomes are arranged end to end. These molecules are cleaved into monomers with single stranded ends as they are packaged. Although end-cutting usually accompanies packaging, it can take place at a low rate *in vitro* in the presence of the λ gene *A* product, gpA. As gpA is not a component of the finished virion, it must be lost after cutting has taken place. The gpD protein is added to the capsid at this stage. End-cutting takes place at specific sites on the DNA (called *cos* for "cohesive sites"), so that λ, unlike T4, has a unique DNA sequence in the virion. One consequence is that, in λ, genetic variations such as deletions, additions, or substitutions that alter the DNA length result in packaging of unit genomes of other than wild type length.

That the normal substrate is a concatemer is inferred from the fact that closed monomer circles are not packaged, either *in vivo* or *in vitro*. With a rolling circle mode of replication, packaging can take place from the tails, while the circles continue to spin out additional DNA without themselves being packaged.

Tail assembly requires the products of 11 λ genes, which can interact to form functional tails *in vitro*. Assembly begins with the fiber protein, gpJ, which, after interacting with several other proteins, causes polymerization of the principal tail component gpV. Finished tails contain about 135 molecules of gpV and a few molecules each of gpJ and gpH.

Lysis

Liberation of λ from the cell at the end of the productive cycle requires two λ proteins, gpS and gpR. Protein gpS appears to digest or damage the cytoplasmic membrane (similar to the t gene product of phage T4). Protein gpR is an enzyme, endolysin, which, like the lysozyme of T4, digests the rigid mucopolypeptide layer of the cell envelope. Endolysin differs from a true lysozyme in that it causes the hydrolysis of a peptide bond, whereas lysozymes split between adjacent N-acetylglucosamine residues. Mutations that inactivate gene *S* allow intracellular phage development to continue for several hours, resulting in very high yields of phage per infected cell.

The Pathway to Lysogeny

The events described up to this point take place in those infected cells that produce virus. It is implicit in the temperate nature of λ that only a fraction of cells enter the productive cycle, whereas other cells survive and multiply as lysogenic bacteria. In a lysogen, prophage is inserted into the chromosome and repressor prevents transcription of most of the viral genome. Both the establishment of repression and insertion into the chromosome are specifically controlled by λ genes.

The fraction of infected cells entering the lysogenic cycle varies with cultural conditions. It increases with multiplicity of infection, concentration of magnesium ion, and intracellular concentration of cyclic AMP (Fry, 1959; Grodzicker et al., 1972). Under optimal conditions, more than 90% of the cells become lysogenic. In its potentiality for stable commitment to one of two alternate pathways against a constant genetic background, the temperate phage-infected cell bears a formal similarity to other differentiating systems.

Immediately after infection, the cell can go in either direction—toward virus production or lysogeny—and the probability of choosing one over the other can be altered by varying external conditions. Once formed, a lysogen is not destabilized by any of the conditions that influence the establishment of lysogeny. Two separate questions can be asked: First, how does the infected cell decide to become lysogenic rather than to enter the productive cycle? Second, after the decision has been made, how is the lysogenic condition maintained?

Insight into the first question has come from the study of mutations that strongly decrease the frequency of lysogenization. Such mutations cause formation of clear, rather than turbid, plaques. Clear plaque mutations fall into four groups: cI, cII, $cIII$, and c_y. The cI gene, as mentioned earlier, determines the structure of the repressor protein. The products of cII and $cIII$ genes are needed to establish, but not to maintain, repression. These mutants form lysogens only rarely (unless complemented by a coinfecting phage), but the lysogens, once formed, are stable.

In mixed infection c_y mutants complement cII and $cIII$, but not cI, mutants. However, genetic mapping places them on the other side of the cro gene from cI itself. As their phenotype resembles that of promoter mutations, their locus is called the promoter for repressor establishment (pre). It is not presently known whether pre is the origin of repressor message or a site for antitermination of message starting further to the right. A small leftward message starting from near gene O is found both in infected cells and in in $vitro$ transcription of λ DNA (reviewed in Szybalski, 1974).

Genes $cIII$ and cII belong to the early leftward and rightward operons, respectively. Following infection and transcription of gene N, these genes are among the first to be included in the extended transcript. In cells destined for lysogeny, these products stimulate transcription of the cI-rex operon, from the pre sites. As repressor accumulates, it binds to op_L and op_R, thus stopping further transcription of genes of the productive cycle.

In so doing, repressor also blocks transcription of the cII and $cIII$ genes, whose products are thus eventually not available to induce cI transcription. The cI gene continues to be transcribed, however, because of a second promoter (called "promoter for repressor maintenance," prm), located next to op_R. Transcription from prm is repressible by high concentrations of repressor (Ptashne et al., 1976), whereas low concentrations of repressor may stimulate transcription. Repressor thus participates directly in its own maintenance.

The *prm* initiation site ("GUA$_{ppp}$" in Fig. 9-4) lies between the *cl* gene and a strong ribosome binding site. Hence messages originating at *prm* are translated less efficiently than those coming from *pre*, in keeping with the idea that less repressor is required to maintain the repressed state than to establish it.

Besides *N, cI, cII,* and *cIII,* another gene, *cro,* influences the establishment of repression. The *cro* product binds to the same sites as repressor itself, and inhibits transcription from these sites.

The ability of gp*cro* to inhibit *cI* transcription from *prm* allowed demonstration of the fact that λ development, like that of a differentiating cell, can be channeled into mutually exclusive steady states. In normal λ infection, the cell chooses between two pathways. One of these, the lysogenic pathway, leads to a stable state that is perpetuated indefinitely during cellular replication. The other alternative, the productive pathway, comprises a chain of events that culminates in lysis and phage production. However, a prophage carrying appropriate mutational blocks in early steps of viral development can achieve a stable state following selection of the productive pathway. The essential thing is to inactivate those genes whose expression in a prophage is lethal to the cells. Two such lethal functions are known: replication (which is controlled by genes *O* and *P* and is lethal when induced on an inserted prophage) and some processes of unknown nature controlled by genes (including gene *kil*) of the early leftward operon. If these genes have been inactivated by mutation, or if their expression is curtailed by the absence of gpN, a lysogenic strain can survive and grow indefinitely without repression—as, for example, when a lysogen of a *cIts* mutant is maintained at 43°C.

If an *N⁻P⁻cIts* lysogen, growing at 43°, is shifted back to 30°, where repressor is stable, it continues for many generations in a derepressed or "anti-immune" state. This condition requires gp*cro*; an *N⁻cro⁻P⁻* lysogen regains immunity within an hour or two after shifting down. The critical function of gp*cro* is to inhibit spontaneous transcription originating at prm. This gives the *cI-cro* interaction the characteristics of a bistable switch. In cells where *cI* is transcribed, repressor is made, which stimulates its own synthesis and represses transcription of *cro*. On the other hand, where *cro* is transcribed, its product prevents repressor formation. Thus a cell growing at 30° may be in either of two states, depending on its previous history, and can pass from one state to the other only with great difficulty. This experimentally contrived situation presumably results from the same interactions that follow upon infection with wild type λ and cause a cell to become locked into either the productive or the lysogenic pathway.

Lambda lysogenizes by inserting its DNA into the host chromosome. In insertion, host and viral DNA break and rejoin at specific sites. Viral DNA is in its circularized form at the time of insertion. The end result (Figure 9-8) is that prophage DNA becomes intercalated between host genes. The prophage is thus a linear segment of the bacterial chromosome, but one whose gene order is cyclically permuted from that in the virion. Evidence for this mode of insertion comes from crosses between lysogenic strains, deletion mapping of the chromosomes of lysogenic bacteria (Figure 9-9) and examination of DNA heteroduplexes between phage and prophage (Figure 9-10).

If the viral and host sites participating in insertion were physically

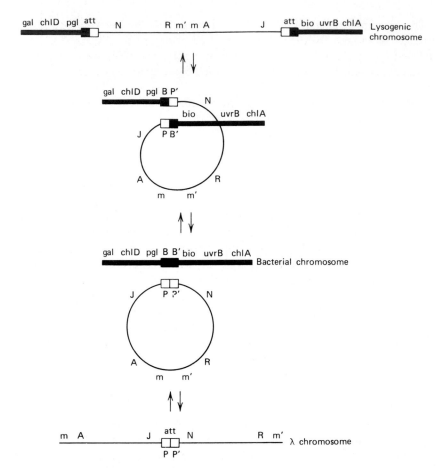

FIGURE 9-8. Insertion and excision of λ. The bottom line represents λ DNA as it occurs in the virion. After infection and circularization, λ DNA is broken at a specific site (P.P'), and rejoined to bacterial DNA at a specific site (B.B'), thus splicing λ DNA into the continuity of the host chromosome. In excision, the final step is reversed, regenerating the bacterial chromosome and the circular λ DNA molecule, which may then replicate and generate progeny virions. Bacterial genetic symbols are gal, a cluster of three genes whose products function in the metabolism of the hexose sugar galactose; chlD and chlA, genes whose products are components of the nitrate reductase system, absence of which renders the cell resistant to chlorate under anaerobiosis; pgl, structural gene for phosphogluconolactonase, bio, cluster of five genes coding for enzymes that function in the biosynthesis of the vitamin biotin, uvrB, gene whose product functions in repair of DNA damage such as that caused by ultraviolet light. From Hershey, The Bacteriophage λ (1971).

identical, insertion might result from conventional recombination between homologs. In fact, however, the actual similarity between host and viral sequences at the point of joining is too slight to be detectable by heteroduplex formation, and mutations that abolish ordinary recombination (such as rec^- mutations of the host) have no influence on the process. Insertion relies rather on a viral protein (integrase) that catalyze genetic exchange at the specific site. The integrase gene (int) lies immediately adjacent to the site of insertion.

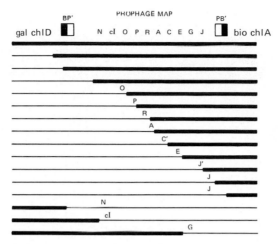

FIGURE 9-9. *Deletion mapping of* λ *prophage. Each line of the figure represents a different bacterial isolate selected as a chlorate-resistant mutant of a lysogenic strain. Thus each deletion (whose extent is indicated by the solid bar) eliminates either* chlA *or* chlD, *or both. Deletions marked* C' *or* J' *terminate within genes* C *and* J, *respectively. Data from Adhya et al.,* Proc. Nat. Acad. Sci. **61,** *956, (1968).*

Such site-specific exchange has the advantage that the virus, by regulating the amount of integrase, can control the time of insertion. Maximum expression of *int* takes place after infection, and depends on leftward transcription from op_L. Once repression is established, this transcription ceases. In a lysogenic cell, *int* is transcribed, albeit at a much lower rate, due to transcription from a promoter close to the *int* gene itself.

When a lysogenic cell is derepressed, phage production depends on the ability of prophage DNA to reverse the insertion process and come out of the bacterial chromosome. Excision from the chromosome requires not only integrase but also another phage-coded protein (excisionase) coded by a second viral gene (*xis*), that is located next to *int*. Transcription of *xis* is exclusively from op_L.

The difference in catalytic requirements for excision and insertion reflects the fact that phage and bacterial sites are nonidentical. We can designate the phage attachment site as "P.P'" and the bacterial site as "B.B'", where "." indicates the crossover point, "P" and "P'" refer to DNA sequences immediately to its left and right, in the phage; and "B" and "B'" are the corresponding sequences in the bacterium. The insertion reaction then becomes

$$P.P' + B.B' \rightarrow P.B' + B.P'$$

whereas excision can be represented by

$$B.P' + P.B' \rightarrow P.P' + B.B'$$

The nucleotide sequences of these four sites have recently been determined (A. Landy and W. Ross, unpublished). The crossover point, indicated here by "." actually comprises a 15-nucleotide segment that is common to all four sites.

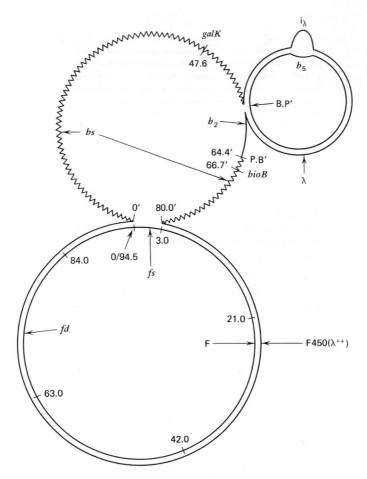

FIGURE 9-10. Diagrammatic representation of three-way heteroduplex formed between single strands of DNA from the fertility factor F, F' gal (λ), and the λ double mutant λb₂b₅. Single strands of F and F' gal (λ) were prepared from covalently closed molecules by gentle treatment with endonucleases, to nick at least one of the two polynucleotide chains. F' gal is a derivative of F generated by abnormal excision of the F factor, so that a segment of E. coli DNA including the λ insertion site has become fused into F. (Cf. Figure 9-15.) When bacteria harboring F' gal are infected with λ, the λ can insert into the F', in the same manner that it normally inserts into the chromosome. The λ b₂ mutation is a deletion that terminates exactly at the insertion site, whereas b₅ is a substitution in the immunity region (like imm434). The b₅ substitution loop serves as a cytogenetic marker, which is the expected physical distance from the insertion site. Based on Sharp et al., J. Mol. Biol. **71**, 499 (1972).

The phage can potentially regulate not only the extent of the reaction, but its direction as well. Exactly how λ decides for insertion following infection and for excision following induction are not understood in detail.

When λ lysogenizes E. coli K12, the λ prophage always inserts at the same site on the bacterial chromosome, between the *gal* and *bio* genes, and in the same orientation. This site is thus preferred by the integrase enzyme. Other, less preferred potential insertion sites are revealed upon infection of bacterial strains from which the preferred site has been eliminated by a genetic deletion (Shimada et al., 1973). Of particular interest are sites that lie within known bacterial genes. For example, in various lysogens λ is inserted into the *galT*, *trpC*, and *ilvA* genes. The genes into which λ has become inserted make no functional product and distal genes of those operons can no longer be transcribed from their normal promoters. Following derepression of such lysogens the prophage may be excised. If the culture is derepressed transiently (as by heating a *cIts* lysogen for a few minutes and then returning it to low temperature for growth) some "cured," nonlysogenic survivors are formed by the action of integrase and excisionase. In these cured derivatives the functions of the interrupted genes are restored. This indicates that insertion and excision are very precise processes, neither of which destroys or mutates DNA in the vicinity of the insertion site.

Specialized Transduction

Excision is generally, but not invariably, precise. The most readily recognized products of imprecise excision are molecules that include some viral DNA (and therefore can be packaged into virions) and some bacterial DNA adjacent to the prophage (that can be detected by the phenotype it confers on recipients bearing appropriate mutations). For example, in a lysate prepared by inducing a lysogen, about one λ particle in 10^5 includes the nearby *gal* operon. When a gal^- recipient infected with such a lysate is placed on medium containing galactose as carbon source, some of the cells that received such particles grow up into Gal^+ colonies.

This ability of phage prepared from a lysogen to transfer specifically those genes close to the prophage attachment site is called *specialized* (or *restricted*) *transduction*, to distinguish it from the *general transduction* carried out by phages such as P22 or P1 (to be discussed later). Specialized transduction results from two distinct types of abnormal particle, each stemming from a different type of aberration from the normal growth pattern of the phage:

IMPRECISE EXCISION

As indicated in Figure 9-11, imprecise excision generates a new type of genome (part viral, part host) that can undergo all the steps of normal λ development such as replication, packaging, and lysogenization, provided only that a wild type λ is present in the same cell to supply the products of any viral genes that have been deleted from the transducing phage. Lysogenization by such a transducing phage produces Gal^+ transductants that are diploid for the *gal* genes. Such strains give rise, by internal recombination during bacterial growth, to haploid Gal^- derivatives.

PACKAGING FROM THE LYSOGENIC CHROMOSOME

In occasional cells, DNA packaging precedes excision. Normal packaging requires two *cos* sites, but from a chromosome with a single *cos* site, rare

TABLE 9-2 GENETIC CONSTITUTIONS OF DEFECTIVE λ BIO PHAGES

λ bio isolate	Genetic markers								
	Phage					Bacterial			
	Nam7	cI47	Oam8	Oam125	Qam21	bioA4	bioC23	bioC18	bioD19
M3-29	−	+	+	+	+	+	+	+	+
M30-7	−	−	+	+	+	+	+	+	−
M42-2	−	−	+	+	+	+	−	−	−
R30h-2a	−	−	−	+	+	+	+	+	+
E5a-20	−	−	−	−	+	+	+	−	−

Cells lysogenic for different λ bio phages (isolated as transductants on a $bioA^-$ recipient) were derepressed, infected with the indicated viral mutants, and scored for production of wild type recombinants ($+ = am^+$ or cI^+ recombinants formed). Lysates containing each λ bio phage were used to infect bacterial strains carrying the indicated bio^- mutation, to see whether Bio^+ bacterial recombinants were produced. The results establish that the genes are arranged on the lysogenic chromosome in the same order as the columns of the table. They also show that there is no fixed relationship between the point of breakage in viral and bacterial DNAs. Data from Kayajanian (1968).

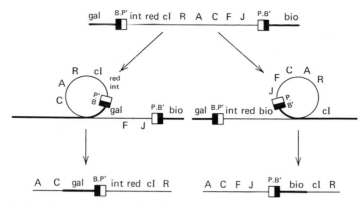

FIGURE 9-11. *Genesis of* λ gal *and* λ bio *transducing phages, by rare abnormal excision from a* λ *lysogen. From Hershey,* The Bacteriophage λ (1971).

virions are recovered in which one end lies at a *cos* site and the other within host DNA. The DNA of such virions cannot circularize, replicate, or lysogenize, but it can cross over with the recipient chromosome to replace recipient genes with their homologs from the donor.

Because they cannot replicate, specialized transducing particles produced by abnormal packaging have only a transient existence. On the other hand, those arising by improper excision can be propagated to give a large number of identical DNA molecules carrying a defined segment of the host genome. The λ*gal* phages obtained from ordinary lysogens are always defective as viruses and can be propagated only in the presence of a co-infecting helper phage. This is because the DNA length from the *gal* operon to λ gene *J* exceeds the packaging capacity of the λ head. Starting with hosts in which some bacterial DNA between *gal* and λ has been removed by previous deletion, plaque-forming λ *gal* phages are obtained. Plaque-forming λ variants carrying the *bio* genes (Figure 9-11) are readily obtained from wild type lysogens.

Specialized transducing variants can also be derived from strains carrying λ in sites other than the preferred one. In that case, the host genes that become incorporated into λ are those adjacent to the prophage of the particular lysogen used.

Specialized transducing derivatives of λ and other temperate phages have found several important uses. First, because each transducing phage contains a specific bacterial segment, these phages comprise a convenient source of large quantities of host DNA sequences for molecular studies. For this purpose, transducing phages arising by abnormal excision have, in some cases, been supplanted by those generated through enzymatic joining of viral and bacterial DNAs (Thomas et al., 1974). Second, because each isolate contains a connected segment of the lysogenic chromosome, they provide a deletion map of viral and host genes (Table 9-2). Third, hybridization with transducing phages allows localization of messenger RNAs along the λ chromosome. Fourth, the normal regulatory circuitry of the virus can be probed by studying the behavior of host genes that have become fused into viral operons. Derivatives of the λ-related phage φ80 in which the host *trp*

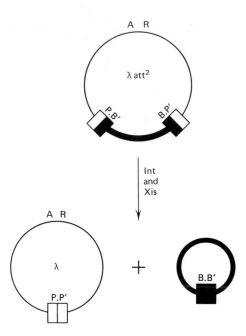

FIGURE 9-12. *Excision of bacterial DNA from the* λ att² *phage. From Hershey,* The Bacteriophage λ *(1971).*

genes replace part of the early leftward operon have been especially helpful (Franklin, 1974).

Another use has been in studies of the insertion and excision reactions themselves. The *int* and *xis* proteins can break and join DNA at the base sequences they recognize, whether they encounter these sequences in the host chromosome or in a virus. Thus integrase can catalyze recombination between two λ phages at their attachment sites. Likewise, if one crosses a λ *gal* phage with a λ *bio* phage, wild type λ is formed by the same reaction that takes place in normal excision.

A λ derivative (λ *att²*) has been isolated in which some bacterial genes are bracketed by attachment sites with the same structures (P.B′ and B.P′) normally found at the two prophage ends, as well as a similar variant where the two attachment sites are of the type (P.P′ and B.B′) whose recombination normally requires integrase alone. When these phages grow in the presence of active *int* and *xis* genes, a large fraction delete the DNA between the two attachment sites by internal recombination (Figure 9-12). The change is easily monitored because the shortened phage is especially resistant to chelating agents. DNA from *att²* phages has been used to assay integrase and gp*xis in vitro* (Nash, 1975).

Plasmid Formation

In principle, there are two ways in which a latent prophage might be perpetuated in a lysogenic cell line: It might insert into the chromosome, as λ does, or it might replicate as a separate, extrachromosomal element. An established genetic element that replicates separately from the bacterial chromosome is called a *plasmid*. As will be discussed in the next chapter, plasmid formation is characteristic of phage P1, as well as various nonviral elements. Wild type λ cannot be maintained as a plasmid, because independent replication comprises one stage of a temporal sequence that leads inexorably to cell death. However, certain mutations, or combinations thereof, block the sequence in a manner 'that allows continued replication of viral DNA within growing cells. What all these mutants have in common is that the *OP* operon is intact and that its rate of transcription is limited by the absence of gpN and the presence of gp*cro* (Berg, 1974). N^- mutants can form plasmids, but the plasmids that have been best characterized are certain variants that have deleted 80 to 90% of λ DNA and that multiply as small circular DNA molecules. These are called λ *dv* (defective virulent) because they were originally obtained from a virulent mutant of λ.

Other Temperate Phages

Various temperate phages establish stable lysogenic complexes by means that differ from those used by λ. The variations may lie either in the manner by which viral functions are regulated, or the method by which prophage is transmitted from parent to progeny.

Phage P2

Phage P2 resembles λ in many aspects of the lysogenic cycle (reviewed by Bertani and Bertani, 1974). The two phages are considered to be unrelated, because they have no homology detectable either by *in vitro* hybridization or by recombination *in vivo*. Like λ, P2 lysogenizes by insertion into the chromosome and shuts off most viral genes with a single repressor protein. Unlike λ, P2 repressor is not inactivated by irradiation or other inducing agents.

The P2 *int* gene lies next to the site of insertion into bacterial DNA, and is more effectively expressed following infection than in a lysogenic cell. However, this result is achieved in a quite different manner than it is in λ. In λ, rapid formation of integrase requires leftward transcription from op_L. In a lysogenic cell, this transcription fails because repressor is bound at op_L. On superinfection by another λ phage, the repressor generated by the prophage will repress the *int* gene of the superinfecting phage as well as that of the prophage. In P2, on the other hand, the *int* gene of a superinfecting phage is expressed even when repressor is present. However, the *int* gene of the prophage fails to function appreciably even when repressor is absent (as when a thermoinducible lysogen is heated). Apparently the *int* gene and its promoter lie on opposite sides of the insertion site, so that insertion into the chromosome automatically splits the operon and stops transcription.

One consequence of this method of *int* control is that P2 cannot form stable lysogens in which two or more prophages are inserted next to each other. In λ, tandem double lysogens frequently are found following mixed infection with genetically marked phages. However, such a lysogen of P2 is intrinsically unstable because, where the two prophages join, the complete *int* operon is restored and is not repressed. In double lysogens formed by P2, the second prophage always lies at some site far removed from P2's preferred site.

As in λ, the P2 *int* gene can promote recombination at the insertion site between two genetically marked phages. The quantitative effect is much more striking in the case of P2, because the frequency of general recombination is very small. In an ordinary mixed infection, about 80% of the genetic recombination between terminal markers on the P2 chromosome is due to exchange at the insertion site, which thus appears as a long empty distance on a formal recombination map.

Phage P22

Phage P22 is a temperate phage of *Salmonella typhimurium*. Because λ grows only in *Escherichia coli* and P22 only in *Salmonella*, it was assumed for many years that the two phages were unrelated. Recently, however, use of *E. coli–Salmonella* hybrids has allowed introduction of both viruses into a common host, where in fact they can form viable genetic recombinants (Botstein and Herskowitz, 1974). The homology between the two phages is apparently confined to certain genome segments within the "early" region between *N* and *Q*. The immunity region of P22 can be crossed into λ and appears to be identical to that of the λ-related coliphage 21.

Phages λ and P22 are quite distinct morphologically. Moreover, they differ in their mechanism of DNA packaging. Whereas λ packages DNA segments terminating at *cos* sites, P22 resembles T4 in having a "headful" cutting mechanism that generates cyclically permuted, terminally redundant molecules in the virion. Whereas T4 DNA molecules seem to have random termini, P22 packaging is manifestly nonrandom. Packaging appears to start at a fixed point on the P22 genome, and to proceed sequentially down the concatemer, so that successive molecules are progressively shifted in phase from the initial one. This interpretation is strongly supported by studies on the distribution of DNA endpoints with P22 variants in which the genome length (and therefore the degree of phase shift) has been altered by deletion or insertion mutations (Tye et al., 1974b).

Given a fixed packaging length, a deletion mutation that shortens the genome increases the extent of terminal redundancy. In the same way, an insertion can reduce terminal redundancy. When an insertion is large enough to increase the genome length above the packaging length, nonredundant molecules of less than genome length are packaged. The virions thus formed are individually unable to form plaques or to carry out a complete infectious cycle following single infection. In multiple infection, virions that cover each others' deficiencies can cooperatively initiate a productive cycle. In this respect, they resemble the isometric particles of T4, which cooperatively generate wild type T4 by recombination. The two cases differ in the consequences of cooperative infection. The T4 small particles can generate wild type T4 genomes, which are then packaged into virions; whereas P22 insertion phages recombine to yield a genome that is still oversized, so that the infected cell liberates the same mixture of fractional genomes as in the previous cycle (Tye et al., 1974a).

The headful cutting mechanism allows P22 occasionally to package bacterial, rather than viral, DNA, as detected by its ability to transduce bacterial genes to appropriate recipients. In fact P22 was discovered by its ability to transduce (because it was present as a prophage in a strain tested for genetic exchange), and transduction was discovered with P22 (Zinder and Lederberg, 1952).

The existence of virions containing host DNA implies that a unique DNA sequence present only on viral DNA cannot be an absolute requirement for packaging. The fraction of bacterial DNA packaged is rather small, amounting to less than 1% of the cellular DNA present before infection; but this fraction increases to 20 to 50% in certain viral mutants (Schmeiger and Backhaus, 1973). Any group of closely linked host genes can be packaged, although of course an individual particle contains only a P22 length segment, amounting to about 1% of the total host genome. This kind of transduction is called *general,* as distinguished from the specialized transduction seen with λ.

The regulation of repressor synthesis and of early viral development in P22 is similar to that of λ, as might be expected from the partial homology between these two phages in the DNA segment determining these functions. Like λ, P22 has one primary repressor gene (*c2*) and two auxiliary genes (*c1* and *c3*), whose products turn on repressor synthesis after infection.

Phage P22 has an additional regulatory element that is absent from λ. A P22 gene (called *ant* for "antirepressor") determines a protein product that can neutralize repressor and thus initiate a productive cycle (Botstein and Susskind, 1974). In a lysogenic bacterium, antirepressor synthesis is itself repressed by the product of another gene (called *mnt* for "maintenance") acting at an operator site v_2. Antirepressor plays no known role in the P22 life cycle. Phage P22 antirepressor neutralizes repressors of various other temperate phages, including λ. Thus P22 infection of bacteria lysogenic for these phages can induce development of endogenous virus.

Phage Mu-1

All the temperate phages discussed thus far become inserted into the bacterial chromosome by genetic exchange between a unique site on the viral chromosome, and one of a limited number of specific sites on the host chromosome. Phage Mu-1 is special in that it can insert with equal ease at a wide variety of host sites, probably at any point on the host DNA (Howe and Bade, 1975). Mu-1 insertion thus frequently interrupts the continuity of some bacterial operon, and constitutes a mutation of that operon. Mu-1 insertion generally does not entail any loss of bacterial DNA at the insertion site, but sometimes deletions are generated as part of the insertion event.

The Mu-1 virion contains a linear double-stranded DNA molecule of about 40 kb. The base sequence is unique rather than permuted, and there is no detectable terminal redundancy. The DNA of all virions is identical, except in two respects. At one end of the molecule, several kb of host DNA are covalently linked to the viral DNA. There are also 100-200 base pairs of host DNA at the other end of the molecule. These are apparently random segments of host DNA; hence artificial duplexes formed by reassociation of single polynucleotide chains almost always are unpaired at one end. Near the same end, there is another segment (G-loop) that frequently undergoes genetic inversion, so that it appears unpaired in some, but not most, artificial duplexes (Figure 10-1). The G segment is flanked by a short inverted

FIGURE 10-1. Structure of phage Mu-1 DNA, as revealed by electron microscopy of duplexes formed by reassociation of complementary virion strands. The variable end is composed of random segments of host DNA. The G segment undergoes frequent inversion and contains short internal inverted repetitions of base sequence. From Howe and Bade, Science **190,** *624, (1975).*

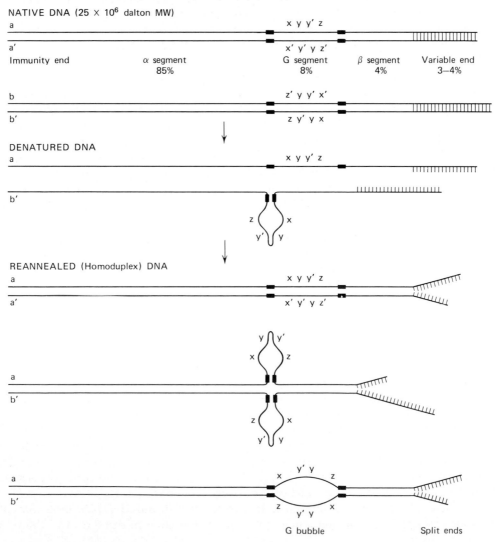

repetition (shown as a heavy bar in Figure 10-1), and contains an internal tandem inverted repetition (labeled yy' in Figure 10-1) as well. This internal structure generates the various kinds of duplexes illustrated in Figure 10-1.

The bacterial DNA seen as split ends in electron microscopy of artificial duplexes may be related to the ability of Mu-1 to insert at random. Unlike λ or P2, Mu-1 does not undergo cyclic permutation, but becomes inserted in the same linear order observed in the virion. The terminal bacterial DNA is not inserted as part of the prophage. Within the infected cell, small circular molecules are found in which Mu-1 DNA and bacterial DNA are covalently joined, which may represent intermediates in normal replication. However, no circular molecules of Mu-1 DNA alone have been observed. Thus at all stages of its life cycle, Mu-1 DNA is inserted between host sequences.

The G-loop inversion seems to play no essential role in either viral development or insertion. Phages bearing deletions that prevent the inversion by removing one or both termini of the G-loop region can still form plaques and lysogenize. However, this is true only of molecules frozen in one of the two possible orientations. Those fixed in the opposite orientation are nonviable. After one cycle of lytic growth, the regions are all in the same orientation. The inversion can take place while the prophage is inserted in the chromosome, even in recombination deficient hosts.

Despite the absence of a demonstrated role in the life of the virus, the G-loop inversion is intriguing as an example (like prophage insertion) of a chromosomal rearrangement that takes place at high frequency, apparently under specific genetic control. The invertible G segment found in phage Mu-1 is also present in the DNA of phage P1 (to be discussed next). As these two phages are otherwise nonhomologous, this suggests that, besides being able to invert, the G loop can be transposed from one replicon to another.

Phage P1

Phage P1 is a large phage (about 90 kb of double-stranded DNA) with cyclically permuted, terminally redundant DNA. It has become the standard phage for use in general transduction of *E. coli* genes.

P1 does not insert its DNA into the host chromosome, but rather perpetuates itself as an independent plasmid. In a P1-lysogenic culture, there is about one P1 genome per host chromosome. Faithful transmission of prophage to all descendants of a lysogenic cell implies some regular segregation process whereby prophage copies are partitioned to each daughter at cell division. The mechanism of segregation is unknown, but is not unique to phages like P1. Certain nonviral plasmids (to be discussed at the end of this chapter), as well as the host chromosome itself, must carry out similar processes.

P1 has a repressor gene (*c1*) that resembles in many respects the *cI* gene of λ; for example, viral development is induced by heating a P1 *c1ts* lysogen (Scott, 1970). The regulatory circuitry of P1 should have some unique features, however. Phages that insert their DNA into the chromosome have two potential regulatory states, in which autonomous replication is either turned on, or else turned off completely. However, in P1 the alternatives are that replication is extensive (in productive infection) or limited (in the plasmid state).

Conversion of Bacterial Properties by Phage

A lysogenic bacterium is immune to lysis by relatives of its prophages. This immunity is a new property of the bacterium. Similarly, some prophages alter the response of the bacterial host to other, unrelated phage (*prophage interference* and *prophage-controlled phage modification*). For example, phage λ grown on *E. coli* K12 is restricted in its ability to multiply in bacteria lysogenic for prophage P1; but if it succeeds in growing on K12 (P1) all progeny λ phage can multiply on P1-lysogenic bacteria. This ability is lost after a single growth cycle on K12 (P1) cells. And we have already described how prophage λ itself changes the lysogenic bacteria by making them nonpermissive for the *rII* mutants of phage T4.

Immunity and modifying effects are both examples of new properties conferred by the genome of a prophage onto its host bacterium. In some instances, the prophage also plays a role in determining bacterial properties that have no obvious relation to the life cycle of a phage, for example, colonial morphology or pigmentation. In these cases one speaks of *conversion phenomena*. The two most remarkable examples of conversion are the changes in somatic antigens of *Salmonella,* discovered by Iseki and Sakai (1953), and the appearance of toxigenicity in *Corynebacterium diphtheriae* (Freeman, 1951).

Salmonella strains have in their cell wall a lipopolysaccharide which constitutes the "somatic" O-antigen (as distinct from the flagellar antigens). The antigenic specificity of the lipopolysaccharide is due to the polysaccharide moiety, which consists of a core with attached side chains (Figure 10-2). The side-chain structure is characteristic for each *Salmonella* serotype and, on the basis of serological cross-reactions, a strain is assigned a certain antigenic formula 1, 4, 5, 12, or 3, 10, 15; each number in the formula corresponds to an antigenic determinant that reacts with a specific test serum. Some of the antigenic determinants are controlled by phages. For example, *S. anatum* has the antigenic formula 3, 10, but after infection with phage ϵ^{15} the serological determinant 15 appears and 10 disappears. Bacteria that become lysogenic for phage ϵ^{15} retain the new antigens 3, 15 as long as the prophage is present and return to the 3, 10 form only if they lose the prophage. In fact, all *Salmonella* strains with 3, 15 antigens found in nature carry an ϵ^{15} phage.

A few minutes after infection of *S. anatum* with phage ϵ^{15}, all bacterial cells, even those cells that will soon undergo lysis, show a rapid appearance of antigen 15 (Uetake et al., 1958). In those infected cells that become lysogenic the antigen 10 is not destroyed, but is diluted out during successive cell divisions. The converting activity of phage ϵ^{15} brings about three biochemical changes (Robbins and Uchida, 1962); (1) production of an enzyme that builds polysaccharide chains with a β-linkage between galactose and mannose; (2) repression or inhibition of a bacterial enzyme that generates chains with an α-linkage between these two sugars; and (3) repression of a cellular enzyme that acylates the galactose residues (see Figure 10-2). Several mutants of phage ϵ^{15},

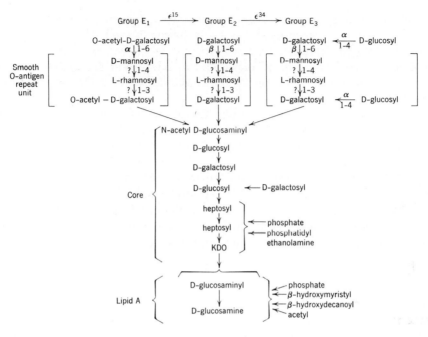

FIGURE 10-2. *Probable chemical organization of the lipopolysaccharides of* Salmonella *groups* E_1, E_2, *and* E_3.

variously defective in one or another of these biochemical functions, have been isolated.

The converting action of phage ϵ^{15} on *Salmonella* polysaccharides is only one example of a large class of phenomena. Other examples are listed in Table 10-1.

TABLE 10-1 EXAMPLES OF CONVERSIONS OF SOMATIC ANTIGENIC DETERMINANTS OF *SALMONELLA* BY BACTERIOPHAGES

Salmonella group	Phage	Conversion of O-antigenic formula
B	P22 or	$4,12 \rightarrow 1,4,12$
		$4,5,12 \rightarrow 1,4,5,12$
B	27	$4,12 \rightarrow 4,12,27$
C_1	14	$6,7 \rightarrow (6),(7),14$
C_2	98	$6,8 \rightarrow 6,8,20$
E_1	ϵ^{15}	$3,10 \rightarrow 3,15$
E_2	ϵ^{34}	$3,15 \rightarrow 3,15,34$
K	6–14	$18 \rightarrow 6,14,18$

Modified from Robbins and Uchida, 1962.
Antigenic determinants affected by conversion are italicized.
Determinants weakened (reduced in amount or reactivity) by conversion are in parenthesis.

The converting functions are not essential for the phages because mutants that have lost the converting ability grow perfectly well. Yet the fact that the polysaccharides of *Salmonella* are responsible for the adsorption of phages to lysogenic bacteria suggests a possible evolutionary role of the converting genes of the phage in determining the range of bacteria that specific phages can use as hosts.

The role of phages in determining the ability of *C. diphtheriae* to produce the diphtheria toxin has been established. Toxigenic bacteria carry prophages of the β group, and nontoxigenic bacteria can be made toxigenic by lysogenization (Groman, 1959). The toxin is coded by the phage genome, and has been synthesized in a cell-free protein-synthesizing system from *E. coli*, directed by β phage DNA (Murphy et al., 1974). Toxin production is regulated by host metabolism in a manner not yet fully understood. In particular, it is strongly inhibited by excess iron in the medium. Toxin appears to play no essential role in viral development, as β phage can form viable nontoxigenic mutants.

Defective and Satellite Phages

Viruses can give rise by mutation to genomes that are functionally incomplete, incapable of carrying out one or more steps in viral development. Such variants tend to be eliminated by natural selection. However, since prophage replication does not require genes of the productive cycle, some mutations that would be lethal to a virulent phage can be perpetuated indefinitely in the lysogenic state.

Defective variants arise from known viruses by all the standard mutational processes familiar to geneticists. As we have already mentioned, phage λ can undergo point mutations (as in conditionally lethal mutants), deletion mutations (sometimes extending beyond the prophage termini into bacterial DNA), and substitutions (as in specialized transducing phages). Defective lysogens of these kinds have been useful in experimental studies of gene function and genome organization.

One method of obtaining defective lysogens is to treat a lysogenic culture with some mutagenic agent, such as ultraviolet light, and then screen the survivors for individuals that have lost the ability to produce phage. Defective lysogens are distinguished from survivors that have been cured of their prophage altogether because the defective lysogens retain at least some viral genes; for example, unless the *cI* gene has been inactivated, they are immune to superinfection by phage of the carried type. Some systematic, but technically difficult, studies of gene function in λ were carried out on defective lysogens before conditionally lethal mutations were discovered.

Besides defective lysogens originating in the laboratory from known phages, many bacteria isolated from nature seem to harbor defective prophages. This condition can be manifested by the presence in culture supernates of objects resembling virions or components thereof (such as empty heads and free tails); or of lysis with production of such objects following exposure to inducing agents. Such strains are considered defective lysogens because they produce virus-like objects but not plaque-forming particles. Where structures resembling complete virions are seen, an obvious question is whether they might not in fact be able to form plaques on some host unavailable in the laboratory, or under some conditions not tested, or even whether they might infect susceptible cells and multiply on them, but with such kinetics that visible plaques are not detected. Doubtless the defectiveness of some natural lysogens will turn out to be apparent rather than real. In other cases, however, there is good reason to believe that, although morphologically complete particles are observed, they are not agents of infectious spread. For example, all known isolates of the spore-forming aerobic bacterial species *Bacillus subtilis* lyse on exposure to inducing agents, and liberate DNA filled particles of similar appearance (Garro and Marmur, 1970). However, most of the DNA packaged into such particles is bacterial DNA synthesized prior to infection, and every segment of the bacterial genome is packaged to some extent. Mutants of *B. subtilis* that are deficient in ability to form such particles have been analyzed; mutations that impair the ability to form heads, tails, or both all lie close together on the bacterial chromosome, suggesting that the determinants of packaging are not widely scattered (Thurm and Garro, 1975).

As mentioned earlier, extensive packaging of host DNA is also observed in certain nondefective mutants of bacteriophage P22. The *B. subtilis* prophages are somewhat reminiscent of that situation. However, the *B. subtilis* phages fail to form plaques. Also, their widespread occurrence suggests that their present lack of specificity in packaging is of some antiquity.

The production by a bacterial strain of phage tails (whether or not connected to heads) is sometimes manifested by the fact that these tails can attach to the surfaces of susceptible bacteria and kill them. Operationally, such lethal substances are classified as bacteriocins; that is, as proteins elaborated by one bacterial strain that can kill bacteria of related strains (Hardy, 1975). The killing may be similar in mechanism to that effected by the DNA-free "ghosts" of phage T4 prepared by osmotic shock of whole phage particles.

Bacteriocins include not only large structures resembling phage tails or whole phage, but also simple small proteins such as colicin Ia (molecular weight 75,000) or colicin E3. Colicin Ia is thought to act on the cell membrane. Colicin E3 causes one species of ribosomal RNA molecule to be cleaved into specific fragments. Because some of these bacteriocins adsorb to the same sites on the cell surface as certain phages, an evolutionary relationship with proteins of the phage base plate has sometimes been suspected.

Like lysogenic bacteria, bacteriocinogenic strains need to be protected from the lethal substances their own cells produce. Bacteriocinogenic bacteria are specifically immune to the bacteriocins they themselves elaborate. This immunity is not caused by a repressor, but rather (at least in the case of colicin E3) by a second protein found in the colicinogenic strain that neutralizes the colicin.

Production of phage-related substances is one manifestation of defective lysogeny. The presence in bacterial strains of viral genes controlling early or regulatory functions has sometimes been detected also by the ability of these genes to recombine with a superinfecting phage and replace their homologs there. For example, when phage P2 is grown on the B strain of *E. coli* and then plated onto another host (such as *Shigella dysenteriae*), a few of the plaques formed contain phage (called P2 Hybrid Dismune) that derive most of their genes from the original P2, but in which a DNA segment including the repressor gene and nearby operator sites has been replaced by analogous genes. P2 Hybrid Dismune is thus related to P2 in the same manner that λimm434 is related to λ: Bacteria lysogenic for P2 Hybrid Dismune are immune to superinfection by another P2 Hybrid Dismune particle, but susceptible to infection by P2; and vice versa. *E. coli* B uninfected by P2 does not liberate any phage spontaneously, but it is immune to infection by P2 Hybrid Dismune. When P2 infects *E. coli* B, some of the survivors have been "cured" of this immunity (and of the ability to generate P2 Hybrid Dismune). Thus *E. coli* B must harbor a defective prophage related to P2 in the same manner that 434 is related to λ. Similar observations have revealed that the K12 strain of *E. coli*, from which λ was isolated, also carries a defective, λ-related prophage that can recombine with λ to contribute genes functionally equivalent to *Q, S,* and *R,* and to the recombination (*red*) genes (Strathern and Herskowitz, 1975).

Among those defective prophages that have arisen as variants of known phages, many, such as the specialized transducing phages and the point mutants of λ, can be packaged into viral particles provided a nondefective virus is present to supply missing gene functions. Production of infectious particles thus depends entirely on coinfection with a helper phage.

Certain natural viruses likewise require a helper virus in order to carry out a productive infectious cycle. Some of these may be defective relatives of the helper. In other cases, the relationship between helper and helped virus is too slight to be detectable either by genetic recombination or by nucleic acid hybridization. A virus dependent on an unrelated helper is called a *satellite virus.* Satellite viruses are found in all the major viral groups.

The satellite phage P4 forms plaques only on bacteria lysogenic for P2 or related phages. Phage P4 is a double-stranded DNA phage of 7×10^6 daltons, whereas P2 is 22×10^6 daltons. It is a temperate phage, and P4 can infect and lysogenize by insertion even when P2 is absent. But plaque formation by P4 depends absolutely on P2. Furthermore, if any of the 17 genes of the P2 prophage that are needed for P2 virion formation or lysis has been inactivated by mutation, P4 fails to grow. Phage P4 activates transcription of the late genes of P2 and uses their protein products to package its own DNA and liberate virions from the infected cell. Replication of P2 prophage is not induced by P4. The quantity of protein derived from transcribing the single prophage copy is apparently sufficient. The P2 genes involved in replication and early regulation are not needed for P4 growth. Apparently P4 provides these functions with its own genes. Yet the behavior of P4 on hosts carrying partial deletions of P2 prophage indicates that some early gene product of P2 stimulates transcription of late genes of P4. Each of the two viruses can thus turn on the late genes of the other. These reciprocal regulatory interactions indicate that P4 is extremely well adapted to exploiting P2 and suggest a relationship of long duration between the two (Barrett et al., 1974).

Comparison with Nonviral Cell Components

Most of the objects just discussed are considered to be defective phages because of their numerous similarities to, or genetic interactions with, infectious agents. At this point we emphasize that there is no strict logical

distinction between a defective provirus and a nonviral host component—
even if that component exhibits some properties common to viruses.

As a group, temperate phages have several typical attributes, some of
them shared by other viruses as well: Virion formation, cellular lysis,
autonomous replication, and insertion into the chromosome are among the
most obvious. There is no logical evolutionary reason why any or all of these
aspects of viral development should not occur separately from the others,
serving functions other than those seen in the viral life cycle. A variety of
bacterial elements, not obviously virus-related, exhibit autonomous replication, insertion, or both (Table 10-2).

Genetic elements that replicate separately from the bacterial chromosome
are called *plasmids* (Schlessinger, 1975). All plasmids thus far characterized are
composed of double-stranded DNA and exist intracellularly as closed circular
molecules. Some of these (such as the fertility factor F indigenous to *E. coli*
K12) can promote their own intercellular transfer, through the agency of small
appendages called pili or fimbriae, which are coded by genes of the plasmid.
Many other plasmids do not carry determinants for pilus formation; but can
themselves be transmitted when pili are present. Autotransmissible plasmids
like F resemble viruses in their ability to spread rapidly by infection
throughout a population of cells. They differ only in the absence of a free
extracellular phase in their life cycle.

Plasmids are of two sorts with respect to replication control. Those
plasmids under *stringent* control resemble phage P1 in that the plasmid is
maintained at about one copy per bacterial genome, and regularly partitioned
to the daughter cells in each division cycle. Plasmids under *relaxed* control are
more like λ*dv*: There are many copies per cell, which replicate, on the average,
once per cell generation. Plasmid copies are distributed at random during cell
division. Complete absence of the plasmid from either daughter cell is
statistically unlikely because the average number is large. Stringent plasmids
are somewhat analogous to the auxiliary chromosomes of certain eukaryotes,
whereas relaxed plasmids are more like cytoplasmic elements.

Many of the drug resistance determinants commonly associated with
plasmids are also translocatable from the plasmids onto the bacterial chromosome or onto other plasmids or phages. Sometimes the resistance genes are
flanked by insertion sequences of types known to be independently translocatable, in reverse orientation with respect to each other. This is the case with
the tetracycline resistance (T_c^R) element included in Table 10-2. Although the
precise mechanism of translocation is not understood, the resistance determinants seem to achieve mobility within the genome and within the population
by becoming associated with other elements (insertion sequences and autotransmissible plasmids, respectively) that can themselves change location
readily.

Specific insertion sequences are also found on the fertility factor F at the
site of its insertion into the bacterial chromosome. Generation of Hfr donor
strains, which in mating can transfer the entire bacterial chromosome rather
than just the F factor, is accomplished by insertion of F into the chromosome.
This insertion apparently entails recombination between insertion sequences
on both plasmid and host. Excision of the F factor from the chromosome is
sometimes imprecise, generating derivatives of F (called F') in which some

TABLE 10-2 MOBILE GENETIC ELEMENTS OF *ESCHERICHIA COLI*

Category	Examples	DNA length (kb)	Autonomous replication	Insertion and Excision	References
Temperate phages	λ	47	In productive cycle of infection only. Certain mutants (λN^-, λdv) can become established as plasmids.	Unique site on phage and chromosome. Mediated by specific viral genes (*int*, *xis*)	Hershey, 1971
	P1	90	In productive cycle and as established plasmid. 1-2 copies per host genome, stringent control.	Unknown with wild type P1. Certain deletion mutants (P1 cryptic) can be inserted.	Scott, 1970 Ikeda and Tomizawa, 1968
Plasmids	F*	95	As established plasmid. 1-2 copies per host genome, stringent control.	Inserts by recombination between ISs common to plasmid and host.	Schlessinger, 1975
	R6*†	97	As established plasmid. Stringent control.	Unknown	Hu et al., 1975. Clowes, 1972.
	RTF component of R6*	84	As established plasmid. Stringent control	Insertion into chromosome unknown. The RTF and r plasmid components of R6 associate and dissociate by insertion and excision	Clowes, 1972
	r plasmid from R6†	13	As established plasmid. Relaxed control.		
	ColE1	6.4	10–15 copies per host genome, relaxed control	Unknown	Schlessinger, 1975.
Insertion Sequences (IS)	IS3	1.4	Unknown	Present at several sites on *E. coli* chromosome, translocatable to new sites.	Saedler and Starlinger, 1972.
	Tc^R element†	10.9	Unknown	Termini of inserted form are inverted repetition of IS3. Translocatable to new sites.	Ptashne and Cohen, 1975. Kleckner *et al.*, 1975.

* Conjugal plasmids.
† Carriers of antibiotic resistance determinants.

bacterial genes have been incorporated into F, just as they can be incorporated into specialized transducing phages (Figure 10-3). Agents like λ and F that can replicate either autonomously or as part of the chromosome have been termed *episomes*.

Some appreciation of the comparative properties of the elements listed in Table 10-2 seems essential to an understanding of the place of temperate phage in the natural world. The possible evolutionary relationships among some of these objects will be discussed in the final chapter of this book.

Plasmids carry a variety of genetic determinants besides those required for their own replication or transfer. Genetic mechanisms are known whereby almost any gene of the chromosome can become associated with a plasmid. However, those genes that are found on plasmids of natural isolates seem typically to be those that are needed under conditions encountered only occasionally by their bacterial host. The determinants for clinically important resistance to antibiotics and various other deleterious agents are almost all located on plasmids. In bacteria of the genus *Pseudomonas*, which are noted for their ability to oxidize a wide variety of organic compounds, many of the genes for specific metabolic pathways are located on plasmids that are carried by some but not all Pseudomonads (reviewed in Schlessinger, 1975). Because these plasmids can be transferred among cells as needed, they increase the effective genome size of the species without placing the burden of replicating all their genes on every individual thereof. The determinants for bacteriocin production are also generally carried on plasmids.

FIGURE 10-3. *Formation of F′* gal *element by abnormal excision from an Hfr strain, in which F is inserted in the chromosome.*

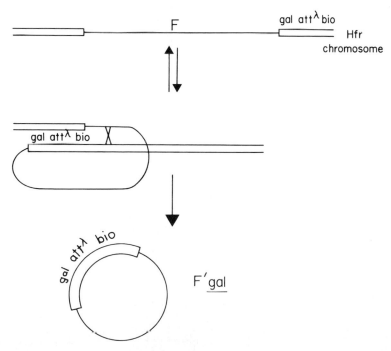

Chromosomal insertion is carried out by temperate phages like λ, plasmids such as F, and certain small genetic elements (*insertion sequences*) that can become inserted at various sites on host DNA but are unable to replicate autonomously (reviewed in Starlinger and Saedler, 1972). Insertion sequences are detected by their ability to translocate themselves within a bacterial genome. Insertion within a gene inactivates its function and generally interrupts transcription of the operon. Several species of insertion sequences indigenous to *E. coli* K12 were first detected as the causative agents of completely polar mutations within known operons. Insertion and excision of insertion sequences do not depend on the host *rec* genes. They thus comprise a specific type of DNA breakage and joining analogous to prophage insertion, although genetic determinants analogous to the *int* and *xis* genes of λ have not yet been identified.

Introduction to Animal Cell
11 Biology

Role of Cell Culture in the Study of Animal Viruses

Until the 1950s the study of animal viruses was stimulated mainly by the importance of certain viral diseases of humans and animals (for example, smallpox, yellow fever, influenza, poliomyelitis, and foot-and-mouth disease of cattle). In this early era the existence of many different kinds of animal viruses, causing varied types of disease, was proved, narrow host ranges for many viruses were observed, and vaccines against viral diseases were developed. The experimental tool for the propagation of viruses in most of the early work was the whole laboratory animal. The difficulties in handling large enough numbers of animals for precise quantitative work were a serious obstacle to progress. Cultivation of certain viruses in the embryonated chicken egg had obviated these difficulties to some extent, but still the isolation of individual cells infected with viruses was not possible.

In 1949 Enders and his associates (Enders, et al., 1949) demonstrated that viruses could be regularly propagated to high titer in explanted cells and bits of tissue growing in culture.

In 1952, Dulbecco (1952) described a plaque assay using cultured cells that allowed accurate counting of the infectious Western equine and poliomyelitis particles. The products of an encounter between a single virus particle and a single cell could be isolated and the concepts that had been developed from bacteriophage studies concerning cell-virus interactions could be tested in individual animal cells. The widespread use of explanted tissues for virus propagation plus the plaque assay for animal cell-virus systems almost overnight opened up the field of quantitative animal virology.

The first cultured cells to be used were fresh isolates, *primary cultures*, from trypsinized tissues such as chicken embryo fibroblasts and minced monkey kidney and testicle tissue. These cultures could not be grown indefinitely, however, and the need to return continually to primary tissue isolations caused a certain amount of irreproducibility. Techniques for the *continuous culture* of cells suitable for virus work were first developed in Earle's laboratory (Sanford et al., 1948). By the late 1940s strains of mouse cells that would grow indefinitely in culture had been developed and, in addition, growth of a culture from a single mouse fibroblast was achieved (Sanford et al., 1959). Continuous cultures of cells from a wide variety of vertebrates and invertebrates have since been established; for example, cell cultures from insects, fish, frogs, many rodents, pigs, dogs, cats and various primates including man are all available.

Major steps in making cell cultures simpler and therefore more widely useful were: (1) the use of antibiotics that did not harm animal cells, but prevented bacterial contaminations; (2) the isolation from a human carcinoma of a cell line (HeLa) which could be grown indefinitely in culture and was susceptible to a wide range of human viruses (Gey et al., 1952; Scherer et al., 1955); (3) the development by Eagle (1960) of a simple, defined medium for the growth, not just survival, of many different cell cultures; and (4) the refinement of techniques for growth of single mammalian cells (Puck et al., 1957).

Two aspects of virology profited equally from using cell cultures and the quantitative plaque assay. Clinical and epidemiological investigations uncovered hundreds of new viruses; so many, in fact, that some remain unrelated to known diseases (Jackson and Muldoon, 1975). Large-scale industrial production allowed vaccines to be produced in cell cultures. Poliomyelitis virus vaccine was administered to several hundred million people by the early 1960s, and now the disease has diminished greatly throughout the world. New vaccines for other viruses, measles, for example, were produced in cultures and were widely accepted after the success with poliovirus (Naficy and Nategh, 1972). The basic virologist also profited greatly from large-scale virus production because virus purification now became practical; reliable procedures were worked out for growing and purifying virtually any virus. Another important advantage of the new techniques was the availability of cells that were homogeneous in origin and could be simultaneously infected in sufficient amounts to allow biochemical studies.

The next several chapters are devoted to the details of animal virus interactions with their hosts, animal cells. The knowledge about the biology and chemistry of animal viruses has accumulated so rapidly that comprehensive coverage of the animal viruses is not possible in a volume devoted to general virology. It is necessary to select for detailed description a few virus-cell systems that provide a sufficiently broad example of the many molecular strategies used by the animal viruses in their growth. Because of the potential impact on the overwhelmingly important medical problem of cancer, and the contribution to the study of integration of viruses with host cells, the tumor viruses are discussed separately in Chapter 16.

Before describing the details of virus growth such as adsorption, penetration and release of nucleic acid, replication, and the effects on host cells and organisms, we briefly discuss the types of relationships that exist between animal viruses and animal cells and consider the facts and techniques of animal cell biology that are most needed to discuss the study of virology.

Characterization of Animal Cell-Virus Relationships

Studies of whole animals revealed that certain viruses—for example, herpes virus and the tumor viruses such as Rous sarcoma—can be isolated from infected animals over protracted periods of time, whereas other viruses such as poliovirus apparently come, cause an infection, and then disappear. It was not until cell culture systems were made practical, however, that the relationship between individual cells and viruses could be experimentally approached.

Two extreme types of interactions have been identified in cell culture, with some gradations in between. Rapid cell death and degeneration, the so-called *cytolytic* or *cytocidal reaction*, results after infection with many types of viruses, the picornaviruses (for example, polio, coxsackie), the togaviruses (for example, Eastern and Western equine encephalomyelitis) and certain myxoviruses (such as influenza) being among the most destructive. Other viruses that regularly produce cytocidal effects, but are less rapid and extensive in their action, are poxviruses, adenoviruses, and reoviruses.

At the other end of the spectrum from the cytolytic viruses are the tumor-producing viruses such as polyoma virus of mice, simian virus 40 (SV40) of monkeys, and Rous sarcoma virus (RSV) of chickens, all of which have the capacity to infect and change appropriate host cells into tumor cells. The ability of the transformed cell to grow is augmented, not destroyed, and its biologic characteristics are grossly altered. A long-term virus-cell association is established involving the integration in the host cell chromosomes of virus-specific DNA (Sambrook et al., 1968).

Such an integrated relationship is similar to lysogeny in bacteria but no generally accepted term has arisen to describe the animal cell case. In whole animals such infections were originally called *latent*; in cultured cells the term *moderate* was at one time proposed; and with the recent establishment of molecularly linked virus and cell DNA, the term *integrated* has been frequently used. Because much effort was expended to demonstrate the integration of the viral DNA, that term will be used in all cases where linked cell-virus nucleic acids have been found.

The action of cytocidal viruses on host cells, although eventually destructive in character, does not always occur immediately. A weak cytocidal interaction can lead to a situation, at least in cultured cells, where a balance between cell growth and virus growth is reached. SV5, a paramyxovirus from monkeys, achieves this balanced situation in monkey kidney cells (Choppin, 1964). Rubella virus infections of embryonic tissue also may lead to continuing growth of both virus and cells even in the presence of antiviral antibody (Rawls, 1968). Other cases of balanced virus and cell growth require a small amount of antiviral antiserum plus a weakly cytocidal virus to prevent the virus from killing all the cells (Walker, 1968). Carrier cultures may have counterparts in chronic animal infections, but are to be distinguished from cases where virus-cell nucleic acid integration has occurred. Finally an important cause in establishing the *carrier culture* state in some cases (vesicular stomatitis virus infection, for example; see Chapter 13) is the existence of *defective interfering* particles. These *DI* particles prevent normal viral replication in most cells, thereby suppressing the cytolytic infection for most but not all cells.

Composition, Structure, and Macromolecular Synthesis in Animal Cells

In recent years investigators with widely different backgrounds have become interested in the study of the biology of animal cells. Those whose interest has evolved through previous microbiological experience with bacteria and viruses, or the biochemistry of lipids, nucleic acids, and proteins tend sometimes to overlook the morphological and organizational complexity of animal cells, whereas workers trained as cell biologists tend to question the feasibility of generalizing results obtained with a bacterium or a virus to the animal cell. As usual in such situations, the middle ground has a great deal to offer.

Many of the most fruitful current ideas about biosynthetic mechanisms and cellular control processes, two of the key areas in cellular biology, have arisen from work with bacteria and viruses. The degree to which these ideas can be tested and proven or rejected in animal cells is still under study.

This brief discussion of animal cell biology does not attempt to provide the uninitiated with an extensive acquaintance with the mysteries of the animal cell. Instead, it points out certain facts about the structure and composition of animal cells that are important in the processes of initiation of viral infection, viral multiplication, and the destructive effect of viruses on their host cells. The emphasis is on cultured animal cell systems and particular tissues that seem to offer the most promising opportunities for the exploration of animal cells with the tools of molecular biology.

The cells of a multicellular organism, examined by light microscopy, differ widely from one tissue to another. It may, therefore, seem naive to attempt to describe "the animal cell." However, many animal cells share three basic features: (1) the content of RNA, DNA, and protein per cell is similar; (2) a structurally similar surface ("plasma") membrane consisting of proteins, frequently glycoproteins, embedded in a phospholipid bilayer; and (3) most cells contain the same structurally identifiable organelles and particulates— nuclei, nucleoli, mitochondria, ribosomes, golgi bodies, lysosomes, microtubules, microfilaments—some of which may be particularly outstanding in specialized cells (for general discussion see Novikoff and Holtzman, 1975).

By expanding our discussion of these areas of similarity, it becomes easier to see why experimental findings with different animal cells may properly and profitably be related.

COMPOSITION OF ANIMAL CELLS

Animal cells are generally quite large—a chick fibroblast in culture is about 10^{-7} mg (dry weight), and a HeLa cell, the first human cell to be cultured indefinitely, about 10^{-6} mg. For comparison, the dry weight of an *E. coli* cell is about 2 to 4×10^{-10} mg. The most accessible individual cells in the animal fall in the range between 10^{-6} and 5×10^{-8} mg dry weight per cell. This includes cells from the liver, spleen, gut, lung, kidney, bone marrow, and connective tissue. Exceptions are muscle fibers (syncytia formed from myoblast fusion) and some cells of the central nervous system, which can be much larger in total mass than other cells of the animal body. Associated with the large size of animal cells is the large mass of virus-specific material that a single cell can produce. For example, a single HeLa cell can produce 1 to 2 $\times$ 10^5 adenovirus particles, about 10^{-7} mg., which is equivalent to the mass of several hundred bacteria.

About 60% of the dry weight of the average animal cell is protein (equivalent to 8×10^{13} protein molecules of average size) and a variable proportion (5 to 20%) is lipid. RNA comprises about 10% of the cell mass (equivalent to about 4×10^6 ribosomes and 6×10^7 tRNA molecules) and DNA is about 5% (Table 11-1). Stored carbohydrates such as glycogen can exist in amounts up to 10% of the dry weight. The amount of DNA in a normal diploid human cell is about 5 picograms (5×10^{-12} g), which is equivalent to almost 5×10^9 base pairs, If the size of the average gene is taken to be 1000 base pairs (see Chapter 4), the number of possible genes in the *haploid* human cells is 5×10^6. There are several reasons why this number is probably too high by at least a factor of 10 (Bishop, 1974). First there is no strict correlation between DNA content per haploid nucleus ("genome size") and the apparent genetic complexity among animals (Figure 11-1). For example, all insects or all amphibians might be expected to be approximately equally complex within their phyla but the genome size of individual species in these phyla varies over a 20-fold range. In addition some amphibians and fishes contain 10 to 20 times as much DNA as humans. Clearly it would seem that all DNA in all organisms must not be devoted to coding for proteins.

TABLE 11-1 CONSTITUENTS OF A HeLa CELL

		Total mass per cell
DNA		15 picograms*
RNA		30 picograms
Protein		400 picograms
Lipid		~50 picograms
		Number per cell
Cytoplasm	Ribosomes	4×10^6
	tRNA molecules	6×10^7
	mRNA molecules†	7×10^5
	Protein molecules‡	4×10^8
	Mitochondria	~100's
Nucleus	Ribosomal precursor RNA molecules§	6×10^4
	hnRNA molecules‖	1.6×10^5
	"Small nuclear" RNA molecules	1.8×10^4

* HeLa cells are hypotetraploid; the normal diploid human DNA complement is about 5 picograms/cell
† An average chain length of 1500 nucleotides was assumed
‡ An average chain length of 400 amino acids was assumed
§ There are about 5 times as many 32S as 45S rpre-RNA molecules
‖ An average chain length of 6000 was assumed

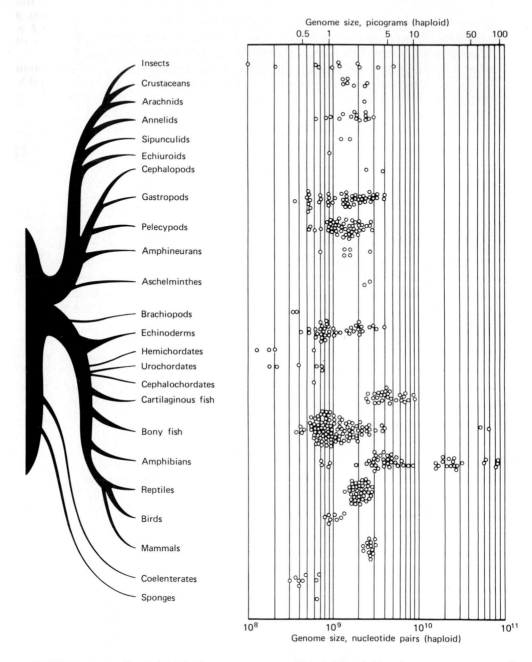

FIGURE 11-1. *Haploid DNA content per cell in animal kingdom. Note range of DNA content within phyla (insects, mollusks have wide range; mammals fairly narrow range) and very weak correlation with supposed phylogenetic complexity. From Britten, and Davidson* Quart. Rev. Biol. **46,** *111 (1971).*

Second, the total number of genes in *Drosophila melanogaster* is estimated to be 5000 both by "saturating" regions with mutations and simply by observing the number of bands, each thought to be one gene, in salivary chromosomes (Judd, 1975). *Drosophila* has about $1/20$ as much DNA as humans, so if the estimate of 5000 genes for *Drosophila* is correct and the "extra DNA"/ structural gene DNA ratio the same, humans would contain about 100,000 genes.

A third argument attempts to take account of "genetic load" (Muller, 1967; Bishop, 1974). Most mutations are thought to be deleterious because they affect the functioning of proteins. By measuring mutation rates in the human population in specific sites, for example, hemoglobin genes, an estimate of the total "target size" of the genome can be obtained and the number of mutations that can be "supported" by the population at large, without disappearance of the species, can be calculated. From these considerations population geneticists conclude that at maximum only a few percent of total DNA could be kept "correct" from generation to generation or our genetic load would be too great.

Whatever the genetic coding potential for the large amount of DNA contained in animal cells, the organization of the DNA within each cell must be of great importance. Microscopic studies of chromosomal structure have been pursued for many years, and this complex problem is now beginning to yield to biochemical approaches. The great majority of eucaryotic DNA seems to be complexed with histones which exist in a constant configuration termed a "nucleosome". Four histone molecules form a tetramer that alternately covers 200 base pairs of DNA leaving 30 to 40 bases uncovered and susceptible to nuclease attack (Axel et al., 1974; Kornberg, 1974). Recent results suggest that DNA actively undergoing transcription is not protected by histones (Weintraub and Gourdine, 1976). In addition, there are a substantial number of "non-histone" proteins associated with DNA. At the moment little is known about the functioning of these DNA-associated proteins, although they undoubtedly play key roles in transcription, replication, and packaging of the DNA.

This general DNA-protein structure at the level of individual genes, and in fact for vast stretches of DNA, does not mean that all chromosomes are alike, however. Even in phyla where the total cell DNA content is fairly constant, the chromosome number may vary considerably among species, for example $n = 6$ for one species of Indian deer and $n = 46$ for another similar species (Wurster and Benirschke, 1970). Such variation in chromosome number in the face of constant amounts of DNA means that chromosome size also varies widely. Within the cells of a single species chromosome size can vary greatly (Figure 11-2). The prevailing belief is that each eucaryotic chromosome represents a single DNA molecule; strong evidence for this conclusion is available only for *Drosophila* and yeasts (Kavenoff and Zimm, 1973; Burke et al., 1975).

Technical difficulties make the precise location of single genes within chromosomes and the study of genetic expression in eucaryotic cells much harder to achieve than with bacteria. Animal viruses range in size from a few genes up to at most a hundred or so genes and also go through a regulated program in their growth cycle. Many virologists hope that a detailed picture of

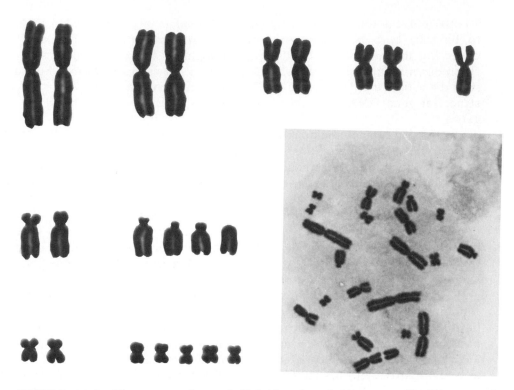

FIGURE 11-2. Karyotype of pseudodiploid cultured hamster cells (Photograph courtesy of J. Biedler). Note approximate 10-fold variation in lengths of chromosomes. Stain was acetoorcein, magnification 1764 for individual chromosomes and 882 for "squashed" cell (lower right).

the operation of these simpler genetic entities within the same cellular milieu may point to principles of molecular organization and regulation that function for cell genes.

In addition to macromolecules, there are in the cell small molecules that remain free of the macromolecules when the latter are precipitated with cold acid. These materials are operationally defined as the *cell pool* (or acid soluble pool) and consist of myriad compounds. Among the major constituents are: amino acids; ribo- and deoxyribonucleotides, chiefly as nucleoside triphosphates, adenosine triphosphate being the most abundant; intermediates of energy-yielding reaction pathways; vitamins and cofactors; lipid precursors; polyamines and, of course, inorganic anions and cations. Knowledge of the dynamics of the cell pool, particularly of free amino acids and nucleotides, is essential to the understanding of many biochemical experiments with viruses whose components are constructed from the soluble pool during virus replication. The soluble amino acid pool, which represents as much as $1/20$ of the total amino acid content of the cell, exchanges every 1 to 2 minutes with the amino acids in the medium. This exchange requires a great energy expenditure because the concentration of most amino acids within the cell is from 5 to 100 times that in the surrounding medium (Levintow and Eagle, 1961). Because of this rapid exchange, "pulse-labeling" experiments with radioactive amino acids can be effectively performed.

Ribo- and deoxyribonucleoside precursors to nucleic acids, however, become phosphorylated once inside the cell and cannot be exchanged back outside the cell. Nevertheless, because the deoxynucleotide pools contain an amount of material sufficient for only a few minutes of DNA synthesis, pulse chase experiments with radioactive deoxyribonucleosides, frequently thymidine, can be effectively carried out.

The ribonucleotide pools, however, are equivalent to the nucleotides consumed during several hours of growth, and are not exchanged with the medium (Mandel, 1964). Thus a truly effective pulse-chase with RNA precursors is not possible. In most cells, the best that can be achieved is a dilution of the labeled intracellular ribonucleotide pool so that the rate of labeling is decreased or maintained at a constant level (Brandhorst and McConkey, 1975). In planning and interpreting experiments where labeled precursors of protein, DNA, or RNA are employed to study viral macromolecular synthesis these characteristics of the acid soluble pools must be borne in mind.

ANIMAL CELL STRUCTURE AND ORGANIZATION

Animal cells have no rigid outer cell wall comparable to bacteria or plant cells. The flexible plasma membrane is, therefore, the only surface that retains or admits necessary intracellular molecules and the only barrier to the entry of particles such as a virus.

With the advent of the electron microscope, not only was the nature of plasma membrane suggested to be a thin lipid bilayer with associated proteins, but a fascinating array of distinctive organelles was discovered to be present in slightly varied form in virtually every eucaryotic (nucleated) cell (Figure 11-3 and 11-4). Many of these newly recognized structures were also membrane bound. In addition to the membrane-enveloped nucleus, which had been observed from the earliest days of light microscopy, circumscribed membranous structures such as mitochondria, "dense" bodies, Golgi bodies, and vast networks of intracytoplasmic membranes, termed the rough and smooth endoplasmic reticulum, were observed. Common nonmembranous structures such as microfilaments, microtubules, and centrioles were also observed in virtually all cells (Porter et al., 1946; Novikoff and Holtzman, 1975).

Deciphering the function of the various cellular organelles made visible by the electron microscope required cell breakage and at least partial purification of the structures so that biochemical examination could be made. This task was begun by Albert Claude and his associates (Claude, 1946; deDuve, 1959) and has been so successful that not only has the general role in cell physiology of all the organelles been established, but many proteins that comprise the structures have been purified and the individual roles of these proteins at least partially outlined.

The mitochondria have been shown to be the center of energy transduction and ATP production via the pathways of metabolic oxidation coupled with phosphorylation. Their intricate double-layered membranes are studded with highly complex structural elements which are, in fact, the enzymes of

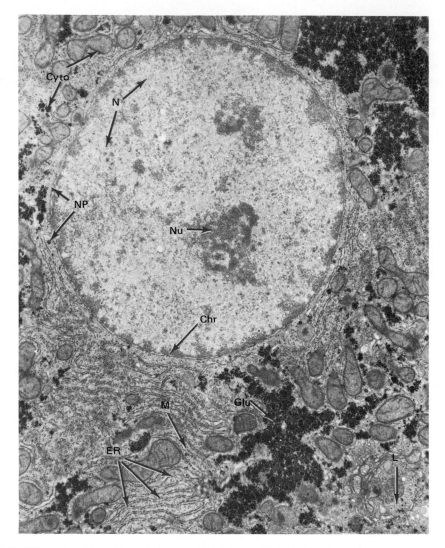

FIGURE 11-3. Electron micrograph of liver cell (×25,000) nuclear and cytoplasmic junction shown (N, Cyto). Marginated chromatin (Chr), nuclear pores (NP) and nucleolus (Nu) indicated. Prominent cytoplasmic structures are L, lysosome, M, mitochondrion, Gly, glycogen, ER, endoplasmic reticulum. (Photograph courtesy of Dr. E. Holtzman).

oxidative-phosphorylation. Abnormal mitochondria are observed in many virus infections and decreased ATP production may be the cause for cessation of virus growth in dying cells.

The "dense bodies" were found to contain hydrolytic enzymes and were renamed "lysosomes," or lytic bodies (deDuve, 1959). A complex system of movement of ingested material through various stages of digestion has been charted in this system. Also old cell fragments are degraded for excretion or for recycling within this intracellular digestive system. Cellular destruction

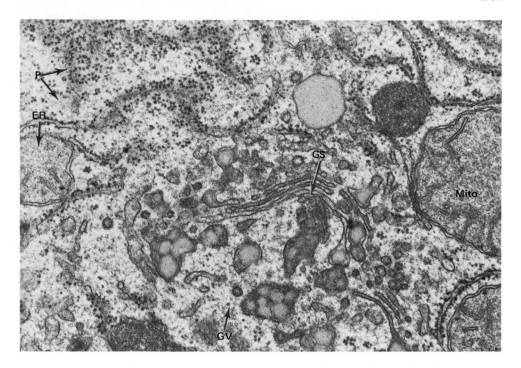

FIGURE 11-4. *Electron micrograph of liver cell (×50,000) Polyribosomes (P) associated with endoplasmic reticulum (ER) very clear. Vesicles and saccules of Golgi apparatus (GV, GS) and mitochondrion are also marked. Photograph courtesy of Dr. E. Holtzman.*

during virus infection may involve release of lysosomal components into the cell cytoplasm with resulting autodigestion.

The endoplasmic reticulum, coupled with functional polyribosomes, is now known to produce proteins for export (Jamieson and Palade, 1968). These proteins contain a lipophilic common amino acid sequence which is responsible for their solution in and passage through the endoplasmic reticulum membrane (Blobel, 1975). The export proteins are then packaged in vesicles within the Golgi bodies, after which a one-way movement to the cell surface is climaxed by fusion of the vesicle packages with the plasma membrane and release outside the cell. It is highly likely that this system is pirated by enveloped viruses that insert their own proteins into the plasma membrane (see Chapter 13).

The microtubules and microfilaments received considerable attention recently because, upon purification, these elements turned out to be simple in molecular structure. The microtubules are composed mainly of a single polypeptide type called *tubulin*. The microfilaments contain *actin* similar to that found in skeletal muscles as well as a *myosin* similar to that found in smooth muscles. The microtubules and microfilaments are thought to be important in maintaining a cytoskeleton underneath the plasma membrane and by contractile properties to be responsible for cell movement (Wessells et al., 1971; Pollard and Weihing, 1973).

These cell structures have an intimate role to play in the process of manufacture of virus particles and in some instances are the targets of destruction during virus-induced cell damage.

Perhaps the cell component that still is less completely understood, and the one that the study of virology might best help to understand, is the plasma membrane. For many years it was assumed on the basis of quantitative lipid determinations (Gorter and Grendel, 1925) and low-angle X-ray scatter data (Schmitt, 1955) that the plasma membrane was a bilayer of lipid with some type of structural association with protein (Davson and Danielli, 1943). The early electron microscope studies of the plasma membrane seemed to confirm this notion, but as more studies were done, confusion set in as to how much of the microscopic image was due to protein and how much was due to lipids. Various extraction procedures seemed capable of removing all the lipid without too greatly changing the electron microscope image, while, on the other hand, pure lipid bilayers produce microscopic images with considerable similarity to the image produced by the plasma membrane. Various alternatives to the lipid bilayer including crystalline lipid structures were discussed. The alternatives to lipid bilayers, such as micelles with central H_2O cores, were thought particularly reasonable to account for some of the physiological properties of membranes such as selective permeability (Luzzati and Husson, 1962).

In the past several years, the prevailing opinion of the structure of the plasma membrane has shifted back to lipid bilayers, which are liquid under physiological conditions, with proteins of various configurations inserted in and through the membrane (Figure 11-5). A good descriptive name for this structure of the plasma membrane is the *fluid mosaic* model (for review see Singer and Nicolson, 1972; Bretscher and Raff, 1975; Edelman, 1976). Among the recent important experiments in this area are: (1) Plasma membranes of packed cells can be frozen and fractured; a replica of the fractured surface can be made and electron micrographs of these fractured surfaces reveal a division between the two lipid layers (Branton, 1969). (2) Proteins that existed in the red blood cell membrane were found to have exposed sections which could be digested away with proteolytic enzymes, leaving portions of the molecule

FIGURE 11-5. *Diagram of possible interactions of surface proteins and glycoproteins. Redrawn from Edelman, Science* **192**, *218, (1976).*

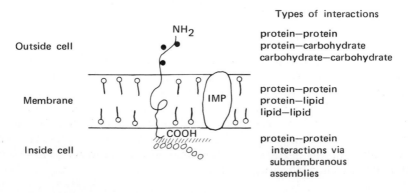

embedded in the membrane. The membrane associated portion is very rich in hydrophobic amino acids while the exposed section is relatively poor in these amino acids (Marchesi et al., 1972). (3) Perhaps the most important findings that led to the fluid mosaic model were made in studies of the mobility of specific binding sites of many substances, notably antibodies (Frye and Edidin, 1970; Taylor et al., 1972) and plant lectins (Nicolson, 1971) in the membranes of living cells. These binding sites, reactive to fluorescent antibodies, for example, were found to be initially distributed over all of the cell surface. After a time (30 minutes or so) the fluorescent bound marker coalesced into "patches" and was then observed to migrate to the cell pole (Figure 11-6). This phenomenon, which was called "capping," depends on the multivalent nature of these binding agents plus the mobility of the binding site within the cell membrane; the attached agent plus the attached cell surface binding site were thereby recruited into clusters with many cross-linking bonds, Clearly, the membranes must be in a fluid state to allow such mobility of the associated antigen-antibody or the complex of lectin-lectin binding sites (Edelman, 1976).

Many of the studies on cell membranes have counterparts in the study of virus membranes, and, in fact, viruses provide some of the purest membrane preparations available for detailed study. In addition, the specific interaction of viruses with cell membrane components, the basis for the first step in virus infections, and the biosynthesis and insertion of virus membrane proteins

FIGURE 11-6. Reaction of fluorescent antiglobulin serum with immunoglobulin on surface of mouse spleen lymphocytes. Fluorescence first shows binding over all the cell surface followed by movement within the fluid membrane of antigen-antibody complexes (patches) leading to final polar distribution ("capping"). Photograph courtesy Dr. G. Edelman.

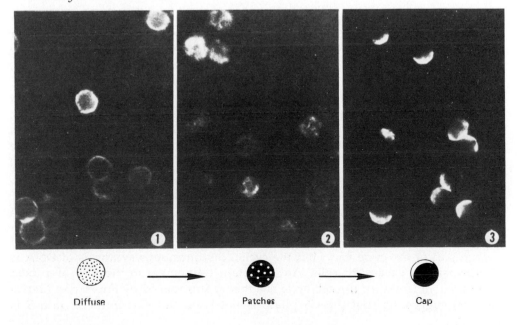

Diffuse   Patches   Cap

into infected cell membranes offer important tools for better understanding cell membrane structure and synthesis.

Although the structure of membranes is more generally agreed upon than it was several years ago, the function of membranes—how they admit or prevent the entry of foreign substances—is considerably less clear and this will be reflected in the discussion of virus entry into cells.

GROWTH AND MACROMOLECULAR SYNTHESIS IN ANIMAL CELLS

A thorough treatment of the current molecular biological understanding of eucaryotic cell growth and division is far beyond the scope of this chapter. However, a presentation of certain features of mammalian cell function is needed to understand properly the importance and the design of certain virological studies to be discussed later. Moreover, many of the more complicated cellular functions may first be understood in miniature by studying viruses.

The most important connection between the preceding discussion of the size and structure of animal cells and the present consideration of cell biosynthesis is the recognition of the physical compartmentalizations of the animal cell. Such separations do not exist in bacteria. Protein synthesis is primarily, if not exclusively, conducted in the cell cytoplasm, while the cell DNA and mRNA manufacturing operations are in the nucleus. Transport—the physical translocation of molecules—within the cell is of great importance and is very poorly understood at present. In this area virological studies play an especially important role because many viruses appear to use cellular channels of communication between the nucleus and the cytoplasm during their growth. For comparison, some viruses utilize the cytoplasm as their exclusive site of replication.

CELL DNA SYNTHESIS   In contrast to bacteria, which synthesize DNA continuously, eucaryotic cells synthesize DNA only at a special phase of their cell cycle. This was first demonstrated by Howard and Pelc (1953) for the dividing cells in onion root tips. A variation of their experiment, performed with HeLa cells which double every 20–24 hours, is shown in Figure 11-7. Cells were exposed to tritiated thymidine for 30 minutes; only about a third of the cells became labeled and none of the cells in mitosis became labeled. Labeled mitotic cells appeared after a lag of about 2 to 3 hours, with a maximum number of labeled mitoses appearing about 8 hours after exposure to label. After about 17 hours the number of labeled mitotic cells declined again to a very low level. Thus a specific DNA-synthesis period, the S-phase, exists. Between the S-phase and mitosis as well as between mitosis and the S-phase, there are two "gaps" in DNA synthesis. The first gap after mitosis is termed G1 and the gap between S-phase and mitosis was termed G^2. A similar cell cycle has been found for all eucaryotic organisms. The lengths of each part of the cycle vary, but the cyclic, discontinuous synthesis of DNA is universal for eucaryotic cells. An important refinement to the idea of a fixed program of times for the cell cycle is a theory introduced by Smith and Martin (1973) suggesting that the G1 phase is not fixed but exit from G1 into S is

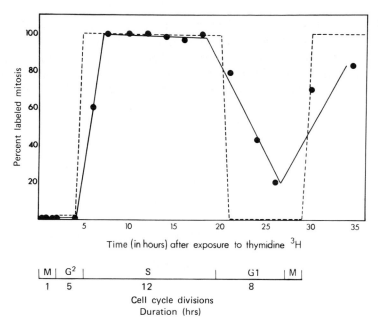

FIGURE 11-7. The cell cycle in cultured mammalian cells. HeLa cells were exposed to ³H thymidine and labeled mitotic cells scored thereafter. The first mitotic cells (M) were unlabeled as were succeeding samples until about 6 hours after labeling. After 18 hours, the frequency of labeled mitotic cells began to decrease, reaching a minimum 26 hours after labeling. The divisions of the cell cycle and their duration are inferred from this experiment: the cell doubling time was 26 hours and mitosis was visually observed to be 1 hour or less; G^2, a gap in DNA synthesis before mitosis is therefore about 5 hours long; some part of the cell DNA is synthesized over a span of 10–12 hours, the S-phase, because all mitotic cells from ~6–18 hours after ³H exposure were labeled. The decline in labeled mitosis after 18 hours signifies another DNA synthesis gap, G1, during the cycle because the doubling time of the culture was 26 hours. Thus G1 is about 8 hours in duration. Redrawn from Baserga and Weibel, Int. Rev. Exp. Path. **7,** 1–31, (1969).

probablistically determined. This idea predicts variabilities in individual cell cycle times which were actually observed.

CHROMOSOME ORGANIZATION AND GENETIC EXPRESSION IN EU-CARYOTES   A second notable difference in bacterial cells and animal cells is the organization of the genes and the relationship, unknown at the moment, between this organization and gene expression. Genetic studies of lower eucaryotes (yeast and molds) have shown that the clusters of related genes (operons) that are coordinately regulated in bacteria either do not exist, or are much rarer than in bacteria (Metzenberg, 1972). The 5000 discrete bands observable in the polytene salivary gland chromosomes of *Drosophila* average about 30,000 base pairs, but it appears likely that each observable band represents one gene. Thus there may be an extraordinarily large "extra" amount of DNA compared to that encoding structural information for proteins. What, if any, function this "extra" DNA performs is unknown.

Molecular biological studies of the question of gene organization and gene expression have taken two main forms: (1) study of the arrangement of sequences within the DNA in hope of identifying structural genes and regulatory sequences (Davidson and Britten, 1973) and (2) examination of the synthesis of RNA in the cell nucleus in an attempt to discover how mRNA is formed (Darnell, 1975).

The studies of DNA structure have centered on reassociation studies of denatured DNA. Britten and Kohne (1968) made the important discovery that unlike bacterial DNA where most sequences are represented once, mammalian DNA contained regions that were repeated exactly or nearly exactly many times. After denaturation of the DNA, these repetitious sequences reassociated much faster than the majority of the DNA (Figure 11-8). By calibration of reassociation rates using phages, bacterial DNA, and homopolymers, they estimated that the DNA of many organisms contained three kinds of DNA: (1) sequences that were repeated many thousands of times and perhaps even more (2) "intermediate" repeat sequences that were present many hundreds of times and included inexact sequence matches, and (3) about 50 to 75% of the

FIGURE 11-8. *Demonstration of rapidly reassociating and slowly reassociating fractions of calf thymus DNA* ($\bullet, \bigcirc, \triangle$) *at three concentrations for various times. The % of reassociation is plotted as a function* C_0t. *The* (+) *symbols are labeled E. coli DNA whose reassociation is unaffected by the calf thymus DNA. From experiments in which DNA that had been reassociated to various* $C_0t's$ *were prepared and reanalyzed a division was made into "foldback" or spontaneously reassociating DNA plus satellite DNA, the most repetitious fraction; intermediate repetitive, and unique DNA. Prepared from data of Britten and Kohne, Science* **161,** *529 (1968) and Britten et al. pp. 187 in "Problems in Biology: RNA in Development, U. of Utah Press, Salt Lake City (1969).*

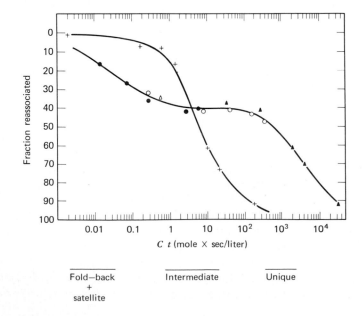

DNA was present as sequences that were unique in the genome. As discussed earlier, it seems unlikely that all this "unique" DNA actually codes for structural proteins.

One class of highly repetitive DNA exists in the centromeric region of chromosomes (Figure 11-9) and incidentally is not transcribed into RNA (Flamm et al., 1969; Pardue and Gall, 1970). This type of highly repetitious DNA includes the so-called satellite DNA. This name refers to the observation that some, but not all, of these highly repetitious centromeric DNA regions have a different average buoyant density from the majority of the DNA and can be separated by equilibrium density sedimentation from the main band of DNA as a "satellite" band. Sequence data on a number of these "satellites" indicates a simple, highly repetitive core structure built up, for example, of repeating hepta- and pentanucleotides for very long stretches with occasional other regions inserted at regular intervals (Southern, 1970; Peacock and Brutlag, 1975). Thus, it is not clear whether to describe such a sequence as repeating each 5, 7, or 12 nucleotides or whether the repeat should be described according to the inserted base that breaks the pattern every several hundred or more. This type of occasional variant base (or bases) is undetectable by reassociation analysis (Botchan et al., 1974), but can be observed when satellite DNA is cut with restriction enzymes.

The less rapidly reassociating "intermediate repeat" type of DNA has also been separated and sequences from this DNA are represented in both nuclear

FIGURE 11-9. *Autoradiograph of a mouse tissue culture preparation after cytological hybridization with radioactive RNA copied in vitro from mouse satellite DNA. The RNA has bound to the centromeric heterochromatin of the chromosomes. Photograph courtesy of Dr. J. Gall.*

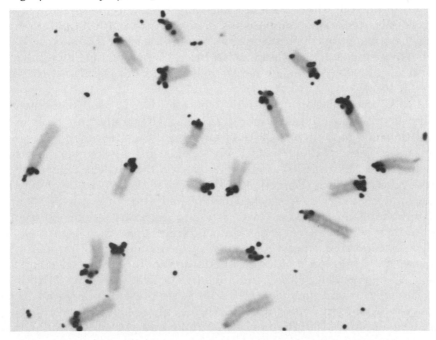

RNA and to a lesser extent in mRNA. Shearing the DNA to graded sizes shows the linkage between repetitive and nonrepetitive DNA and a general pattern for many types of genomes has emerged: 200 to 300 nucleotide sequences of repetitive DNA are interspersed with single-copy regions 1000 to 2000 nucleotides long. Also regions exist that seem to be devoid of repetition for many thousands of bases (Davidson and Britten, 1973).

Another type of unusual DNA sequence has recently received attention (Wilson and Thomas, 1974). When DNA from animal cells is denatured, a certain fraction returns to double-stranded form so rapidly that the reassociation does not appear to take place between two completely separate strands. This "spontaneously reassociating" or "fold-back" fraction is composed of intrastrand hybrids, inferring an inverted repeated region of DNA which allows Watson-Crick base-pairing within a *single strand* of DNA. This fraction exists in short, 200 to 300 nucleotide inverse-repeats as well as regions several thousand nucleotides long. The nucleotide sequences in this fold-back class fall into sequences that are very frequently repeated in the genome as well as regions that may be "single copy", that is occur at most a few times per haploid genome (Perlman et al., 1976).

The meaning of any of the described structural features of the DNA is far from clear at the moment. For example, it is not known how much of the "single copy" or "unique" DNA actually codes for protein sequences. It appears that much of the mRNA from cultured cells comes from the regions of interspersed unique and intermediate repeat DNA, but some mRNA may come from other regions (Davidson and Britten, 1973). How regulation of mRNA formation is accomplished cannot at the moment be inferred from the evidence on DNA structure.

The other type of work that should contribute to the eventual understanding of gene regulation in eucaryotes is the study of nuclear RNA synthesis and mRNA formation. Here the emphasis has been on the characterization of mRNA molecules and comparisons between the mRNA and the nuclear molecules, many of which belong to a class of nuclear RNA termed hnRNA, for heterogenous nuclear RNA, which has a higher average molecular weight than does mRNA. (For review see Darnell, 1968, 1975; Derman, Goldberg, and Darnell, 1976).

Unlike the situation in bacteria where mRNA is used immediately after synthesis, posttranscriptional modification of RNA transcripts prior to mRNA utilization is a prominent feature of eucaryotic cells (Figure 11-10). At the 3' terminus of most mRNA molecules a segment of polyadenylic acid approximately 200 to 250 nucleotides long is added in the nucleus before the mRNA reaches the cytoplasm (Darnell, 1975; Sawicki et al., 1977). At the 5' terminus of the mRNA molecule a blocked methylated structure, $m^7GpppNmpNp$ or $m^7GpppNmpNmp$ termed a "cap," is found in most if not all eucaryotic cells and animal virus mRNA molecules. Without the "cap" structure, mRNA molecules bind ribosomes very poorly and thus the rate of protein synthesis depends on the presence of a cap. In addition to the caps, internal methylated bases, N^6-methyladenylic acid and 5-methylcytidylic acid exist (Shatkin, 1976).

The poly(A) and methylated structures are present in large nuclear molecules as well as cytoplasmic mRNA. Furthermore, several types of sequences that derive from transcription of cell DNA are also present within both mRNA

FIGURE 11-10. Features of eukaryotic mRNA.

Explanation of Symbols

> Cap, A blocked methylated 5' oligonucleotide with general structure
> m⁷GpppNmpNp or m⁷GpppNmpNmpNp
> where N = any of 4 nucleotides and Nmp is 2'0 methyl-ribose

NC 1, Untranslated noncoding region between cap and initiation at first AUG;
probably varies from ~10 to ~100 nucleotides and is rich in A and U
residues.

> i, AUG is initiating codon in eukaryotic as well as prokaryotic systems.
> Coding region, averages 1500 nucleotides for total mRNA from cultured
> cells.

> t, UAA, UAG, UGA are terminators in eukaryotic as well as prokaryotic
> cells.

NC 2, 3' untranslated noncoding region at least 50–150 nucleotides in mRNAs
so far sequenced; strong sequence conservation evident in 3' regions of
hemoglobin genes; A_2UA_3 sequences present in this region in every
eukaryotic mRNA (~10 examples) so far sequenced.

A_{20}-A_{200} Symbol for polyadenylic acid which is about 200–250 nucleotides
initially but becomes shorter with age.

> Summary from Gatlinburg Conference on "mRNA: Structure and Func-
> tion," Darnell, Prog. Nuc. Acid Res. **19**, 493 (1976).

and hnRNA molecules. The shared sequences, which presumably code for protein, include: (1) virus-specific, high molecular weight nuclear RNA and lower molecular weight, virus-specific mRNA (Darnell, 1975); (2) silk fibroin mRNA and a nuclear molecule several thousand nucleotides longer that also contains the coding sequences of silk fibroin mRNA (Lizzardi, 1976); (3) hemoglobin-specific nuclear molecules several times the size of hemoglobin mRNA (Ross, 1976; Curtiss & Weissmann, 1976); (4) an overlap in sequences between the total mRNA and hnRNA from cultured cells that can be detected by molecular hybridization (Herman et al., 1976); and (5) oligonucleotides are present in hnRNA in the form of double-stranded loops and at least a part of these sequences can be detected in cytoplasmic mRNA by hybridization with the nuclear double-stranded regions (Naora and Whitelam, 1975). Implicit in all of the evidence cited above is the cleavage of a higher molecular weight nuclear precursor molecule in the formation of at least some mRNA molecules. What remains unproven in any of the studies is the actual precursor role of the very large hnRNA in mRNA formation. That the true transcripts for mRNA might be much nearer mRNA size, as appears to be true for silk fibroin mRNA, is still a possibility. In any event, posttranscriptional modification, for example, poly(A) and methyl group addition, is necessary for mRNA formation, and it appears quite possible that RNA cleavage occurs to remove at least a small RNA segment.

The suggestion that mRNA might derive through processing of a higher molecular weight precursor came from an earlier finding that both rRNA and tRNA in mammalian cells are, in fact, derived from higher molecular weight RNA precursors (Figure 11-11; for review see Darnell, 1968, 1975). In mammalian cells pre-rRNA is 14,000 nucleotides long (45S RNA) and undergoes a series of specific cleavages within the nucleolus.

Pre-tRNA in mammalian cells has thus far only been detected as units about 50 nucleotides larger than the ~80 bases of the finished molecule. However, the cleavage of pre-tRNA to tRNA by extracts of HeLa cells was the first demonstration of RNA processing *in vitro* (Mowshowitz, 1970). Recent work with bacterial cells has revealed that here, too, processing occurs in the production of rRNA, tRNA, and, in at least one case, mRNA . The early mRNA from T7 bacteriophage DNA is produced by cleavage of a single high molecular weight RNA transcript into specific mRNA segments (see Chapters 4 and 8).

Finally, the most important point about posttranscriptional modifications during eucaryotic mRNA formation remains totally obscure at present. Does the cell use these posttranscriptional steps to regulate mRNA levels? Can the cell choose to add poly(A) or cap differently under different conditions? Although transcriptional control may finally prove to be the most important or even sole locus of regulation in higher cells, it must be remembered that posttranscriptional regulation is a possibility in eucaryotic cells.

The final resolution of the problems of the regulation of gene expression in mammalian cells clearly may be aided by the study of mRNA production

*FIGURE 11-11. The processing of 45 S pre-rRNA from HeLa cells. Top portion of diagram shows the 45 S molecule, the addition of methyl groups to 28 and 18 S regions of the molecule, and cleavage products derived from the 45 S molecule, together with molecular weights. Bottom portion is a tracing of an electron micrograph from Wellauer and Dawid, Proc. Nat'l. Acad. Sci. USA **70**, 2827, (1973) of a 45 S molecule showing characteristic secondary structure pattern.*

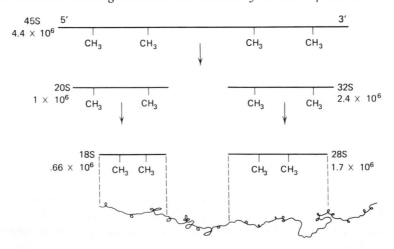

from DNA viruses, such as adenovirus, that enter the cell nucleus and there produce mRNA molecules which function in the cytoplasm (see Chapter 14).

THE RNA POLYMERASES OF EUCARYOTIC CELLS RNA synthesis in eucaryotic cells is not carried out by a single enzyme as apparently is the case with bacteria. Three forms of enzyme, I, II, III or A, B, and C, are found in a variety of eucaryotic cells and tissues (Roeder and Rutter, 1970; Chambon, 1975). Pre-rRNA is synthesized in the nucleolus by polymerase I and polymerases II and III are located outside the nucleolus. Polymerase II is thought to be responsible for hnRNA and mRNA production, while small RNA molecules, at least pre-tRNA and 5S RNA, are formed by polymerase III. The number of functional molecules of polymerase I and II was recently reported (Cox, 1976) to be 2000 and 10,000 per cell, respectively. These relative amounts of the two polymerases agree with the estimated ratio of pre-rRNA and hnRNA synthesis (Soeiro et al., 1968).

Only polymerase III has thus far been demonstrated to correctly initiate its RNA product in a cell free system. Both polymerase I and II will complete RNA chains in isolated nuclei and both are active with various DNA templates in a purified form, but discrete identifiable products with proper termini have not yet been synthesized with these purified enzymes.

Modifications of RNA polymerase are important in the regulation of RNA synthesis, both in bacteriophage replication and bacterial sporulation (see Chapter 4), and such regulatory events may occur in eucaryotic cells, both normal and virus-infected. No evidence on this point is yet available.

The Use of Inhibitors of Macromolecular Synthesis in Mammalian Cells

Many key discoveries in molecular biology were made by studying bacteria bearing mutations that prevent specific macromolecular synthesis: for example, cells unable to make, say, arginine, stop making protein abruptly when placed in medium lacking arginine; cells unable to make thymine stop DNA synthesis quickly in a medium lacking thymine. With such cultures it was found that DNA synthesis was unnecessary for protein synthesis, but protein synthesis was required to begin a round of DNA synthesis. Such specific mutants are not available in mammalian cells. However, the interrelationship of protein, RNA, and DNA synthesis in the production of host or viral macromolecules remains of great interest. Animal cell biologists have therefore resorted to the use of various pharmacologic agents that inhibit, more or less specifically, protein, DNA, or RNA synthesis. No less than with bacterial mutants, care must be exercised to be sure that any effect observed after administration of an inhibitor is due to the primary site of action of the inhibitor. With judicious use and careful interpretation, studies with the inhibitors of macromolecular synthesis have yielded very valuable information on the interdependence of protein and nucleic acid synthesis in both normal and virus-infected eucaryotic cells.

INHIBITION OF PROTEIN SYNTHESIS

Interference with protein synthesis could theoretically occur at initiation, elongation, or termination of a polypeptide chain because these three stages of protein synthesis all involve specific mRNA codons and proteins specific to each step (see Chapter 4). Agents that affect initiation and elongation are in fact available; none is yet known that specifically affects termination. *Pactamycin* is perhaps the most commonly used drug to limit *initiation* (Cohen, et al., 1969). If the concentration used is too high, pactamycin also inhibits chain elongation. Stopping initiation without stopping chain elongation results in the completion of nascent chains and disappearance of polyribosomes. Thus radioactive amino acids introduced after inhibition with pactamycin allows labeling only of the —COOH terminal region of the remaining incompleted chains. This has been a valuable technique in ordering the position of the RNA sequences coding for specific polypeptides.

Edeine is another inhibitor that blocks formation of the complex between the ribosomal subunit and the mRNA. Edeine probably has a greater specificity for inhibition of initiation than does pactamycin and may be used more in the future (Pestka, 1969). A useful *in vitro* inhibitor of chain initiation is histinol, an analogue of 1-histidine which causes histidine tRNA to become uncharged. This leads secondarily to inhibition of initiation (Vaughan and Hanson, 1973).

Many experiments simply demand a quick halt in total protein synthesis, regardless of what step in inhibited. *Cycloheximide* is frequently the drug of choice in this case because chain *elongation* is prevented (>95% in one minute of exposure); the ribosomes are "frozen" in position for as long as the drug is present, and when the drug is removed protein synthesis promptly resumes (Baglioni et al., 1969). Cycloheximide is a member of a class of inhibitors called glutarimides (also including streptovitacinA and acetoxycycloheximide), which act by combining with some factor(s) responsible for translocation of the peptidyl tRNA complex on the ribosome. *Emetine,* an alkaloid derived from the ipecac plant, has a similar drastic and complete action in blocking elongation but emetine is irreversible (Grollman, 1966).

Puromycin, an analogue of the 3' terminus of amino-acyl tRNA, was one of the earliest specific protein synthesis inhibitors and can act at any site from the first peptide bond to the last (Pestka, 1969). Puromycin substitutes for the incoming aminoacyl tRNA and receives the growing polypeptide, which now becomes peptidyl-puromycin instead of peptidyl-tRNA, with the result that the growing chain is prematurely terminated and the ribosome leaves the mRNA chain.

In addition to compounds that act on the protein synthetic machinery to reduce or block the synthesis of proteins, another useful mechanism for exploring the need for ongoing protein synthesis is incorporation of amino acid analogues with the resulting production of faulty polypeptides. Such drugs as parafluorophenylalanine, ethionine, canavanine, and azetidine-2-carboxylic acid, which are, respectively, analogues of fluorophenylalanine methionine, arginine, and proline are incorporated into proteins which are thereby rendered nonfunctional. (Jacobson and Baltimore, 1968).

INHIBITION OF RNA SYNTHESIS

Theoretically a drug might inhibit the initiation of RNA synthesis by combining with the nucleic acid site, either RNA or DNA, where the new RNA chain must begin or by combining with the RNA polymerase before or after it is bound to the initiation site. In bacteria a group of antibiotics, *rifamycin* and derivatives, has been found that does, in fact, combine with free bacterial RNA polymerases to prevent chain *initiation*. No effect on chain elongation is evident (Sippel and Hartman, 1968). Recent evidence suggests that *DRB* (5,6-dichloro ribofuranosylbenzimidazole) may act mainly on chain initiation by the RNA polymerase responsible for hnRNA synthesis in insect cells and cultured mammalian cells (Egyhazi, 1975; Sehgal et al., 1976). Such a drug has vast potential usefulness in helping to localize the point of origin and direction of transcription of both cellular and viral RNA molecules related to mRNA formation.

The most generally used inhibitors of RNA synthesis have been agents that stop *RNA chain elongation*. Again, two mechanisms are possible, binding to the DNA template or binding to the RNA polymerase. *Actinomycin D* intercalates in the DNA helix binding firmly by polypeptide side chains to

$$5'. . . GpCp. . .3'$$
$$| \quad |$$
$$3'. . . CpGp. . .5'$$

dinucleotides and thus preventing RNA polymerase progress down the chain (Reich et al., 1961; Sobell, 1974). This drug is by all means the most widely used RNA synthesis inhibitor and its highly specific binding structure makes it likely that at least for some minutes after it is administered to cells its only action is to limit RNA synthesis. Another drug that binds to DNA via intercalation is *ethidium bromide*, which is particularly interesting because of its rather specific ability to inhibit mitochondrial RNA transcription without affecting nuclear RNA metabolism (Zylber et al., 1969). Mitochondrial DNA is a covalently closed circular DNA and it might be expected that other DNAs that are transcribed in this form would also be susceptible to ethidium bromide. *α-Amanitin*, a principal poison from the mushroom, *Amanita phalloides* (Wieland, 1968) acts directly on RNA polymerase II, the polymerase which appears to be responsible for hnRNA and mRNA production. Penetration into the cell is poor, however, and α-amanitin has been used mainly in cell-free studies involving purified polymerases and isolated nuclei that carry out RNA synthesis (Roeder and Rutter, 1970).

A large group of analogues of the natural ribonucleotides exists which affect RNA synthesis in various ways by acting to repress formation of natural triphosphate precursors, being incorporated with resulting deficient RNA, or acting as competitive inhibitors for incorporation (Suhadolnik, 1970). For example, *azaserine* inhibits enzymes of purine production while high doses of adenosine repress formation of pyrimidines (Ishii and Green, 1973) by feedback repression. *Azaguanine* (Perry, 1965) and *fluorouracil* (Milcarek et al., 1974), which are incorporated into pre-rRNA, prevent proper processing to rRNA. *3'Deoxyadenosine* (cordycepin) acts either as a chain terminator by being incorporated thus preventing further chain elongations, probably the case with pre-rRNA (Siev et al., 1969), or by competitive inhibition with polymerases, as likely occurs in the posttranscriptional addition of poly(A). (Sawicki, et al., 1977). The literature of nucleoside inhibitors is vast and the reader is referred to Suhadolnick (1970) for a good treatise on compounds investigated through 1970.

INHIBITION OF DNA SYNTHESIS

Since the mechanism of *initiation* of chain synthesis for DNA is unclear at the moment there could be no known specific inhibitors for chain initiation. A number of agents, which simply block ongoing DNA synthesis, can also be used to prevent the beginning of the "S" phase of DNA synthesis. This allows cells not in "S" phase to progress through G2, mitosis, and G1, halting just before "S" phase. *Aminopterin*, which blocks one carbon transfer and consequently TTP synthesis, *FUDR* (Fluorodeoxyuridine), which inhibits TTP production probably by feedback inhibition (Littlefield, 1966), high doses of thymidine, which block dCTP synthesis (Xeros, 1962), and *hydroxyurea*, which depletes the deoxypyrimidine pools (Young et al., 1967), all work to stop DNA synthesis by depriving the cell of precursors. *Arabinosyl cytosine (ara-C)* probably acts both by being incorporated and rendering DNA ineffective as well as by depleting the pool of normal pyrimidine deoxyribosides (Suhadolnick, 1970).

The halongenated pyrimidines IUDR (iododeoxyuridine) and BUDR (bromodeoxyuridine) are incorporated into DNA with variable effects. IUDR usually renders the DNA ineffective, but BUDR substitution is compatible with continued DNA function. However, BUDR incorporation causes cells to be sensitive to killing by exposure to ordinary light (Kao and Puck, 1968).

Animal Viruses: Adsorption and Entry 12 Into the Cell

The Stages of Early Infection: Definition

This chapter deals with the initiation of virus infection of animal cells—the initial attachment of viruses to their host cells, the steps leading to entry into the cells, and the beginning of events directed by virus nucleic acid. The specificity of this initial phase of viral infection is especially important because the host range of many viruses is probably determined at this level; if a particular cell is unable to adsorb a virus particle, it is safe from any damage that virus may cause.

The first steps in infection for all types of viruses have traditionally been referred to as *adsorption*, *penetration*, and *uncoating*. Adsorption is an appropriate term for all virions and denotes the initial cell-virus contact which frequently is at first weak, *reversible adsorption*, then progresses to a stronger linkage, *irreversible adsorption*. Penetration and uncoating are unsatisfactory terms if they are used to apply to all of the different events that occur when the many varieties of animal viruses enter cells. Penetration is misleading because it implies an active process on the part of the virion during entry, which has not been demonstrated; rather in many cases it is more likely that without further activity by the attached virion, physico-chemical complementarity between the virus surface and *receptor* molecules on the cell surface mediates the changes necessary for virus entry.

What is meant by virus entry? The key event that begins the intracellular phase of virus infection is the appearance on the *cytoplasmic* side of the plasma membrane of either a viral nucleic acid or a functional viral nucleoprotein. The departure of the virions from the cell periphery (for example, by engulfment), even though the plasma membrane has not been crossed, is clearly not entry into the cell, although this event has been widely observed and has been described as penetration.

275

Although almost the entire virion of some viruses must cross the plasma membrane to cause infection, for most others only a subviral unit or the nucleic acid alone need enter. If entry into the cytoplasm of the whole virion does occur in some cases, it may not lead to productive infection but rather to digestion of the particle. Finally, in many cases access to the interior of the cell may *at the same time* release the nucleic acid or functioning nucleoprotein, thus "penetration" and "uncoating" may be part of one entry process. In this chapter we will therefore treat the early steps of infection in two stages, *adsorption* and *entry*.

One final concept that virologists have used to describe the early events of infection requires definition. Sometime between attachment and the beginning of virus-specific synthesis, that is, during "entry," the virion becomes so altered that it cannot be recovered as a stable infectious particle. Its nucleic acid may be recovered as infectious material but with a much lower efficiency of infection. This loss of virion integrity has been termed *"uncoating"* or *"eclipse"* and the phase of infection between adsorption and the appearance of new virus has been termed the *eclipse phase*.

Knowledge about the molecular basis of animal virus adsorption and the subsequent activation of nucleic acid function has increased considerably in recent years, but still is less detailed than about the entry of the DNA of some bacteriophages (for example, T-even phages) into bacteria. It seems likely that no animal virus possesses a complex infection mechanism comparable to the T-even phages for introduction of its nucleic acid into host cells. In fact, agents with diverse surface characteristics (for example, adenoviruses, polioviruses, myxoviruses, and pox viruses) may differ greatly in the manner in which they gain effective access to the cell interior.

The molecular events between virus attachment to the cell and the beginning of virus-directed synthesis occur very rapidly and cannot yet be reproduced step by step with isolated virus and cell components. We are dependent, therefore, on various experimental tricks to define the stages of the entry process. Before describing the entry of specific viruses we will outline these experimental definitions of various stages of early infection.

EXPERIMENTAL METHODS OF INVESTIGATING VIRUS-CELL INTERACTIONS

Success in the study of the first step of infection, specific adsorption, depends on the accuracy with which the virus particles can be assayed. After Dulbecco (1952) introduced the plaque assay for animal viruses, two types of quantitative assays for adsorption became available. The most direct assay of adsorption, which is applicable at all multiplicities of infection (m.o.i.) simply measures the initial virus input, and the virus unattached to cells after varying times of exposure. The virus removal per unit time per cell can be calculated as the attachment rate constant for the particular virus-cell pair under study (see Chapter 2). The use of radioactive virus allows a simple assay of cell-associated and free virus by scoring radioactivity rather than infectivity. A second technique measures the attachment rate by mixing cells and virus at low m.o.i. and determining, after various exposure times, the number of infected cells. This is accomplished with suspension culture cells by plating the exposed cells in the presence of other susceptible cells which will serve as indicators of infections arising from single infected cells ("infectious center" assay). If the exposed cells were already attached as a "monolayer" in a Petri dish, the rate of adsorption to the monolayer can be simply scored by removing excess virus at various times and scoring the subsequent appearance of plaques. With these quantitative assays for infection, the effects of environmental changes—temperature, pH, and composition of medium—on virus adsorption can be accurately determined.

In contrast, disappearance of the virus into the cell and/or disruption of the virus particle, can be measured only in an indirect manner. Among the techniques and criteria useful in following the fate of cell-associated virus during entry ("penetration" and "uncoating") are: loss of accessibility of virus to antiviral antibodies; recoverability of viral infectivity from the cell surface by physical or chemical treatment of the virus-cell complex; isolation of cell receptor material capable of specifically inactivating virus infectivity; the effect of temperature, metabolic inhibitors, and surface binding agents on the entry process (Lonberg-Holm and Philipson, 1974). A powerful tool in studying the fate of adsorbed virions is the electron microscope (Dales, 1973). It must be remembered, however, that the ratio of physical particles to infective particles is high for many animal viruses so that visual determination of infectious particles is impossible. Morphologic observations must generally be interpreted in combination with biochemical evidence.

THE GENERAL PICTURE OF ANIMAL VIRUS ADSORPTION

Work with a variety of animal viruses, both enveloped and nonenveloped, allows the following general picture of virus attachment (see Lonberg-Holm and Philipson, 1974 for review). Virions collide randomly with sites on the cell surface and about one in every 10^3 or 10^4 collisons leads to physically complementary union between a site on the cell and a site on the virion, probably with the aid of ions in the medium (Valentine and Allison, 1959; Koch, 1960; Koch and Feher, 1973, Ogston, 1963). The surface site on the virion may be a projection of a special virus protein such as the spike of the enveloped toga-, myxo-, and paramyxoviruses, and the fiber protein projecting from the icosahedral adenovirus, or the virus-binding site may be an individual structural protein on the virion surface or a mosaic of several capsid proteins (as might be the case with the picornaviruses). The receptor site on the cell is probably in all cases a surface protein, frequently a surface glycoprotein. Different specific sites exist on the cell for different viruses, but overlap between viruses make it possible to describe "families" of viruses related according to binding sites. Viruses that share binding sites may or may not be otherwise related (Lonberg-Holm, Crowell and Philipson, 1976). A specific type of binding site on a susceptible host cell may be present in as many as 10^4 to 10^5 copies.

Once attachment of the virion has occurred, infection is not by any means assured. The initial binding can be "loose" or "reversible," that is, the virion can leave the cell surface; some of the cell-associated virions, however, proceed to a tighter "irreversible" binding. The suggestion has been made that this tighter binding involves multiple-site attachment, possibly by recruitment of additional binding sites within the fluid mosaic membrane. Once the particle is firmly bound, the pathways vary widely for different virions. The following section will outline some of the experimental support for this general picture of the initial animal virus-animal cell attachment.

NATURE OF INITIAL CELL-VIRUS BOND

Animal viruses associate with charged surfaces such as silica gels, calcium phosphate gels, and treated cellulose (Puck et al., 1951; Levintow and Darnell, 1960). Perhaps because of these properties, it was suggested some time ago that the first stage in cell-virus interaction was electrostatic. Just as the binding of viruses to charged substrates depends on conditions of ionic strength and pH, so does the binding of viruses to cells. For example, adenovirus type 7 binds best to erythrocytes between pH 5.5 and 8.7, a range where a positively charged arginine on the virion and an ionized carbonyl group on the cell have been suggested as participants in an interaction (Neurath et al., 1970). The range of pH optima for specific virus-cell interactions may be quite narrow, for example, Coxsackie B4 attaches best between pH 3.0 to 3.5 to HeLa cells, but conditions for interaction are quite diverse when many different cell-virus pairs are considered, for example, pH adsorption optima exist from 3 to 8.7 (Lonberg-Holm and Philipson, 1974).

Monovalent cations in the medium promote maximum attachment of several viruses—polio (Holland and McLaren, 1959), NDV (Sagik and Levine, 1957), adenovirus (Neurath et al., 1970), and influenza (Hirst, 1965). In addition, divalent cations have also been implicated in binding for some infections. The rate of attachment of Coxsackie A9 and human rhinovirus type 2 to HeLa cells is increased by Mg^{++} or Ca^{++} and inhibited by the presence of the chelating agent, EDTA (ethylene diamine tetra-acetate) (Lonberg-Holm and Korant, 1972).

A final bit of evidence that electrostatic interactions may occur between cells and viruses comes from the now classical demonstration that a neuraminidase from *Vibrio cholera*, the receptor-destroying enzyme, RDE, would render cells unable to bind influenza virus. The action of RDE removes a negatively charged group, N-acetyl neuraminic acid (NANA) from the cell surface, decreasing the affinity for influenza virus (Gottschalk, 1959). Either the electrostatic or chemically specific binding to NANA apparently mediates influenza binding.

Thus electrostatic bonds appear important in allowing proper cell-virus contact. It is not clear, however, whether the ions act as salt bridges or play a role in helping achieve the correct complementarity between virus and cell proteins or sugars on the glycoproteins of either virus or cell.

RATES OF ADSORPTION

The speed of interaction of various viruses with their host cells differs widely. For example, Sindbis, an enveloped togavirus, is completely adsorbed to a monolayer of host cells virtually as soon as exposure is afforded (Purifoy and Sagik, 1968). Poliovirus, while adsorbing sufficiently rapidly to permit synchronous infections, has a measureable adsorption constant of 4 to 5×10^{-9}

cm^3 min^{-1} $cell^{-1}$ to HeLa cells (Darnell and Sawyer, 1960) and 2 to 3×10^{-8} to monkey kidney cells (Bachtold et al., 1957). Two naked icosahedral viruses, polio and adeno, have lower attachment rates when adsorption is performed in the cold, while many enveloped viruses seem to still combine well in the cold. Finally, when poliovirus-HeLa cell complexes are warmed there is frequent dissociation of the initial union (Joklik and Darnell, 1961). All of these results—speed, variable tightness of binding of viruses, and the temperature effect—suggest that the stabilization of the initial electrostatic cell-virus union requires multivalent attachment of virions to the cell. To achieve a multivalent attachment for poliovirus, for example, would almost certainly require the recruitment of additional receptors into the region of the initial attachment site. Only about 10^4 receptors per cell exist for polio (Lonberg-Holm and Philipson, 1974) and if they are randomly distributed over the several hundred square microns of surface area per cell, a single particle of 0.02 microns in diameter might well not have more than one receptor in the vicinity of the first cell-virus bond. The myxoviruses, on the other hand, may stick to all glycoproteins that contain a sialic acid, an interaction that might not necessitate a recruitment of sites from a great distance, and therefore firm binding might occur in the cold. A recruitment of virus binding sites would be analogous to "capping" or "patch" formation described for antibodies to cell surface proteins (see Chapter 11). Many electron micrographs show that, soon after attachment, virions are partially surrounded by very closely applied plasma membrane, a configuration that could result from multisite attachment to a number of protein units within the flexible plasma membrane causing the membrane to wrap around the virion (Dales, 1973). The idea of multiple site attachment is attractive, but not yet proven.

NATURE OF VIRION PROTEINS INVOLVED IN ATTACHMENT

Electron micrographs of some, but not all, types of animal virions reveal regularly arranged thin filamentous surface protein structures about 2 nm in diameter and from 10 to 30 nm in length (Figure 12-1). Clear evidence exists in at least some cases that these fibers, or "spikes" as they have been termed, are the mediators of initial cell-virus binding. As such, they represent some of the most highly studied and best understood proteins of any contained in animal cell virions.

Most viruses that possess spikes are enveloped; myxoviruses (influenza), paramyxoviruses (NDV), rhabdoviruses (VSV), togaviruses (Sindbis), as well as some of the less commonly studied virus groups such as coronoviruses and probably arenaviruses, all of which contain spikes (Dales, 1973; Figure 12-1). Vaccinia and herpes viruses are enveloped, but lack visible spikes; they might, however, contain projections too short to visualize. In most cases, virion spikes are probably composed of one or two types of polypeptide either or both of which are glycosylated. The sugar moiety varies from virus to virus and also may vary for a particular virus depending on the host cell that produced the virion (Burge and Strauss, 1970).

Perhaps the spikes from the myxoviruses (Laver, 1973) are chemically the best understood of any virus proteins (Figure 12-2). In 1941 both Hirst (Hirst, 1941) and McClelland and Hare (1941) discovered that influenza virus could

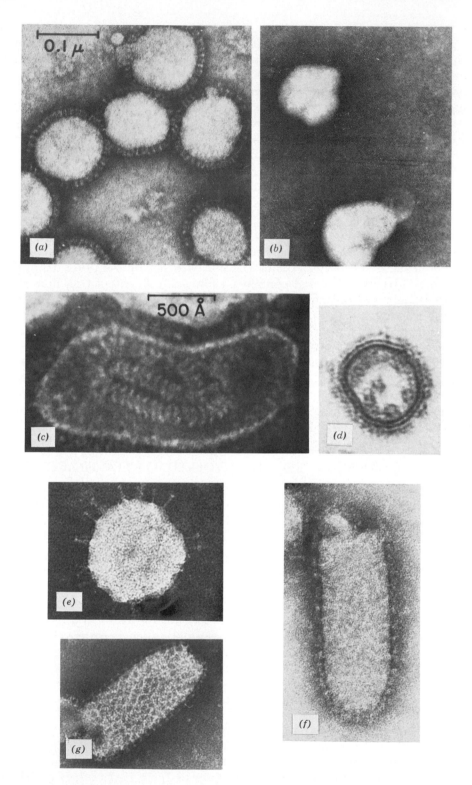

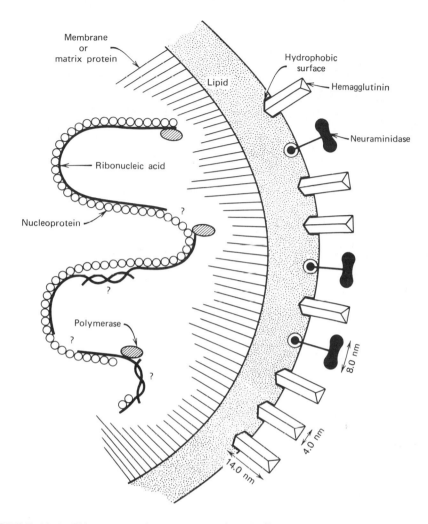

FIGURE 12-2. *Diagrammatic structure of an influenza virus particle. From W. G. Laver,* Adv. Virus Research **18,** *51–103, Academic Press (1973).*

adsorb to and cause red blood cells to clump or agglutinate. It is now known that this hemagglutination reaction is due to one of the two "spike" proteins of influenza virus. A large number of animal viruses have the capacity to hemagglutinate a wide variety of different red blood cells (RBC), each virus characteristically favoring attachment to certain cells. This test has provided a

FIGURE 12-1. *Surface projections of animal viruses. Electron micrographs reveal surface projections that probably function in cell attachment. (a) Influenza, (b) Influenza virus digested with bromelain, a protease; such particles will no longer attach to cells; (c) Simian virus 5, a paramyxovirus, (d) Simian virus 5 showing bilayer membranes, virion is just above cell membrane, (e) avian infectious bronchitis virus, a coronavirus, (f) vesicular stomatitis virus, (g) rabies virus. Photographs courtesy of Drs. P. Choppin, a–d, and R. Simpson, e, f and g).*

TABLE 12-1 HEMAGGLUTINATION BY ANIMAL VIRUSES

Virus	Red blood cell type Conditions and comments	Hemadsorption
Myxoviruses	Chicken, guinea pig, human, and others; monovalent ions necessary, reaction proceeds at 4°, spontaneous elution at 37° in wide pH range; Receptors destroyed.	Positive
Paramyxoviruses NDV, mumps	Fowl, human, guinea pig, 40° optimum; elution occurs; receptors destroyed.	Positive
Measles	Monkey, 37°, no elution	Positive
Togaviruses	Young chicken or goose cells preferred, fairly sensitive to pH (6–7 required), inhibited by normal lipids in serum.	Positive
Rhabdoviruses	Goose, 4°	Positive
Coronaviruses	Human	
Bunyawera	Goose, neutral pH critical	
Picornaviruses Coxsackie, Echo	Group 0 human cells; virion has hemagglutinating capacity, reaction occurs in cold, some spontaneous dissociation occurs at 37° without affecting cells ability to be reagglutinated	Negative
Polyoma	Guinea pig cells most commonly used	Negative
Adenoviruses	Variety of cells hemagglutinated. Reaction proceeds at 4° up to 37°	Negative
Parvoviruses	Human, guinea pig, mouse, 4°	Negative
Reoviruses	Human cells agglutinated by all types; only type 3 agglutinates ox cells; RDE destroys ox cell receptor, but virus does not.	Negative
Pox viruses	Many cell types agglutinated, adult chicken commonly used, optimum temperature 37°, no elution; hemagglutinin separable from infectious particle and not made by all strains.	Positive

very important means of quick, quantitative assay for viruses that do hemagglutinate. It has proved especially important in clinical studies where antibodies of convalescent patients can be detected by combining with virus in the so-called *hemagglutination inhibition* test. Table 12-1 lists the hemagglutinating viruses and conditions for demonstrating hemagglutination. An additional interesting use of the reaction of RBC with virus proteins is the so-called hemadsorption reaction; cells containing virus spikes at their surface adsorb RBC just as virus particles do (Shelakov et al., 1958).

Hirst also discovered (1950)that when the influenza virus-RBC complex is warmed to 37° the virus elutes from the cells and the cells are no longer easily agglutinable. The chemical basis for the overall reaction involves two spike proteins (Figure 12-2): one binds to the red cell and the second, a viral enzyme, can destroy the binding site on the cell (Compans and Choppin, 1974). The multiple hemagglutinating spikes in a single virion are able to bind two cells together and, acting in concert, many virions can create multiple bridges between cells leading to the agglutination.

The cell receptors contain neuraminic acid and the virus enzyme is a neuraminidase. Upon warming the virus cell complexes, the enzyme catalyzes the removal of N-acetylneuraminic acid (Figure 12-3) and the bridges between red blood cells disappear (Gottschalk, 1959; Laver, 1973). In nature, where the virus encounters mucopolysaccharides in nasal and respiratory tract secretions, it may use this enzymatic capacity to get to the plasma membranes of cells. The spikes with neuraminidase activity may also play a role in adsorbing to susceptible cells because mice can be protected from influenza virus by inhalation of RDE or immunization against neuraminidase (Stone, 1948; Schulman et al., 1968). The function of the spike proteins and their identity with the hemagglutinating property of influenza virus can be diagrammatically illustrated by treating the particles with proteolytic enzymes. The spikes are no longer visible and the treated particles fail to hemagglutinate (Compans et al., 1970; Figure 12-1).

Paramyxovirus virions and the spike proteins from these virions can also

FIGURE 12-3. *Site of action of neuramidase.*

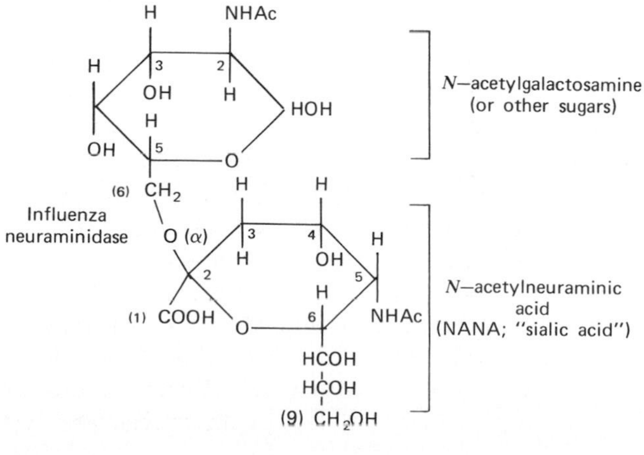

agglutinate red blood cells, but in this case the same spike proteins contain both the hemagglutinin and neuraminidase activity (Scheid et al., 1972).

The spikes of Sindbis virus, a togavirus, are also known to be the organs of cell attachment because, as was shown with influenza virus, treatment of the virions with the proteolytic enzyme bromelain removes the spikes viewed by electron microscopy and the particles can no longer adsorb to cells (Compans, 1971). Gel electrophoretic analysis of the Sindbis virion proteins after such digestion revealed that the surface glycoprotein(s), but not the core protein, were destroyed by the bromelain treatment.

Rhabdoviruses possess a spike protein that has also been purified and has a hemagglutinating activity (Arstila et al., 1969).

The only icosahedral viruses yet described that contain protein projections similar to the spikes of enveloped virions are the adenoviruses. The fiber of adenoviruses is a multimeric protein, probably consisting of two or three identical linearly polymerized polypeptides plus a globular terminal protein that could possibly be a different polypeptide (Philipson et al., 1968). The fibers can adsorb to susceptible cells and, in fact, can seriously depress host protein metabolism (see Chapter 15). Removal of the spikes and the base to which they are joined renders adenovirions unable to attach to cells (Laver et al., 1969; Prage et al., 1970).

Adenovirions can cause hemagglutination as can the isolated fibers so long as the fibers are aggregated. Monomers of fiber protein thus apparently do not possess multiple binding sites and the intracellular bridges required for hemagglutination do not form.

The adenovirus particle contains fibers at each of 20 vertices, and each vertex probably contains only a single cell binding site. To immobilize such a large particle it is thought likely that firm attachment of the virion requires cooperative binding of several fibers of the same virion to cell membrane sites that are recruited into the local area around the particle (Lonberg-Holm and Philipson, 1972). Figure 12-4 shows adenovirus, with its characteristic cubic shape intact, which must be bound in at least two places to the cell membrane.

The sites of cellular attachment for viruses that lack spikes is a matter of conjecture. As will be discussed more thoroughly in a later section, all of the surface structure of icosahedral virions like the picornaviruses can be considered part of one large polymeric protein network (Talbot and Brown, 1972; Lonberg-Holm and Philipson, 1974); thus binding sites may exist in a polypeptide mosaic at the surface of such virions and the conversion of "loosely" bound to "tightly" bound particles may involve movement within the membrane to bring more than one binding site in contact with the appropriate surface structure. Support of the complex nature of the binding site for these nonenveloped icosahedral viruses comes from the finding that picornaviruses can be subtly altered in their structure by acid treatment (Korant et al., 1972) or exposure to cells or cell extracts (Crowell and Philipson, 1971). The particles are still similar in size and, although they have lost the smallest internal viral protein (Lonberg-Holm and Butterworth, 1976), their surface proteins are all still intact. Such particles, however, no longer bind to the cell, an indication that the structural alteration has either caused the binding site on a single surface polypeptide to become hidden, or, just as

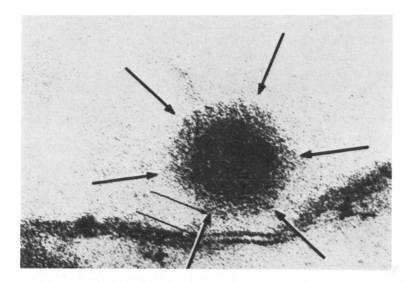

FIGURE 12-4. *Effect of adenovirion attached to cell surface. Icosahedral shape of particle still evident (arrows) and effect on membrane structure underlying particle evident even though particle separated from cell by distance equal to fibers which exist on particle surface. Photograph courtesy of Dr. D. Brown.*

likely, a complex binding site created from several polypeptides has been destroyed by the structural change in the particles. In line with the conclusion that a changed surface architecture is responsible for removing the cell binding site, such particles also are antigenically changed, the so-called D → C transition (Lonberg-Holm and Yin, 1973).

SPECIFICITY OF CELL RECEPTORS

Perhaps the most important aspect of cellular receptors for animal viruses is that infection and virus disease are largely determined by the ability of viruses to bind to the cells of a particular animal species. In addition, within a given species, the expression of different genes in different cell types and within the same cell types at different stages of development can result in the formation of a different sets of specific virus receptors.

For example, poliovirus does not find receptors on nonprimate cells while cultured primate cells of a variety of types—monkey kidney cells, HeLa cells (human carcinoma cells) and human diploid fibroblasts—all have poliovirus receptors (McLaren et al., 1959). In humans, poliovirus can infect cells of the nasopharynx, gut, and the anterior horn cells of the spinal cord and cause poliomyelitis (the dread "infantile paralysis"). However, poliovirus finds no receptors in a variety of other tissues, as tested by the inability of extracts to bind and inactivate the virus upon subsequent exposure to susceptible cells (Holland, 1963). Other viruses appear to attack a wider range of tissues, for example, encephalomyocarditis virus has receptors in heart muscle, brain tissues, and other tissues as well (Kunin, 1964). Some viruses may have receptors in virtually every tissue, for example, measles virus may infect epithelial cells of nasopharynx and lungs, as well as spread via lymphatics and possibly the blood stream to skin and subcutaneous tissues, and, finally, may also infect the central nervous tissue.

Tissues that have receptors at one time during the life of an animal may not always possess such receptors; for example, Coxsackie B1 and B3 virus, which can affect humans at all ages causing nasopharyngeal, skeletal muscle, and heart muscle infections, can only infect newborn mice, not adult mice (Kunin, 1964). On the other hand, Coxsackie A receptors are absent in most mouse embryo tissues except differentiating myoblasts correlating with the fact that muscle infection, producing "myositis," is the primary disease caused by Coxsackie A in mice (McLaren et al., 1960).

Receptors for various viruses are divisible into "families" (Figure12-5, Table 12-2). HeLa cell fragments bind all three serotypes of poliovirus, detected by the reduction of infectivity for intact cells of a virus suspension. Pretreatment of the cell fragments with killed poliovirus type I decreases the binding ability of the cell fragments for all three poliovirus types, but not for the binding of Coxsackie B, NDV, or equine encephalitis viruses (Quersin-Thiry, 1961; Quersin-Thiry and Nihoul, 1961). Recent quantitative measurements of relatedness between cell receptors for different viruses have been carried out (Lonberg-Holm et al., 1976). Saturating amounts of virions of one

*FIGURE 12-5. Test for overlap in adsorption sites of various virions. Radioactive virions were adsorbed to cells in presence of excess unlabeled virions and fraction of labeled virions attached to cells measured (radioactive virion/competing virion). Virions used were coxsackie virus, type A21 (Cox A21), coxsackie virus, type B3 (Cox B3) human rhinovirus, type 14 (HRV-14), adenovirus, type 2 (Ad-2) and poliovirus, type 2 (Polio 2). Redrawn from K. Lonberg-Holm et al., (1976 Nature **259**, 679).*

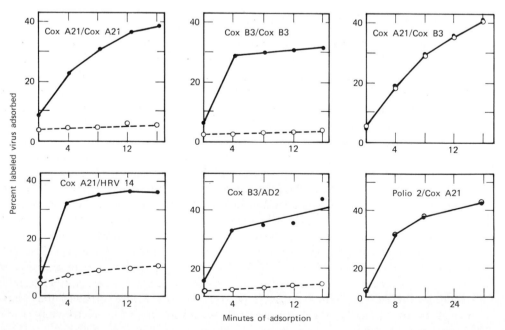

TABLE 12-2   VIRUS RECEPTOR FAMILIES IN CULTURED HUMAN CELLS

Family	Radioactive viruses tested	Excess inhibitors			
		HRV-2	HRV-14	AD-2	Polio-2
I	HRV-2	+	−	±	−
	1A	+	−		
	1B	+	−		
II	HRV14	−	+	−	−
	3	−	+		
	5	−	+		
	15	−	+		
	39	−	+		
	41	−	+		
	51	−	+		
	Cox A21	−	+	−	−
III	Cox B 3	−	−	+	−
	Adeno 2		−	+	−
IV	Polio 1				+
	2	−	−	−	+
	3				+

HeLa cells were tested for adsorption of radioactive virus (viruses tested) in the presence of excess of listed viruses.
HRV = human rhinovirus or Cox = coxsackie virus (see Figure 12-5).

type were mixed with cells to determine the effectiveness in blocking sites for a second radioactively labeled virion. Human rhinovirus type 2 blocked sites for types 1a and 1b. Human rhinovirus type 14 inhibited the binding of human rhinovirus types 3, 5, 15, 39. However, the two rhinovirus groups did not cross-react. In addition, shared receptors were found for otherwise unrelated viruses; for example, adenovirus, type 2 blocked Coxsackie B3 adsorption and partially blocked human rhinovirus type 2. Extension of this type of experiment is of obvious value in learning about the number, specificity, and distribution of cell surface glycoproteins.

The cellular genes responsible for different virus receptors can be located on different chromosomes. Hybrid cell lines that contain chromosomes from both mouse and human cells lose human chromosomes upon continued culture. The disappearance of particular chromosomes or chromosome groups is correlated with the loss of susceptibility to particular human viruses. For example, human chromosome 19 was identified in this way as having the gene for poliovirus receptors (Medrano and Green, 1973; Miller et al., 1974).

Thus receptors of many different types must exist, coded for by specific genes, but it is also true that some viruses share the same cell receptors for attachment.

PROBABLE GLYCOPROTEIN NATURE OF RECEPTORS

Receptors in the membrane of virus-susceptible cells are in a dynamic state. Receptor material is constantly lost to the medium surrounding cells and if cells are treated with various agents, trypsin or neuraminidase, for example,

TABLE 12-3 PROPERTIES OF VIRIONS RELATED TO ATTACHMENT AND ENTRY INTO CELL

Virion characteristics			Components that must enter cells		
Naked	Enveloped	Spikes	Infectious nucleic acid	Nucleocapsid plus enzymes to make mRNA	Structurally altered nucleocapsid
Picornaviruses		−	+		
Papovaviruses		−	+		
Parvoviruses			+		
Reoviruses			+		
	Togaviruses	+		+ (Capsid must be partially digested)	+ (? Enzymes are capsid proteins)
Paramyxoviruses	Paramyxoviruses	+		+	
	Myxoviruses	+		+	
	Rhabdoviruses	+		+	
	Retroviruses	+		+ (Enzymes to make DNA for integration)	
	Vaccinia	?		+ (Core structure more complicated than just helical nucleocapsid)	
Adenoviruses		+	± (DNA has associated covalently-bound protein)		+ (?)
	Herpes	?			
	Bunyaweravirus	+	} unknown		
	Arenaviruses	+	} unknown		
	Coronavirus	+	} unknown		+ (?)

they temporarily lose virus receptors. Receptors are regenerated, however, with a characteristic time for each different type of molecule, for example, 6 hours for half-regeneration of adenovirus type 2 after subtilisin digestion (Philipson et al., 1968), 1 to 2 hours for poliovirus receptors after trypsin treatment, and in the same cells the return of Coxsackie B receptors requires at least 5 hours (Levitt and Crowell, 1967). Regeneration of N-acetylneuraminic acid-containing receptors cannot be accomplished by simple readdition of sugar moiety; the entire glycoprotein must be resynthesized (Marcus and Hirsch, 1963).

In the foregoing discussion the nature of the cellular receptor has been assumed to be a surface glycoprotein because of trypsin and/or neuraminidase removal of receptor activity. Purification of a cellular receptor has been achieved in only one instance so far. Human red blood cells contain a variety of antigenic glycoproteins, which form the chemical basis for the various blood groups. The glycoprotein responsible for MN blood group reactivity is apparently the influenza virus receptor protein (Marchesi et al., 1971).

All hemagglutination, however, is clearly not the result of attachment of viruses to the same glycoprotein or even to sialoglycoproteins. Many icosahedral viruses, share with the viruses-containing spikes the capacity to adsorb to and hemagglutinate red blood cells (Table 12-1). The chemical nature of the sites for the "nonspiked" viruses differs from that of myxovirus receptors, however, because removal of sialic acid does not affect binding. The binding sites for many of these viruses may be to other types of glycoproteins, however, because various sugars are capable of preventing hemagglutination by enteroviruses (Lerner, et al., 1965) and reoviruses (Lerner, 1968).

Entry Into the Cell and "Uncoating" of Animal Viruses

THE VARIETY OF THE PROBLEM

Once a virion is firmly bound to the surface of a susceptible cell the next step(s) in infection result in the entry into the cell of part or all of the virion and the beginning of virus-specific protein or mRNA synthesis. The initial binding may follow similar principles for many viruses; the actual entry and activation of the virus genome probably follows many different pathways (Table 12-3).

Enveloped viruses and naked viruses clearly encounter different physicochemical problems during cell entry. It has long been considered that some "membrane melting" or fusion process is a possible mode of entry for the enveloped virion (Hoyle, 1962). However, the only mechanism known by which relatively large protein structures such as naked virions might enter the cell was phagocytosis, a variant of which, "viropexis," was hypothesized to result in virus entry (Fazekos de St. Groth, 1948).

In recent years another important mechanical necessity with respect to entry processes has come to light (Table 12-3). In some cases the only component of the virion that actively participates in making new virus material is the nucleic acid, whereas in other cases the virion nucleic acid plus a virion-associated RNA or DNA polymerase are involved in replication and therefore must enter the cell. Potentially, therefore, the problem of crossing the cell membrane exists for naked nucleic acid on the one hand and nucleoproteins on the other.

Some viruses that lack virion polymerases clearly need have only their nucleic acid enter the cell, a fact most strikingly demonstrated by the infectivity of isolated RNA from picornaviruses (for example, polio), togaviruses (such as Western equine) and of isolated DNA from polyoma or SV40 viruses. However, there are large DNA viruses lacking virion polymerases (for example, herepesvirus and adenovirus give extremely low yields of infectious nucleic acid). It may be that these viruses still require some part of their virion structure to facilitate the initiation of replication. Thus here also a nucleoprotein may possibly require entry to the cell.

Finally, there are at least two general types of viruses with virion polymerases: (1) those virions with polymerases that can be activated by "stripping" that is, physically disrupting or partially disrupting a virion lipoprotein envelope (for example, vaccinia, RNA tumor viruses, or NDV) and (2) those viruses that may require proteolytic digestion of a part of their capsid structure in order to have a virion polymerase activated (for example, Reovirus, blue tongue virus of sheep, or cytoplasmic polyhedrosis virus of insects).

It seems clear that the molecular events of virus entry must be very different for different viruses although the end result of initiation is always the same—the beginning of virus protein or mRNA synthesis. This result is achieved whether particles are enveloped or nonenveloped or carry a polymerase or do not carry a polymerase.

EXPERIMENTAL APPROACHES TO STUDYING ENTRY

There are two major experimental approaches for examining the entry process of animal viruses—the morphologic and the biochemical approaches. In the former, sections of cells are examined with a high-resolution electron microscope in an attempt to localize the invading particles and to determine the time and place of disappearance of the characteristic features of the virion (Dales, 1973). The biochemical approach mainly involves radioactively labeled virions, virion components, and infectious particles to determine when the physical integrity of the virion is lost (Lonberg-Holm and Philipson, 1974).

On the one hand the electron microscopic approach has led some observers to the conclusion that "viropexis" (Fazekos de St. Groth, 1948), apparently a rather nonspecific engulfment of attached particles, is responsible for getting the virion within the confines of the cell, but not yet within the cytoplasm. According to this model, once inside a vesicle, enzymatic action leads to release of virus into the cytoplasm (Dales, 1973).

On the other hand, the biochemical approach has emphasized the possibility of physical interactions between plasma membranes and virus surfaces which, by the nature of their interactions (? allosteric), leads to physical changes in the virion and in the membrane (structural rearrangement of nonenveloped viruses and membrane fusion with enveloped viruses) with the result that entry is accomplished in a specific manner. This type of entry is envisaged both at the cell periphery or from within a vesicle. Although no definitive statements are possible yet about a single case, let alone all cases of virus entry, the second set of conclusions seem to have the better of the argument at the present time. The wide variety of problems encountered in entry of animal virus genomes is best illustrated with details from specific cases.

ENVELOPED VIRUSES

MYXOVIRUSES AND PARAMYXOVIRUSES   The mRNA of these enveloped viruses is formed by a virion polymerase. Therefore, the end result of virus entry in this case must, as a minimum, admit a nucleoprotein. In purified disrupted virions, the virion polymerases act while the RNA is in a helical ribonucleoprotein form, suggesting that the entire helical nucleocapsid must enter tne cell (see Chapter 13). Early morphological observations (Hoyle, 1962) showed cell fragments could "fuse" with the envelope of influenza virus with the discharge of helical nucleocapsid material. Electron micrographs taken during the first few minutes of infection also suggested "surface melting" between cell membrane and virus envelope accomplished entry of myxoviruses (Morgan and Rose, 1968) and paramyxoviruses (Morgan and Howe, 1968; Figures 12-6 and 12-7). Nevertheless, many particles are seen apparently still intact in intracellular vesicles, and adherents of "viropexis" conclude that it is these particles that are responsible for infection. In this latter case, however, it is entirely possible that membrane fusion of the virion membrane with the membrane of the intracellular vesicle could be responsible for nucleocapsid release.

It seems particularly likely for paramyxoviruses that cell membrane-virus envelope fusion is the mechanism of entry. This group of viruses possesses a powerful hemolytic action and a capacity to fuse different cells. Sendai virus is best known for this property and, after ultraviolet-irradiation, Sendai virions have been widely used to promote heterokaryon formation between cells (Okada, 1958). Sendai virus is normally grown in embryonated chicken eggs (Homma, 1971; Scheid and Choppin, 1974). Such virus is not only infectious for chicken cells, but also causes the production of particles in a strain of bovine kidney cells. However, the particles produced by the bovine cells are not infectious for either cell type. The glycoproteins normally seen in the egg-grown virus are absent from the bovine kidney cell-grown virus and in their place is a higher molecular weight glycoprotein (Figure 12-8). Treatment of the abnormal virions with proteolytic enzymes not only cleaves the presumed precursor glycoprotein to the normal-sized glycoprotein, but also restores the infectivity of the bovine kidney-grown virions. Coincident with the restoration of the infectivity, another activity is regained—the ability of Sendai virus to promote cell fusion. Thus it appears that infection is possible only when these virions are in a state where cell fusion also can be promoted, suggesting thst such fusion activity is necessary for infection. This phenomenon of particles becoming infectious after surface glycoprotein cleavage could be of major importance in determining host ranges of viruses for animal species or for cells of a particular animal. The cells possessing the proper proteases will be both the most susceptible to infection and the most productive of infectious virus.

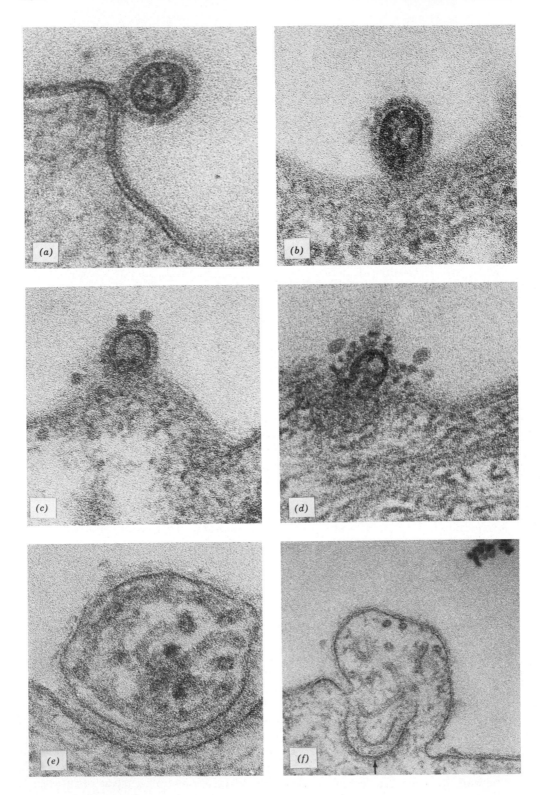

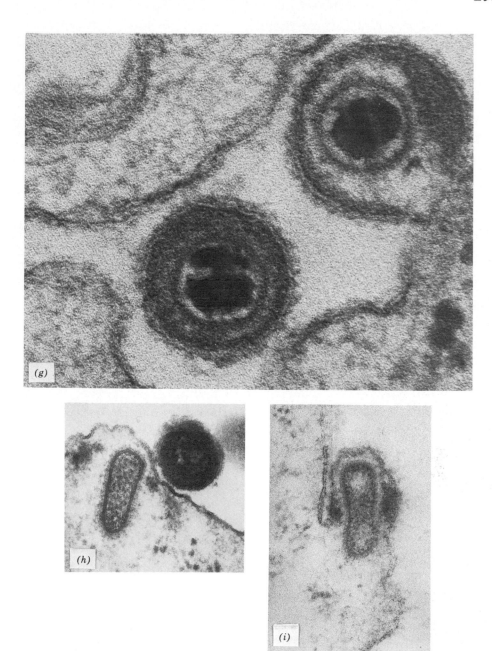

FIGURE 12-6. *Electron micrographs exhibiting apparent virion entry by fusion with cell membrane. (a–d) Influenza particles attached and entering chick embryo fibroblasts (180,000×). (e–f) Attached Sendai virion with membrane beginning to envelop particle; after fusion particle has released helical nucleocapsid (200,000 ×). (g) Herpes particles between microvilli at surface of a HeLa cell. Particle at right has fused with membrane; if microvillus were closed off it would appear as if particle at left were engulfed (260,000 ×). (h–i) Vaccinia virions that have entered L cells by apparent fusion with plasma membrane; lateral bodies of virion still associated with membrane. Photographs courtesy of Dr. C. Morgan. (a g) and Dr. S. Dales (h–i).*

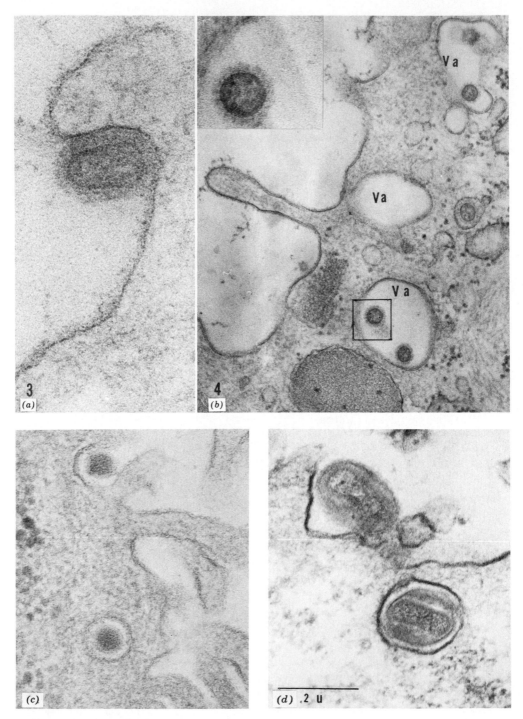

FIGURE 12-7. Entry of virions by apparent "viropexis." Particles appear to be inside cell without apparent morphologic change. (a and b) Influenza virion attached at cell surface and present in vesicles within cell; inset higher magnification of a vacuole. (c) Adenovirion attached at surface and another within vesicle. (d) Vaccinia virion apparently intact within a cytoplasmic vesicle; another virion is attached at cell surface. Photographs courtesy of Dr. S. Dales.

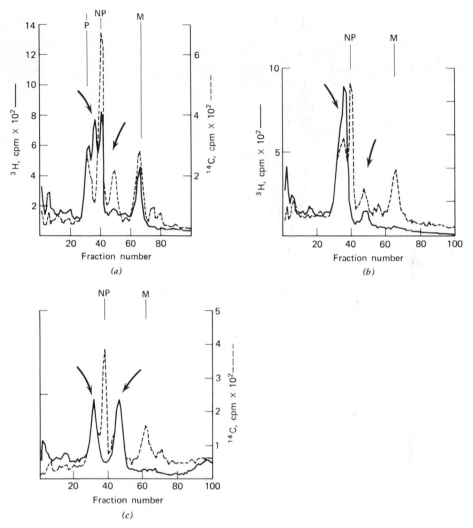

FIGURE 12-8. *Sendai virus proteins involved in fusion and infectivity.* (a) *Co-electrophoresis of amino acid labeled polypeptides of Sendai virus grown in chick embryo cells (----, * ^{14}*C label) or in bovine kidney cells (——, * ^{3}*H label). The chick virus is infectious; the bovine virus is not. The proteins marked P (polymerase), NP (nucleocapsid protein) and M (membrane protein) are the same in the two samples. The two large arrows (↘) point to the differences (presence or absence of polypeptides in the two samples). (b and c) Sendai virions from bovine cells labeled with amino acids (----) or glucosamine (——), which labels the sugar of glycoproteins, were compared before (b) and after (c) trypsin digestion. The large glycoprotein (arrow) present in the bovine cell-grown virions is cleaved to yield a normal sized glycoprotein like that in chick embryo grown virions. The conversion of this protein restores the ability to cause both cell fusion and infectivity. Redrawn from Scheid and Choppin, Virology* **57,** *475 (1974).*

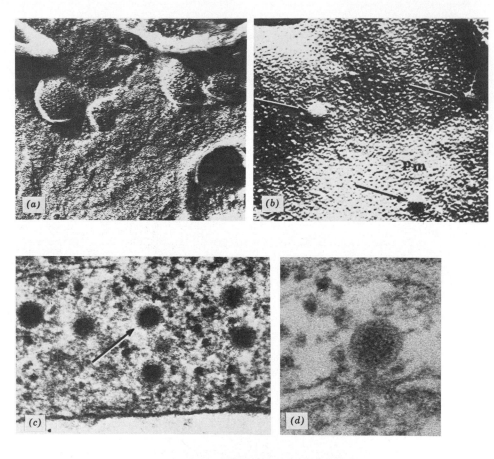

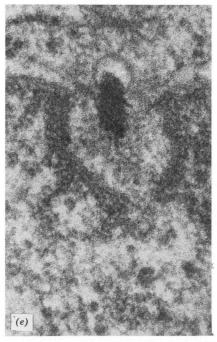

VACCINIA   The entry of vaccinia virus into the cytoplasm requires that a relatively huge organized core structure—essentially the virion lacking only the lipoprotein envelope—be admitted to the cytoplasm. The early mRNA is transcribed by virion associated DNA-dependent RNA polymerase contained in this core structure (Kates and McAuslan, 1967; Munyon et al., 1967). Within a few minutes of mixing cells with vaccinia virions, "core" structures are apparent within the cytoplasm (Figure 12-7; Dales, 1973). When the m.o.i. is quite high many particles end up in intracellular vesicles. It is not known how many particles ever emerge from such vesicles nor is it known whether such vesicles form when the m.o.i. is quite low. Recent electron microscopic evidence has identified a number of particles on the cell surface engaged in a fusion process with the plasma membrane. In any event, it seems possible that some form of cell membrane-virus envelope fusion is responsible for cores emerging from intracellular vesicles, just as it is possible that some digestive process leads to membrane rupture and admission of cores.

An elaborate scheme proposed some years ago by Joklik (1968) for vaccinia virus "uncoating," which involved an induced cell protein to make DNA available for transcription has now been shown to be incorrect. Disruption of the envelope and subsequent activity of the virion polymerase, the first-discovered virion polymerase (Kates and McAuslan, 1967), signals the beginning of vaccinia-specific synthetic activity. Host cell macromolecule synthesis is not required in this process (see Chapter 14).

HERPES   Herpes is an enveloped DNA virus with a complicated icosa-hedral double shell of protein surrounding a very large DNA genome (96 × 10⁶ daltons). Micrographic evidence of free, intracellular nonenveloped nucleocapsids, as well as vesicles containing still-enveloped particles, has been described (Morgan et al., 1968a). Regardless of the mechanism of nucleocapsid entry, the DNA must finally arrive in the nucleus, because it is here that cellular RNA polymerases and other enzymes are responsible for herpes mRNA formation (Wagner and Roizman, 1969). Empty nucleocapsids have been visualized in the cytoplasm with the DNA presumably having entered the nucleus.

NONENVELOPED (NAKED) VIRUSES

ADENOVIRUSES   The beginning of adenovirus-specific synthesis in infected cells requires a chain of events similar to that for herpes, the difference lying in the initial modes of attachment. Both require the entry of DNA into the nucleus from a nucleocapsid structure that is delivered into the cytoplasm. Whereas herpes has an envelope, adneovirus, as discussed previ-

FIGURE 12-9. *Stages of adenovirus entry into cells.* (a *and* b) *Freeze fracture of membrane with adenovirus embedded within plasma membrane (Pm). Photographs courtesy of Dr. Dennis T. Brown.* (c) *Adenovirions within cytoplasm; hexagonal outline less distinct perhaps because of fiber-penton removal. Photographs courtesy of Dr. D. T. Brown.* (d *and* e) *Adenovirus DNA apparently being discharged at nuclear border at or near a nuclear pore. Photographs courtesy of Drs. S. Dales* d) *and C. Morgan* (e).

ously, is bound to the cell by projecting fibers or spikes (Figure 12-9). Adenovirus has also been described as subject to "viropexis" by some workers (Chardonnet and Dales, 1970) or as being quickly admitted through the plasma membrane by others (Morgan et al., 1968b). Evidence favoring direct penetration comes from the freeze-fracture technique of plasma membrane analysis performed on infected cells (Brown and Burlingham, 1973). In these experiments, the lipid membrane is physically fractured between the lipid bilayers revealing embedded adenovirus particles (Figure 12-1). Even in the reports describing adenovirus entry by "viropexis," each vertex in turn appears to be bound so that the particle is internalized by what could easily be numerous specific virus-cell bonds. The first intracellular stage is a subviral particle minus its fibers and the penton base units to which the fibers are attached (Lonberg-Holm and Philipson, 1969). The resulting structure is clearly "looser" than native virions because 70% of the DNA can be digested by DNAse at this stage. A second intracellular stage can be detected as the hexon capsid structure is lost, leaving a DNA-internal protein complex. The DNA complex and the hexon-capsid can be separated by isopycnic banding in CsCl. Finally, the DNA must be delivered to the nucleus from which virus mRNA later emerges. Morgan et al. (1968b) and Chardonnet and Dales (1972) demonstrated that particles at the nuclear border engaged in what could be delivery of the DNA into the nucleus (Figure 12-9). It is not known whether the DNA-protein complex detectable in extracts of infected cells must be released at the nuclear border in order to succeed in entering the nucleus.

POLIOVIRUS AND RHINOVIRUSES Chemical treatment of polio and rhinovirus produces a number of altered forms of virions and subviral particles. Study of the relationship of these changed particles to similar structures that occur during entry into the cell has been very instructive. Both poliovirions (Summers et al., 1965) and rhinovirions (Lonberg-Holm and Korant, 1972) contain four types of polypeptides VP1, VP2, VP3, of 25 to 35,000 daltons and VP4 of about 6000 daltons. Polypeptide VP4 is probably internal, as manifested by its inability to become labeled with iodine after lactoperoxidase treatment of whole particles (Lonberg-Holm and Butterworth, 1976), although it can be iodinated when capsids are disrupted. Initial attachment of poliovirus to HeLa cells at 4° results in loss of attached particles in about 50% of cases when the cell-virus complex is warmed to 37°, (Joklik and Darnell, 1961). The eluted particles are structurally changed and are unstable in 3M CsCl in which they are normally stable for years. The structural alteration of eluted virions of Coxsackie B3, a related picornavirus, was associated with an almost quantitative loss of VP4 (Crowell and Philipson, 1971). Thus a physical change in the structure leads to a loss of an apparently internal protein. Further evidence on the changed nature of the eluted particles was obtained by antigenic analysis. Fully infectious poliovirus combines with antibodies termed "D" reactive; killed virus and eluted virus combines with antibodies termed "C" reactive (Katagiri et al., 1968).

The stages of dismantling of bound infectious particles was first examined by the loss of virion-bound light-sensitizing dye (neutral red) as a measure of virion integrity (Wilson and Cooper, 1962; Schaffer and Hackett, 1963). Whole virions can be recovered until structural changes of adsorbed virions allow escape of the dye. These structural changes occur during the first hour of infection but occur *after* the antibody sensitivity of the invading viruses is lost. Loss of antibody sensitivity does not necessarily correlate with entry into the cell, of course, because binding of virion protein to many cell surface sites might preclude antibody access to the virion.

Recent studies have concentrated on careful assessment of the structural changes secondary to chemical treatment of poliovirus and rhinoviruses, a picornavirus family that has a naturally less stable capsid structure. These changes have then been compared to the physical and biological changes that occur to virions after adsorption to cells (Figure 12-10).

Rhinoviruses are very sensitive to acid pH (which incidentally renders them unable to survive stomach acidity), losing infectivity almost entirely after brief exposure to pH 5 (Hamre, 1968). Particles treated in this manner degrade into two components that sediment more slowly in zonal sedimentation analysis than native particles (Korant et al., 1972). The slowest sedimenting structures, the B particles, contain VP1, 2, and 3 but have lost VP4 and the viral RNA; the A particles, which sediment between the B particles and the native virions, still contain RNA but have lost VP4 (Figure 12-10). Coincident with these structural changes, the rhinovirus particles also undergo an antigenic change, equivalent to the D to C conversion mentioned earlier for eluted poliovirus particles (Lonberg-Holm and Yin, 1973). Similar A and B particles, lacking VP4 and either containing or lacking RNA, have also been prepared by alkaline treatment of poliovirus (Katagiri et al., 1971).

To explore the possible physiological meaning of the so-called A and B particles, a comparison between these changed virions and adsorbed particles was made both during a normal infection and during virion attachment in the presence of compounds that block or reduce virion infectivity. Loosely attached rhinovirus particles can be recovered from cells by simply washing extensively with EDTA-containing buffer, while loosely bound poliovirus can be freed from cellular receptors by treatment with the nonionic detergent NP40. Tightly bound rhinovirus is released by NP40, while tightly bound poliovirus requires treatment with SDS for release from cellular components. The released particles can be monitored in two ways, (1) as radioactive structures with characteristic sedimentation behavior, and (2) as infectious particles, which obviously require a normal ability to readsorb to cells. Even when the exposure of cells to human rhinovirus type 2 was limited to 7 minutes at 34.5°, 68% of the adsorbed virus underwent eclipse; that is, was no longer infectious in a second round of infection, and after 45 minutes had passed almost all of the cell associated virus was eclipsed (Lonberg-Holm, 1975). In the case of adsorbed poliovirus, the eclipse was slightly less rapid, but was also largely complete within 37 minutes. In both these experiments almost all the adsorbed, recovered, noninfectious particles were either in the A or B configuration; that is, they had suffered a change in the physical properties of their capsids. That this change was mediated by activity in the plasma membrane of the cells was strongly suggested by an experiment using concanavalin A (Con A). This plant lectin is known to bind to a variety of cell surfaces that possess glycoprotein sites containing an α mannoside linkage. When a cell is saturated with Con A, the movement of previously free antigenic components of the plasma membrane is prevented; the membrane

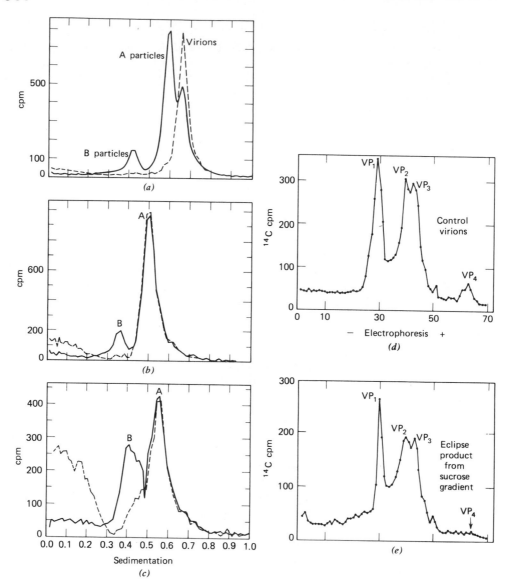

FIGURE 12-10. "Eclipse products" of rhinovirions reisolated from HeLa cells. (a) Sedimentation of purified virions (V, - - -) compared to altered A and B particles released from infected cells. (b and c) Progression in adsorbed particles from A to B with time; adsorption for 20 minutes is compared to 45 minutes. In addition particles were labeled with amino acids (———) and ³²P (- - -) to demonstrate that A particles contain RNA while B particles do not. (d and e) Gel electrophoresis of proteins for purified virions (d) and from A particles (e) demonstrating loss of VP4 (B particles have same proteins as A particles). Redrawn from Lonberg-Holm and Korant, J. Virol. **9,** 29 (1972); and Lonberg-Holm and Noble-Harvey, J. Virol. **12,** 819 (1973).

proteins are "frozen." Presumably, the turgidity of the membrane has increased because of the many cross-links between the attached multivalent Con A molecules. When HeLa cells were treated with Con A at the same time the poliovirus or rhinovirus was adsorbed, virus adsorption was decreased. That which did adsorb was not eclipsed nearly so rapidly and A particle formation was significantly inhibited. In a presumably different way, treatment of infected cells with a thiopyrimidine also impeded conversion of particles to the A form and blocked eclipse (Lonberg-Holm, 1975b). Thus conversion to A particle form is synonymous with eclipse. All of these studies lend support to the proposal that tight binding of the virion to the cell membrane (probably through multivalent attachment) is associated with a structural change that loosens up the virus capsid, allowing the RNA to penetrate the cell membrane.

A final very suggestive experiment on the conditions for entry of poliovirus RNA has been performed. Cells were infected in the presence of a protein synthesis inhibitor, pactamycin, which at the right concentration acts primarily on initiation of protein synthesis. The final conversion of particles from A (RNA-containing, structurally altered) to B (lacking RNA, same protein structure as A) was substantially inhibited by pactamycin (Butterworth et al., 1976). The A to B conversion might be a measure of entry of RNA into the cell. Since the first thing an entering RNA of a picornavirus must do inside the cell is to be translated, an association with ribosomes may complete both the entry of RNA and the final conversion of the particle structure. Whether or not this hypothesis is proven to be true, the experiment with pactamycin is suggestive and demonstrates that a highly coupled set of events must be studied to carry the analysis farther.

REOVIRUSES Entry of the reoviruses into the cell cytoplasm in an effective form presents a qualitatively different problem from any other virus group (Joklik, 1975). The complicated double-shelled, icosahedral structure is superficially analogous to herpes or adenovirus, but here the similarity ends. The genome is double-stranded RNA and virus-specific synthesis is initiated by the action of an RNA-dependent RNA polymerase in the virion. Activation of this polymerase in purified reovirus virions can be accomplished by chymotrypsin treatment, which removes two specific polypeptides. The same virion polypeptides are removed from the virions in the infected cell. The entry of reovirus by phagocytosis into what appears to be lysosomal vesicles that contain digestive enzymes would appear to offer the proper environment to produce an effective subviral particle (Dales, 1973). The problem remains that true lysosomal vesicles contain RNAse, so some provision must be made either to shield the cores and emerging mRNA or have the cores admitted across the membrane prior to activation of the virion mRNA polymerase.

Conclusion

Binding sites for viruses in cell membranes are demonstrably specific. The physical details of crossing the cell membrane and entering the proper cell compartments for initiation of replication are too poorly understood to claim that here too specificity must occur. However, because the various viruses have such widely differing needs in terms of what structures must be delivered inside the cell, it seems logical that the entry processes will be found to be specifically mediated by viral protein: membrane protein complementarity of some sort rather than leaving such events to chance engulfment.

Animal Virus Multiplication: The RNA
13 Viruses

Although the multiplication of animal viruses and bacterial viruses involves the same operations—transcription and replication—the actual mechanisms are often so different that the two types of viruses are best considered separately. These differences are partly a result of the differences between their hosts but also reflect evolutionary specializations that have no obvious relation to the nature of the host cell. There is no reason to believe that animal viruses evolved from bacteriophages (or vice versa) and attempts to draw parallels between phages and animals viruses can be misleading.

The term animal viruses will refer here only to vertebrate viruses. General usage excludes insect (invertebrate) viruses (Chapter 18) from this category, although many viruses can grow both in insect and in vertebrate cells. Also numerous plant viruses can grow both in insect and in plant cells (see Chapter 17). There are presumably great similarities among the plant, insect, and animal viruses, and in the future an integrated picture of the viruses of higher cells will probably emerge. No viruses are known that grow both in bacteria and in any higher cell.

One of the striking differences between bacterial and animal viruses is the time required for a one-step growth cycle. For animal viruses, the fastest growth cycles are 5 to 6 hours, some require days, and many viruses establish only persistent infections in which the cell is not killed but rather it and its progeny produce virus indefinitely (see Figure 13-1). The long growth cycle, compared to the shorter cycle of most phages, is probably a function of the relative size of the two host cells. A mammalian cell has a volume about 1000-fold that of *E. coli* so the virus-specific proteins have a larger volume in which to float around. The elementary processes are no slower in higher cells than in bacteria—RNA and protein synthesis, at least, proceed at about the same rates—but morphogenesis and the rates of acceleration of virus-specific processes are slower.

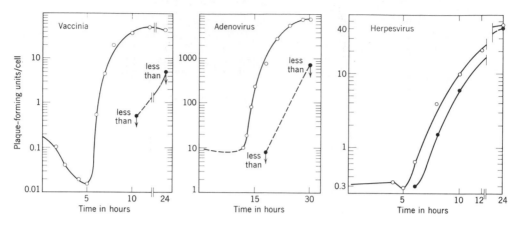

FIGURE 13-1. *One-step growth curves of various viruses in cell cultures. All data are from experiments in which cell cultures were infected with sufficient virus so that every cell in the culture was infected. The open circles (-○-) represent the total virus in the culture; the closed circles (-●-) represent virus spontaneously released from the infected cells. All infectivity measurements were by plaque assay and the results are presented as the average numbers of plaque-forming units per cell assuming that all cells yielded virus. (Sources: poliovirus: Darnell et al., 1961; Newcastle disease virus: Franklin et al., 1957; reovirus: Gomatos et al., 1962; vaccinia: Salzman, 1960; adenovirus: Green, 1962; herpesvirus: Kaplan, 1957). From Darnell, 1965.*

Many specializations of animal viruses relate to specific aspects of eukaryotic cell architecture. Most DNA viruses make their DNA in the cell's nucleus. Proteins for all viruses are made in the cytoplasm and many viruses have their mRNAs designed to make polyproteins that are cleaved to the functional proteins by cellular proteases. The enveloped viruses use the machinery that the cell has for membrane biogenesis. The tumor viruses make themselves part of the cell's chromosomes. Oddly, no viruses appear to have a home in the cell's mitochondria.

Animal viruses have developed numerous highly specialized functions that make each class of viruses a separate organism. There are at least four different general strategies by which the RNA of various viruses is transcribed and replicated. Each class of DNA virus has a characteristic way of replicating its DNA. The hierarchies of proteolytic processing vary widely and must serve diverse functions. The variety of mechanisms employed by the various animal viruses is one of the most intriguing aspects of virology.

When animal viruses infect cells, two general outcomes are possible. Either the infected cells is killed, but produces a large yield of virus (a *lytic* interaction) or the cell continues to multiply while making small amounts of virus. Cultures of growing cells that produce virus are called *persistently infected*. Almost all animal viruses can, under special conditions, generate persistently infected cultures but many viruses rarely lyse cells and usually establish a persistent relationship. There is, however, no phenomenon known among the animal viruses that is an exact parallel of lysogeny although in

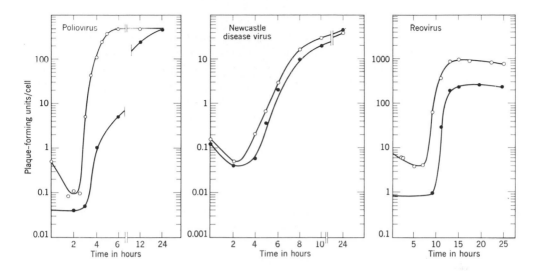

some cases integration of cellular and viral DNA does occur. A variety of other mechanisms other than lysogeny exist for nonlytic, productive host-virus interactions (see Chapter 15).

This chapter first presents the outlines of a scheme for organizing the myriad animal viruses according to their modes of transcription and replication. Each of the classes of animal viruses is then discussed and their characteristic properties are emphasized.

There are five separable events that generally occur during a successful lytic infection of cells and involve the functions of virus-specified proteins. These events may take place simultaneously or serially through the course of a one-step virus growth cycle; their timing depends on the specific properties of the virus. The events are: *inhibition of cellular functions* (discussed in Chapter 15), *transcription* of viral mRNA, *replication* of the viral genome, *morphogenesis* of the virion, and *release* of the virus from the cell.

The transcriptive events that synthesize viral mRNA are usually initiated very soon after infection and continue throughout the course of the viral life cycle. For most DNA viruses, sequential synthesis of two or more classes of transcripts occurs. For most RNA viruses, the evidence for qualitative changes in the transcripts made during the course of an infection cycle is at best slight.

Except for those RNA viruses for which replication and transcription are the same event, replication usually commences some time after transcription. Replication may then continue for a short time, generating a pool of progeny molecules that are later integrated into progeny virions. More usually, however, replication continues through the whole later part of the viral growth cycle.

Morphogenesis of progeny virions is the final intracellular event of the viral life cycle. Where complex virions are formed, an individual virion may require hours for completion; for simple virions a few minutes may suffice.

The mechanism of release of virions from infected cells varies markedly depending on specific factors for the viral life cycle. Viruses that bud from the cell surface as the final step of maturation never have a cell-associated infectious form (see Newcastle disease virus and herpesvirus in Figure 13-1) (Franklin, 1962). The picornaviruses, such as poliovirus, apparently encode a cell-lysis function analogous to that of bacteriophages, allowing them to be released in a burst at variable times after intracellular virion synthesis ceases (Lwoff et al., 1955; Amako and Dales, 1967). Some DNA viruses escape very slowly from infected cells and seem to depend on nonspecific cellular disintegration for their release (see vaccinia and adenovirus in Figure 13-1).

Viral Genetic Systems: Classification of Animal Viruses

To discuss the multiplication of individual animal viruses, it is convenient to group them in classes that have common strategies of behavior (Baltimore, 1971b). Extensive enough understanding of the molecular biology of most animal viruses has accrued over the last 15 years to allow meaningful groupings. For simplicity, and because of the central role of mRNA in viral multiplication, the relationship of the virion nucleic acid to the mRNA provides the best initial distinction among viruses. In this way, six classes of animal viruses can be distinguished. The pathways of replication and transcription used by different viruses may be called *viral genetic systems.* Homology of viral genetic systems is the best way we have to determine the evolutionary relatedness of different viruses.

The Watson-Crick base-pairing rules imply that for any given RNA molecule, a sequence of bases complementary to that RNA can be written. As a useful convention for classifying viruses, one defines viral mRNA as "plus" RNA and its complementary sequence "minus" RNA. Six classes of animal viruses can be defined by the structural relationship between their mRNA and the nucleic acid in the virion (Baltimore, 1971a). Of course, bacteriophages, insect viruses, and plant viruses could be placed in such a classification but at present the system is most useful in thinking about animal viruses.

Figure 13-2 diagrams the relationship of the virion nucleic acid of the six classes of viruses to their mRNA. Class I viruses are the *double-strand* DNA viruses such as vaccinia virus. They make their mRNA in the same way that

FIGURE 13-2. Relationship of virion RNAs to mRNAs. The various classes of animal viruses are defined by their transcriptional mechanisms. Examples of each class are provided. Adapted from Baltimore, 1971a with permission of the publisher.

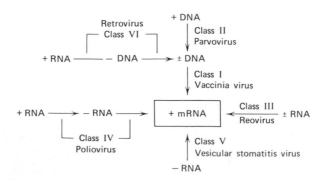

cells make their mRNA: a double-stranded DNA genome acts as template for the synthesis of the mRNA. The Class II viruses are *single-strand* DNA viruses. Their mRNA may be the same strand as their DNA so an intermediate synthesis of minus-strand DNA is needed to be template for mRNA synthesis.

The other classes are viruses with an RNA genome. The Class III viruses are the *double-strand RNA* viruses, such as reovirus, for which double-stranded RNA acts as template for the asymmetric synthesis of mRNA. It happens that the virions of all Class III viruses so far described have *segmented genomes* (multiple chromosomes), each of which encodes a single polypeptide.

Class IV viruses are also called *positive-strand* RNA viruses: for these viruses the genome RNA has the same polarity as their mRNA. Two types of Class IV viruses can be distinguished. Class IVa viruses, typified by poliovirus, are viruses for which a single mRNA molecule serves as a template for all viral protein and all of the proteins are made as a single *polyprotein* which is then proteolytically cleaved to generate the functional proteins. For Class IVa viruses all the mRNA has the same length as the genome RNA. Class IVb viruses, also called togaviruses, synthesize at least two forms of mRNA in the cell: one is the same size as virion RNA while the other is a fragment of virion RNA.

The Class V viruses are known as *negative-strand RNA viruses*. For these viruses, the virion RNA is complementary in base sequence to the mRNA. Thus the virion contains a template for making mRNA but does not carry any sequences that encode proteins. Two subdivisions of Class V can be distinguished. Class Va viruses have a single molecule of RNA as their genome and they synthesize a series of mRNAs from this single template strand. In all known cases, these mRNAs are monocistronic. Class Vb viruses have segmented genomes, each molecule of which is a template for synthesis of a single mRNA. The mRNAs made by Class Vb viruses may be able to encode polyproteins or may be monocistronic.

Class VI viruses are also known as *retroviruses*. They are the most unusual RNA viruses we know because they direct the formation of a DNA that acts as the template for making their mRNA. The mRNA and virion RNA are of the same polarity and some at least is of the same length. Many fascinating consequences result from this remarkable viral genetic system.

Key properties of many of the known types of animal viruses are shown in Table 1-1. Included there are all the viruses discussed in this chapter as well as a number that are not mentioned.

Replication

The replication of each class of viruses is as specialized as their transcription. Class I and II could be extensively subdivided according to the replication patterns of their DNA. Rather than using class designations it is easier to use the common names for the DNA viruses because, unlike the RNA viruses, there are a limited number of names. Thus, the parvoviruses, papovaviruses, adenoviruses, herpesviruses, poxviruses each have their own special replication schemes, DNA sizes, and DNA morphologies, which are described in Chapter 14.

Each of the classes of RNA viruses has a single style of replication linked to its style of transcription. For Classes IV and VI, replication is no different from transcription or at most very slightly different. For Class III, the double-stranded genome must be reformed from plus strand RNA in a mechanism specific to the replication of those viruses. For Class Va viruses, a replication mechanism completely separate from the transcriptional one forms new genome RNA. The replication process involves making a full-length plus-strand RNA from which the full-length minus-strand RNA is copied. Replication of Class Vb viruses is not yet understood.

The Virion Polymerases

Many animal viruses carry in their virions molecules of a nucleic acid polymerase. In some cases the need for this is evident: uninfected cells have no known RNA-dependent RNA polymerase or RNA-dependent DNA polymerase that viruses could use to initiate their infection cycle. Thus, the negative-strand RNA viruses, the double-strand RNA viruses, and the retroviruses appear to have no choice; along with the infecting RNA they must put into the cell those polymerase molecules necessary to begin the growth cycle. For the negative- and double-strand viruses, these polymerase molecules are RNA-dependent transcriptases that must make the first round of mRNA; after that, new polymerase molecules encoded by the mRNA can accelerate the infection process. For retroviruses the polymerase is a DNA polymerase (reverse transcriptase), able to make at least one complete DNA transcript of the genome RNA. Ultimately the cell's RNA polymerase appears to do the actual work of transcribing the mRNA. Poxviruses also have a virion polymerase; they are cytoplasmic DNA viruses that bring into the cell their own DNA-dependent RNA polymerase thus circumventing the need for the equivalent cellular enzyme to function outside of its usual nuclear location.

Defective Interfering Particles

Most, if not all, animal viruses spawn defective particles if the viruses are propagated by multiple cycles of growth at a high multiplicity of infection. The defective particles are produced because their deleted function is complemented by genes of a co-infecting complete virus. A fraction of the defectives can interfere with growth of the wild type virus; these are called *defective interfering (DI) particles* and the virus that spawns them is called the *standard virus*. The distribution and properties of DI particles has been discussed in a number of review articles (Huang and Baltimore, 1970; Huang, 1973; Huang and Baltimore, 1977). For most animal viruses discussed here we will not mention their DIs but interested readers are referred to the reviews.

Positive-strand viruses: Picornaviruses (Class IVa)

Rather than scanning the various virus classes in numerical order, we first look in detail at the replication of positive-strand RNA viruses because they provide a good general prototype of animal virus multiplication (Baltimore, 1969; Levintow, 1974). Poliovirus is the most intensely studied of these viruses, which are known collectively as picornaviruses. The class includes Mengovirus and EMC virus (mouse picornaviruses), rhinoviruses (the viruses of the common cold), and foot-and-mouth disease virus.

THE VIRION

The virion of poliovirus consists of a single molecule of RNA embedded in an icosahedron composed solely of protein molecules. To produce virus stocks in the laboratory or to study the physiology of poliovirus infections, cultures of primate cells—such as the HeLa cell—are infected with virus at high multiplicity and the infected cells are then incubated at 37°. Over a 6 hour period each cell will produce approximately 10^5 virions, but because of the inefficiency of poliovirus infection these will score as only a few hundred plaque-forming units of virus. Virions are easily purified from cellular debris because they resist disruption by the detergent sodium dodecyl sulfate, which solubilizes almost all other nucleic acid-protein complexes in the cell (Mandel, 1962). After treatment with sodium dodecyl sulfate a cell lysate can be fractionated directly by a combination of rate zonal centrifugation and isopycnic banding to give chemically pure virus.

VIRION PROTEINS

When virions of picornaviruses are disrupted and the proteins are separated by polyacrylamide gel electrophoresis in the presence of sodium dodecyl sulfate, four main bands of viral protein are found (Summers et al., 1965). From the ratio of protein to RNA and the molecular weights of the RNA and protein it can be calculated that each of the polypeptides occurs in the virion approximately 60 times (Rueckert, 1971; Stoltzfus and Rueckert, 1972), once for each of the 60 crystallographic subunits in the virus (Finch and Klug, 1959).

VIRAL RNA

The virion RNA is a single, infectious molecule (Colter et al., 1957; Alexander et al., 1958) with a molecular weight of approximately 2.6×10^6, corresponding to a length of 7500 bases or a coding capacity of approximately 2500 amino acids (Granboulan and Girard, 1969). Virion RNA does not self-anneal and therefore represents a pure single-stranded RNA with no contamination by the complementary strand. At the 3'-terminus of the RNA is a run of approximately 75 nucleotides of adenosine (Armstrong et al., 1972; Yogo and Wimmer, 1972). This 3' terminal poly(A) is heterogeneous in length even when cloned virus is analyzed (Spector and Baltimore, 1975a). The 5'-terminus is pUp linked via its 5' phosphate to a protein (Flanegan et al., 1977; Lee et al., 1977). As will be described below, intracellular 35S poliovirus RNA has a free 5'-terminus of pUp.

Many cellular mRNAs as well as viral RNAs, have poly(A) at their 3' end. The only function known for the poly(A) is long-term stabilization of the RNA

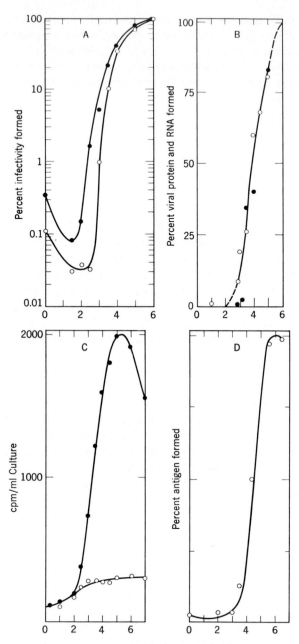

FIGURE 13-3. *Kinetics of formation of poliovirus RNA and protein studied by four types of experiments. (A) Growth curve of whole virions (○) and viral RNA (●) measured by their infectivity. (B) Radiochemical determination of viral growth. A culture infected with poliovirus was divided into a number of samples and either* [14]*C-adenosine or* [14]*C-valine was added at intervals during the course of infection. Virus was purified from each sample and its relative radioactivity determined. A decrease from the maximum radioactivity in a virus sample indicates the prior synthesis of a corresponding amount of viral protein (●) or viral RNA (○). Data from Darnell et al., (1961). (C) Use of actinomycin to measure viral RNA synthesis.*

to intracellular degradation (Brawerman, 1974). The poly(A) on poliovirus RNA has an absolutely required function because its removal—by ribonuclease H treatment of viral RNA hybridized to poly(dT)—renders the RNA noninfectious (Spector and Baltimore, 1974). The poly(A) of poliovirus RNA is not required for translation and therefore probably serves a function not served by the poly(A) of cellular mRNAs. That function may be to initiate replication of the RNA because the 3'-end of the RNA is where replication must begin.

Mengovirus and EMC virus RNA have somewhat shorter poly(A) chains at their 3' ends than does poliovirus RNA (Spector and Baltimore, 1976). Along with a number of other picornavirus RNAs, they also have an internal stretch of poly(C) approximately 100 nucleotides long, the exact location of which is not known (Brown et al., 1974).

THE GROWTH CYCLE: TIME COURSE OF RNA AND PROTEIN SYNTHESIS

The time-course of synthesis of viral RNA and protein can be followed in a number of ways, all of which show a close coordination between the times of synthesis of RNA, protein, and infectious particles (Figure 13-3). Neither by measuring infectious RNA production nor by measuring labeled viral RNA can a sizable pool of encapsidated virion precursor RNA be demonstrated (Baltimore et al., 1966).

The curves of accumulation of viral RNA and protein after infection indicate a 2 hour lag before any synthesis can be demonstrated (Figure 13-3). Using large amounts of labeled precursors and fractionation methods that can identify small amounts of RNA, one can show that viral RNA synthesis actually is initiated within the first hour of the infection cycle and probably between 15 and 30 minutes after infection (Baltimore et al., 1966). Even before the first viral RNA is produced, there must be synthesis of viral protein because a newly made virus protein is required to synthesize viral RNA molecules.

The time-course of synthesis of total viral RNA and protein gives no information about the time required to synthesize an individual virion. The time required for a newly made molecule of RNA or protein to become part of a virion can be determined by pulse-labeling cells during the phase of the infection cycle when virions are being actively formed. Such experiments have

---

Actinomycin, a compound that inhibits DNA-dependent RNA synthesis, was added to a culture, which was then infected with poliovirus. ^{14}C-uridine was then added and total acid-precipitable radioactivity assayed (●) at intervals after infection. As a control, actinomycin-treated uninfected cells were also tested (○). (Zimmermann et al., 1963). (D) Immunological measurement of virus capsid protein. Infected cells were exposed to ^{14}C-labeled amino acids immediately after virus infection. Newly formed poliovirus protein was detected by first treating cellular extracts with rabbit antiserum against purified poliovirus and then precipitating with sheep antiserum to rabbit γ-globulin. The total amount of newly formed poliovirus protein was determined by assaying ^{14}C in the precipitate. (Scharff and Levintow, 1963).

shown that newly made RNA molecules can become part of finished virions within 5 minutes of their synthesis whereas newly made protein molecules require at least 20 minutes before becoming integrated into virions (Penman et al., 1964; Baltimore et al., 1966). The reasons for the longer lag for protein molecules will become clear after the discussion of viral morphogenesis given below.

SITE OF SYNTHESIS OF VIRAL RNA

Autoradiography has shown that in uninfected cells labeled uridine is first incorporated in the nucleus but in virus-infected cells only the cytoplasm is ever labeled even with very short pulses (Franklin and Baltimore, 1962). Furthermore, enucleated cells can carry out the entire virus life cycle (Pollack and Goldman, 1973) and work with metabolic inhibitors has shown that neither synthesis nor function of DNA is involved in poliovirus multiplication (Simon, 1962).

Fractionation of infected cells that have been pulse-labeled for a short time with an RNA precursor has shown that the site of viral RNA synthesis is a structure called the *replication complex* that can be recovered from the smooth membranes of the cell (Girard et al., 1967; Caligueri and Tamm, 1970a,b). This complex is completely separate from poliovirus-specific polyribosomes (Huang and Baltimore, 1970) and no evidence has been found for coupling of viral RNA and protein synthesis. The replicative intermediate RNA, described below, is part of the replication complex.

The replication complex contains an enzymatic activity that is not found in uninfected cells and is able to synthesize 35S viral RNA in a cell-free system (Baltimore and Franklin, 1963; Baltimore, 1964). The activity appears in infected cell cytoplasm in parallel with the acceleration of the rate of viral RNA synthesis. It is not clear which viral proteins are responsible for the new RNA-dependent RNA synthetic system of infected cells (Lundquist et al., 1974). A virus-specific activity is also apparently responsible for synthesis of the 3'-poly(A) of the RNA (Dorsch-Hasler et al., 1976).

STRUCTURE OF INTRACELLULAR RNAs

Using combinations of centrifugation, gel electrophoresis, and differential solubility, it is possible to demonstrate and purify three forms of intracellular virus-specific RNA. These forms are *single-stranded RNA,* which is plus strand RNA of the size and general structure of the virion RNA; *double-stranded RNA,* consisting of a strand of plus RNA and a strand of minus RNA (Montagnier and Sanders, 1963; Baltimore, 1966); and *replicative intermediate RNA,* consisting of a minus strand of RNA and incomplete molecules of plus strand RNA (Baltimore and Girard, 1966; Baltimore, 1968). These three types of RNA molecules appear with somewhat different kinetics through the course of the infection cycle (Figure 13-4). Although they are generally analyzed as radioactive molecules, it is possible to demonstrate directly their presence in infected cells by electrophoretic separation of RNA through gels of agarose followed by staining of the RNA with ethidium bromide. Methods that can be used to

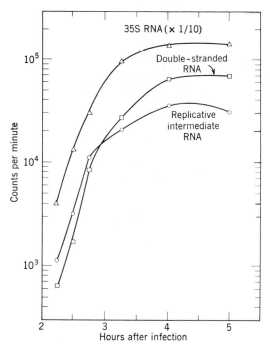

FIGURE 13-4. Accumulation of various types of radioactive virus-specific RNA in poliovirus-infected HeLa cells labeled with ³H-uridine. "35S RNA" indicates single-stranded, complete viral RNA chains (ordinate values to be divided by ten for this curve). Redrawn from Baltimore and Girard, 1966.

produce pure preparations of the three forms of virus-specific RNA are described in the flow chart in Figure 13-5 (Spector and Baltimore, 1975a).

Poliovirus mRNA is infectious and therefore is not significantly different from virion RNA. It differs from virion RNA only in its 5'-end where the virion RNA has a protein but the mRNA has just pUp (Hewlett et al., 1976; Nomoto et al., 1976; Fernandez-Munoz and Darnell, 1976).

Double-stranded picornavirus RNA sediments at approximately 20S compared to the 35S sedimentation rate of single-stranded viral RNA. The rigid configuration of a double-stranded polynucleotide causes it to sediment more slowly than single-stranded RNA in spite of its having twice the molecular weight. The most characteristic property of double-stranded RNA is that it resists digestion by ribonuclease (Montagnier and Sanders, 1963).

The replicative intermediate RNA was named because it is labeled by RNA precursors before any of the other virus-specific RNA in the cell. It also rapidly comes to a steady state as labeling is continued, consistent with the idea that it is the site of synthesis but not accumulation of viral RNA.

The structure of replicative intermediate RNA has been investigated in many ways but it is still not completely clear what state it has in the infected cell. There is reason to believe, as shown in Figure 13-6, that in the cell there are long regions of minus strand RNA that are not occupied by growing plus strand (Oberg and Philipson, 1971; Thach et al., 1974). However, after

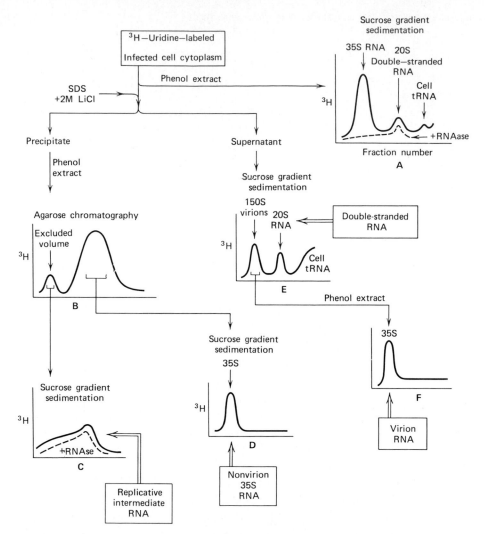

FIGURE 13-5. *A scheme for purifying the RNA molecules involved in poliovirus replication. Starting with infected HeLa cells labeled from 3 to 4 hours after poliovirus infection with ³H-uridine, a cytoplasmic fraction is prepared by homogenization. A phenol extract of the cytoplasm yields the RNA molecules, displayed by sedimentation through a gradient of sucrose, shown in pattern A. The radioactivity that resists digestion by ribonuclease, and is therefore double-stranded, is portrayed by the dashed line. It consists of heterogeneous replicative intermediate RNA and a discrete, pure 20S double-stranded RNA.*

Fractionation of the extract with sodium dodecyl sulfate and 2 M LiCl produces a precipitate and a supernatant. Extraction of the precipitate with phenol followed by agarose chromatography yields an excluded peak and an included peak (pattern B). The excluded volume is seen to be pure replicative intermediate by sedimentation (pattern C). The included peak is all of the 35S RNA that is not in virions (pattern D).

The 2M LiCl supernatant contains two forms of virus-specific structures separable by sedimentation; they are virions and 20S double-stranded RNA (pattern E). From the virions, 35S RNA can be extracted (pattern F). Modified from Spector and Baltimore, 1975a.

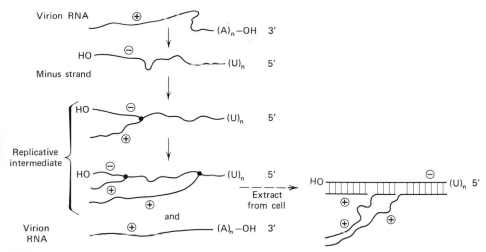

FIGURE 13-6. Schematic representation of poliovirus-specific RNA synthesis in the cytoplasm of infected cells. The collapsed form of replicative intermediate RNA produced by phenol extract of the cytoplasm is also schematically presented. Modified from Baltimore, 1976.

extraction from infected cells, the minus strands become covered by nascent plus strands producing the collapsed replicative intermediate structure shown in Figure 13-6. Consistent with that structure is the finding that if replicative intermediate RNA is treated with ribonuclease, the ribonuclease-resistant fraction sediments at about the rate of double-stranded RNA. One can separate from replicative intermediate RNA the nascent polynucleotides by heating the molecules to a temperature below the melting temperature of double-stranded RNA (Baltimore, 1968). Analysis of the nascent chains has indicated that there are approximately five growing chains on an individual replicative intermediate complex (Baltimore, 1969).

Girard (1969) showed that replicative intermediate RNA is the precursor to both double-stranded RNA and plus strand RNA. He labeled replicative intermediate RNA for a short time *in vitro* and then chased the label with a large excess of unlabeled nucleoside triphosphates. During the chase period both double-stranded and single-stranded 35S RNA were labeled at the expense of the loss of label from the replicative intermediate RNA.

DYNAMICS OF RNA SYNTHESIS

Figures 13-3c and 13-4 show the accumulation of RNA in infected cells on both linear and logarithmic scales. It is evident that there are two phases of viral RNA synthesis. One phase occupies the time from the initiation of infection until 2.5 to 3 hours after infection and involves an exponential accumulation of both replicative intermediate RNA and 35S RNA. After this period, a linear rate of 35S RNA ensues for about an hour. The exponential phase corresponds to the time when the number of replicative intermediate molecules is multiplying while in the linear phase that number remains approximately constant. Most of the final yield of viral RNA is made during the hour of linear synthesis when 2×10^5 molecules of viral RNA are formed or about 3000 chains per minute.

The time to make an individual molecule of RNA can be determined in a manner analogous to that used by Dintzis (1961) for determining the time required for reticulocyte ribosomes to synthesize a chain of hemoglobin. As shown in Figure 13-7, the procedure requires separating the growing chains on the template from the finished chains released from the template and labeling the infected cells for times approximating the synthesis time. The time at which one-half of the incorporated radioactivity is in finished chains is then the mean synthesis time for an RNA molecule. For poliovirus growing in HeLa cells this time is about 1 minute (Darnell et al., 1967; Baltimore, 1969).

Given a synthesis time of 1 minute, an overall rate of RNA production of 3000 chains per minute, and about five growing chains per replicative intermediate molecule, an infected cell must have about 600 replicative intermediate molecules during the linear phase of RNA synthesis. In a cell that was infected with a single particle, a doubling time for the rate of RNA synthesis of 10 to 15 minutes (Figure 13-4) gives about 2 hours to generate 600 templates from a single template. Adding to this a lag at the beginning of the infectious cycle for penetration of the infecting viral RNA and a factor for the turnover of replicative intermediate molecules that continually occurs during the infection cycle (Baltimore, 1968), it is understandable that the exponential phase lasts 2.5 to 3 hours.

FIGURE 13-7. Determination of the time for synthesis of one poliovirus RNA molecule following the principle of Dintzis (1961). When ³H-uridine is added to poliovirus-infected HeLa cells, radioactivity will appear first in growing, template-associated RNA chains. After completion of some chains, radioactivity appears in 35S viral RNA not attached to the template. When a period of time has elapsed that corresponds to the average synthesis time of one viral RNA molecule there will be equal amounts of radioactivity on the template and off it.

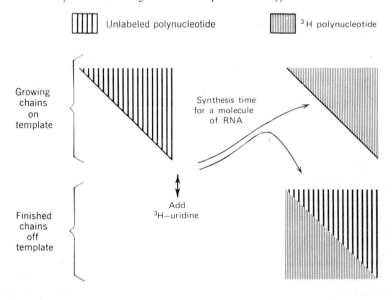

We can see from this analysis that the various numbers describing the rates of poliovirus-specific RNA synthesis and accumulation of poliovirus-specific RNA molecules come together to produce a self-consistent picture of the dynamics of poliovirus RNA synthesis in infected HeLa cells.

INHIBITORS OF POLIOVIRUS RNA SYNTHESIS

Certain chemicals inhibit poliovirus replication in as yet mysterious ways. The best known of these is the simple molecule, guanidine hydrochloride; a less well-known but equally specific inhibitor is 2-(α-hydroxybenzyl)-benzimidazole (Tamm and Eggers, 1967). Poliovirus mutants exist that are either resistant to the inhibitory effect of guanidine or dependent on it for their replication, indicating that the compound interacts with a virus-specific protein. Although the ultimate effect of guanidine is to grossly inhibit viral RNA synthesis, it is not clear that the compound works directly on the RNA synthetic machinery (Caliguiri and Tamm, 1968; Baltimore, 1969). Some viral RNA is made after guanidine addition to infected cells but this RNA apparently is blocked from forming polyribosomes or new virus particles (Huang and Baltimore, 1970). The RNA made in the presence of the drug is intact and contains a normal complement of poly(A) at its 3'-end and pUp at its 5'-end. At the moment, the ability of guanidine to block effective polyribosomal or virion RNA synthesis makes it a very useful compound for the elucidation of events during viral replication, but its mode of action remains uncertain.

SYNTHESIS OF VIRAL PROTEIN

Like all proteins, poliovirus proteins are synthesized by ribosomes reading an mRNA strand. The mRNA for the synthesis of viral proteins is 35S RNA; no smaller mRNA can be detected in infected cells. Except for the mRNA, all of the components needed to make viral proteins appear to be derived from the host cell. The viral proteins are synthesized on polyribosomes that are extremely large, consisting of as many as 35 to 40 ribosomes per unit. These polyribosomes completely replace the smaller, host cell polyribosomes (Figure 13-8).

The proteins made in infected cells span a wide size range from those as large as 100,000 daltons down to very small ones. The distinctive pattern of virus-specific proteins found after electrophoretic separation in polyacrylamide gels (Figure 13-9) contains the three larger virion proteins as well as many nonvirion proteins (designated N or, originally, NCVP for noncapsid viral proteins) (Summers et al., 1966).

The sum of the molecular weights of all the polypeptides seen in infected cells greatly exceeds the coding capacity of the viral genome, hence many of the proteins must be related to each other. Evidence for a relationship first came from pulse-chase experiments using radioactive amino acids which showed that protein N1 is formed rapidly during a pulse and then disappears during a chase while the virion proteins are not formed during the pulse but appear during the chase (Summers and Maizel, 1968; Holland and Kiehn, 1968; Jacobson and Baltimore, 1968b). Fingerprints of tryptic digests of N1 and of the virion proteins show that N1 is cleaved to form the virion proteins (Jacobson et al., 1970). A general term for a polypeptide such as N1, one that

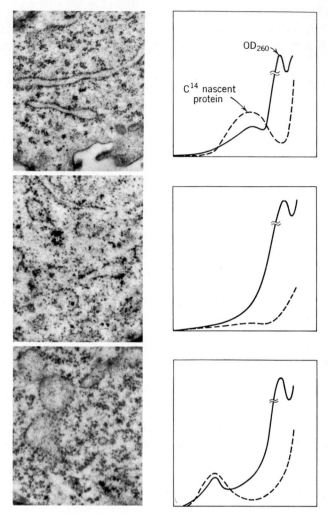

FIGURE 13-8. *Electron micrographs showing the distribution of ribosomes in poliovirus infected HeLa cells. (a) Normal cell in which ribosomes are in small clusters with no obvious association with membranes; (b) virus-infected cells about 2 hours after infection, showing loss of normal polyribosomal clusters; (c) large polyribosomes appear at the time of virus synthesis, frequently in proximity to membranous structures. The graphs near the figures show the polyribosome patterns in sucrose gradient of extracts of cells labeled for 5 minutes with* ^{14}C *amino acids.*

Solid lines: OD_{260}, *corresponding to ribosomal material; broken lines: radioactivity in nascent proteins. Photographs courtesy of Dr. Theodore Borun; sucrose gradient patterns redrawn from Penman et al., 1963 and Summers et al., 1965.*

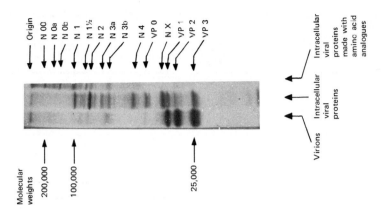

FIGURE 13-9. *Poliovirus proteins displayed by electrophoresis through polyacrylamide gels in the presence of sodium dodecyl sulfate. The proteins are separated according to their molecular weights. They were labeled with ^{35}S-methionine and visualized by autoradiography. The lane marked "virions" shows only three of the four virion proteins; the fourth is so small (about 7000 daltons) that it migrated off the end of the gel. The other lanes show total intracellular virus-specific polypeptides made in the absence or presence of amino acid analogues. The analogues cause polypeptides larger than 100,000 daltons to accumulate: such proteins are usually cleaved to smaller size during their synthesis but the analogues prevent the cleavage. Adapted from Villa-Komaroff et al., 1975 with permission.*

gives rise to two or more proteins by cleavage, is a *polyprotein*. The extent of polypeptide cleavage during poliovirus protein formation only became apparent when infected cells were treated with a series of amino acid analogs (Jacobson and Baltimore, 1968b). These analogs, as well as other compounds (Korant and Butterworth, 1976), can prevent cleavages, generating viral polypeptides as large as 200,000 daltons in infected cells (Figure 13-9).

Studies with inhibitors of proteolysis have led to the conclusion that the total poliovirus genome is transcribed into a single, long polyprotein that is cleaved in a number of steps to form the functional viral proteins. A general model for the formation of protein in all picornaviruses is shown in Figure 13-10; this model is widely accepted although Abraham and Cooper (1975) have raised the possibility that one of the viral proteins could be made as a separate translation unit.

Two distinct types of polypeptide cleavages occur during the generation of the poliovirus proteins. One type is called *nascent cleavage* because it occurs soon after the ribosome has formed the peptide bond to be cleaved: N1, NX, and N1.5 are produced by nascent cleavages (Figure 13-10). The second type of cleavages, *cytoplasmic cleavages*, have characteristic half-times of 10 minutes or longer and so must occur after the protein is finished and released from the ribosomes. A third type of cleavage, one linked to the maturation of the virion, will be described below.

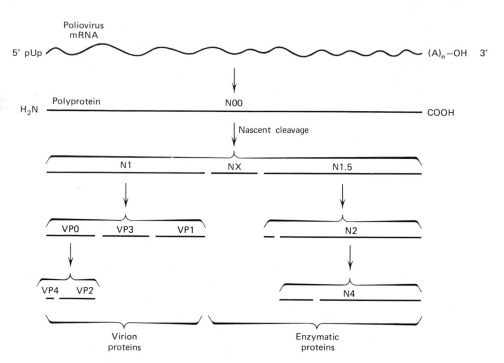

FIGURE 13-10. *Synthesis of poliovirus proteins. Picornaviruses make their functional proteins by translating their mRNA into a continuous string of amino acids
(the polyprotein, N00). During synthesis, "nascent cleavages" take place that
generate the primary products, N1, NX, and N1.5. N1 is further cleaved to form the
virion structural proteins while NX and the products of N1.5 are probably involved
in intracellular functions such as synthesis of viral RNA.*

It is possible to order the genes that encode the various poliovirus
proteins along the viral RNA because of the unique mode of synthesis of these
proteins. An inhibitor of initiation of protein synthesis will allow the
ribosomes that are on the mRNA to finish their synthesis, but will not allow
any new ones to enter the mRNA. In this way, during the transient period of
synthesis after inhibition of initiation, a gene near the 3'-end of the RNA will
be translated by every ribosome on the mRNA while a gene near the 5'-end of
the RNA will only be translated by a few ribosomes. Hence the relative yields
of proteins made following an initiation block, compared to the yields under
normal conditions, is a function of the map position of the protein. The order
of poliovirus proteins shown in Figure 13-10 was deduced using the inhibitor
pactamycin (Taber et al., 1971; Summers and Maizel, 1971) and more recently
by high-salt inhibition of initiation (Saborio et al., 1974).

Why has poliovirus evolved such a cumbersome mechanism for the
synthesis of its proteins, one which does not even allow the virus to regulate
the ratio of enzymatic proteins to structural proteins? One hypothesis,
strongly supported by evidence from other systems, is that eukaryotic mRNAs
are restricted to having one functional initiation site for protein synthesis on
any one RNA molecule (Jacobson and Baltimore, 1968b). Such a restriction

clearly does not exist in bacteria: the classic example of a polycistronic mRNA with separate initiation sites is the RNA of RNA bacteriophages (Chapter 8). There is as yet no solid biochemical explanation why a eukaryotic mRNA could not have multiple initiation sites but no viral or cellular mRNA with multiple, functional sites has been found.

Strong support for a model of translation of viral RNA starting at a single initiation site and using cleavage mechanisms to generate the virion proteins has been provided by the *in vitro* translation of picornavirus RNA. Among the products of *in vitro* translation there is a polypeptide of the same size as the one thought to be precursor to all of the others (Villa-Komaroff et al., 1975). Many of the other *in vivo* products can also be identified in the pattern of *in vitro* products, even if extracts of uninfected cells are used for synthesis, indicating that the nascent cleavages are carried out by cellular rather than virus-specific proteins. The cytoplasmic cleavages do not occur in the *in vitro* system and may involve virus-specific proteins. In fact, one of the EMC capsid proteins may be a protease (Lawrence and Thach, 1975).

MORPHOGENESIS OF THE VIRION

A series of intermediate structures between the N1 polyprotein and the finished virion can be identified and a sequence of events during morphogenesis has been deduced (Figure 13-11). The first step, shown from work on EMC virus, is the aggregation of N1 to form a pentamer sedimenting at 14S (McGregor et al., 1975). Cleavage of N1 follows and the cleaved pentamers of VP0, VP1, and VP3 then aggregate to form the 74S *procapsid* (Jacobson and Baltimore, 1968a). This step, which generates a particle of the size and shape of the virion, has been demonstrated *in vitro* and is accelerated by the presence of membranes (Perkin and Phillips, 1973).

Viral RNA appears to be added to the procapsid to form the 125S *provirion* (Fernandez-Tomas and Baltimore, 1973) but provirion could also arise by aggregation of pentamers around the viral RNA. Complete 150S virions then appear in which VP0 is reduced to a very minor component and VP2 plus VP4 become evident. Fingerprints of tryptic digests of the virion proteins has shown that VP2 and VP4 are cleavage products of VP0; this cleavage of VP0 changes the provirion into the virion (Jacobson and Baltimore, 1968a; Jacobson et al., 1970).

POLIOVIRUS DEFECTIVE INTERFERING PARTICLES

After high multiplicity serial passage of poliovirus the pure virions begin to become contaminated with slower sedimenting particles containing shorter RNA molecules. These particles can interfere with infection by standard virions and thus have been designated poliovirus defective interfering particles (poliovirus DI) (Cole et al., 1971).

Poliovirus DI can be highly purified and shown to contain the same proteins as standard poliovirus. It infects sensitive cells and initiates an infection cycle in which a full yield of DI RNA is produced. Some poliovirus proteins are also produced but protein N1 is totally absent, indicating that the deletion in DI RNA is in the N1 region (Cole and Baltimore, 1973a). All

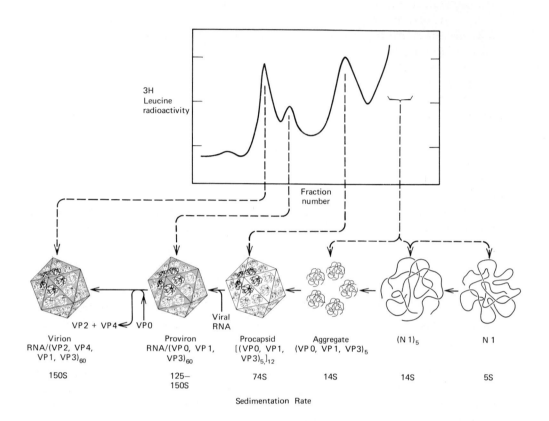

FIGURE 13-11. Morphogenesis of the poliovirion. The upper pattern shows the amino acid-labeled structures in an infected cell separated by centrifugation through a sucrose gradient (Fernandez-Tomas and Baltimore, 1973). The various structures are shown diagramatically below along with their sedimentation coefficients.

precursors of N1 are also missing but shortened versions of N1 and its precursors can be found.

The mechanism by which poliovirus DI interferes with the growth of standard virus has been determined (Cole and Baltimore, 1973b). When cells are infected by poliovirus, approximately 2×10^5 molecules of progeny RNA are formed whether the cells are infected with a mixture of DI and standard virus or by either one separately. The two types of RNA molecules appear to multiply independently so that if cells are infected by equal multiplicities of standard virus and DI they make equal amounts of the two RNAs. One consequence of this equality is that half of the RNA made in a coinfected cell does not have the gene for synthesizing capsid protein so the amount of capsid protein formed will be 50% of normal. The second consequence is that because DI RNA can be incorporated into virions as efficiently as standard RNA, the procapsid made in infected cells is shared by the two types of RNA molecules, reducing the yield of infectious particles again by one-half. These two levels of interference are independent of one another and multiply together so that the yield of standard virus from coinfected cells is the square of the percentage of standard virus in the inoculum. Hence cells infected by equal multiplicities of the two kinds of particles produce only one-fourth of the yield produced by standard virus infection alone.

Poliovirus DI provides a model for DI particles in other viral systems. It may be more than a laboratory curiosity because large concentrations have been found in poliovirus vaccines (McLaren and Holland, 1974) and poliovirus DI may be at least partially responsible for the efficacy of the vaccine.

POLIOVIRUS GENETICS

Poliovirus, and other picornaviruses, undergo genetic recombination (Ledinko, 1963; Cooper, 1969; Cooper et al., 1975). A linear map of viral mutants has been established combining data on temperature-sensitive mutations and on mutations defining resistance to agents known to be lethal to virus growth. Because the proteins carrying the mutations are not known, the map cannot be compared to the RNA map derived by inhibition of protein synthesis.

Complementation between viral temperature-sensitive mutants is either not measurable or extremely low (Cooper, 1969). One-way complementation of DI particles by standard virus is, however, very efficient because the standard capsid protein can package the DI genome (Cole and Baltimore, 1973b). Complementation by coat protein would be predicted from the high efficiency of phenotypic mixing between picornavirus strains (Ledinko and Hirst, 1961). The failure of two-way complementation by temperature-sensitive mutants of clearly different physiology implies that the noncapsid functions may be *cis*-active and thus not be able to complement other mutations.

Togaviruses (Class IVb)

Those positive-strand viruses that utilize two different sizes of messenger RNA are all members of a group of viruses called the togaviruses, a name indicative of the envelope that surrounds the core of the virion. The synthesis of this envelope is discussed in a later section. Here, only the RNA and protein synthetic mechanisms used by this class of viruses are discussed (Pfefferkorn and Shapiro, 1974).

Before considering the molecular biology of togaviruses it is interesting to recall how this virus group was recognized (see Fenner et al., 1974; Casals, 1971). Field workers noticed that many of the viruses that infect vertebrates are carried by tick or mosquito vectors. Those that are human pathogens are generally endemic to a different animal species and only infect humans when the appropriate arthropod carrier happens to bite a human being. These viruses became collectively known as the "arthropod-borne" viruses, a name that became shortened to arboviruses. It later became evident that the arboviruses are not a uniform biochemical grouping because viruses from various classes can be carried by arthropods. Arthropod-carried viruses usually multiply in their insect vector and therefore must have the special ability to grow both in vertebrate and in insect cells.

A major fraction of the arboviruses fall into the biochemical grouping of the togaviruses. Two subgroups of togaviruses, Group A and Group B, distinguished by serology, are now called *alphaviruses* and *flaviviruses*. At least two non-arthropod-borne viruses are togaviruses: rubella virus and lactic dehydrogenase-elevating virus (Brinton et al., 1975). This discussion deals with the properties of only the two best-characterized togaviruses, Sindbis and Semliki Forest, both alphaviruses.

THE VIRION

Togaviruses, like other virus groups to be discussed later, have as an integral part of their virions a lipoprotein envelope derived from the plasma membrane of the infected cell. Inside the envelope is an icosahedral core containing the viral RNA. Envelope synthesis is described in a later section of this chapter. The virion has only four virus-specific polypeptides. Three of these, called E1, E2, and E3, with molecular weights, respectively, of about 50,000, 50,000, and 10,000, are glycoproteins that are on the surface of the virus embedded in the lipid bilayer that forms the envelope (Garoff et al., 1974). The fourth protein, called C, has a molecular weight of about 30,000, is not glycosylated, and is the protein component of the nucleocapsid core.

A single RNA molecule constitutes the genome of togaviruses. It sediments at about 42S and has a molecular weight of approximately 4.2×10^6 daltons. The RNA is infectious by itself (Wecker, 1959) and will act as a messenger RNA in *in vitro* cell-free systems; togaviruses are therefore classified as positive-strand viruses. The 3' end of the RNA is polyadenylated (Eaton and Faulkner, 1972) and the 5' end has a capping group, so the togavirus RNA structure is the classic structure of a mammalian messenger RNA (see Chapter 11). The 5' terminus of togavirus genome RNA, however, differs from the terminus of most mammalian messenger RNAs in that the only methyl group at the 5' end is in the terminal guanosine while the adenosine at the other end of the 5'—5' triphosphate bond has no methyl groups (Mefti et al., 1976).

INTRACELLULAR RNAs

The pattern of intracellular single-stranded RNA synthesized in togavirus-infected cells shows some fundamental differences from picornaviruses. Although the genome RNA is 42S, the major intracellular RNA species in infected cells is a 26S RNA of 1.8×10^6 daltons (Simmons and Strauss, 1972a). If the intracellular RNAs are separated by polyacrylamide gel electrophoresis, two other RNA species can be resolved between the 42S and 26S RNA: 38S RNA and 33S RNA. Separation of the T1 ribonuclease-generated oligonucleotides of these four RNA species has shown that the 26S RNA sequences are contained within the 42S RNA (Kennedy, 1977). Each of the minor species is a conformational variant of one of the major species, the 38S RNA being indistinguishable from the 42S RNA by primary structure analysis and the 33S RNA being indistinguishable from the 26S RNA (Kennedy et al., 1977).

Electron microscopy has shown that the 42S RNA can be circularized (Hsu et al., 1973) and the complementary sequences holding together the circular configuration have been isolated and sequenced. They constitute a terminal double-stranded region of about 140 nucleotides except that the 3' poly(A) is not included in the double strand.

Both the 42S RNA and the 26S RNA appear to serve as mRNAs in infected cells because both can be recovered from virus-specific polyribosomes (Simmons and Strauss, 1974b).

VIRAL PROTEIN SYNTHESIS

The role of the 26S RNA has become clear through analysis of protein synthesis in infected cells and in cell-free systems. It appears to be an mRNA encoding a polyprotein that consists of the amino acid sequences of all of the proteins of the virion. Demonstration of this polyprotein has been difficult, however, because the type of pulse-chase experiment that revealed the precursor of the picornavirus capsid protein failed to detect any polyprotein in a togavirus-infected cell.

The first evidence for linkage of two virion polypeptides was the demonstration of a 60,000 dalton intracellular precursor to E2 and E3, called PE2 (Schlesinger and Schlesinger, 1972). A 100,000 dalton polypeptide was later found in some host cells whose peptide composition corresponds to the combination of PE2 and E1. The final evidence for the full-length polyprotein came from a temperature-sensitive mutant of Sindbis virus that produces a 130,000 dalton protein containing the peptides of all four of the virion proteins (Schlesinger and Schlesinger, 1973). It appears that nascent cleavages give rise to C, PE2, and E1 and that a final cleavage of PE2 to E2 plus E3 is coincident with their appearance on the cell surface. Knowing that the proteins come from a polyprotein their sequence in the polyprotein can be derived using the same types of mapping procedures used for picornaviruses. The order (5' to 3' on the mRNA) is C–E3–E2–E1 (Clegg, 1975; Lachmi et al., 1975).

In vitro, the 26S mRNA displays only a single initiation site and only directs synthesis of polypeptides related to the virion proteins (Clegg and Kennedy, 1975; Cancedda et al., 1975). If 26S RNA is the mRNA for the capsid polyprotein, what then does the 42S mRNA do?

Although it has been difficult to resolve any proteins in infected cell extracts other than those of the virion, high-resolution polyacrylamide gel electrophoresis combined with autoradiography can reveal faint bands of other proteins (Kennedy et al., 1976). Use of a temperature-sensitive mutant that produces little capsid protein under nonpermissive conditions makes these intracellular polypeptides more evident (Lachmi and Kaariainen, 1976). There may be as many as four of them, possibly all derived from a single polyprotein. At least two may be involved in an RNA replicase activity that can be isolated from infected cells (Kennedy et al., 1976).

The *in vitro* translation products of 42S RNA are consistent with the idea that it synthesizes a polyprotein separate from that synthesized by the 26S RNA. Most initiation on the 42S RNA comes from a site separate from that found on 26S RNA (Cancedda et al., 1975) and the material made from 26S RNA has been reported to be absent from the products of the 42S RNA

(Simmons and Strauss, 1974a). Because the 26S RNA represents the 3' end of the 42S RNA (Kennedy, 1977), the 42S RNA must contain the initiation site for the synthesis of the capsid polyprotein although that site must be unavailable to ribosomes in the 42S RNA (Figure 13-12). A similar situation exists for some plant viral mRNAs (Shih and Kaesberg, 1973; Hunter et al., 1976). Such situations strongly suggest that eukaryotic mRNAs cannot display more than one initiation site for protein synthesis, even if a second initiation site is present on the RNA and can be made functional in a different configuration.

VIRAL RNA SYNTHESIS

The synthesis of both the 26S and 42S RNAs requires that there be a minus strand in infected cells. Such a full-length minus strand can be found as part of both a double-stranded RNA and an apparent replicative intermediate.

FIGURE 13-12. Synthesis of togavirus RNAs and proteins. From a full-length copy of the viral genome (minus strand) there are two types of RNA copied. One is the virion RNA, the other is 26S RNA. Virion RNA serves both as precursor to virions and as mRNA. It has two sites for initiating protein synthesis (AUGs), one of which is available to initiate synthesis of a polyprotein that is precursor to the enzymes for viral RNA synthesis and the other of which is not available for initiation at all on 42S RNA. The other RNA made is a 26S mRNA that represents the 3' end of the 42S RNA and has the second initiation site available for making a polyprotein that is precursor to the virion proteins.

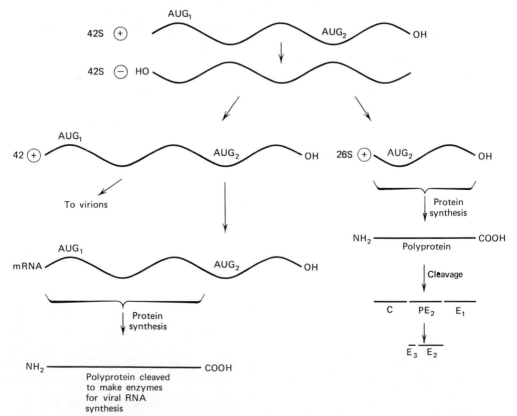

No replicative intermediate RNA or double-stranded RNA containing a minus 26S RNA can be found in infected cells, hence the 26S RNA must be synthesized from a minus 42S RNA (Simmons and Strauss, 1972b; Bruton and Kennedy, 1975). The exact mechanism by which it is synthesized is not known; there could be processing of 26S RNA from 42S RNA or an independent initiation site on the 42S RNA. A tentative picture of the togavirus life cycle is shown in Figure 13-12.

It is interesting to consider why togaviruses might have evolved this complicated dual mRNA mechanism as compared to the simpler scheme used by the picornaviruses. An advantage that the togaviruses gain is evident from observations on certain mutants of togaviruses that cause RNA synthesis to be temperature-sensitive. If cells infected with these mutants are first incubated for a while at the permissive temperature and then shifted to the nonpermissive condition, the RNA synthetic enzymes made under permissive conditions continue to work at nonpermissive conditions and the cells yield virus (Pfefferkorn and Burge, 1967). No new RNA synthesizing activity need be formed during the latter part of the infection cycle. The virus has thus achieved an early/late dichotomy, which picornaviruses do not have. Having a dual mRNA system, the virus can also separately control the levels of RNA synthetic activity and of capsid proteins. Furthermore, capsid proteins do not compete with ribosomes for newly made RNA, as happens with picornaviruses, because the mRNA function of the 42S RNA is no longer relevant at the time virions are being made. The flexibility inherent in separate controls may be the basis for the ability of these viruses to grow both in their arthropod vectors and in vertebrate hosts. As we see in the next section, the negative-strand viruses, which have evolved mechanisms that allow a complete dichotomy between replication and transcription, have an even broader host range.

Negative-Strand Viruses (Class V): Vesicular Stomatitis Virus

Negative-strand viruses fall into three major morphological categories: rhabdoviruses (Wagner, 1975; Howatson, 1970), paramyxoviruses (Choppin and Compans, 1975), and orthomyxoviruses (Compans and Choppin, 1975). In terms of biochemical strategy, the rhabdoviruses and paramyxoviruses are very similar and they constitute most of the well-characterized viruses of Class Va. This section focuses on one rhabdovirus, vesicular stomatitis virus (VSV), because it has been studied particularly thoroughly. VSV is a mild pathogen of cattle and is of minor economic or health concern. It grows rapidly to high titer in cell cultures and is lethal to the cells. Most other rhabdoviruses and paramyxoviruses usually establish persistently infected cell cultures rather than killing cells and are therefore more difficult to study.

Following the discussion of VSV, orthomyxoviruses are considered briefly. These viruses, of which the best-known are the human influenza viruses, are segmented, negative-strand viruses (Class Vb).

VSV, like the togaviruses, is enveloped but the virion has a characteristic bullet shape (see Figure 3-22). The name rhabdovirus derives from "rod" signifying the asymmetry of the particle. The bullet shape reflects the shape of the cylindrically coiled nucleocapsid which contains a single RNA molecule of about 4×10^6 daltons. This RNA has none of the hallmarks of an mRNA; it is not infectious, lacks 3'-poly(A) and is not capped at its 5' end. It functions as a template from which mRNAs are copied and is therefore a minus strand. The nucleocapsid is a very stable structure in which the RNA is completely ribonuclease-resistant (Soria et al., 1974); nucleocapsids are infectious but have a very low specific infectivity (Szilagyi and Uryvayev, 1973).

There are five different protein molecules in the VSV virion and these are the only viral proteins found in infected cells. Figure 13-13 shows the separation of these proteins by electrophoresis and their approximate positions within the virion. The most abundant virion polypeptide, the N protein, is the major protein of the nucleocapsid. The nucleocapsid also has two minor proteins, called L and NS, that are involved in RNA synthesis. Between the lipoprotein envelope and the nucleocapsid there is a space filled by the M protein and outside of the lipid bilayer the G protein is organized into a series of spikes that cover the external surface of the virus particle.

The paramyxoviruses are not bullet-shaped but rather have a loosely spherical appearance with a less-ordered nucleocapsid than that of the rhabdoviruses.

FIGURE 13-13. The proteins of vesicular stomatitis virus. A polyacrylamide gel electropherogram of the virion proteins is shown and each protein is named by a letter and its molecular weight shown (K = 1000). Below an idealized picture of the virion is presented with the locations of the proteins in the virion indicated. The L and NS proteins appear to be enzymes involved in replication but their exact functions are uncertain. The molecular weight of NS is probably lower than indicated by its electrophoretic mobility (Knipe et al., 1975). The particle is shown as a baccilliform structure but electron micrographs usually show the virions to be bullet-shaped (see Figure 3-22) apparently because the region where the terminus of the budding particle is sealed often ruptures during purification and preparation for electron microscopy.

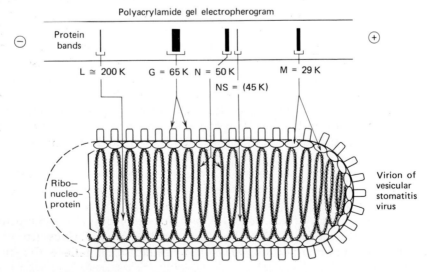

INTRACELLULAR RNA SPECIES

A major clue to the nature of negative-strand viruses came from an examination of the RNA species formed in infected cells (Bratt and Robinson, 1967). If infected cells are labeled with ^{3}H-uridine and the virus-specific RNAs are separated by electrophoresis, five capped, poly(A)-containing RNA species that anneal to virion RNA can be resolved (Rose and Knipe, 1975; Figure 13-14). These RNAs can be isolated from polyribosomes and fingerprinting shows them to be different from one another. The 5'-terminal structure on each of the mRNAs appears to be identical in nucleotide sequence but there is heterogeneity in the methylation of all of the termini (Rose, 1975). The basic structure is m[7]G5'ppp5'ApApCpApGp . . . (Rhodes and Banerjee, 1976). Thus, the five RNAs have the characteristics of eukaryotic mRNAs. The final proof that they are mRNAs came with the demonstration that they will direct synthesis of the five VSV proteins in cell-free systems (Knipe et al., 1975; Both et al., 1975b).

The negative-strand viruses thus utilize monocistronic mRNAs that are complementary to virion RNA for synthesis of their proteins. This strategy is fundamentally different from that of the positive-strand viruses on two counts: use of a complementary copy of virion RNA, rather than virion RNA itself, as mRNA and synthesis of individual proteins from individual mRNAs rather than synthesis of polyproteins followed by cleavage.

FIGURE 13-14. The vesicular stomatitis virus (VSV)-specified RNAs. An electropherogram of the RNAs made in an infected cell is shown. Before analysis, the RNAs were converted to double-stranded form by hybridization to VSV virion RNA followed by digestion with ribonuclease. The RNAs are identified by the proteins they encode as shown by cell-free protein synthesis and ultraviolet light inactivation (Knipe et al., 1975; Freeman et al., 1977). A preparation of reovirus double-stranded virion RNA was analyzed in parallel to provide size markers. The sizes of the classes of reovirus double-stranded RNA are approximately: L1–L3 = 2.3–2.7 × 10⁶, M1–M3 = 1.3–1.6 × 10⁶, S1–S4 = 0.6–0.9 × 10⁶. Adapted from Freeman et al. (1977) with permission of the publisher

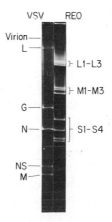

VIRION TRANSCRIPTASE

The enzyme responsible for transcribing the mRNAs of negative-strand viruses from the virion RNA is contained within the virion. One consequence of its existence is the lack of infectivity of naked virion RNA. The transcriptase is easily demonstrated *in vitro*; detergent treatment removes the lipoprotein envelope from the virion releasing the nucleocapsid which can then synthesize the five mRNAs when provided with the four ribonucleotide triphosphates (Baltimore et al., 1970; Moyer and Banerjee, 1975; Both et al., 1975a). Purified virions are also able to polyadenylate the mRNAs (Banerjee and Rhodes, 1973) and to generate the 7-methylguanosine-containing cap at their 5′ end (Rhodes et al., 1974). The mRNAs synthesized by the virion direct synthesis of four of the five virion proteins in cell-free systems (Both et al., 1975a; Ball and White, 1976). The inability of 28S RNA made *in vitro* to code for L protein, which is made by 28S mRNA from infected cells, has not been explained.

All three of the proteins of the VSV nucleocapsid are required for transcriptase activity (Emerson and Yu, 1975). The L and NS proteins appear to be responsible for the act of polymerization but the RNA must be coated by the N protein to be a functional template.

MECHANISM OF TRANSCRIPTION

There are two ways the transcriptase could read the genome RNA to produce the mRNA. Either a single polymerase molecule could start at one end and copy continuously down the template or there could be individual entry points at the beginning of each of the mRNA molecules. By the use of ultraviolet light inactivation of *in vitro* transcription, these two models are distinguishable because synthesis of the individual mRNAs will either be inhibited coordinately or sequentially. Ball and White (1976) showed that inhibition is sequential, indicating that the polymerase molecules start to traverse the template at only one site. This experiment, which used a coupled transcription-translation system, also provided an order for the genes of the VSV genome: 3′-L-G-M-NS-N-5′.

Lack of complementarity of the terminal sequences of the N mRNA and the virion RNA suggested to Abraham et al. (1976) that there might be an RNA transcribed before the N mRNA (a "leader sequence"). They have found a short, unmethylated RNA segment made early during the *in vitro* reaction that has the appropriate 5′ terminal nucleotide sequence and appears to represent such a leader (Colonno and Banerjee, 1976; Figure 13-15).

Although it is clear that in copying the genome, a polymerase molecule starts at the leader and sequentially copies the genome, it is not known if a long poly-mRNA is formed and then cleaved or if the polymerase stops and restarts at each junction between transcripts. The poly-mRNA model is supported by the unique biochemistry of the VSV capping reaction. Two of the three phosphates of the triphosphate bridge derive from the terminal GTP suggesting that a monophosphate, possibly left after cleavage of the mRNA from a poly-mRNA, was present at the 5′ end of the mRNA before the cap was added (see Chapter 4).

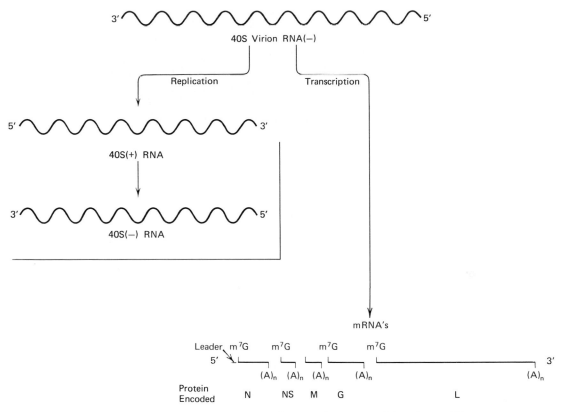

FIGURE 13-15. *The genetic system of vesicular stomatitis virus. The separate pathways of replication and transcription are indicated. Messenger RNAs are shown in the order they map along the genome RNA.*

A remarkable number of separate enzymatic activities are involved in the transcription process that generates the VSV mRNAs. All of these activities are present in purified virions of VSV although it is not certain that they are all carried out by virus-specific enzymes. Trace amounts of host cell enzymes could be in the virions. If they are all virus-specific processes, then the L and NS protein must have a number of different enzymatic sites to carry out the different reactions. It is conceivable that the N protein that coats the entire length of the virion RNA might also have enzymatic function.

REPLICATION OF RNA

As part of its life cycle, VSV must both make more virion RNA (replicate) as well as transcribe mRNA. Synthesis of virion RNA (40S minus-strand RNA) requires a template of 40S plus-strand RNA and such a molecule has been identified in infected cells (Morrison et al., 1975). The 40S RNA in cells is a mixture of some plus-strands and an excess of minus strands. Appropriate temperature-sensitive mutants maintained at nonpermissive temperature can make mRNA but not 40S RNA of either polarity, showing that 40S plus RNA is not a precursor to mRNA. If poly-mRNA exists as precursor to the functional mRNAs, it must be cleaved before it is completed. Thus, two separate RNA synthetic systems are involved in the VSV life cycle: a replication system that generates 40S plus and minus strands and a transcription system that produces mRNA (Figure 13-15). The products of these systems also differ: both 40S RNAs occur in cells exclusively as nucleocapsids but the mRNAs are not in nucleocapsids (Soria et al., 1974).As mentioned earlier, all of the detectable virus-specific proteins in infected cells are also found in the virion. Therefore replication must be carried out utilizing only the same viral proteins that are needed for transcription, although cellular proteins not involved in transcription might be involved in replication.

By appropriate manipulations it is possible to totally separate transcription from replication. Certain temperature-sensitive mutants, at the nonpermissive temperature, can carry out one process in the absence of the other and inhibition of protein synthesis in infected cells can stop replication without affecting transcription (Perlman and Huang, 1973).

The existence of separate transcription and replication processes represents a step of complexity beyond that shown by any of the positive-strand viruses. Possibly as a consequence of the increased complexity, negative strand viruses, and specifically rhabdoviruses, have a very wide host range. There are rhabdoviruses that grow both in plant cells as well as in cells of their insect vectors and others that grow both in insect cells and in vertebrate cells.

TIMING OF INTRACELLULAR EVENTS

One consequence of the life cycle of negative-strand viruses is that it is possible to distinguish two time-periods within the cycle. During the early stage of the infection only input genomes are acting as templates for transcription. This "primary transcription" occurs even if cells are infected in the presence of inhibitors of protein synthesis (Perrault and Holland, 1973). As much as 10% of the normal amount of mRNA can be made by primary transcription (Huang and Manders, 1972). "Secondary transcription" begins when new nucleocapsids start to be formed in the infected cell. There must also be stages of primary and secondary replication.

MUTANTS

Temperature-sensitive mutants of VSV are easily derived and complement well (Flamand and Pringle, 1971; Cormack et al., 1973). A large fraction of all mutants isolated fall into a single complementation group, Group I, the product of which is involved in RNA synthesis and appears to be L protein (Hunt et al., 1976). Individual Group I mutants may be defective either in transcription or replication, indicating that the L protein is involved in both processes but that the sites involved are individually mutable (Perlman and Huang, 1973/4). Mutants affected in the N protein (Ngan et al., 1974) cease replication at the nonpermissive temperature, suggesting that it may play a critical role in controlling the replication pathway.

Group III mutants make RNA normally at the nonpermissive temperature but appear to be defective in the M protein (Lafay, 1974). Group V mutants affect the G protein that normally occurs on the outside of the envelope of the virus and allows the virus to absorb to cells. Cells infected with Group III or

Group V mutants make no virion-like particles at the nonpermissive temperature. Group V mutants with a heat labile G protein can be complemented by other enveloped viruses. For instance, retroviruses can donate their envelope protein to Group V mutants producing a VSV particle with the tropism of the retrovirus. Such particles in which the envelope is derived from one virus and the internal components from another are known as *pseudotypes* and are formed between many types of enveloped viruses.

DEFECTIVE INTERFERING PARTICLES

A characteristic feature of the VSV life cycle is the generation of DI particles during virus growth (Huang and Wagner, 1966). These are shorter than normal bullet-shaped particles containing only a portion of the viral RNA, that interfere so effectively with the growth of standard virus that continual passage of a VSV stock rapidly leads to a preponderance of DI particles (Stampfer et al., 1971). The step in the virus life cycle at which interference occurs is the replication of the nucleic acid (Palma et al., 1974).

OTHER NONSEGMENTED NEGATIVE-STRAND VIRUSES

The other viruses of Class Va follow the same pattern of transcription and replication as VSV. They all have a single long genome RNA of 4 to 5 × 10⁶ daltons that is copied into five to seven mRNAs (Choppin and Compans, 1975; Roux and Kolakofsky, 1975).

Many paramyxoviruses have two envelope proteins rather than the one associated with VSV. One protein is a "fusion factor" that causes the virus to fuse to cells and can cause different cells to fuse to each other (Scheid and Choppin, 1974). The ability of killed Sendai virus to fuse cells has been used extensively for production of hybrid cells (Okada, 1972). The fusion factor is also responsible for the ability of these viruses to lyse red blood cells and probably is responsible for virus entering into host cells (Chapter 12). The second envelope protein of paramyxoviruses has two activities (Scheid et al., 1972): binding to cell surfaces, as indicated by its ability to agglutinate red blood cells, and neuraminidase activity, possibly to allow the virus to release itself from mucopolysaccharides.

SEGMENTED NEGATIVE-STRAND VIRUSES (CLASS Vb)

There are a number of viruses that carry the minus strand of RNA within the virion but which have segmented genomes. Those characterized segments are monocistronic antimessenger RNAs (Palese and Schulman, 1976), although some segments might be polycistronic. The best known such virus is influenza virus, an orthomyxovirus (Etkind and Krug, 1975). Other virus types that use the same strategy include the bunyaviruses and arenaviruses (Ranki and Pettersson, 1975; Carter et al., 1974). These viruses have helically organized nucleocapsids; in fact, all helically organized enveloped viruses found thus far are negative-strand viruses. The nucleocapsid of orthomyxoviruses is different from that of the nonsegmented viruses because it is ribonuclease-sensitive (Pons et al., 1969).

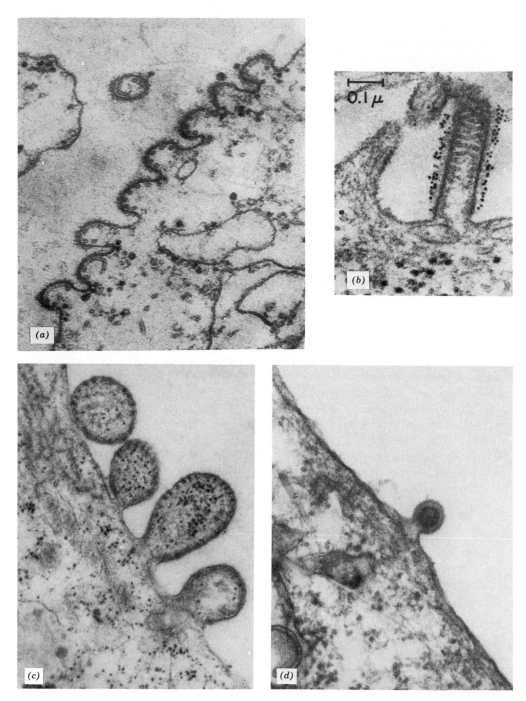

FIGURE 13-16. *Formation of enveloped viruses by budding from the plasma membrane.* (a) *A row of eight paramyxovirus particles (simian virus 5) budding from the surface of a monkey kidney cell. The circles within the buds are cross-sections of the nucleocapsid that abuts the inner face of the plasma membrane and is organized in coils* ×80,000. *Taken from Compans et al., 1966 with permission.* (b) *A filamentous form of simian virus 5 budding from the surface of an infected cell.*

Because of the segmented nature of the Group Vb viral genomes, they should show high levels of recombination between genes located in different segments. This is the case for influenza virus (Simpson and Hirst, 1968).

Influenza virus probably has eight different segments of RNA as its genome (Ritchey et al., 1976). It is not known whether they are packaged randomly into the virus particles or whether there is a specific mechanism for selecting eight different ones. Hirst and Pons (1973) have presented evidence that the packaging may be random and that many influenza virions may lack a full complement of RNA molecules.

The segmented nature of the influenza virus genome may have important consequences in the epidemiology of its disease. Influenza viruses of very different antigenicity arise with approximately an eight year periodicity. It has been suggested that there is a continual shifting of RNA segments between animal influenza viruses and human influenza viruses, allowing for the insertion of new antigenic types into the human population (Webster et al., 1973).

Viral Envelopes

The togaviruses, negative-strand viruses, and retroviruses (see Chapter 14) have a common feature, a lipoprotein *envelope* surrounding their ribonucleo-protein core. They derive this envelope in a common manner: the ribonucleo-protein associates with the under surface of a region of altered plasma membrane and as the ribonucleoprotein exits from the cell it is enveloped by this altered membrane. This process is called *budding* and the nascent particle as it forms in continuity with the plasma membrane is called a *bud*. Buds are easily recognized in thin section electron micrographs as regions of character-istically altered plasma membrane (Figure 13-16). A number of review articles describing present knowledge about envelope structure and the budding process are available (Choppin et al., 1972; Simons et al., 1974; Lenard and Compans, 1974).

VIRION STRUCTURE

There are three major classes of structural proteins in enveloped viruses: *glycoproteins, matrix proteins,* and *nucleocapsid proteins*. The macrostructure of

---

The coiled nucleocapsid is visible in the upper part of the bud. The budding virion is labeled with a ferritin-conjugated antibody against the virus (black dots are the ferritin). The absence of viral protein in the region of plasma membrane not included in the bud is apparent from the lack of ferritin away from the bud. ×90,000. Taken from Compans and Choppin, 1971 with permission. (c) Budding measles virus particles from a persistently infected cell line. The four particles seen are in late stages of budding. The circles at the edges of the buds are cross-sections of the nucleocapsid. The black dots in the particles are possibly ribosomes. Micrograph courtesy of S. Rozenblatt and C. Moore. (d) Budding C-type retrovirus. Continuity of the cell's plasma membrane with the viral envelope is evident. Micrograph courtesy of C. Moore.

the virion is determined by a closed lipid bilayer surface that surrounds the nucleocapsid. The outside of the lipid bilayer is covered by glycoprotein, the inside is coated with matrix or nucleocapsid protein (see Figure 13-13).

The lipid in the viral envelope is entirely derived from the host cell—no virus-specific lipid metabolism has been demonstrated. The composition of the virion lipid closely reflects the lipids of the cell's plasma membrane: it consists mainly of cholesterol, glycolipids, and phospholipids. There is significant, species-specific variation in the types of lipid in plasma membrane and the virus made by a given cell reflects the lipid composition of that cell (Klenk and Choppin, 1970). Different enveloped viruses derived from the same host cell are very similar in lipid composition but minor differences between viral membranes and host membranes may indicate some specificity of interaction between the viral proteins and certain lipids.

The glycoproteins of different viruses have specific properties as well as common characteristics. They are all located on the exterior surface of the virion and can be removed from the virus by protease treatment. Because no other viral proteins are cleaved by protease treatment of intact particles, the glycoproteins appear to be the only ones that extend out of the bilayer. The protease does not remove the entire glycoprotein; a "foot" of highly hydrophobic polypeptide remains with the virion, apparently buried in the lipid and inaccessible to the protease (Utermann and Simons, 1974).

The foot of the togavirus glycoprotein probably extends through the bilayer to the interior of the virion because some glycoprotein peptides are inaccessible to a hydrophilic reagent that cannot penetrate membranes unless membrane integrity is destroyed. Also, the glycoprotein can be cross-linked to the nucleocapsid protein by a reagent that is only one-fourth the thickness of the lipid bilayer (Garoff and Simons, 1974). The amount of protein actually embedded in lipid must be small because low-angle X-ray diffraction shows that the center of the bilayer approximates the density of pure lipid (Harrison et al., 1971) and no intramembrane particles can be seen after freeze-fracture electron microscopy (Brown et al., 1972).

If the results on togavirus glycoproteins are representative of all viral glycoproteins, these proteins can be classified as integral membrane proteins, that is proteins embedded in the lipid bilayer by a hydrophobic domain (Singer, 1971). The togavirus glycoprotein, and presumably other viral glycoproteins, is a transmembrane integral protein because it has hydrophilic regions on either side of the bilayer (Nicholson, 1976).

Like almost all proteins on the exterior surface of plasma membranes the viral glycoproteins have carbohydrate chains attached to the exterior hydrophilic portion of the protein. The function of the carbohydrate is not known but may be to increase the hydrophilic nature of the protein. The terminal sugar on the carbohydrate is sialic acid except in those viruses that have a virion-bound neuraminidase. No virus-specific glycosyltransferases are known, hence the structure of the carbohydrate must reflect the action of cellular enzymes modifying specific sites on the viral proteins (Grimes and Burge, 1971).

The matrix proteins appear not to be integral membrane proteins. The VSV matrix protein is freely soluble and is not intimately bound into membranes (Knipe et al., 1977b). Matrix proteins do not have carbohydrate;

thcy are presumably in the class of peripheral membrane proteins (Singer, 1971, 1974) and are bound to the lipid bilayer either by ionic association with the hydrophilic portion of the membrane or by interaction with the interior portion of the glycoprotein.

The togaviruses and bunyaviruses lack a matrix protein, their virions containing only glycoproteins and a nucleocapsid. The nucleocapsid protein appears to abut the interior face of the lipid bilayer occupying the same position in the virion as that of the matrix protein in myxoviruses, paramyxoviruses, and rhabdoviruses. The togavirus nucleocapsid protein has characteristics of a peripheral protein, not an integral protein (Helenius and Soderlund, 1973). In viruses with a matrix protein, the nucleocapsid probably does not interact with the lipid bilayer but rather interacts with the matrix protein.

The structural location of the two or three major proteins of the enveloped viruses strongly suggests that their architecture results from protein-protein and protein-lipid interactions, but exactly what interactions are most critical has yet to be shown. No bipartite interaction among the three proteins has been clearly established. The cross-linking experiments indicate that the glycoprotein and nucleocapsid proteins of togaviruses are closely apposed but do not establish the existence of a linkage.

VIRION ASSEMBLY

The first step of virion synthesis is formation of the individual proteins. Each of the three protein classes appears to be made as a separate entity, often on a separate mRNA (Knipe et al., 1977b).

The glycoprotein is formed on a membrane-bound mRNA and is never free in the cell. It matures by passage from the rough endoplasmic reticulum to smooth membranes and possibly Golgi and then to the plasma membrane. It gains its carbohydrate while passing through the intracellular membranes and then is placed onto the cell surface where it presumably floats freely in the fluid lipid bilayer (Knipe et al., 1977a,c).

The matrix and nucleocapsid proteins are made on free polyribosomes and the nucleocapsid protein rapidly binds to RNA. How the matrix protein is handled intracellularly is uncertain: after cell fractionation this protein is often found in association with all cell membranes (Hay, 1974). It may be a soluble protein that easily aggregates and sticks to membranes (Knipe et al., 1977b).

Formation of the nascent virion bud appears to occur by a cooperative process in which the nucleocapsid, matrix protein, and glycoprotein coalesce. This is most evident for VSV since in the earliest stage of formation of the bullet-shaped particle only the tip of the bullet is formed while the rest of the nucleocapsid remains unstructured and the surrounding plasma membrane shows no alteration. It is thought that during the coalescence, glycoprotein floating free in the membrane diffuses to the nascent bud and is trapped either by self-interaction or, more likely, by interaction with proteins beneath the membrane (Garoff and Simons, 1974; Simons et al., 1974). Bud formation is also presumably nucleated by such transmembrane interactions.

Virions of enveloped viruses contain no detectable host proteins, although their coat is derived from the host plasma membrane (Pfefferkorn and Clifford, 1964). Garoff and Simons (1974) suggested that the close packing of viral glycoprotein on the nascent bud sterically excludes host proteins from the virion surface. Whatever the mechanism involved, it is presumably a consequence of membrane fluidity that the virus can manage to exclude host membrane proteins from the region of the bud.

FUNCTION AND PROCESSING OF GLYCOPROTEINS

The glycoproteins present as structural elements of the virion also serve critical functional roles. They are the sites of virion attachment to cells and often have highly specific receptor affinities. They can be neuraminidases and can also be responsible for cell fusion and for red blood cell lysis. All of these functions are thought to play significant roles in the cell target specificity and the pathologic properties of the viruses.

As part of the maturation of viral glycoproteins there is often a proteolytic cleavage. One of the two togavirus glycoproteins is cleaved before it reaches the exterior surface of the cell (Sefton et al., 1973). On the contrary, the protein in myxoviruses that agglutinates red blood cells and appears to attach the virus to susceptible cells is cleaved only after it reaches the cell surface and perhaps only after virion budding has begun. Cleavage of this protein greatly enhances the infectivity of the virus (Lazarowitz and Choppin, 1975). Similarly, the fusion factor of paramyxoviruses is only activated by cleavage on the external face of the membrane or in the virion, and that cleavage is required for infectivity (Scheid and Choppin, 1974).

Double-Stranded RNA Viruses (Class III)

Double-stranded RNA viruses have been found in molds, higher plants, insects, and vertebrates. All such viruses lack lipid and have two capsid layers, an inner core and an outer shell surrounding it. The inner core contains multiple segments of double-stranded RNA along with variable amounts of small oligonucleotides that appear to serve no genetic function. Reviews of these viruses include Gauntt and Graham (1969), Verwoerd (1970), Joklik (1974), and Shatkin and Both (1976).

The most extensively studied double-stranded RNA viruses are the human reoviruses which cause no known human disease—although reovirus-like agents appear to cause infantile gastroenteritis (Kapikian et al., 1976)—but are frequently isolated from humans and grow easily in laboratory conditions. Some evidence is also available about selected plant and insect double-stranded RNA viruses (Joklik, 1974).

MULTIPLE MONOCISTRONIC VIRION RNAs

The first evidence that reoviruses might contain double-stranded RNA came from an unlikely source: it was observed that the staining characteristics of intracellular viral RNA with acridine orange were more like the properties of DNA than of RNA (Gomatos et al., 1962). Reovirions, however, clearly contained RNA. When it was discovered that the RNA of the virions is ribonuclease-insensitive, shows a sharp melting temperature, and diffracts X-rays like a double strand, it became clear that the genome must be double-stranded RNA (Gomatos and Tamm, 1963; Langridge and Gomatos, 1963).

The RNA of reovirus, and of all other double-stranded RNA viruses, is present not as a single long filament but as a defined spectrum of about 10 individual RNA molecules (Figure 13-14). Molecular hybridization has shown that the size classes consist of unique nucleotide sequences (Bellamy and Joklik, 1967; Watanabe et al., 1967). Reovirus contains exactly 10 RNA segments and the primary protein products made in infected cells consist of 10 polypeptides (Both et al., 1975c). The correspondence of size between the polypeptides and the RNAs leaves little doubt that the double-stranded RNAs are each monocistronic RNAs encoding a single polypeptide. A number of the polypeptides are proteolytically processed after their synthesis. It is not certain what determines the ability of the virion to package a complete set of RNAs; at some stage in the life cycle of the virus there may be a connection between them, although probably not a covalent one.

THE VIRION TRANSCRIPTASE

Because double-stranded RNA will not act as mRNA, the RNA segments of the genome must either be copied or separated to generate mRNA. If the segments generate mRNAs by transcription, then the necessary enzyme must be in the virion, because copying of double-stranded RNA is not thought to occur in uninfected cells.

The core of the reovirion is, in fact, an RNA synthetic factory. Whole virions will not make RNA, but synthesis of all 10 mRNAs can be activated either by treating virions with a protease or by other methods of disruption (Shatkin and Sipe, 1968; Borsa and Graham, 1968). Activation of the core RNA polymerase can also be demonstrated soon after infection of cells, even if the infection is performed in the presence of an inhibitor of protein synthesis. Protease treatment *in vitro* strips the whole outer shell from the virion whereas the *in vivo* activation process leaves one of the outer shell polypeptides almost intact (Joklik, 1972). The core particles have an intrinsic infectivity about 10^{-5} that of the virions.

Synthesis of mRNA by virion cores is a conservative process; only newly formed, single-stranded RNA molecules are extruded from the cores while both strands of the double-stranded RNA remain intact inside (Skehel and Joklik, 1969; Levin et al., 1970). The synthesis is an asymmetric one; all single-stranded RNA formed is of the same polarity as intracellular mRNA. During the entire synthetic process the core remains impermeable to ribonuclease, indicating, as electron microscopy also shows, that the cores remain intact (Bartlett et al., 1974) (Figure 13-17). The cores can manufacture RNA for days in an *in vitro* system. It is possible in this way to generate huge amounts of reovirus mRNA that can direct synthesis of the reovirus-specific proteins in a cell-free system (McDowell et al., 1972).

RNA STRUCTURE

The ends of the virion RNA and mRNA have been chemically analyzed. The mRNAs have no 3' poly(A) nor does the plus strand of the virion RNA

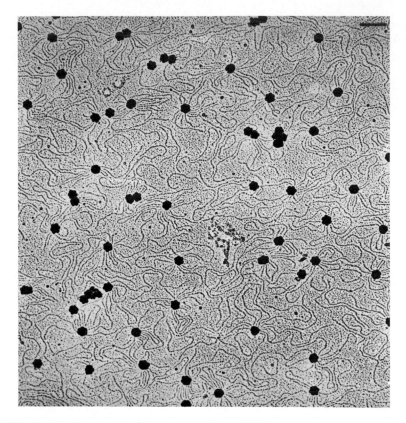

FIGURE 13-17. *Extrusion of newly made mRNA by reovirus cores. As the cores synthesize RNA, multiple RNA molecules are extruded from them. In this electron micrograph, many of the cores show six protruding spikes. There are a total of 12 such spikes per core which are really hollow cylinders that probably provide the channels for extruding the RNA. Reprinted from Bartlett et al. (1974) with permission.*

have poly(A) (Stoltzfus et al., 1973). These are the only known mammalian viral mRNAs without poly(A). Both the mRNA and the virion plus RNA are capped at their 5′ ends with the structure m^{7}G5′ppp5′GmpCpUp (Furuichi et al., 1975a,b; Chow and Shatkin, 1975). The minus strand RNA in the double-stranded RNA is not capped, is not polyadenylated, and has at its 5′ end a diphosphate, probably generated from an original triphosphate by a phosphohydrolase in the virion (Kapuler et al., 1970; Borsa et al., 1970). The nucleotide structures at the four ends of each double-stranded RNA segment appear to be identical in each of the segments. The sequences in the two strands are exactly complementary at the ends indicating that the RNA molecules do not have single-stranded tails.

VIRION ENZYMES

The whole machinery for manufacturing the mRNA is found in the virion, hence the virion core must have the RNA polymerase activity as well as the enzymes that cap and methylate the 5' terminus. Even the enzyme that ribose-methylates the guanosine in the cap is a virion enzyme (Shatkin, 1974). Newly made mRNA is extruded from the core; electron microscopy has shown cores with 10 or more nascent chains in the process of extrusion (Bartlett et al., 1974; Figure 13-17). The core has 12 hollow projections from its surface that may provide channels through which the nascent RNA molecules leave the core. Extrusion is presumably an active process and may involve the nucleoside triphosphate phosphohydrolase activity of the core.

None of the enzymatic activities of the virion has ever been isolated in a soluble form. In all probability the core of the virion is a tightly integrated system for synthesizing, capping, methylating, and extruding RNA. The "structural" proteins of the core are probably also its enzymatic proteins; the highly ordered symmetric structure of the core may be necessary to the functioning of the proteins.

REPLICATION

Thus far only the synthesis of mRNA has been considered. But to generate new reovirions it is also necessary that minus strand RNA be formed. In infected cells, the plus strand is made much earlier than the minus strand (Schonberg et al., 1971). All minus strand synthesis occurs in a particulate fraction of the cell which can incorporate labeled nucleotides into minus strands *in vitro* (Acs et al., 1971; Sakuma and Watanabe, 1972).

The process of minus strand synthesis is sensitive to pretreatment of the particles with ribonuclease but when the process is complete, the newly made, double-stranded RNA is insensitive to ribonuclease even under salt conditions in which free double-stranded RNA would be digested. The structures containing the newly made RNA are similar to the core of the virion, suggesting that morphogenesis of the core is coordinated with synthesis of the minus strand.

In all probability the structure that makes the minus strand of RNA is a partially completed core structure. To form the precore, 10 single-stranded RNAs must come together with the appropriate proteins. Recognition of the difference between the 10 RNA molecules of the virion therefore occurs when they are single-stranded, but how the recognition occurs is not known. Zweerink (1974) has been able to fractionate the precores by size and has shown that the slowest-sedimenting ones make predominately minus strands of the smallest double strands while the fastest-sedimenting ones make mostly the minus strand of the largest double strands. Replication therefore appears to be a sequential process rather than being a coordinated copying of all mRNA strands at once.

Temperature-sensitive mutants of reovirus have been isolated; they fall into a number of freely recombining classes. This behavior suggests that the single-stranded RNA segments are drawn from a common pool of RNA to form the precore structure within which the double-stranded RNA is ultimately finished (Fields and Joklik, 1969; Joklik, 1974).

THE LIFE CYCLE

Disruption of cores after infection occurs even if protein synthesis has been inhibited and must be caused by preexisting, presumably lysosomal enzymes (Silverstein and Dales, 1968). The virion is never degraded past the core stage and the input cores can reacquire an outer shell later in infection (Silverstein et al., 1972).

The primary transcription phase lasts a few hours following which synthesis of double-stranded RNA becomes evident and the rate of mRNA synthesis greatly accelerates (Joklik, 1974). Most of the viral mRNA appears to be made from progeny double-stranded RNA. Progeny virions are formed over a 10 hour period starting about 6 hours after infection. The entire life cycle takes place in the cytoplasm.

Animal Virus Multiplication: DNA Viruses and Retroviruses

14

Because RNA viruses have genetic systems with no apparent counterparts in normal cells, the major problems they face involve production of enzymes that can carry out the processes of transcription and replication. DNA viruses, on the contrary, direct mRNA synthesis in the same way that the normal cell does and replicate their genome in a formally similar manner to the replication of the cellular genome. Furthermore, the DNA of most DNA viruses is transcribed and replicated in the nucleus where the cell normally carries out these functions.

The similarity of basic processes in cells and in DNA viruses suggests that a DNA virus could multiply without providing any new enzymatic functions in the infected cell. The provision of capsid proteins might be sufficient for viral growth and the genome of the virus would thus consist only of capsid protein genes. Such a DNA animal virus does in fact exist but most DNA viruses lead a more complicated life. They run the gamut of size and complexity: the smallest has a DNA of only 1.5×10^6 daltons, the largest has 100 times more DNA. As the viral genomes get bigger they get more complex both in having more genes and in having more complicated mechanisms of DNA replication.

Since a tiny DNA virus can replicate perfectly well, one might wonder why the large DNA viruses have evolved at all. One advantage is that as the genome gets larger the virus becomes less dependent on the cell. Specifically, the large DNA viruses can grow in cells that are not in S phase whereas the smallest DNA viruses require cells to be in S phase. The intermediate DNA viruses encode a protein that can drive the host cell to enter S phase. Many virus-specific enzymes appear to duplicate host functions; in all likelihood, the ability to encode these functions allows the viruses to grow in cells in which the functions are not being expressed.

All groups of DNA viruses except the simplest ones have oncogenic

343

representatives. Their oncogenic potential is probably a by-product of the growth mechanisms used by the virus. The viruses that induce cells to enter S phase may cause malignancies by virtue of that ability; they could override normal cellular controls and maintain cells in a continuously growing condition. Discussion of the oncogenic potential of DNA viruses is left for Chapter 16. The present chapter deals only with DNA viruses as productive, infectious agents. The last group of viruses considered in this chapter is the retroviruses, viruses that use both RNA and DNA in their life-cycle. These can also be oncogenic, a property that is taken up in Chapter 16.

Parvoviruses

Parvoviruses are probably the simplest of all viruses. They have only 1.5×10^6 daltons of single-stranded DNA in their genome and even that is not completely used to encode their single product, a capsid protein molecule. The growth of this tiny parasite presumably involves virtually only cellular functions (Hoggan, 1970, 1971; Berns, 1974; Rose, 1974).

There are two general classes of parvoviruses, the *autonomous* and the *defective* viruses. The autonomous virises, all known ones being viruses of rodents, can use host cell enzymes for replication, transcription, and any other needed functions. The defective parvoviruses are dependent on coinfection of cells with an adenovirus, which provides some critical functions. There is no known uninfected cell in which the defective parvoviruses can grow.

The autonomous parvoviruses cannot grow in stationary cells; only cells replicating their own DNA, that is, cells in S phase, will support their multiplication (Tattersall, 1972). This limitation is expressed in the types of cells that become infected in diseased animals. Parvoviruses cause abnormal fetal development as well as defects of growing tissues in the newborn animal. They also cause intestinal problems, presumably because of growth in the rapidly dividing cells of the intestinal crypts.

Defective parvoviruses neither require S-phase cells for growth nor will they grow in S-phase cells without an adenovirus helper. Only adenovirus can fulfill their requirements; herpesviruses support some aspects of the defective virus growth but not production of complete particles (Atchison, 1970). For this reason, the defective viruses are all called adeno-associated viruses (AAV).

One striking difference between autonomous and defective parvoviruses is that the genome of the autonomous viruses is a unique single strand of DNA, but the DNA found in virions of AAV consists of equimolar quantities of complementary single-stranded DNA molecules (McGeoch et al., 1970; Berns and Rose, 1970). Annealing of the DNA isolated from AAV readily converts it all to double-stranded molecules.

Parvoviruses are about the size of a ribosome (20 nm in diameter). They contain no lipid but have in their capsids three different sizes of polypeptides, the largest of which is 90,000 daltons. The tryptic peptides of the smaller proteins are subsets of the tryptic peptides of the 90,000 dalton polypeptide. It is believed, therefore, that the viral mRNA encodes only 90,000 daltons of capsid protein (Tattersall et al., 1976).

RNA SYNTHESIS

Only a single virus-specific mRNA has been identified in parvovirus-infected cells. This mRNA is about 9×10^5 daltons and therefore has the appropriate coding capacity for the synthesis of the 90,000 molecular weight polypeptide (Carter and Rose, 1974). No higher molecular weight precursor of this RNA has been identified. The molecular weight of the genome is larger than the size of the mRNA, suggesting that there might be coding capacity for a second mRNA, but no such second mRNA has been identified. Since the total RNA from infected cells is unable to protect more than 80% of the genome DNA, the single 9×10^5 dalton mRNA may be the only one formed. For the nondefective parvoviruses, a cellular RNA polymerase must be responsible for the synthesis of the mRNA. The defective ones, however, make no RNA unless there is coinfection by adenovirus, suggesting that an adenovirus-specific protein plays a role in transcription from the AAV genome.

VIRION DNA STRUCTURE

AAV DNA has two properties that indicate the existence of repetitious structures at the ends of the molecules. One is that the ends of the single-stranded molecules can associate with each other to generate single-stranded, circular DNA molecules (Koczot et al., 1973; Berns and Kelley, 1974). There must be inverted terminal repetitions in the DNA to provide the necessary hydrogen bonding specificity to form the circles. The second property is the ability of double-stranded AAV DNA to form circles, which indicates the presence of terminal noninverted repetitions (Gerry et al., 1973). Therefore, AAV DNA has both terminal repetitions and inverted terminal repetitions.

One structure that is consistent with these two properties is that the ends have identical sequences that can fold back on themselves to form "hairpins" (Figure 14-1). Denhardt et al. (1976) have provided evidence that such structures can exist in AAV DNA. If the foldback structures at the two ends of the molecule are identical, then they are also self-complementary, and the molecules should form circles, linear dimers, and dimeric circles as well as higher order polymers; such structures have been seen in the electron microscope (Koczot et al., 1973; Berns and Kelley, 1974; Figure 14-2).

The structure of the DNA of autonomous parvoviruses is different from that of AAV (Figure 14-1). In the one well-studied case, that of minute virus of mice, there is a stable hairpin structure only at the 5' end and no inverted terminal redundancy can be demonstrated (Bourguignon et al., 1976). However, the 3' end appears also to have a weakly associating hairpin structure because the molecule can be replicated by DNA polymerases that require a hydrogen-bonded 3'-end for initiation.

DNA REPLICATION

Cells infected by either type of parvovirus accumulate double-stranded forms of DNA that appear to be intermediates in replication (Tattersall et al., 1973; Siegl and Gautschi, 1976; Straus et al., 1976). A large fraction of such

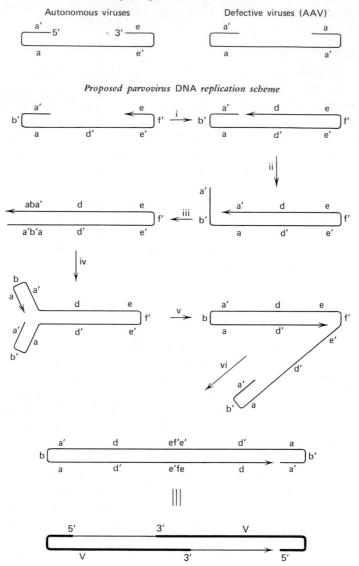

FIGURE 14-1. Structure and replication of parvovirus DNA. The upper portion diagrammatically represents the structures of the DNA from autonomous and defective (adenovirus-associated) parvovirus. The designations x and x' are meant to represent DNA sequences that have exact Watson-Crick base complementarity. In the autonomous viruses, sequences a and a' represent the stable 5' hairpin, sequences e and e' represent a transient 3' hairpin and d' represents 95% of the length of the molecule. On the model of replication of autonomous viruses pictured in the lower portion of the figure, the 3' hairpin is shown initiating DNA synthesis (i). When the growing strand reaches the 5' hairpin, it is dissociated and replicated (ii and iii). The 5' hairpin then reforms generating a complementary 3' hairpin (iv) that can reinitiate replication (v), ultimately producing a covalent dimer (vi) consisting of two virion strands and two complementary strands. In the bottom diagram, V signifies the two virion DNA's in a dimer duplex. Repetition of steps (iv), (v), and (vi) would generate a tetramer and continued iteration could produce higher multimers. Adapted from Tattersall and Ward (1976) by permission.

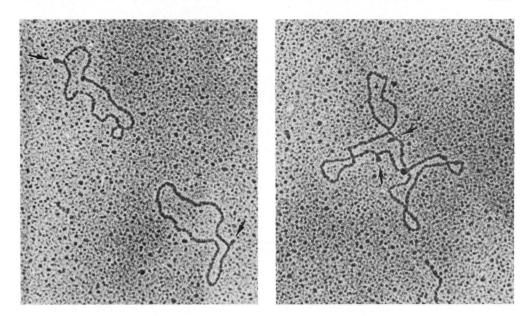

FIGURE 14-2. Circles of double-stranded DNA from adenovirus-associated, defective parvovirus. The left panel shows monomer circles and the right shows a dimeric circle. In both, "pan-handles" are marked with arrows. They are about 100 base pairs long and in the dimer are separated by about 180°. The pan-handles presumably represent the self-associating regions of the molecules that cause them to be circles. Reprinted from Berns and Kelly (1974) by permission.

double-stranded molecules cannot be irreversibly denatured, suggesting that the two strands are covalently linked. Linear dimeric double-stranded DNA molecules that spontaneously renature have also been observed (Straus et al., 1976). The monomeric and dimeric double-stranded DNAs appear to involve plus and minus strands linked end to end: one of each in the monomer and four alternating plus and minus strands in the dimer.

These observations have suggested a model of replication for parvovirus DNA (Straus et al., 1976; Tattersall and Ward, 1976). Figure 14-1 shows the model proposed for the autonomous parvoviruses. The model generates a duplex dimeric molecule (or higher polymer) from which specific nuclease action must excise the virion DNA strands. Tattersall and Ward (1976) have suggested that the DNA may be packaged as it is peeled off of the replicating form possibly by displacement synthesis. Figure 14-3 shows a proposed packaging scheme for newly made molecules of DNA that will lead to incorporation of a unique strand into the virion of the autonomous viruses.

These models would have to be modified for AAV in view of the evidence that AAV has identical 3' and 5' ends. This difference is slight, however, for AAV the plus and minus strands could both initiate replication in the same manner as the virion DNA of the autonomous viruses. Because the strands are not distinguishable at their ends, they could both be packaged. This is, in less detail, the model of Straus et al. (1976) and explains why AAV virion DNA

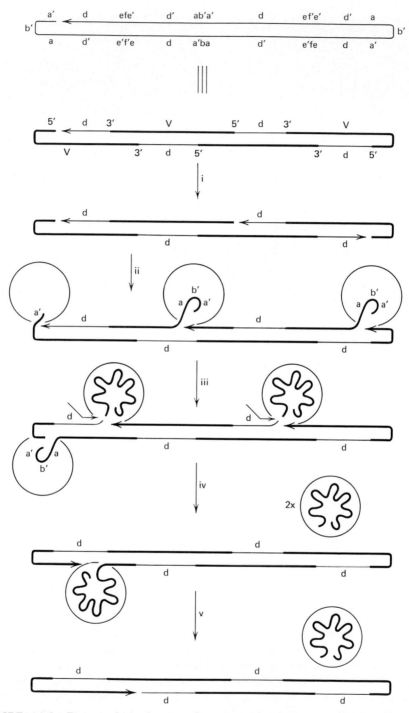

FIGURE 14-3. *Proposed mechanism for parvovirus DNA packaging by displacement synthesis from a tetrameric intermediate. The progeny DNA strand, V, is indicated throughout as a heavy line. The sequence d represents the approximately 4100 nucleotides of the complementary strand that is not packaged by the autonomous parvoviruses. The circle represents the capsid of an immature virion. Adapted from Tattersall and Ward (1976) by permission.*

consists of both plus and minus DNA while the virion DNA of the autonomous viruses is unique.

Because parvoviruses are dependent on cellular enzymes for their replication, the mechanism of replication they use may have a close relationship to that used by cells and may provide detailed information about it.

Viruses often have proteins that carry out multiple functions, so that, despite the fact that parvoviruses may encode only a single protein, that protein might also play some role in DNA synthesis. Rhode (1976) has in fact isolated mutants affected in the coat protein which make decreased amounts of single-stranded progeny DNA. The protein may be involved in progeny DNA synthesis, but not in the synthesis of the double-stranded DNA intermediates of replication.

Papovaviruses

The papovavirus group is best-known for its two widely studied oncogenic members, SV40 and polyoma, which are both lytic viruses that grow in a narrow range of mammalian cells. Their oncogenicity, as expressed by cellular transformation *in vitro*, is generally studied in cells of species in which the viruses do not carry out a complete lytic cycle. In the present chapter we consider only their lytic cycle (Sambrook, 1972; Tooze, 1973; Levine, 1976; Salzman and Khoury, 1974).

There are more members than just SV40 and polyoma in the papovavirus group. The name papovavirus derives from three prototypes: rabbit *pa*pilloma virus, *po*lyoma virus, and simian *va*cuolating virus-40 (SV40). These specific viruses do not appear to be indigenous to human beings, although SV40 may infect humans occasionally, but three papovaviruses are widespread in the human population: JC, BK, and human wart virus. JC virus has been associated with a progressive neurological degeneration in humans (Padgett et al., 1971). BK is often recovered in the urine of people whose immune system has been suppressed but it has not been associated with a human disease (Gardner et al., 1971). Human wart virus, like many animal papilloma viruses, causes only benign overgrowths of the epidermis.

Only the properties of SV40 and polyoma virus are considered here. Papilloma viruses have been studied only as physical entities because they do not grow well in tissue culture. They have a somewhat larger DNA molecule than SV40 and polyoma (Watson and Littlefield, 1960).

THE VIRION

The virion of papovaviruses consists of only protein and DNA. The capsid protein is composed mainly of one major polypeptide, VP1. Two minor polypeptides, VP2 and VP3, which consist largely of identical amino acid sequences, are also present (Fey and Hirt, 1974). Aside from these virus-specific proteins, the virion has cellular histones associated with the viral DNA (Frearson and Crawford, 1972); only the H-1 histone is absent (Louie, 1974).

The DNA of papovaviruses is a single, closed-circular duplex molecule weighing 3 to 3.5×10^6 daltons. The DNA is not significantly larger in coding capacity than the 1.5×10^6 dalton single-stranded DNA of the parvoviruses. The papovavirus genome, however, encodes not only the virion proteins, but also an early protein that plays a significant role in the viral life cycle.

Papovavirus DNA gently released from the virion is recovered as a circular molecule with approximately 20 adhering particles which apear to be homologous to the histone-containing particles associated with cellular DNA. Present evidence suggests that the structure of the papovavirus chromosome is very similar to that of the DNA of normal cells (Louie, 1974).

VIRAL DNA

The closed-circular duplex form of papovavirus DNA is a very constrained molecule. The lack of an end requires that any perturbation of the number of turns of the two single-stranded DNA molecules around each other must be accompanied by three-dimensional alterations in the DNA structure. Binding of histones to the DNA apparently provides such a perturbation (Germond et al., 1975) so that after deproteinization the naked DNA molecule is slightly deficient in the turns necessary to accommodate the Watson-Crick base-paired structure of the DNA. This deficiency in turn is taken up by a "supercoiling" in isolated molecules of either SV40 or polyoma DNA (Vinograd et al., 1965). A nick in either strand of the DNA releases the constraints that give rise to the supercoiling and allows the molecule to relax.

Multiple forms of papovavirus DNA can be produced by a combination of single-strand nicks, double-strand cuts, and denaturation. These interrelations are illustrated in Figure 14-4 (Vinograd and Lebowitz, 1966). Single-stranded

FIGURE 14-4. *Model of various forms of polyoma or SV40 DNA. The native molecule (I) is a supercoiled, covalently closed circular double helix. It can be converted to II by a single-strand break and to III by a double-stranded break. Upon denaturation, I becomes a rapidly sedimenting tangled form, II gives one circular and one linear strand and III gives only linear single-strands. Redrawn from Vinograd et al., 1965.*

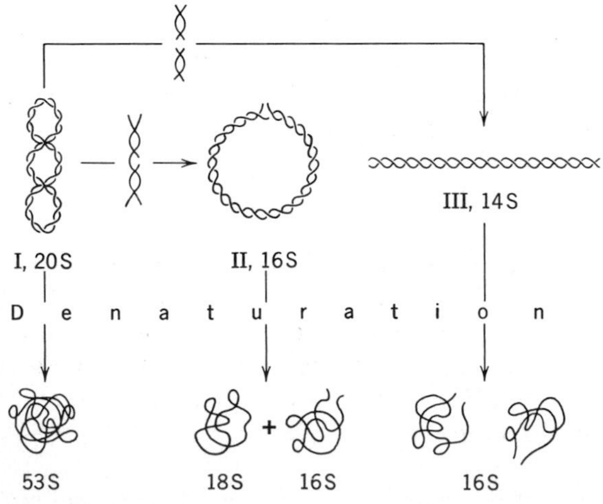

circles sedimenting at 18S, single-stranded linear 16S molecules, double-stranded linear 14S molecules, and nicked circles sedimenting at 16S can all be generated from the 20S circular duplex. Possibly the most unique molecule is a double-stranded collapsed molecule generated by denaturing the native DNA. This molecule sediments at 53S because, although it is denatured, the topological constraints on the molecule prevent it from coming apart into free single strands.

MAPPING THE DNA

Using restriction endonucleases, it is possible to cut papovavirus DNA into a large number of defined fragments (Danna and Nathans, 1971). Figure 14-5 shows a map of the physical relationship between the fragments of SV40 DNA generated by different restriction endonucleases. The position of fragments relative to one another was determined by recutting partial degradation products as well as by analyzing the fragments generated by serial application of different restriction enzymes (Danna et al., 1973). The map is oriented relative to a zero position defined by the single cut produced by the EcoRI endonuclease and positions are conventionally denoted by fractional distances from that point (Morrow and Berg, 1972).

MAPPING MUTANTS ONTO THE DNA

Four complementing classes of temperature-sensitive mutants, A, B, C, and D, have been isolated for both polyoma viruses and SV40 (Chan and Martin, 1974). The map positions of these are shown in Figure 14-6. The B and C mutants appear to be in the same gene, suggesting the occurrence of intragenic complementation.

Two procedures have been used to map the temperature-sensitive mutants. In one, heteroduplex molecules are made between a circular single-stranded DNA of the mutant and a fragment of wild type DNA generated by a restriction endonuclease. If the fragment overlaps the mutant gene, cells infected with the heteroduplex molecule yield wild type virions; otherwise only mutant progeny are produced (Lai and Nathans, 1974).

The second mapping method uses complete heteroduplex molecules between mutant and wild type DNA. The mismatched sequence at the point of mutation generates a small single-stranded region that can be cut by S_1 nuclease of *Aspergillus oryzae*, a nuclease that will specifically degrade single-stranded regions of DNA. The nuclease cuts both strands, modifying the pattern of fragments produced by a restriction enzyme and thus providing a map position for the mutant relative to the position of the restriction enzyme sites (Shenk et al., 1974).

TIME COURSE OF REPLICATION

The complete growth cycle of papovaviruses requires a few days from the time of infection to the time of maximum yield. There are three distinct phases in the life cycle of these viruses (Figure 14-7). During the first phase, until about 8 hours after absorption of the virus to cells, all that is evident is

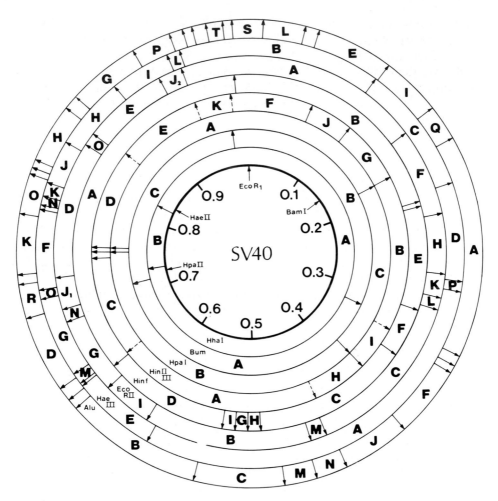

FIGURE 14-5. Cleavage of SV40 DNA by various restriction enzymes. The inner disk shows the map positions as decimal fractions of a circle starting from a zero point defined by cleavage with the EcoR₁ *restriction endonuclease. The positions of cleavage by other enzymes that cut only once is also indicated at the edge of the inner disk. The outer rings represent cleavage sites by a variety of endonucleases that make multiple pieces. Provided by J. Sambrook. An equivalent map of polyoma virus has been generated (Griffen et al., 1974).*

movement of viral DNA to the nucleus. The second or *early phase* begins with synthesis of early mRNA and early protein. The early protein is mainly if not entirely a protein called "T-antigen." Effects on the host cell are also evident during this phase of the life cycle: there is induction of a series of cellular enzymes of DNA metabolism and the beginning of virus-induced synthesis of host cell DNA (Hartwell et al., 1965).

During the *late phase* of the growth cycle, lasting from about 12 hours after infection until the end, synthesis of viral DNA, late mRNA, and late protein,

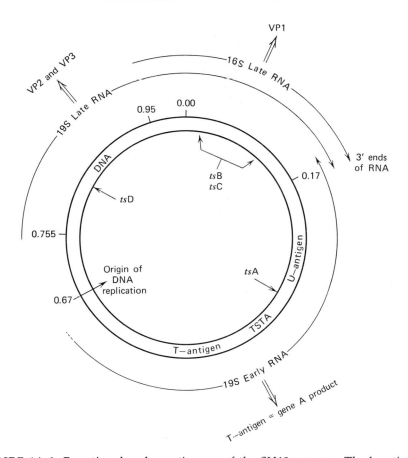

FIGURE 14-6. *Functional and genetic map of the SV40 genome. The locations of the four groups of temperature-sensitive mutants is shown as is the site of initiation of viral DNA replication. The regions from which three mRNAs are copied is indicated; the late mRNAs have been mapped by electron microscopy of RNA-DNA hybrids (May et al., 1976). The proteins made by each of the mRNAs are also shown. The region of DNA producing the 19S early RNA is responsible for three separable effects as seen in specific adenovirus-SV40 hybrid viruses: U-antigen synthesis, tumor-specific transplantation antigen (TSTA) formation and T-antigen synthesis. The subregions responsible for eliciting these antigens are indicated within the annulus representing the DNA (Lebowitz et al., 1974).*

as well as virion maturation, occur together. Host DNA synthesis and that of early viral products also continues for much of the late phase.

THE EARLY PROTEIN: T ANTIGEN AND GENE A

The first protein made after infection of cells by a papovavirus is the so-called T-antigen. This antigen is defined by either immunofluorescence or complement fixation using the serum of an animal that carries a tumor

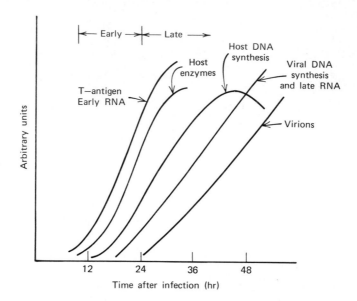

FIGURE 14-7. *Time-course of the events following SV40 infection of monkey cells.*

induced by the specific papovavirus (Black et al., 1963). Polyoma and SV40 viruses make distinct T-antigens, each located in the nucleus of infected cells.

Antisera that are specific for T-antigen will precipitate mainly a polypeptide of about 90,000 daltons from infected cells (Tegtmeyer et al., 1975) and T-antigen has been partially purified (Carroll et al., 1974; Jessel et al., 1975). Multiple forms of T-antigen can be detected by sedimentation rate, indicating that either it self-associates or has affinity for host proteins (Osborn and Weber, 1974). The T-antigen of SV40 binds tightly to any double-stranded DNA; on SV40 DNA it appears to bind most efficiently near or at the site where DNA replication starts (Reed et al., 1975).

Because T-antigen is the earliest protein made in infected cells, mutations in the gene coding for T-antigen should behave as early mutations. Two classes of apparent early SV40 mutants are known: the *ts*A and *ts*D groups. The *ts*D mutations appear to prevent the earliest stage of the life cycle, by inhibiting the complete uncoating of the viral DNA, but they are really in a late gene (Robb and Martin, 1972; Martin et al., 1974). Only the *ts*A group of mutations defines a truly early function which could be carried out by T-antigen.

The SV40 T-antigen from cells infected by *ts*A mutants shows a number of differences from wild type T-antigen by complement fixation (Alwine et al., 1975) and by DNA binding (Tenen et al., 1975); at the nonpermissive temperature it also does not form aggregates like wild type T-antigen (Osborn and Weber, 1974). Finally, it appears to regulate its own synthesis, so that at nonpermissive temperature much more of the specific 90,000 dalton polypeptide is made than at permissive temperature or by wild type virus (Tegtmeyer et al., 1975).

These studies have led to the provisional conclusion that the gene defined by the *ts*A mutants encodes a 90,000 dalton polypeptide that displays T-antigenicity. A protein of this size would require at least half of the coding capacity of the SV40 genome and would require an mRNA of some 900,000 daltons for its synthesis. During the early stage of infection, in fact, the only virus-specific mRNA found in infected cells has a size of about 900,000 daltons, a sedimentation coefficient of 19S, and arises from the lower half of the DNA map (Weinberg and Newbold, 1974; Figure 14-6). The tsA mutations are grouped within a small section of this *early region* of the viral DNA.

The major physiological consequence of a tsA mutation at nonpermissive temperature is a block in initiation of viral DNA synthesis (Tegtmeyer, 1972). Impairment of the induction of cellular DNA synthesis may also occur (Chou and Martin, 1975) and T-antigen may play a critical role in cellular transformation by papovaviruses (see Chapter 16). The affinity of T-antigen for DNA and the effects of tsA mutations on DNA synthesis suggest that T-antigen functions by interacting with and modifying DNA so that it can be replicated. Whether the role of T-antigen in transformation involves this same action is not known.

The proposal that one early papovavirus protein carries out all of the early functions satisfies much of the relevant data but there are at least two observations that are not well-explained by the model:

1. SV40-infected cells have on their surface a virus-specific antigen called TSTA (tumor-specific transplantation antigen). It appears as an early product of lytic infection (Girardi and Defendi, 1970). Because it is localized on the cell surface, while T-antigen is nuclear, it seems unlikely to be part of T-antigen. It could therefore be a separate early protein. A second antigen, called U-antigen, is also found in infected cells, appears separate from T-antigen, and is made early (Lewis and Rowe, 1971).

These three early antigens have been mapped onto the early region by their appearance after infection with adenovirus-SV40 hybrid viruses containing different portions of the SV40 early region (Figure 14-6). While both TSTA and U-antigens might be parts of T-antigen or be specific modifications of cellular macromolecules, their occurrence requires explanation before the model of a single A gene product can be accepted.

2. In polyoma virus there exist host-range mutants that were selected to grow only on cells expressing some polyoma functions; these now have been found to grow on certain uninfected cells as well (Benjamin, 1970; Benjamin and Goldman, 1974). These mutations, called hr-t because they affect both host range and transformation as described in Chapter 16, define an early gene that appears to complement the polyoma tsA mutations. They map in the early region of the genome but quite separate from the tsA mutants (Feunteun et al., 1976). Polyoma either has two early genes or an efficient intracistronic complementation can occur between hr-t mutants and tsA mutants.

INDUCTION OF CELLULAR DNA SYNTHESIS

Soon after T-antigen becomes evident, a wide range of enzymes of DNA metabolism start to increase in concentration in the infected cells (Hartwell et al., 1965). The induction requires new protein synthesis and at least 10 enzymes are involved including DNA polymerase, DNA ligase, and thymidine kinase (Eckhart, 1968). All of the enzymes appear to derive from the host cell because there is no place on the viral genome for encoding them. Thymidine kinase has been shown to be a host enzyme because when cells lacking the enzyme are infected by the virus no new thymidine kinase is produced (Basilico et al., 1969). The induced DNA polymerases are also host enzymes (Narkhammer and Magnusson, 1976).

Soon after the enzymes of DNA metabolism increase, host cell DNA synthesis begins in infected cells (Vogt et al., 1966). The virus apparently sets in motion a chain of events that first involves the stimulation of a variety of enzymes and then—possibly as a consequence of one or more of the newly made proteins—the initiation of replication of host cell DNA. This chain of events is evident only when resting cells are infected by the virus; then the entire complement of cellular DNA is replicated (Gershon et al., 1965). Cellular histone synthesis is also induced by the virus (Winocour and Robbins, 1970).

LATE PROTEINS AND THEIR mRNAs

As indicated previously, the capsid consists of three polypeptides, two of which are related. These proteins are made only after viral DNA synthesis begins in infected cells. At this time, two new RNAs appear in the cytoplasm of the infected cell (Weinberg et al., 1972); the translation of these mRNAs *in vitro* has identified them as the mRNAs for the capsid proteins. These mRNAs are related to each other much as the two mRNAs of togavirus-infected cells are related: the larger mRNA includes the sequences of the smaller mRNA (Kamen and Shure, 1976). The larger mRNA, which sediments at 19S, codes mainly for VP2 while the smaller 16S mRNA codes for the major capsid constituent VP1 (Prives et al., 1975). How VP3 arises is unclear. The two RNAs map together to form the *late region* of the viral genome (Figure 14-6). VP1 is encoded by the region of the genome in which the *ts*B and *ts*C mutations occur, showing that these mutations must affect the function of that protein. The *ts*D mutations, which interfere with the functioning of viral DNA after infection, are all localized within the region that encodes VP3. The lesion caused by *ts*D mutations is not known, but possibly VP3 is bound to the DNA (Griffith et al., 1975).

TRANSCRIPTION AND PROCESSING OF RNA

Three mRNAs are responsible for the synthesis of the papovavirus-specific proteins: the early 19S mRNA, the late 19S mRNA, and the late 16S RNA. Which strand of the viral DNA encodes the mRNAs and what controls the switch from early to late synthesis?

Separation of the strands of SV40 DNA was made possible by the fortuitous circumstance that *E. coli* RNA polymerase copies one strand much more readily than the other (Westphal, 1970). Hybridization of the early and late RNAs to the individual strands of DNA has shown that the two classes of mRNA arise from the opposite strands and consequently must be copied in different directions (Sambrook et al., 1973; see Figure 14-6). The strand of DNA encoding the early mRNA is called the E-strand; the other is called the L-strand.

Although these observations suggest that the switch from early to late transcription involves a change of the DNA strand that is transcribed, there is evidence to suggest that both strands of DNA are copied all of the time but that some selection process—possibly involving processing of the initial RNA transcripts—is the central controlling event in what types of mRNAs reach the cytoplasm and are responsible for viral protein synthesis (Aloni, 1974). Because these questions are controversial and unsettled, it is not profitable to discuss at great length the evidence supporting various models (see Acheson, 1976). One fact, however, is clear: late after infection the whole L-strand of the DNA is transcribed and RNA molecules even longer than the size of a single DNA strand can be found in the cell's nucleus (Acheson et al., 1971). It is probable, but unproven, that the late mRNAs derive by cleavage from a very long L-strand transcript. The cleavage from 19S to 16S appears to occur in the cytoplasm (Aloni et al., 1975).

A final issue is whether the viral DNA is integrated into host cell DNA at the time it is transcribed and whether free viral DNA molecules have promoters for transcription or if transcription might start from cell promoters. Little evidence exists about early RNA synthesis, but evidence suggests that late RNA is transcribed from free viral DNA molecules and consequently from a viral promoter (Mousset and Gariglio, 1976).

DNA SYNTHESIS

Viral DNA synthesis and virion maturation both take place in the nucleus of the cell. The study of viral DNA synthesis was greatly facilitated when Hirt (1967) discovered that exposure of cells to sodium dodecyl sulfate and high salt leads to precipitation of chromosomal DNA but solubilization of low molecular weight DNA such as viral DNA.

The topology of viral DNA synthesis has been studied by two techniques. One technique is a variant of the procedure of Dintzis (1961) in which DNA is briefly labeled and the distribution of label in *finished* molecules is examined using restriction enzymes to divide the DNA into different portions (Danna and Nathans, 1972). The second method is electron microscopy of replicative intermediates that can be purified because they have a characteristic sedimentation rate and buoyant density. Cutting the replicative intermediate molecules once with a restriction enzyme identifies a defined point on the molecule and allows the visualization of the topology of the DNA synthesis relative to that point (Fareed et al., 1972).

Both techniques have shown that there is a single origin for viral DNA synthesis located at approximately 0.67 on the physical map. The techniques also have shown that DNA synthesis is bidirectional from this point so that a spreading wave of DNA synthesis passes away from the origin in two directions until a point of termination is reached about 180° away from the point of initiation (Figure 14-8). Using deletions it is possible to show that the point of termination is not specific but is simply the point at which the two growing forks meet. The growing DNA molecules separate from each other while they still have gaps in their growing strand; the gaps are later filled and the molecules are converted to closed circular duplexes (Chen et al., 1976).

The replication of a closed circular duplex DNA involves unwinding of

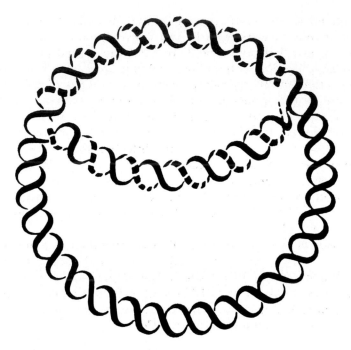

FIGURE 14-8. Diagrammatic representation of replicating SV40 DNA. The salient features of the molecule are: (1) both parental DNA strands (solid lines) are covalently closed, and (2) the two newly synthesized DNA strands (broken lines) are not covalently linked to the parental DNA nor are they linked together. Reprinted from Sebring et al. (1971) with permission.

the two DNA strands from around one another. To accomplish this unwinding, at least one nick must be put into one strand of the DNA, providing a swivel point. Replicative intermediate DNA molecules might be expected to have a nick and not to be supercoiled, yet electron microscopy has shown that there is a supercoiled arm in replicating DNA molecules (Figure 14-9). The necessary nick, therefore, must be a transient one such that at most times replicating molecules are not nicked (Sebring et al., 1971). Cells contain an enzyme that can carry out the nicking and closing cycle and which is likely to be responsible for the release of the strain imposed by uncoiling parental DNA (Champoux and Dulbecco, 1972). This enzyme is in high concentration and has been purified; with no apparent energy requirement it will release the supercoils of a viral DNA molecule (Keller and Wendel, 1974).

Because DNA synthesis is bidirectional, at least one of the strands growing in either direction must be made with a polarity opposite to the polarity of movement of the replicating fork. Okazaki et al. (1968) first suggested that in such a situation there may be small fragments of DNA made and then linked together. Such Okazaki pieces have been found for papovavirus DNAs and, surprisingly, appear to come from both strands of the DNA at either growing fork (Fareed et al., 1973). At all four points of DNA synthesis, discontinuous production of the growing strand must occur. Each of the small fragments of DNA is linked to a 10-nucleotide RNA molecule that is presum-

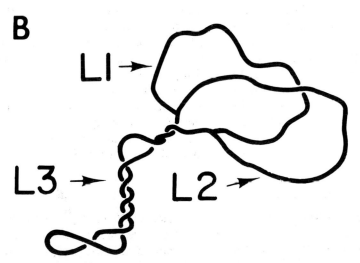

FIGURE 14-9. *SV40 DNA molecule in the process of replication. In the electron micrograph (A), the superhelical arm and the open, replicated loops are evident. In the interpretive drawing (B), the replicated loops are denoted L1 and L2, the superhelical arm is denoted L3. From measurements of a large number of such molecules, L1 = L2 and L3 + L1 = length of an SV40 molecule. Reprinted from Sebring et al. (1971) with permission.*

ably the primer for initiating synthesis of Okazaki pieces (Piaget et al., 1974). It is not known what molecular process initiates the overall replication of a given DNA molecule, the process that requires gene A function.

Adenoviruses

Adenoviruses have six to eight times more DNA than papovaviruses but aside from the increased number of proteins involved, the pattern of their multiplication cycles is similar. The two viruses display early/late control and transcribe their mRNAs from both strands of DNA. Adenovirus DNA is linear, however, and therefore must replicate differently from papovavirus DNA. Adenovirus DNA also shows a high enough rate of genetic recombination to allow formal genetic analysis but papovavirus DNA recombines too infrequently for such studies (Ishikawa and Di Mayorca, 1971).

THE VARIETY OF ADENOVIRUSES

A large number of different adenoviruses have been isolated from a wide range of species; 31 different serotypes of human adenoviruses have already been characterized. The basic molecular biology of these viruses appears to be similar enough not to require distinguishing among them.

The major disease caused by adenoviruses is an acute respiratory tract infection but when certain serotypes of human adenoviruses are injected into hamsters they cause tumors. Almost all of the adenovirus strains will also transform rat fibroblasts in culture but none are associated with human malignancy. Adenoviruses, therefore, are interesting for their characteristic molecular biology, for their ability to cause self-limiting human diseases, as well as for their ability to cause cancer. Reviews of adenoviruses include Philipson and Lindberg (1974); Philipson et al. (1975) and Wold et al. (1976).

The adenoviruses have a very elegant physical structure (Figures 3-2 and 3-29). The 14 or more proteins involved in the manufacture of the virus particle include individual species that make up the hexons, pentons, and fibers on the surface of a virus. The location of certain individual proteins in the virus particle is shown schematically in Figure 14-10.

DNA STRUCTURE

Adenovirus DNA is a linear duplex of 20 to 25×10^6 daltons with ends remarkably similar to those of the defective parvoviruses (adenovirus-associated viruses). If the duplex DNA is melted and the single strands of DNA are allowed to reanneal, circles of DNA are formed, indicating that there are inverted repeat structures at the two ends of the viral DNA (Garon et al., 1972; Wolfson and Dressler, 1972). Inverted repetitions 100 to 140 nucleotides long have been isolated (Roberts et al., 1974). If the double-stranded DNA is digested from its 3' ends by exonuclease III, the 5' terminal single-stranded regions will form hairpins, indicating that an inverted repeat also exists within the first 180 nucleotides from either end (Padmanabhan et al., 1976).

Although pure adenovirus DNA is a linear structure, if adenovirions are

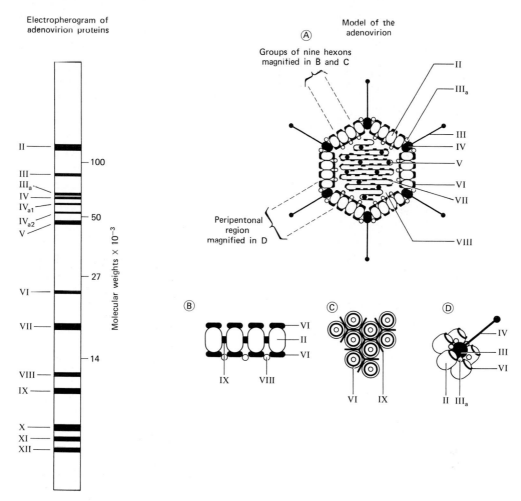

FIGURE 14-10. The adenovirion proteins. On the left is an idealized electrophoretic separation of the virion proteins (reprinted from Anderson et al., 1973 by permission). On the right is a hypothetical model of the location of the virion proteins in the particle. The molar ratio between the internal basic proteins VII and V is 5, which is maintained in the schematic representation. Protein VII may cover about 50% of all phosphates of the DNA. (A) A schematic view of a vertical section of the virion. The different polypeptides are indicated by their Roman numerals as indicated to the left. Magnified views of groups of nine hexons and the peripentonal region are given in B, C, and D. (B) A vertical section through a group of nine hexons showing the tentative location of proteins VI, VIII, and IX. (C) A horizontal view from the outside of the group of nine hexons showing the tentative location of proteins VI and IX. (D) A magnification of the peripentonal region showing the proteins II, III, IIIa, IV, and VI. Reprinted from Everitt et al. (1973) with permission.

Adenovirus DNA map

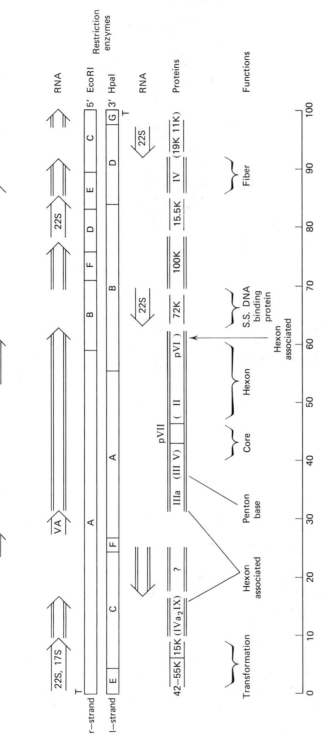

FIGURE 14-11. Adenovirus DNA maps. The DNA is shown in a conventional orientation with the strand that is copied in a rightward direction shown above the other. Restriction enzyme digest patterns are shown for EcoR₁ and Hpa I. The positions encoding the early and late mRNAs and the VA RNAs as well as their direction of synthesis and the strand encoding them, are all indicated. Below, the sizes of proteins made from each region—as determined by translating mRNAs selected on specific DNA fragments—are shown with the virion polypeptides numbered with Roman numerals. The functions of the proteins, where known, are indicated below the proteins. It can be seen that the structural proteins are all made late while a protein involved with DNA replication (the 72K protein) and the proteins that are made in all transformed cells (the 42–55K and 15K proteins) are made early. Adapted from the cover of the Cold Spring Harbor Laboratory Abstracts, SV40 and Adenovirus Meeting, 1976.

gently disrupted and not treated with a protease, some circular DNA can be isolated. The circles are held together by interaction between proteins covalently attached to both ends of the viral DNA (Robinson and Bellett, 1974). These proteins make the DNA much more infectious than naked DNA.

THE GROWTH CYCLE

For the most rapidly growing adenovirus serotypes a full yield of virus requires 24 hours from the time of infection. An early set of mRNAs is made during the first 8 hours of infection and the early proteins include a T-antigen found in both the nucleus and cytoplasm of infected cells (Pope and Rowe, 1964). The late phase of the cycle starts after about 8 hours when viral DNA synthesis begins. The late class of mRNAs is not made if viral DNA synthesis is blocked either by drugs or by appropriate temperature-sensitive mutants (Berget et al., 1976). Both classes of mRNA have poly(A) at their 3' end and a cap structure at their 5' end consisting mostly of $m^7G^{5'}ppp^{5'}m^6AmpCp$ . . . (Wei et al., 1975; Sommer et al., 1976).

Adenovirus encodes a group of RNAs of unknown function, known as "VA" RNAs (virus-associated RNAs). They are molecules about 150 nucleotides long that are synthesized by a different polymerase than the one that synthesizes the bulk of adenovirus RNA (Price and Penman, 1972). Most adenovirus RNA synthesis is sensitive to α-amanatin, indicative of synthesis by polymerase II, but VA RNA synthesis resists the inhibitor and therefore is thought to be synthesized by RNA polymerase III (Weinman et al., 1976). One of the VA RNAs appears in much higher concentration than the others and it has been sequenced (Ohe and Weissman, 1970).

THE DNA MAP

Six separate methodologies have been used to map the organization of adenovirus DNA (Figure 14-11). Because of the amount of detailed information produced in these different mapping procedures, only a brief description of each will be given here.

1. *DNA fragment map.* There are now some 40 restriction enzymes available with a variety of cleavage sites allowing the generation of an extremely fine map of the viral DNA. Examples of the maps of adenovirus DNA generated with various nucleases are shown in Figure 14-11 (Mulder et al., 1974).

Not only can the double-stranded DNA fragments produced by the nucleases be isolated, but the individual fragments can then be denatured and the two strands separately isolated by electrophoresis (Sharp et al., 1974). By hybridizing these DNAs back to the strands of intact viral DNA, it is possible to tell from which strand each of the pieces comes. The orientation of the two intact strands relative to the endonuclease cleavage map was determined by labeling them at their 3' ends and determining which pieces of the DNA contained the label.

2. *Early/late mRNA map.* The early and late mRNAs have been mapped onto the genome by hybridizing the total RNA of cells infected for various times to the labeled, separated strands of the individual DNA fragments. The map shows four blocks of early genes and six blocks of late genes interspersed on the two strands of the DNA, which are denoted *r* and *l* to signify that one is transcribed to the right and the other to the left (Figure 14-11). The VA RNAs have also been mapped in this fashion and found at approximately 0.3 of the genome from its left-hand end (Soderlund et al., 1976).

3. *Separation of mRNA by size.* The individual size classes of mRNA can be mapped onto the DNA by hybridizing labeled RNA molecules to unlabeled DNA fragments (Craig et al., 1974; Flint et al., 1976). The minimum number of mRNAs encoded by the various early and late regions of the DNA can thus be determined. Most of the early regions encode single 22S mRNAs but the early region at the left-hand encodes two RNAs of 22S and 17S.

4. *Translation map.* By hybridizing early and late virus-specific RNA to separated DNA fragments, recovering the RNA, and translating it in cell-free systems, the mRNAs encoding six early proteins and nine late proteins have been mapped (Lewis et al., 1975, 1976). Where a correlation is possible, the map determined by translation gives the same positions for early and late proteins as the map determined by hybridization (Figure 14-11).

5. *Standard genetic map.* A large number of temperature-sensitive mutants have been generated from adenovirus type 5 and these have been mapped relative to one another using standard techniques of genetic recombination (Williams et al., 1974).

6. *Physical map of mutations.* Because adenoviruses types 2 and 5 give rise to different patterns of restriction enzyme fragments—but the viruses are homologous enough to allow recombination between them—by studying the fragment pattern it is possible to determine the site at which recombination has occurred in an interspecific genetic cross. Using adenovirus type 5 temperature-sensitive mutants and temperature-sensitive mutants of a strain of adenovirus type 2 containing a segment of SV40 DNA, certain mutations have been localized within limits on the adenovirus map (Williams et al., 1975).

MECHANISM OF RNA SYNTHESIS

Although the mapping data shows where the different stable RNA species on the adenovirus DNA come from, the mRNA could be made by individual initiations or by processing of longer transcripts.

By labeling RNA for a very short time and hybridizing the labeled RNA to different restriction enzyme fragments, the sequences from the right two-thirds of the adenovirus *r* strand can be identified in long RNA from which later the individual late mRNA appears to be excised (Bachenheimer and Darnell, 1975). The target sizes for ultraviolet light-inactivation of regions of the *r* strand also correlate with their map position showing that two-thirds of the *r* strand is copied as a long RNA molecule (Goldberg et al, 1977). The VA RNAs are initiated *de novo* at 0.3 but the initiation site of the large *r* strand transcript, however, indicates that it is to the left of the VA RNAs, between 0.12 and 0.18, and therefore initiation of VA RNA synthesis and of mRNA precursor synthesis are different events (Weber et al, 1977).

A recent finding with adenovirus mRNAs promises to be of revolutionary importance. Electron microscopy of AD-2 mRNA hybrids with specified segments of AD-2 DNA indicates that individual mRNA molecules are made of sequences complementary to different *non-contiguous* sites on the AD-2 genome. Since the large precursor RNA appears to be the only nuclear product from the right hand 85% of the genome, RNA:RNA ligation or "splicing" may occur during mRNA formation.*

Synthesis of adenovirus mRNA involves a number of different reactions: polyadenylation, capping, processing of long transcripts, selection of certain sequences to be transported from the nucleus to the cytoplasm, and both symmetric and asymmetric transcription. Also, sequences destined for transport change with time and some RNA may even be held after its synthesis for later transport. Thus, many, if not most, of the events thought to be involved in mRNA formation in uninfected cells may have their counterpart in the synthesis of adenovirus mRNA.

VIRAL DNA SYNTHESIS

The mechanism of viral DNA synthesis, which takes place in the cell nucleus, is not known in detail but it is likely to be dictated by the nature of the ends of the viral DNA. The two critical facts about the ends are that they are identical and contain an inverted repeat sequence. The identity of the ends suggests that initiation of DNA synthesis could occur with equal rates at either end of the viral DNA molecule and the existence of a local inverted repeat suggests that a model for initiation like that described for parvoviruses might be applicable.

Using variants of the procedure Danna and Nathans (1970) devised for studying SV40 DNA replication, evidence has been obtained that DNA synthesis does start at either end and proceed toward the middle of the molecule (Horwitz, 1976). Such a process, as seen in Figure 14-12, would generate transient single-stranded regions at either end of the molecule, and such single-stranded regions have been identified (Lavelle et al., 1975).

The cellular enzymes involved in adenovirus DNA synthesis are not known but one adenovirus-specific protein—a 72,000 dalton protein that binds to single-stranded DNA—is thought to be involved (Levine et al., 1974). A temperature-sensitive mutant that apparently affects the structure of this protein makes no adenovirus DNA at the restrictive temperature. In adenovirus type 12, at least three complementation groups have been identified that are involved in the initiation of DNA replication (Shimojo et al., 1974). For adenovirus type 5, two such complementation groups have been identified. Thus, there are probably more proteins to be found that are adenovirus-specific and involved in the replication of viral DNA.

---

* See Berget, S. M., Moore, C., and Sharp, P. A. (1977). PNAS, *74*, 3171-3175; Chow, L. T., Gelinas, R. E., Broker, T. R., and Roberts, R. J. (1977). Cell, *12*, 1–8; Evans, R., Fraser, N. W., Ziff, E., Weber, J., Wilson, M., and Darnell, J. E. (1977). Cell, *12*, 733–739; Klessig, D. F. (1977). Cell, *12*, 9–22.

DNA

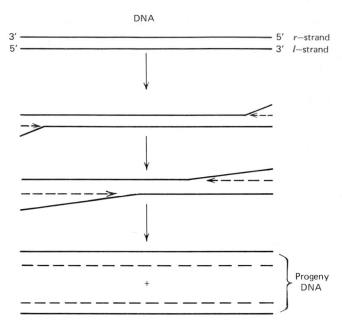

FIGURE 14-12. *Model of replication of adenovirus DNA. The two strands are thought to initiate replication independently using identical sequences at the two ends. The initiation mechanism is not known but probably involves the type of hairpin initiation described for parvoviruses in Figure 14-1.*

INDUCTION OF HOST CELL DNA SYNTHESIS

In growing cells adenovirus does not stimulate host cell DNA synthesis but rather inhibits it (Hodge and Scharff, 1969). When nongrowing cells are infected a different picture emerges. In hamster cells, where adenovirus is not able to induce its own DNA synthesis, infection of nongrowing cells produces a stimulation of host cell DNA synthesis (Shimojo and Yamashita, 1968). If nongrowing permissive human cells are infected with adenovirus there is a transient stimulation of host cell DNA synthesis followed by inhibition (Yamashita and Shimojo, 1969). In both permissive and nonpermissive nongrowing cells an induction of cellular enzymes involved in DNA synthesis occurs after infection with adenovirus (Ledinko, 1968; Suzuki and Shimojo, 1971). It therefore appears that adenoviruses, like papovaviruses, produce a protein that stimulates host cell DNA synthesis in resting cells. Adenoviruses, however, go on to inhibit host cell DNA synthesis before much cellular DNA has been made. This inhibition may be a secondary consequence of the fact that adenoviruses turn off host cell protein synthesis—which is required for host cell DNA synthesis—while papovaviruses do not.

ADENOVIRUS-SV40 HYBRIDS

A series of hybrid viruses between adenovirus and SV40 have been isolated and have been very informative. The formation of these covalent complexes has been observed as a consequence of the ability of SV40 to help adenovirus grow in monkey cells where ordinarily the human adenoviruses will not complete their life cycle.

If monkey cells are infected with both adenovirus and SV40, covalently linked complexes of adenovirus plus SV40 DNA, packaged in an adenovirus coat, emerge along with the parental viruses. Two types of hybrids have been isolated; those *defective* in their adenovirus component (Baum et al., 1966) and those with no loss of essential adenovirus genes (*nondefective* hybrids; Lewis et al., 1973).

To grow the defective adeno-SV40 hybrids, cells must be coinfected with standard adenovirus. Therefore, such virus mixtures yield a two-hit titration curve on monkey cells although the adenovirus component is completely competent for growth on human cells. In the hybrid DNA, the SV40 component may replace as much as 40% of the adenovirus DNA, and the deletion in adenovirus can occur in a number of different places. The extent of SV40 insertion can be greater than two complete SV40 genomes but in some hybrids only a fraction of an SV40 genome is found (Kelly et al., 1974).

The nondefective adeno-SV40 hybrids have a nonessential region of the adenovirus genome replaced by SV40 information. All known nondefective hybrids come from a single original stock and therefore may be closely related (Lewis et al., 1973). In these viruses, about 5% of the adenovirus genome is missing. The SV40 genes in the hybrid include a small amount of the late region of the SV40 genome as well as variable sections of the early region (Lebowitz et al., 1974). Probably as a consequence of the position in the adenovirus genome where the SV40 sequences are integrated, only the early sequences are transcribed into mRNA. Those hybrids that make SV40-specific proteins provide the helper function for adenovirus to grow in monkey cells, but two of the viruses make no SV40 mRNA and will not provide the helper function. The smallest hybrid that can grow in monkey cells makes no TSTA or T-antigen but does make U-antigen; the U-antigen-bearing polypeptide can therefore provide the as yet mysterious function that allows human adenovirus growth in monkey cells.

Herpesviruses

The herpesviruses have diverse growth characteristics but share a common morphology, a common size of DNA, and appear to be part of a homogeneous biochemical grouping. Those that have been studied in greatest detail are the lytic ones, including human herpes simplex virus type 1—the virus responsible for cold sores—herpes simplex virus type 2, and a number of rapidly growing animal herpesviruses (Roizman and Furlong, 1974).

The most widely studied, nonlytic herpesvirus is the Epstein-Barr (EB) virus, which is responsible for infectious mononucleosis and is also recovered consistently from the cells of patients with two forms of cancer, Burkitt's lymphoma and nasopharyngeal carcinoma (Biggs et al., 1972). While herpesvirus types 1 and 2 grow lytically in almost all cell cultures, EB virus infects only primate B-lymphocytes and in these cells the virus is only produced by an occasional infected cell (Roizman and Kieff, 1975).

Herpesvirus encodes at least 49 proteins and their synthesis would utilize most of the coding capacity of the viral genome (Honess and Roizman, 1973). This complexity makes study of herpesvirus physiology difficult.

DNA STRUCTURE

Herpesvirus DNA is a linear duplex of approximately 100×10^6 daltons consisting mainly of unique sequences (Frankel and Roizman, 1971). When the DNA is alkali-denatured only a small fraction is recovered as full-length single-stranded molecules because the majority of the native DNA contains either single-strand nicks or ribonucleotides (Hirsch and Vonka, 1974). The position of the alkali-labile bonds appears to be random, although at one time it had been thought that they occurred at specific places (Wilkie et al., 1974).

Three different approaches have shown that the herpesvirus DNA molecule has a number of repetitive segments. The first indication came from electron microscopic investigation of the structure of those full-length single-stranded DNA molecules released by denaturation of virion DNA. In a fraction of such molecules, both ends hybridize to a single internal position on the molecule, producing a structure with two loops connected by a double-stranded stem (Sheldrick and Berthelot, 1974; Figure 14-13). The stem takes up about 10% of the length of the single-stranded molecule and consists of 6% derived from the side of the larger loop and 4% derived from the side with the smaller loop. Partial denaturation mapping has shown that the two areas of closure differ in base sequence as well as length (Wadsworth et al., 1975).

Another aspect of the structure of viral DNA has been identified by electron microscopy of double-stranded molecules which have been digested slightly from either their 5' ends or their 3' ends. In both cases, circular DNA molecules are produced, indicating the presence of a terminal repetition (Grafstrom et al., 1975; Wadsworth et al., 1976). If digestion by exonuclease is carried further than a few percent from the end, the ability to circularize disappears, suggesting that the ends can snap back on themselves to produce double-stranded hairpin regions (Figure 14-14). This structure is similar to that thought to exist at the ends of both parvovirus and adenovirus DNA and suggests that all of the linear DNA viruses may use similar mechanisms of DNA replication.

In addition to this complexity a unique aspect of DNA structure is revealed by fragmentation of herpesvirus DNA with restriction endonucleases (Hayward et al., 1975). Three types of fragments are produced which occur with molarities of 1, 0.5, and 0.25 per DNA molecule. The sum of the molecular weights of all of these fragments is larger than the total genome. This odd behavior can be rationalized if herpesvirus DNA consists of one long and one short DNA molecule covalently joined in random orientation to each other at the junction of their identical regions (Figure 14-14). The random association of the long and short arms will produce four different types of DNA molecules depending on the orientation of the two arms.

This unique DNA structure has numerous implications for the biology of herpesviruses. For instance, the two arms of the DNA molecule may replicate in independent pools or, alternatively, there may be extensive recombination at the joining sequences generating the assortment of molecules that are

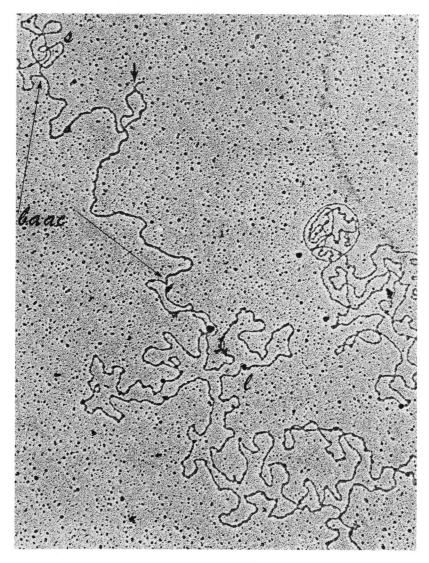

FIGURE 14-13. *Electron micrograph of a self-annealed intact single strand of herpesvirus, type 1, DNA. The duplex region extends between the two long arrows and is marked "baac." The large (l) and small (s) loops are indicated. The heavy arrow marks a spur in the duplex region, presumably indicating one end of the DNA. An interpretive drawing of the molecule is presented in Figure 14-14. Micrograph from Wadsworth et al. (1975), by permission.*

found. Whatever the nature of the replicating molecules, genetic mapping is likely to show two separate linkage groups.

GROWTH CYCLE: DNA SYNTHESIS

The herpesvirus growth cycle is faster than that of any of the small DNA viruses. New progeny virions appear in the nucleus as soon as 6 hours after

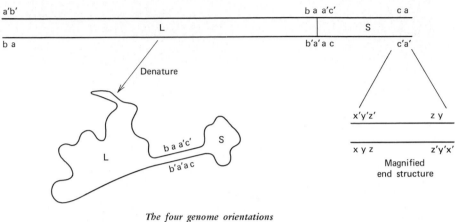

Herpesvirus DNA

The four genome orientations

FIGURE 14-14. *The structures of herpesvirus DNA. One orientation of the double-stranded DNA of the virion is shown at the top. The consequences of melting the double strand and allowing a single strand to self-anneal are shown below: a molecule with a large loop (L), a small loop (S), and duplex region (baa'c'/b'a'ac) is formed. A magnified view of an end is also shown to indicate that within a short distance from an end there is a region of inverted complementary structure. The four different orientations of the genome are shown in the lower half of the diagram. The two independently oriented segments are called "La" and "Sh". Each has two possible orientations. Because all four ends of the two segments are identical (a/a') the joint between the segments is the same in all orientations (aa'/a'a).*

infection and new viral DNA is made there by 3 hours after infection. The ultimate yield of virions can be as high as 10^5 particles per infected cell, although only a small fraction of these will score as plaque-forming units.

The DNA from the herpesvirion is infectious, indicating that the virion contains no required enzymes or other proteins (Sheldrick et al., 1973). The virion, however, is extremely complex; it shows a number of different layers in the electron microscope and is separable into as many as 24 polypeptide chains.

The mechanism by which viral DNA is formed is not known in any detail except that pulse-labeled DNA appears entirely in fragments that are only slowly joined (Frankel and Roizman, 1972). The ribonucleotides and/or nicks in the mature viral DNA may be a result of the incomplete fusion of the fragments that make up the replicating pool.

The enzymes responsible for viral DNA synthesis are at least partly viral in origin (see Roizman and Furlong, 1974). For instance, the virus appears to encode a thymidine kinase; this enzyme is induced by the virus even in cells in which the host thymidine kinase has been mutationally inactivated (Dubbs and Kit, 1964). The viral thymidine kinase has a number of properties that differentiate it from cellular enzymes, including a deoxycytidine kinase activity, and mutants of the viral enzyme have been produced. Another enzyme that may be produced under the direction of the herpesvirus genome is a ribonucleotide reductase. The reductase found in infected cells is much less sensitive to feedback inhibition by dTTP than is the enzyme found in uninfected cells (Cohen, 1972).

A number of different deoxyribonuclease activities have also been found to be increased in infected cells. Furthermore, infected cells contain a new DNA polymerase activity presumably synthesized under the direction of the viral genome (Hay et al., 1976). The full range of enzymes that might be encoded by the viral genome is probably not yet known, and it is also unclear if the virus is able to induce the increased synthesis of any host cell enzymes.

If DNA synthesis is already occurring when cells are infected, virus infection slowly shuts off DNA synthesis. In quiescent cells, the virus does not induce the synthesis of host cell DNA. It would be interesting to know if herpesvirus induced host cell DNA synthesis in a nonpermissive system, but the virus has such a ubiquitous host range that no such nonpermissive system has yet been found.

VIRAL RNA AND PROTEIN SYNTHESIS

The central fact differentiating the RNA control system of herpesvirus from that of other DNA viruses is that all viral proteins apparently can be synthesized even if viral DNA synthesis is completely blocked by inhibitors (O'Callaghan et al., 1968). The early/late control characteristic of almost all other DNA viruses therefore does not seem to exist in herpesvirus. Sequential control of mRNA formation, however, is suggested by experiments in which infected cells are treated with an inhibitor of protein synthesis for various lengths of time after infection. When the inhibitor is removed, different classes of proteins are made depending on the time at which the inhibitor had been added. Such data suggests that there are three classes of herpesvirus proteins—called alpha, beta, and gamma—and each sequentially induces the synthesis of the next and is then inhibited in its synthesis by the next, all independently of viral DNA synthesis (Honess and Roizman, 1974).

Whatever control exists on protein synthesis does not seem to be a reflection of the control of RNA synthesis per se. All of the sequences of DNA that are represented in RNA late after infection appear to be also represented early after infection. Furthermore, if cells blocked in protein synthesis are infected, more than 40% of the viral DNA is expressed in nuclear RNA, while at most 10% can be found in the cytoplasm (Frankel et al., 1973). It has therefore been proposed that most control of herpesvirus protein formation occurs as a consequence of specific processing of RNA sequences that are formed in the nucleus rather than by control of the synthesis of those RNA species.

Synthesis of the core structure of herpesvirions occurs in the nucleus. As the multilayered core exits from the nucleus, it is often enveloped by the nuclear membrane. Cores may also migrate to cytoplasmic and plasma membranes and acquire envelopes by budding through these membranes. Viral proteins are inserted in both the internal and external membranes of the cell so that all membranes provide sites that could yield virus-specific envelopes (Roizman and Spear, 1971). Whether particles that gain envelopes from different cellular membranes are equally infectious is not known.

EPSTEIN-BARR VIRUS

The isolates of EB virus come either from cultured cells of patients with Burkitt's lymphoma or from throat washings of patients with infectious mononucleosis. Most work on the virus has utilized antisera derived from people previously infected by the virus. Such antisera allow three types of viral antigens to be distinguished: antigens that appear early in the growth cycle of the virus (EA); antigens that are part of the viral capsid (VCA); and an EB-specific nuclear antigen (EBNA).

The cells that produce EB virus and the cells that are infectable by EB virus are all peripheral B-lymphocytes. Infection by the virus can be carried out, for instance, by exposing lymphocytes to cell-free extracts from virus-producing cultures (Gerber et al., 1969). Uninfected lymphocytes will not proliferate whereas if a successful infection occurs, a continuously growing culture of lymphocytes is produced. This measure of infection is called a "transformation assay," although it might better be called an "immortalization assay" (Miller et al., 1974).

Only a very small fraction of cells from tumor cell lines or from transformed lymphocytes synthesize EA and an even smaller fraction of cells synthesize both EA and VCA. Many cultures are totally negative for both of these antigen classes. Only cells that produce VCA produce any infectious virus, hence the yields of virus are very low. One line of cells derived from a Burkitt's lymphoma produces an unusually large amount of virus but that virus is unable to transform lymphocytes and can only induce antigen-positive cells in recipient cultures (Miller et al., 1974).

Despite the fact that only rare cells in infected cultures form either EA or VCA, all cells in such cultures are EBNA-positive (Reedman and Klein, 1973). In many cases, treatment of cultures with iododeoxyuridine can greatly increase the fraction of virus-yielding cells (Gerber, 1972) and in cultures with only a few antigen-positive cells, each individual clone of cells has a fraction that are antigen-positive. Therefore, the heterogeneity of behavior of cells in

EB-infected cultures does not imply that some cells have different potentialities than others, it rather implies the occurrence of an induction event that may be either spontaneous or produced by chemical treatment of the cells.

It appears that all cells becoming EA-positive or VCA-positive go on to die. The relation of EB virus to infected cells is therefore not unlike the process of lysogeny carried out by bacteriophages. EBNA, on such a model, might resemble the repressor protein and EA and VCA would be proteins whose synthesis is repressed in most cells but can be derepressed in occasional cells. Such a derepression would then commit the cell to a lytic cycle and to the death of the cell. For EB virus it must be postulated that for some cells the lytic cycle is blocked after EA synthesis because in many cultures no VCA is ever produced. Furthermore, the analogy with phage lysogeny is limited by the fact that EBNA-positive cells can be productively superinfected.

The analogy with lysogeny suggests that the EB virus might be found integrated into host cell DNA. EB-infected cells contain at least four genomes of viral DNA per cell and many cultures show an average of over 100 viral genomes per cell. Most of these genomes are found as covalently closed circular duplex molecules, unlinked to host cell DNA, and it is still a matter of controversy whether there is any integrated EB virus genome in infected cultures or whether all EB viral DNA multiplies as a plasmid (Lindahl et al., 1976).

In infected cultures, EB viral DNA appears to replicate at a specific point early in the S-phase of the cell cycle (Hampar et al., 1974)—at that time the number of EB viral genomes doubles. It is exactly at that time that the cells are sensitive to induction by iododeoxyuridine (Hampar et al., 1973) suggesting that the analog acts by being incorporated into viral DNA.

LATENCY AND HERPESVIRUS

One of the reasons for focusing attention on EB virus is its ability to establish a repressed, persistent state in cell cultures. Although this appears very different from the behavior of herpes simplex viruses in cell cultures, the behavior of EB virus may be an important clue to understanding the *in vivo* behavior of all herpesviruses.

It is a common observation of afflicted people and one of the oldest puzzles of virology that herpesviruses can maintain a latent infection for many years. Herpes simplex type 1 can be activated from its latency by a wide range of stimuli producing "cold sores" when it appears. From infected animals or human cadavers herpesviruses can be recovered from nerve ganglia, suggesting that somehow the virus can maintain an association with ganglion cells (Stevens, 1975). There is as yet no idea how this relationship is maintained because once ganglion cells are placed in culture they rapidly start producing virus. The ability of EB virus to remain latent in B-lymphocytes may be a model that makes evident a generalized ability of herpesviruses which has so far eluded workers who deal with lytic herpesvirus in cell cultures. It should also be noted that another herpesvirus, called herpes zoster, is the cause of childhood chicken pox but can remain for many years in a latent state and then produce in adults the disease called shingles (see Chapter 15).

Poxviruses

All of the DNA viruses discussed thus far have their DNA made and mature their virions in the nucleus of the infected cell. Poxviruses multiply exclusively in the cytoplasm of the cell and therefore face very different conditions than those faced by nuclear DNA viruses.

There are many different types of poxviruses, the most important human one being smallpox virus. Those that have received the most detailed analysis are vaccinia virus and its animal relatives, rabbitpox virus and cowpox virus (Joklik, 1968; Moss, 1974). All poxviruses have a common antigen (Woodroofe and Fenner, 1962). Poxviruses are not the only cytoplasmic DNA viruses but other such viruses are less well-studied and will not be discussed here (see McAuslan and Armentrout, 1974).

AUTONOMY OF POXVIRUS GROWTH

The cytoplasmic localization of poxvirus-specific processes has been indicated by electron microscopy of infected cells. The best evidence, however, comes from the ability of poxviruses to carry out most of their life cycle in enucleated cells such as those produced by treating cells with cytochalasin B. Infection of such fragments leads to virus-specific DNA synthesis and the synthesis of many viral proteins although no progeny virions are made (Prescott et al., 1971; Pennington and Follette, 1974).

Poxviruses therefore relocate the functional center of the cell from the nucleus to the cytoplasm. One might expect this feat to require a large amount of virus-specific information and the synthesis of a large number of virus-specific proteins. Consistent with such an expectation, poxviruses have the largest DNA of any animal virus and during the course of their growth cycle there is induction of multiple enzymatic activities. To grow in the cytoplasm, poxviruses must solve problems more like those faced by RNA viruses than those faced by nuclear DNA viruses. In fact, like some RNA viruses, poxviruses begin transcription of their DNA using a virion enzyme and the virion also contains all of the enzymes necessary to process that RNA into a functional mRNA.

VIRAL DNA

Poxviruses have a double-stranded DNA of approximately 150×10^6 daltons. Such a DNA contains sufficient information to encode 75 or more proteins of which only a fraction has been functionally identified.

When viral DNA is melted much of it renatures instantaneously, suggesting that it may be cross-linked. An explanation of this phenomenon came from electron microscopy of totally denatured vaccinia virus DNA molecule which was shown to be circular (Geselin and Berns, 1974). Apparently at either end of the double-stranded viral DNA there are covalent links so that when the molecule is denatured both strands stay together in a huge circle. Aside from this unique feature of vaccinia virus DNA, little is known about the structure of poxvirus DNAs.

THE VIRION

The virion consists of a complicated series of layers and substructures (see Figure 3-23). Thirty different virion proteins have been resolved by electrophoresis (Sarov and Joklik, 1972a); two of these are glycosylated but the only sugar in the carbohydrate is glucosamine (Moss et al., 1973). Detergent treatment of the virion releases a core structure that contains at least 17 of the polypeptides. During the infection process a somewhat more complicated core is released from the virion, containing the same 17 proteins as well as the glycoproteins. The glycoproteins are therefore not on the outside of the virion, as found with enveloped viruses, but are rather part of an internal membrane structure (Sarov and Joklik, 1972b). The outside of the virus particle contains five proteins that are accessible to iodination in the intact virion and that are released by detergent as well as by digestion with chymotrypsin.

The core released by detergent disruption of virions is able to synthesize RNA (Kates and McAuslan, 1967b; Munyon et al., 1967). The identification of this activity was the first demonstration that a virion can contain a nucleic acid polymerase. The RNA made by cores has 3'-poly(A) as well as a 5'-terminal cap structure (Kates, 1970; Wei and Moss, 1975). The ability of the core to synthesize completely formed mRNA molecules shows that it contains not only a polymerase, but a series of other enzymes and is, in fact, a transcription factory. Disruption of the cores with sodium deoxycholate and a reducing agent releases a number of soluble enzymes including those that modify RNA transcripts at their termini. Eight enzymes have been highly purified from the virions and a number of others have been demonstrated although not yet purified. A few hundred-fold purification is required to produce homogeneous enzymes from the virions and where the concentration of enzyme per virion is known, it is about 100 molecules per core. The known enzymes of the core are as follows:

1. *Poly(A) polymerase.* This enzyme is able to add poly(A) stretches to the 3' end of primer RNA molecules. It uses ATP as a substrate and requires no template for specifying the poly(A) structure. It can use poly(C) as a primer and therefore may not even need a primer oligo(A) on which to initiate poly(A) synthesis (Moss et al., 1975).

2. *Messenger RNA guanylyltransferase and messenger RNA (guanine-7-) methyltransferase.* It is possible to demonstrate these two enzyme activities by using as a substrate mRNA that was made by vaccinia cores in the absence of S-adenosylmethionine (Ensinger et al., 1975). The two activities purify together and the complex containing these activities has two polypeptides. However, the relationship of the polypeptides to the enzymatic activities is not known (Martin et al., 1975). These enzymes will modify any polynucleotide ending in a purine including poly(A) and poly(G). They carry out a coupled reaction of 5' capping using GTP followed by methylation of the linked GMP by S-adenosylmethionine (Martin and Moss, 1975). It may be that by coupling the two reactions, the reversible capping reaction is made irreversible by the addition of the 7-methyl group to the terminal GMP residue.

3. *Messenger RNA (ribose 2'-)methyltransferase.* A soluble enzyme able to methylate the purine residue proximal to the cap has been demonstrated but

not purified (Ensinger et al., 1975). This enzyme is not part of the complex of enzymes that adds and methylates the m^{7}G of the cap (Martin et al., 1975).

4. *Nucleic acid-dependent nucleoside triphosphate phosphohydrolases.* Two separable activities able to hydrolyze a nucleoside triphosphate have been purified from vaccinia virions (Paoletti et al., 1974; Paoletti and Moss, 1974). These enzymes carry out the hydrolysis of only the terminal phosphate from a nucleoside triphosphate. Presumably their activities are coupled to some process occurring in the virion but it is not known what that process is. One candidate is the extrusion of newly made mRNA from the core, a process known to require ATP (Kates and Beeson, 1970).

5. *Deoxyribonucleases.* There appear to be at least three deoxyribonucleases in vaccinia virions, but only one of these has been purified to homogeneity (Rosemond-Hornbeak et al., 1974).

6. *Protein kinase.* A protein kinase able to transfer phosphate residues from ATP to serine or threonine residues in protein has been purified from virions (Kleiman and Moss, 1975a,b). The enzyme is not dependent on a cyclic nucleotide for activation but does show a requirement for protamine as a cofactor. Protamine is not phosphorylated by the enzyme but rather there are two acceptor proteins that have been purified from the virions, one of 39,000 daltons and another of 12,000 daltons (Kleiman and Moss, 1975a).

7. *Nicking-closing enzyme.* An enzyme able to relax the supercoils in a closed circular duplex DNA molecule by making and closing nicks in the DNA has been purified from vaccinia virions (Bauer et al, 1977). Like other chromosomes, the vaccinia chromosome is probably folded into supercoiled regions inside the virion. The function of the relaxing enzyme may be to remove the strain in the molecule that could occur during RNA synthesis. Because extrusion of newly made mRNA appears to be coupled to its synthesis, it is difficult to imagine that the RNA could follow the polymerase around the rigid DNA molecule during synthesis. The constrained nature of the DNA also prevents it from rotating and it is the relaxing enzyme that presumably releases the continually produced strain during RNA synthesis.

INITIATION OF THE INFECTIOUS CYCLE

If cells are infected in the presence of an inhibitor of protein synthesis all of the phospholipid in the virion as well as one-half of the virion protein is released into a soluble form (Joklik, 1966). It appears, therefore, that constitutive enzymes of the cell carry out this early step of the uncoating process. The released core has its DNA in a deoxyribonuclease-insensitive form. If further protein synthesis is not allowed, the core remains inside the cell as a stable structure; if protein synthesis is allowed, the core is further uncoated to release the DNA into a deoxyribonuclease-sensitive form.

The virion transcription system is active in the core and synthesizes early mRNA (Munyon and Kit, 1966; Kates and McAuslan, 1967a; Woodson, 1967). The RNA formed by the cores represents 14% of the DNA sequences and the molecules are smaller on average than those made late in the infection (Oda and Joklik, 1967). It was the observation of synthesis of RNA by the coated genome that led to the discovery of the virion enzyme. This stage of vaccinia virus infection is reminiscent of reovirus infection where the core structure becomes a transcription factory in which the template is resistant to exogenous nucleases. One product of vaccinia core transcription must be the mRNA that encodes the protein responsible for degrading the core.

If uncoating is allowed to proceed to the stage where the core is disrupted, and the DNA becomes sensitive to deoxyribonuclease, then the rate of RNA synthesis rapidly declines. Whether uncoated genomes that have not yet replicated can be transcribed at all is uncertain. No class of "delayed early" mRNA synthesis has been identified but there is evidence from assay of virus-specific enzymes that such a class might exist (see below). Only two classes of virus-specific RNA have been characterized: early (made before DNA synthesis) and late (Oda and Joklik, 1967).

During the early stage of the infection a number of new enzymes appear in infected cells. All of these are thought to be encoded by the viral genome because their appearance requires *de novo* protein synthesis and because they are distinct from cellular enzymes. All of the known enzymes are involved in DNA synthesis and none appear to be packaged in virions. However, some virion structural proteins are made early (Holowczak and Joklik, 1967).

The best-studied of the early enzymes is thymidine kinase. This enzyme is certainly a product of the viral genome because viral mutants can be isolated that lack it (Dubbs and Kit, 1964). The other early enzymes include a DNA polymerase activity that is separable from the host DNA polymerase (Citarella et al., 1972); a number of deoxyribonucleases, one of which has been purified quite extensively (McAuslan and Kates, 1967) and two others of which can be differentiated by their pH optima (Jungwirth and Joklik, 1965); and polynucleotide ligase (Sambrook and Shatkin, 1969). There is also a requirement for continued protein synthesis in order for DNA synthesis to continue; this presumably involves yet a separate protein (Kates and McAuslan, 1967c).

Whether all of these proteins are formed from mRNAs made by cores is not certain. The possibility of "delayed early" enzymes must be considered. For thymidine kinase there appears to be no question; if cells are infected in the presence of cycloheximide, and actinomycin D is added at the time cycloheximide is washed out, the cells make the virus-specific thymidine kinase (Kates and McAuslan, 1967a). The mRNA for the enzyme must therefore have been made by the coated genome. In the same experiment, however, no new virus-specific DNA polymerase was found, suggesting that the polymerase is encoded by a delayed early mRNA. Also if cells are infected with ultraviolet light-irradiated virus, thymidine kinase is induced but neither DNA polymerase nor a virus-specific deoxyribonuclease are formed (Jungwirth and Joklik, 1965). The irradiated virus is not uncoated effectively (Joklik, 1964), suggesting again that there are two phases of early protein synthesis.

At the time when viral DNA synthesis is initiated—the beginning of the late phase—there is a marked decrease in the rate of synthesis of early enzymes. This inhibition of early enzyme synthesis requires viral DNA synthesis and may be an expression of products made from progeny viral DNA. It is not simply a result of the final step of uncoating (Jungwirth and Joklik, 1965). The mRNA for early proteins is very stable because if cells are infected for 2 hours and then actinomycin D is added, thymidine kinase synthesis will continue for at least 18 hours (McAuslan, 1963a,b; Figure 14-15). Studies of early mRNA by chemical methods have also indicated its stability

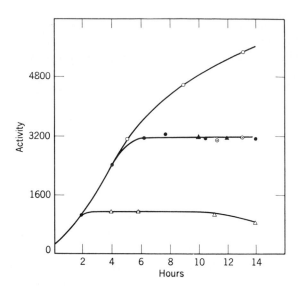

FIGURE 14-15. *The effect of actinomycin D and puromycin on the formation of thymidine kinase in vaccinia-infected HeLa cells. Enzyme activity was measured in total cell extracts after various treatments:* ●, *untreated;* ○, *actinomycin D added 2 hours after infection;* △, *puromycin added at two hours. The ability of actinomycin D to prolong synthesis of the enzyme argues for a translational repression of mRNA function beginning at about 4 hours after infection. The inhibition by puromycin indicates that de novo enzyme synthesis is responsible for the increased enzyme activity. From McAuslan (1963b).*

(Oda and Joklik, 1967). Furthermore, RNA synthesis decreases markedly when uncoating occurs but synthesis of early enzymes can continue indefinitely.

The stability of early mRNA implies that when viral enzyme formation is stopped by the switch to late functions, there must be inhibition of protein synthesis directly at the translational level. This is probably the best case for translation control of specific mRNAs that has ever been found (McAuslan, 1963a,b). Another evidence of translational control is that early transcription appears to continue throughout the growth cycle in spite of the inhibition of early enzyme synthesis (Oda and Joklik, 1967). The translational block to early enzyme synthesis is presumably the result of a protein made under the direction of the late mRNAs, but there has been no recent work on this phenomenon that might help to elucidate it.

THE LATE PHASE

The late phase of the viral growth cycle begins by 1½ hours after infection and extends for up to 24 hours. The major events of the late phase are synthesis of viral DNA, synthesis of late viral proteins, and maturation of virions.

The late viral proteins include two broad classes of proteins: the virion enzymes and the virion structural proteins. Synthesis of the late proteins is paralleled by an increased rate of RNA synthesis and the appearance in the cell of mRNAs that were not present during the early phase of the growth cycle (Oda and Joklik, 1967). The late mRNAs appear to have a very short half-life compared to the half-life of early mRNA (Shatkin, 1963). During the late phase of the growth cycle, synthesis of early RNA continues but this early RNA may come from cores that were not uncoated rather than from free DNA.

The synthesis of viral DNA takes place in discrete regions of the cytoplasm not intermingled with the usual cytoplasmic organelles. The number of such foci of DNA synthesis is proportional to the multiplicity of infection, and at low multiplicity, only one per cell is evident (Cairns, 1960). Viral DNA is induced in infected cells whether or not the cell is synthesizing its own DNA. This is not surprising considering the wide range of enzymes apparently synthesized under the direction of the viral genome. Viral DNA synthesis takes place between 1.5 and 5 hours after infection, and maturation of the DNA into virions requires a further period of up to 10 hours.

Virions mature initially from the surface of the cytoplasmic foci of viral DNA and later diffuse away into the cytoplasm. The initial event is apparently a *de novo* formation of membrane in the cell cytoplasm, followed by enclosure of a "headful" of viral DNA in the membrane. For unknown reasons, the drug rifampicin is able to block morphogenesis at the stage of immature membrane formation (Moss, 1973). Release from the rifampicin block leads to the rapid envelopment of DNA followed by the further stages of maturation leading to the final brick-shaped particle (Sarov and Joklik, 1973). A series of polypeptide cleavages that ordinarily occur are blocked by rifampicin, hence they are probably steps in maturation.

Retroviruses

The retroviruses have been left to last because they have properties of both RNA and DNA viruses. Their virion contains RNA but their intracellular state is as an integrated DNA. The entering viral RNA is actually turned into cellular genes that can be passed from parent cell to progeny cell as stably integrated DNA molecules. There is no DNA virus known to be inherited in this way, because all DNA viruses kill the cells in which they carry out a productive infection. The only time DNA viruses are stably integrated into cells is in nonproductive conditions (see Chapter 16). The retroviruses, by using the nondestructive budding procedure common to many RNA viruses, are able to maintain a productive infection without killing cells. The central issue in retrovirus multiplication is then how they are able to turn themselves from an RNA virus into a DNA gene; this is the process known as reverse transcription because it reverses the standard direction of information flow.

Many different retroviruses exist, and some of them can cause cancer. The best known are Rous sarcoma virus and the leukemia viruses of chickens and mice. Retroviruses are the only RNA viruses known to cause cancer—hence the common designation of them as "RNA tumor viruses"—but many retroviruses cause either no detectable disease or nonmalignant conditions. What unifies the retroviruses is their mode of multiplication. Like any other class of viruses, there are differences among them in the size and number of virion proteins, their detailed morphology, and their host range. Most work on retroviruses has focused on the murine and avian leukemia-sarcoma viruses and most of the following discussion deals with these viruses. They are called "C-type viruses" as part of a morphological classification system that distinguished three types of murine retroviruses: A-type (intracellular particles), B-type (mammary tumor viruses), and C-type (leukemia-sarcoma viruses) (Bernhard, 1960). An excellent book has reviewed the historical aspects of oncogenic retroviruses (Gross, 1970). Reviews of retrovirus biochemistry and genetics include Tooze (1973), Temin and Baltimore (1972), Temin (1973) and Vogt (1976).

VIRAL PROTEINS AND GENES

Retroviruses mature by the budding process described in Chapter 13. They have five to eight virion polypeptides, including one or two glycoproteins in the envelope that surrounds the virion core (Bolognesi, 1974) (Figure 14-16). One of the virion polypeptides appears to be a subunit that polymerizes to form the shell of the core. The RNA of the virus, as well as the DNA polymerase responsible for reverse transcription (the reverse transcriptase), are found inside the core.

FIGURE 14-16. *The organization of the retrovirus genome and location of the viral products in the virion. The gene order shown has been defined only for the avian leukemia viruses but is presumably the same for all retroviruses. Three genes are shown each of which encodes a single polypeptide, but the 60,000–76,000 dalton precursor of the internal proteins (the* **gag** *gene product) is cleaved to form a number of products of 30,000 daltons and less. A fourth gene,* src, *is found in Rous sarcoma virus and presumably encodes a transforming protein.*

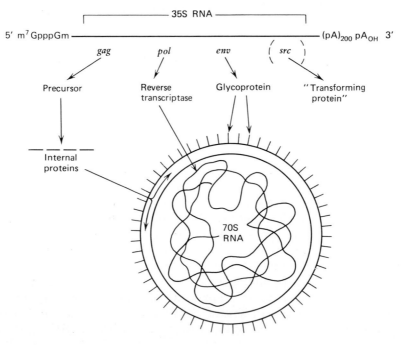

No intracellular viral proteins have been found that are not represented in the virion except for the nonessential transforming protein made by Rous sarcoma virus (see Chapter 16). Also, the sum of the molecular weights of the virion proteins virtually exhausts the coding capacity of the virion RNA so it is not likely that any new, large virus-specific proteins have yet to be discovered.

The known virion polypeptides fall into three classes: envelope polypeptides; structural, nonenvelope polypeptides; and the reverse transcriptase. The three regions of the viral RNA that encode these three classes of polypeptides have been called, respectively, *env, gag* (for *g*roup-specific *a*ntigens, an immunologic characteristic of this group of proteins), and *pol* (Baltimore, 1974). The individual polypeptides have not been delinated well enough to assign names to the genes that encode them but the three class designations are sufficient at the present level of genetic resolution. The *gag* genes are all linked because the internal proteins all arise by cleavage from a single precursor polypeptide (Vogt and Eisenman, 1973).

VIRAL RNA

Two classes of RNA molecules can be extracted from the virion core: a high molecular weight RNA complex sedimenting at 60 to 70S (generally called the 70S RNA) and RNA molecules sedimenting at about 4S. Most of the 4S RNA is cellular transfer RNA. When 70S RNA is completely denatured it generates two types of products: one consists of 2.5 to 3.0 $\times$ 10^6 dalton molecules known as 35S RNA for their sedimentation rate; the second is 4S RNA, again mainly cellular transfer RNA. Electron microscopy of 70S RNA, 35S RNA, and intermediate partially denatured forms has shown the 70S RNA to be made of two 35S subunits joined at their 5'-ends (Bender and Davidson, 1976; Figure 14-17). There are also defined loops in this 35S RNA under intermediate denaturing conditions. RNA from some viruses shows the loops and the dimer configuration very clearly while RNAs from other viruses show features that suggest they have the same underlying structure. What holds the molecules together at their 5' ends is not known but it involves no detectable covalent association. The low molecular weight RNAs may perform some sort of linking function.

By determining the frequency of occurrence of unique oligonucleotides in the viral RNA, it has been shown that the two 35S RNAs in the 70S complex must be identical (Weissmann et al., 1974; Beemon et al., 1976). Each of the 35S subunits has at its 3' end a stretch of poly(A) and at its 5' end a cap of $m^7G^{5'}ppp^{5'}GmpCp$, at least for the avian and murine viruses. The RNA also shares another feature common to molecules that arise from nuclear DNA, it has several N^6-methyl adenosine residues.

MAPPING THE RNA

The location of individual genes on the 35S RNA has been mapped using a three-step procedure. In the first step, ^{32}P-labeled RNA was randomly fragmented and all of the fragments containing poly(A) were purified using an

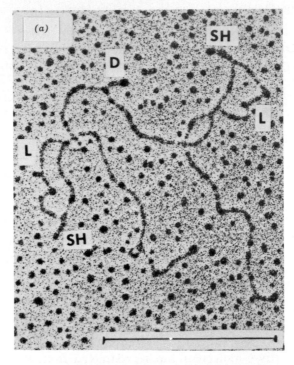

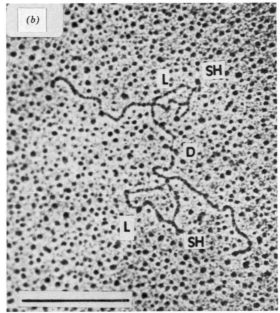

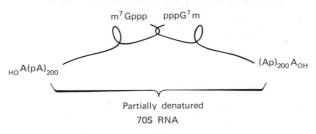

(c)

oligo(dT) column. These poly(A)-containing fragments were then separated by sedimentation, degraded to completion with T_1 ribonuclease, and the distribution of individual T_1-oligonucleotides through the different sizes of molecules was determined. In this way, the position of each unique oligonucleotide relative to the 3'-end was determined (Coffin and Billeter, 1976).

The second step to mapping the genome has been to determine what oligonucleotides are missing in specific deletion mutants. Two types of deletion mutants are available. In one type of deletion, the transforming gene for Rous sarcoma virus, called *src*, is deleted. In other deletions, *env*-specific regions are deleted. Assuming that the genome has 10,000 nucleotides, the *src* deletions lead to disappearance of oligonucleotides that map between 600 and 2000 nucleotides from the 3' end. The deletions in *env* map between 2800 and 5000 nucleotides from the 3' end (Wang et al., 1976a). Thus, these two genes are next to each other in the 3' half of the viral RNA.

Usable deletions do not exist in either *gag* or *pol* so a different procedure had to be used to map them. This procedure involves isolating recombinant viruses between two strains that have different oligonucleotide maps (Joho et al., 1975, 1976; Wang et al., 1976b). Because recombination occurs with high frequency in retroviruses, such recombinants can be readily isolated. By correlating the recombinant markers with recombinant oligonucleotides it is possible to localize genes. Analysis of recombinants has confirmed that *src* is poly(A)-proximal and *env* is next to it. The analysis has also shown that the *pol* gene is in the 5' half of the RNA but its localization relative to *gag* is still uncertain. Because the *gag* gene is the one most efficiently translated from the 35S RNA, it seems likely that *gag* is nearest the 5' end of the RNA and the provisional map in Figure 14-16 shows it in that way. Thus the total genome of Rous sarcoma virus consists of (5' → 3') *gag-pol-env-src*-poly(A).

An alternative way to determine the linkage relationships of the various retrovirus genes is by standard genetic recombination analysis (Vogt, 1971). The recombination rate of retroviruses is so high, however, that only a limited amount of linkage data is available. That data shows linkage of *pol* to *env* and *gag* to *src*; it also shows a *lack* of linkage of *env* to *gag* and *pol* to *src* (Vogt, 1976). These results are consistent with the physical mapping data (Figure 14-16) except that the two ends of the physical map seem genetically linked. One possible explanation of the apparently circular map is that recombination occurs at the stage of a circular DNA provirus (see below).

FIGURE 14-17. Dimeric structure of the 70S RNA from retroviruses. Retroviruses appear to have a dimeric genome, held together at the 5' ends of the individual 35S RNA molecules, with loops about one-quarter of the molecule away from the 5' end. This structure is reproducibly evident in many mammalian retrovirus RNAs but in avian viral RNAs it is only rarely seen, and hence they could be different. (a) Electron micrograph of partially denatured RNA from a baboon endogenous virus; (b) RNA from a virus recovered from a wooly monkey; (c) A schematic representation. In the electron micrographs D = the dimer linkage structure; L = loop; SH = short hairpin structures seen within the loops. Photographs reprinted from Kung et al., 1976 with permission.

MULTIPLICATION CYCLE: PHASE I

The multiplication cycle of retroviruses has two distinct phases that are not homologous to the early/late phases found with most other viruses. Phase I involves synthesis of the provirus and its integration into host cell DNA. Phase II involves the expression of the proviral DNA and the maturation of new virions. Phase I is initiated in the cytoplasm of the infected cell and, because enucleated cells can make normal amounts of DNA, it appears that most of the DNA synthesis process can go on in the cytoplasm (Varmus et al., 1974a). Integration, of course, occurs in the nucleus and so there must be transport of DNA or intermediates of DNA synthesis from the cytoplasm to the nucleus. The maturation of virus occurs at the plasma membrane so virion precursor RNA, after being made in the nucleus, must be transported to the cytoplasm.

There is extensive evidence for the synthesis of a DNA intermediate in the life cycle of retroviruses. The original evidence that such a phase might exist came from the response of retroviruses to specific inhibitors (Temin, 1971). Although most RNA viruses are insensitive to inhibitors of DNA synthesis, retroviruses were found to be very sensitive to drugs like cytosine arabinoside and aminopterin. Growth of retroviruses was also found to be very sensitive to actinomycin D. These and other facts led as early as 1963 to the suggestion that there might be a proviral intermediate in the growth of retroviruses. It was not, however, until 1970, when the virion DNA polymerase was discovered, that the possibility of a DNA intermediate was widely accepted (Temin and Mizutani, 1970; Baltimore, 1970).

There now exists very direct and persuasive evidence of the existence of the provirus. The clearest evidence is isolation from infected cells of a DNA that can transmit the infection to other cells (Hill et al., 1974). Such infectious DNA is present in the high molecular weight DNA of transformed cells. Early after infection, unintegrated infectious proviral DNA can also be found (Smotkin et al., 1975). From dilution kinetics, as well as from studies on circular duplex DNA, it has been concluded that a double-stranded DNA molecule of about 5 to 6 $\times$ 10^6 daltons carries all of the information for synthesis of a retrovirus. This size is consistent with the complete genetic information being found in a single RNA molecule of about 3 $\times$ 10^6 daltons.

A more indirect but informative method for analyzing DNA provirus information is nucleic acid hybridization. Because the provirus is double-stranded, each of the strands can be separately analyzed. To detect DNA complementary to virion RNA (the "minus" strand), labeled virion RNA is usually employed (Baluda and Nyack, 1970; Nieman, 1972). To detect plus strands of DNA use is made of labeled DNA formed *in vitro* by reverse transcriptase. Unintegrated proviral DNA can be analyzed separately from integrated DNA by using, for instance, the Hirt (1967) fractionation method.

Using hybridization and nucleic acid fractionation, three types of DNA molecules have been found in cells infected by murine or avian retroviruses: (1) covalently closed circular duplex DNA of about 5.5 $\times$ 10^6 daltons; (2) linear virus-specific DNA molecules containing complete minus strands but only partially completed plus strands and (3) integrated viral DNA molecules that are covalently bonded to host cell DNA. The interrelation of these three types of molecules is unclear, but it is assumed that the linear DNA is precursor to the circular DNA which in turn is precursor to integrated DNA. The three types of DNA are best considered as independent entities.

CLOSED CIRCULAR DNA.   These molecules can be isolated by cesium chloride-ethidium bromide centrifugation and detected by hybridization (Varmus et al., 1974a; Gianni et al., 1975). The circular proviral DNA of Moloney leukemia virus is infectious and can be isolated chemically pure by taking advantage of the absence of an Eco-R_1 restriction endonuclease site in the viral DNA (Smotkin et al., 1975). No other circular cellular DNA lacks such a site. Electron microscopy has shown that the circular proviral molecules have a molecular weight of about 5.5×10^6 daltons (Gianni et al., 1976; Figure 14-18).

LINEAR DNA.   Most of the DNA made after infection by retroviruses is not present as covalently linked circles. Using a restriction endonuclease that cuts the circular proviral DNA only once, the completely double-stranded portion of this DNA is fragmented into two defined pieces (Smotkin et al., 1976). It must therefore be a single sequence of linear DNA rather than being circular DNA. This DNA is highly infectious (Smotkin et al., 1975). If the linear DNA is denatured, the minus strand is recovered mainly as linear full-length molecules. The plus strand, however, is largely in the form of DNA fragments much smaller than full-length molecules (Varmus et al., 1976). In many of the molecules, the plus strands do not completely cover the minus strand.

INTEGRATED DNA.   The third form of virus-specific DNA is the integrated viral DNA that appears to be covalently attached to host cell DNA. The orientation of this DNA relative to the viral RNA has not been determined, so

FIGURE 14-18. *Electron micrograph of the closed circular double-stranded DNA provirus of Moloney murine leukemia virus. The leukemia virus circles are marked MuLV and are compared with mitochondrial DNA (mito) and SV40 DNA. Reprinted from Gianni et al. (1976) with permission.*

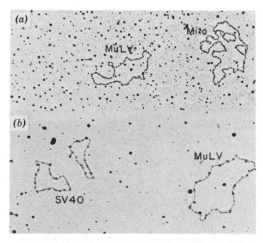

the site at which integration occurs is not known. The integrated DNA, however, must be in a form in which it is able to synthesize the viral RNA, so that the sequence of free linear DNA and integrated DNA is presumably very similar.

In almost all studies, the high molecular weight DNA from infected cells has been found to contain more virus-specific DNA than similar DNA from uninfected cells. Cells of most vertebrate species, however, contain endogenous copies of the DNA of whatever viruses ordinarily infect that species. Such endogenous sequences, which reflect the inheritance of virus-specific DNA by normal cells, make it difficult to demonstrate newly integrated DNA.

The most elegant demonstrations of integration of the provirus have come from studies on the infection of cells that contain no endogenous virus-related sequences. Such a situation can be realized in the infection of either mammalian cells or duck cells by Rous sarcoma virus. In both cases it is possible to demonstrate that infected cells contain virus-specific DNA (Baluda, 1972; Varmus et al., 1973a,b). In all cases, the number of viral DNA copies in infected cells is small (one to five copies per cell).

MULTIPLICATION CYCLE: PHASE II

Having established itself as part of the chromosome, the integrated provirus is available for transcription. There is no indication that it can be transcribed before integration.

The major stable RNAs that have been detected in infected cells are of the same polarity as virion RNA (Coffin and Temin, 1972). Viral RNA is known to be synthesized from a DNA template because of the effects of inhibitors both *in vivo* and *in vitro* (Temin, 1963; Rymo et al., 1974). The total virus-specific RNA of infected chicken or mouse cells represents 0.5 to 1% of the total cellular RNA (Besmer et al., 1975). Two size classes of virus-specific RNA are found in the cytoplasm of infected cells: 35S RNA and a series of lower molecular weight species of RNA, the exact size of which depends on the system examined (Fan and Baltimore, 1973; Bishop et al., 1976). Both high and low molecular weight RNAs have been implicated as mRNAs for the synthesis of viral protein, but the exact relationship between the mRNA and the proteins is not yet known.

As indicated previously, the virion contains five to seven structural proteins including both internal proteins, envelope glycoproteins and reverse transcriptase. All of the internal proteins are encoded by the *gag* genes, are synthesized as a single high molecular weight polyprotein and are formed by cleavage of this polyprotein (Vogt and Eisenman, 1973). This polyprotein precursor is synthesized under the direction of viral RNA in *in vitro* systems (Kerr et al., 1976).

Newly synthesized proteins along with 35S RNA molecules organize themselves at the surface of the cell and bud from the plasma membrane to form the infectious virion. The 70S RNA must be formed by aggregation of 35S subunits during the process of budding because no 70S RNA can be found inside infected cells whereas newly made virions do contain 70S RNA (Fan and Baltimore, 1973; Stolzfus and Snyder, 1975). Neither viral RNA nor reverse transcriptase are necessary components of the virion; particles lacking one or the other of these molecules have been found (Levin et al., 1974; Panet et al., 1975b).

THE REVERSE TRANSCRIPTASE

All retroviruses carry in their virions a DNA polymerase that is necessary for reverse transcription of viral RNA (Temin and Baltimore, 1972). The reverse transcriptase was first detected in virions of avian and murine viruses after partial disruption and incubation with deoxyribonucleoside triphosphates (Baltimore, 1970; Temin and Mizutani, 1970). This reaction is sensitive to ribonuclease, indicating that the template for DNA synthesis is RNA. Because the nascent DNA molecules can be recovered in association with 70S RNA and because the DNA can be hybridized to the 70S RNA, the template for DNA synthesis in the virion must be the 70S RNA. DNA made by disrupted virions can be infectious, although viral RNA is not infectious, showing that reverse transcription by virions is a physiologically significant process.

The reverse transcriptase has been purified to homogeneity from virions (Kacian et al., 1971). An antibody made against it does not inhibit other DNA polymerases. Using such an antibody, a competitive radioimmunoassay shows that there are 70 molecules of enzyme per virion of avian myeloblastosis virus (Panet et al., 1975b). The viral enzyme is not only a DNA polymerase, it is also a ribonuclease. Its ribonuclease degrades the RNA strand of a DNA·RNA hybrid; the enzyme does not degrade either free RNA or double-stranded RNA (Moelling, 1975). Such an enzyme is called a ribonuclease H.

The reverse transcriptase is a primer-dependent enzyme that cannot initiate new chains of DNA *de novo* but can only extend preformed chains (Baltimore and Smoler, 1971). When the enzyme copies 70S RNA in virions, the primer for synthesis of DNA appears to be a cellular transfer RNA species tightly bound to the viral RNA: for avian viruses, the primer transfer RNA is tryptophan tRNA while for murine viruses it is proline tRNA (Harada et al., 1975; Peters et al., 1977). The site at which the primer is bound had been localized near the 5' end of the 35S RNA (Taylor and Illmensee, 1975). DNA synthesis appears to be initiated near the 5' end of the RNA and, because of the polarity of synthesis, the newly formed DNA can only extend from the primer site to the 5' terminus of the molecule. Because much longer DNA molecules can be synthesized in the reverse transcription reaction, it must be possible for the growing DNA chain to extend past the 5' end by restarting near a 3' end of either the same template molecule or a different one (Haseltine et al., 1976).

Not only does the primer tRNA have a special affinity for the 35S RNA but it also binds with high affinity and specificity to the avian reverse transcriptase (Panet et al., 1975a). The reverse transcriptase is therefore a highly specialized DNA polymerase designed to start DNA synthesis by binding to a specific primer which is attached at a specific place on the RNA template. The ultimate product of the reverse transcription reaction *in vitro* is a complete DNA copy of 35S viral RNA formed by a linear extension process starting from the initiation point, jumping the 5' to 3' gap, and then continuing all the way down the 35S RNA back to the 5' end (Figure 14-19).

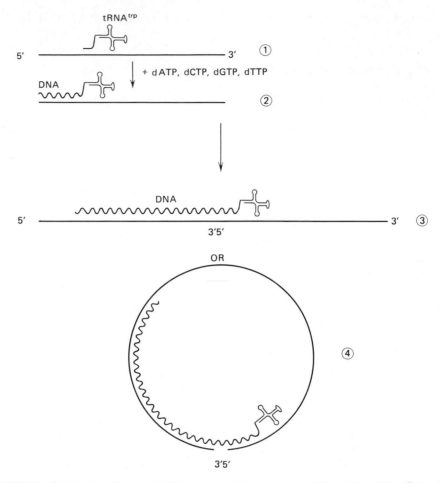

FIGURE 14-19. Synthesis of DNA copy of retrovirus RNA. The tRNA^{Trp} primer is attached to the 35S genome RNA about 135 nucleotides from its 5' end (structure (1)). The first product formed after addition of deoxyribonucleotides is a short DNA attached to the tRNA primer and extending to the 5' end of the template (structure (2)). Two possible models for the next step are shown. In structure (3), two 35S RNA are used as template and the DNA product is extended across the 5' to 3' gap between the molecules and then continues to copy all of the second template strand. In structure (4), a single, circularized template is shown on which the DNA product jumps the 5' to 3' gap and then continues to copy the whole RNA.

That the reverse transcriptase is encoded by viral RNA has been proved by the isolation of both deletion and temperature-sensitive mutants in the enzyme (Hanafusa et al., 1972; Verma et al., 1974). Viruses with temperature-sensitive lesions in the reverse transcriptase make little or no viral DNA after infection (Varmus et al., 1974b). The enzyme is therefore necessary for the synthesis of viral DNA and is encoded by the viral RNA. Despite some claims of reverse transcriptase-like DNA polymerases in uninfected cells, no convincing evidence for the existence of such enzymes has been provided.

RELATION OF VIRUS GROWTH TO THE CELL CYCLE

Retrovirus growth is tightly coupled to the progression of cells through the cell cycle. Newly infected cells must pass through mitosis in order for the synthesis of new virus to begin (Humphries and Temin, 1972). Neither transcription of the infecting genome nor synthesis of viral protein can be detected until mitosis has occurred (Humphries and Temin, 1974). It is unclear whether virus synthesis requires a mitosis *per se* or if RNA synthesis can only be initiated in the G_1 phase of the cell cycle so that mitosis is needed to bring the cell into the G_1 phase. In chronically infected cells virus production will occur only if the cell is growing: resting cells will not produce retroviruses (Paskind et al., 1975).

INHERITANCE OF RETROVIRUSES

The life cycle of retroviruses has been outlined above as a two-stage process. But cells can obtain an integrated viral genome in another way: they may inherit the viral genome rather than having it inserted after infection. In fact, once a cell is infected with a retrovirus, every progeny cell will inherit the retrovirus genome because the virus has effectively turned itself into a set of cellular genes.

When retroviruses become integrated into the germ line cells of an animal, all progeny animals will inherit the virus. In many species of animals proviral DNAs are inherited, presumably reflecting a germ line infection of an ancestral animal (Todaro et al., 1976; Baltimore, 1974; 1976). Nucleic acid hybridization studies have shown that in most species of animals copies of viral genomes are inherited by every animal of the species. It is likely that many of the inherited viral genomes are defective and cannot code for the synthesis of a complete virus particle. This might be expected because inherited viral genomes should accumulate mutations that would ultimately inactivate their ability to produce infectious particles.

Very few of the inherited genomes have been mapped in the cellular chromosomes. In one case, however, it has been possible to localize a viral genome to a specific site (Chattopadhyay et al., 1975). It seems likely that all inherited viral genomes have specific chromosomal locations.

In most cases where cells inherit viral genomes the genomes are transcriptionally silent. Expression of these genomes can often be induced, however, by treatment of the cells with specific chemical agents. The first demonstration was the finding that treatment of AKR mouse embryo cells with either bromodeoxyuridine or iododeoxyuridine induced the synthesis of virus (Lowy et al., 1971). Induction involves activation of viral RNA synthesis and is a consequence of the incorporation of the halogenated pyrimidine into the cellular DNA (Besmer et al., 1975). Induction of virus can also occur spontaneously; all AKR mice, for instance, become viremic soon after birth.

Effects of Animal Viruses on Host Cells and

15 **Organisms**

Virion Manufacture and Host Damage: Separate Processes?

Investigations of the events following virus infection of animal cells have dealt more with the synthesis of new virion components than with effects on host cells. This is largely because of the ease with which the virion-specific material can be followed and the slower development of knowledge and methods that allow a detailed comparison of cellular processes in normal and infected cells. Since viruses are the ultimate cellular parasites, it might seem artificial to separate a discussion of their effects on cells from a discussion of their growth cycle. Yet, even at the conclusion of their growth cycle, animal viruses comprise at most a few percent of the total mass of an infected host cell, an amount of protein and nucleic acid that would contain no more amino acids and nucleotides than that present in the cell pool at the time of infection. Massive destruction and reutilization of cell material would therefore appear to be unnecessary. Thus, the cellular changes observed late in the virus replicative cycle may have nothing to do with the synthesis and assembly of virion protein and nucleic acid.

The separate nature of the events of virion manufacture and the events that specifically affect the cell is evident for animal viruses that replicate and assemble in the cell cytoplasm but exert profound effects on the cell nucleus. Understanding the cell-damaging or cell-modifying viral functions is the key to understanding viral diseases at the molecular level and, in the long run, the basis of any rational therapy of viral infection that might conceivably emerge.

CATEGORIES OF VIRUS EFFECTS ON CELLS

Three effects of animal viruses on their host cells can be distinguished. The most easily recognized is a destructive or *cytocidal* effect characterized by

391

the extensive cell damage of many different cell organelles. Primary damage is probably caused by virus-specific macromolecules which may in turn begin a chain of secondary destructive events mediated by cellular products.

At the other end of the spectrum of virus action is *transformation*, the induction in the virus-infected cell of unlimited growth potential. Transformation appears to result from the permanent integration of all or a part of the virus genome in the cellular chromosomes without causing the death of the cell. In fact, the transformed cell is frequently released from normal growth restraints (see Chapter 16). Some viruses that do not integrate into the cell chromosomes cause effects intermediate between the most cytolytic and the transforming viruses. In these cases, cells continue to function for some time after infection and at least in one case (paramyxoviruses) to grow and divide while producing virus ("persistent infection").

A separate class of cellular reactions, which does not necessarily relate to the ultimate outcome of the cell-virus interaction, can be referred to as *inductive* actions of the infecting virus. Many viruses have the capacity to induce the formation, within the infected cell, of proteins that are not specified by the viral genome, but are apparently cellular products made in response to infection.

Cytolytic Effects of Viruses: Morphological Observations

The early studies of the morphologic effects of viruses on cells were carried out by pathologists who examined those organs of infected animals or humans that were most affected by the viral disease—for example, the brain and spinal cord in cases of encephalitis, or the liver in cases of "infectious" jaundice such as yellow fever. A major difficulty in correlating the histopathologic changes with virus infection was the inability of the light microscope to reveal which cells contained virus and which cells were damaged by secondary reaction. Nevertheless, the early histological work clearly described the morphologic changes associated with virus infection and remains useful in the diagnosis of certain virus diseases. Massive intracellular lesions (Figure 15-1) called *inclusions* or *inclusion bodies* were identified by cellular pathologists. These inclusions occur in either the nucleus or the cytoplasm and their location and staining properties are characteristic for the particular virus that causes them. Some, like the famous Negri bodies found in the cytoplasm of nerve cells infected by rabies virus, are so diagnostic that public health rules still recognize the diagnosis of rabies in the brain of an infected dog based on their presence or absence rather than on the outgrowth of the virus.

Two modern developments have greatly increased the accuracy of description of virus-caused morphologic lesions: first, cultured cells, all of which can be infected, settle the question of which morphologic changes are a direct result of infection and which changes occur because of secondary inflammation. Second, the use of electron microscopy has revealed the contents of inclusion bodies as well as many other more subtle derangements which occur in infected cells (see Table 15-1 and Figure 15-2). Inclusion bodies in some cases are sites of intensive virus manufacture or accumulation; for example, Guarnieri bodies are vaccinia "factories" in the cytoplasm (Cairns, 1960) and

Negri bodies are collections of rabies nucleocapsids (Matsumoto, 1970). Other characteristic inclusions such as the Cowdry type A nuclear changes of Herpes infection are the result of clumped and marginated cellular chromatin. Still other inclusions such as the nuclear inclusions of adenovirus-infected cells are crystalline arrays of virions, or in some cases virus proteins.

The disruptive morphologic changes do not always occur coincident with virus manufacture. Poliovirus growth is largely complete 5 hours after infection, but morphologic changes are minimal at this time compared to the breakdown of cellular integrity that occurs later (Bablanian et al., 1965). Likewise, the major site of virus activity need not represent the major site of cellular morphologic change. Group B arboviruses, which are thought to multiply exclusively in the cytoplasm, cause a pronounced perinuclear membrane proliferation (Matsumura et al., 1971). On the other hand, herpesviruses also cause extensive nuclear membrane changes (Roizman and Heine, 1972), but herpes multiplies in the nucleus and virus-specific protein is inserted into the nuclear envelope.

An interesting instance of morphologic abnormality without obvious connection to virus replication is the association of reovirus virions with the mitotic spindle of the host cell (Spendlove et al., 1963; Dales, 1963). Although interruption of host DNA synthesis is clearly important in eventually stopping mitosis (Tamm, 1975), interaction with spindle fibers could account for the abrupt cessation of mitosis that accompanies some virus infections.

Morphologic effects of viruses on host cell chromosomes are visible in a wide variety of cultured cells infected with any of several different viruses. These changes are probably nonspecific because they resemble the random action caused by ionizing radiation. It was claimed at one time that herpesviruses cause site-specific chromosomal lesions but since the amount of breakage increases when more virus is added to cells, it seems the action is not limited to only a few chromosomal sites (O'Neil and Rapp, 1971; Figure 15-1). Chromosomal breakage in herpes-infected cells is definitely due to a virus product formed early in infection because infected cells treated to prevent virus DNA replication still undergo extensive chromosomal breakage.

The ability of tumor viruses to cause chromosomal breakage and DNA synthesis stimulation and how this may be related to virus DNA integration is dealt with later in this chapter and in chapter 16.

Cytolytic Effects of Viruses: Biochemical Observations

With the knowledge that so many viruses cause extensive cytopathic effects in their host cells, biochemists began to question whether all cellular protein, RNA, and DNA synthesis ceased and in what order cellular syntheses stop. A central and still incompletely answered question is whether inhibition of host cell synthesis is due to specific virus molecules or due to competition for building blocks with virus replicating systems. If virus macromolecules are the basis for inhibition, are virions intrinsically toxic or is synthesis of new virus-specific material required to stop host functions? Partial answers and some general rules are emerging from a variety of experiments. (1) The mechanism of inhibition of protein synthesis is probably different for different viruses.

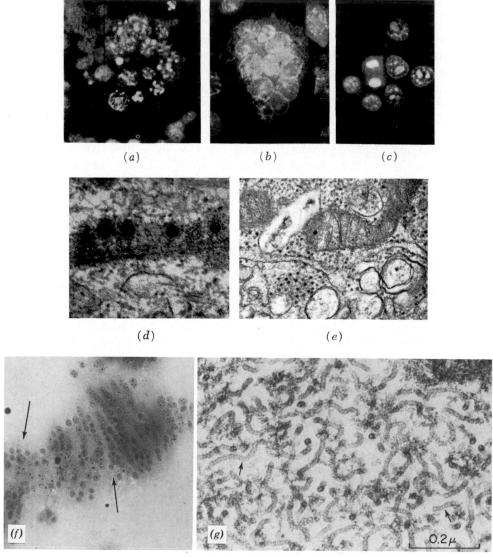

FIGURE 15-1. *Effects of various animal viruses on cells. (a), (b), (c) Cells stained with the fluorescent dye acridine orange. Normal cells are shown in (a). (b) and (c) show the formation of polykaryocytes induced by herpes and measles viruses, respectively. (d) An electron micrograph of reovirus particles associated with spindle fibers. (e) An electron micrograph of a thin section of an L-cell infected with mengovirus, showing a damaged mitochondrion and many cytoplasmic vacuoles. (f) Polykaryocyte induced by infection with simian virus 5 (SV5), light photomicrograph showing pale circumscribed inclusions. (g) Electronphotomicrograph of SV5 inclusion showing nucleocapsids. (h) Poliovirus crystal (electron micrograph) inside a HeLa cell and a view of the total disarray seen in the cytoplasm 5 to 6 hours after infection. (i) Portion of an adenovirus-infected cell showing crystalline array of virus particles within the nucleus (electron micrograph). (j) Nuclear border of cell infected with herpes simplex showing redundant membrane formation and herpes maturation (electron micrograph). Photographs (a,b,c) courtesy of Dr. B. Roizman; photographs (d,e,h,i) courtesy of Dr. S. Dales; photographs (f,g) courtesy of Dr. P. Choppin; and photograph (j) courtesy of Dr. F. Rapp.*

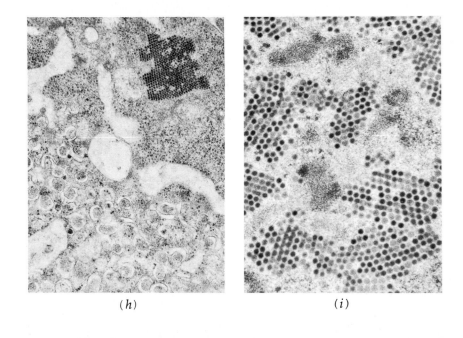

(h)   (i)

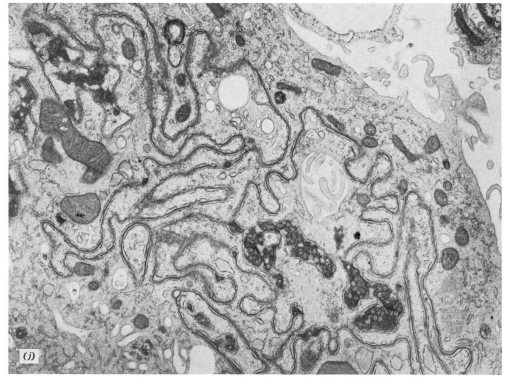

(j)

TABLE 15-1 TYPICAL MICROSCOPIC CHANGES IN CELLS INFECTED WITH COMMON CYTOPATHIC ANIMAL VIRUSES

Virus	Nuclear changes	Cytoplasmic changes	Comment
Small nonenveloped RNA viruses	Pyknosis; nucleus pushed to eccentric position in cell	Large eosinophilic mass which displaces nucleus seen in some cells; generalized increase in basophilia; cytoplasm appears to bubble at cell periphery.	Cell destruction not characterized by specific features; changes can occur even when virus multiplication is blocked. Among the most rapid cytopathic agents.
Togaviruses	Relatively unaffected early	Granular cytoplasmic inclusions seen to be collections of virions that have budded into internal vesicles; hyperplasia of membranes sometimes evident	Hundreds of agents in major groups vary widely in extensiveness of cytopathic and target cell lesions; encephalitis with brain cell destruction in whole animals is main feature with many viruses. Yellow fever virus causes liver cell destruction
Myxoviruses and Paramyxoviruses Influenza, fowl plague	Pyknosis	Cytoplasmic vacuolization contraction and degeneration; electron microscopy of surface reveals viral "budding."	No striking features during degenerative changes
Mumps, NDV	Pyknosis	Distinct eosinophilic, Feulgen-negative cytoplasmic inclusions	Cell fusion in absence of nuclear division frequent with mumps virus
Measles	Cytoplasmic inclusions, also some nuclear ones	Eosinophilic, Feulgen-negative inclusions formed by condensation of smaller clumps of eosinophilic material	Giant cell formation prominent; spindle or stellate cells also produced
Reovirus	Very little affected	Characteristic eosinophilic cytoplasmic masses which progress to occupy most of cytoplasm	Acridine staining of cytoplasmic masses shows a pale green-yellow fluorescence (since single-stranded RNA gives red fluorescence, this indicates double-stranded RNA).

Poxviruses	Margination and coarsening of chromatin	Development of distinctive Feulgen-positive cytoplasmic inclusions that contain virions; number of inclusions increases with time in infected cells; early in cell infection the number of inclusions depends on multiplicity of infection.	Various agents of the group differ in compactness of lesions and extent and speed of cellular destruction
Herpesviruses	Early changes marked by nucleolar displacement and reticular appearance of chromatin followed by granularity and margination of chromatin, leaving characteristic central Feulgen-positive inclusion	Vacuolization, especially of giant cells	Overall response of infected cells can be either (1) recruitment of nuclei into giant cells, or (2) destructive with nuclear fragmentation.
Adenoviruses	Types 1, 2, 5, and 6 cause small eosinophilic inclusions which increase in size and eventually appear as Feulgen-positive bodies surrounded by vesicles; electron microscopy reveals virus crystals. Second group (3, 4, 7) causes formation of Feulgen-positive masses arranged in distinctive rosette pattern; crystals seen in electron micrographs may be protein crystals only.	Few changes recorded	Loss of adhesion to glass observed particularly with Types 1, 2, 5, and 6. Cytopathic effect slow to develop. Plaques well developed only after 9–11 days.
Rhabdoviruses Rabies	Little early change	Pathognomonic eosinophilic Feulgen-negative round or oval inclusions 2–10μ in diameter (Negri bodies) consist of masses of bullet-shaped nucleocapsids.	Nerve cells in brain and spinal cord are targets; most abundant lesions in hippocampus and posterior horn of spinal cord.

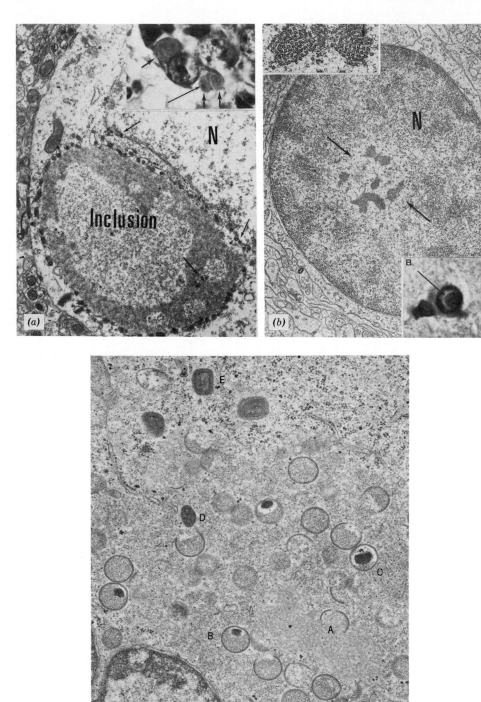

FIGURE 15-2. Effects of animal viruses in cells: inclusion bodies. (a) Intracytoplasmic inclusion of rabies (classic Negri body) shown in light micrograph, inset,
and in an electron micrograph where nucleocapsids and peripheral mature virions
can be seen. (b) Electronmicrograph of intranuclear (N = nucleus) inclusion body of

The extent and timing of inhibition of cell protein synthesis varies with different cytolytic viruses. (2) A block in cellular RNA accumulation frequently results from stopping ribosomal precursor RNA (pre-rRNA) processing with little or no effect on pre-rRNA transcription. Cellular tRNA synthesis is frequently unaffected. Cellular mRNA formation is often impaired but the mechanism(s) of interruption are very poorly understood. (3) The initiation of cell DNA synthesis is often inhibited, but in some virus infections, cells that have entered S phase may complete a round of synthesis and cells that have passed through the S phase may proceed through mitosis. The inhibition of cellular DNA synthesis is probably secondary to the cessation of cell protein synthesis, since continuous protein synthesis is necessary for DNA synthesis to proceed.

The following examples illustrate the experimental evidence underlying the above conclusions.

Protein Synthesis Inhibition

A general test for inhibition of host protein synthesis is to measure incorporation of radioactive amino acids into total protein in infected cells. A decline in incorporation, first noted for poliovirus-infected HeLa cells (Darnell and Levintow, 1960) has been observed for many different virus infections, for example, mengovirus (Baltimore et al., 1963), Sindbis (Strauss et al., 1969), herpesvirus (Sydyskis and Roizman, 1967), and paramyxoviruses (Wheelock and Tamm, 1960). Sometimes, as in adenovirus (Ginsberg et al., 1967) or vaccinia (Becker and Joklik, 1964) infections, total amino acid incorporation is not greatly altered, but an increasing proportion of the new proteins are virus-specific, indicating a true decline in host protein synthesis.

PICORNAVIRUS INHIBITION OF HOST PROTEIN SYNTHESIS

The molecular mechanism for the inhibition of host protein synthesis has been most thoroughly investigated in picornavirus infected cells, particularly in poliovirus-infected HeLa cells. These studies provide a model to consider how host protein synthesis inhibition can occur at the same time virus-specific synthesis flourishes (Figure 15-3).

To completely understand the inhibition of any cellular function may

---

←

measles virus in brain cells of a chronically infected patient with syndrome of subacute sclerosing panencephalitis (SSPE). Inset (B) is light micrograph of inclusion and (A) is a higher power electron micrograph to demonstrate nucleocapsid structure of the measles virion. (c) Electron micrographic view of region of cytoplasm equivalent to Guarnieri body (cytoplasmic inclusion) of HeLa cell infected with vaccinia virus. Central amorphous area of virus specific protein surrounded by all stages of vaccinia virion maturation. (A) membrane starting to form (B) Virion with core and DNA (C) Nucleoid present but not condensed (D) Mature virion. Photographs of (a) and (b) Courtesy of Dr. H. Koprowski and (c) courtesy of Dr. C. Morgan.

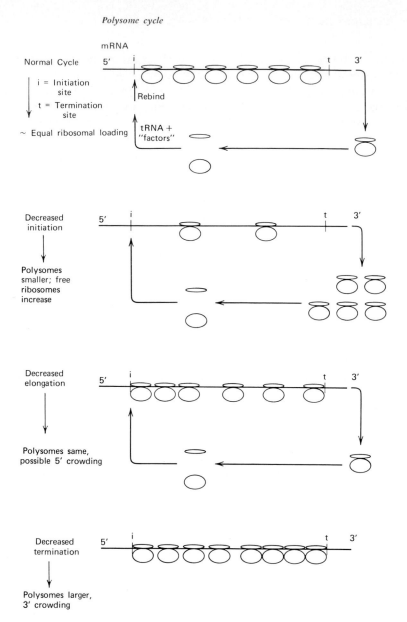

Polysome cycle

FIGURE 15-3. *Polysome cycle during poliovirus suppression of host protein synthesis. The polysome cycle and effects of interrupting it at initiation, elongation or termination are given above. (a) and (b) on opposite page show the effect of poliovirus in decreasing the polyribosomal content of a cell extract, detected by zonal sedimentation analysis and UV absorbancy at 260 nm: (a) was from cells infected only 20 minutes and (b) from cells infected for 65 minutes. (c) shows that the active polyribosomes detected by 1 minute ^{3}H amino acid pulse, also sediment more slowly after 90 minutes of infection. Cell polyribosomes still exist but they contain fewer ribosomes. If such infected cells are treated with cycloheximide, a drug which allows protein synthesis initiation but greatly slows elongation, the cell polyribosomes recruit additional ribosomes, thereby increasing the sedimentation value of the polysomes (d). Redrawn from data of R. Leibowitz and S. Penman (1971) J. Virol. 8, 661.*

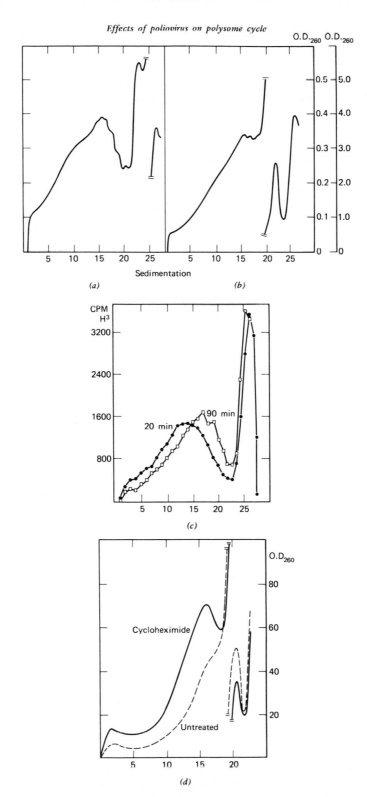

Effects of poliovirus on polysome cycle

require that purified cell components work satisfactorily in cell-free systems. Since intracellular inhibitors may act by altering rates of reactions, not just stopping reactions completely, the requirement for *in vitro* proof of a subtle alteration may be very difficult to meet. Therefore, the study of whole poliovirus-infected cells is first described followed by evidence obtained from cell-free protein synthesis.

1. *Are all host cell proteins affected?* Total amino acid incorporation in poliovirus-infected HeLa cells declines by about 80% within 2 to 3 hours, after which virus synthesis becomes evident. The cellular polyribosomes, which average 5 to 10 ribosomes each, disappear to be replaced by poliovirus polyribosomes, which average 35 to 40 ribosomes each (Summers et al., 1966). Experiments performed in the presence of guanidine, a drug that does not affect host cell protein synthesis (Summers et al., 1965) but stops viral RNA replication show 95 to 98% of host cell protein synthesis to be suppressed. Examination of the remaining host protein synthesis by electrophoresis of the labeled proteins revealed no specific protected class of protein—all cell protein synthesis is stopped.

2. *Is the inhibition due to a viral product?* Ultraviolet inactivation kills the virus' ability to affect host synthesis (Penman and Summers, 1965). Since ultraviolet-irradiation primarily affects nucleic acids, it appears that the entering polio RNA cannot act as an mRNA or cannot be replicated or both. Because guanidine allows virus protein synthesis late in infection, it probably would allow the entering RNA to serve as mRNA, but it does not allow RNA replication. Guanidine, however, does allow the entering viral RNA to stop host protein synthesis. Thus, a product of the entering strand of RNA, probably protein, appears to cause host protein synthesis interruption (Summers et al., 1965). A direct attempt was made to determine whether protein synthesis from the incoming polio RNA strand interrupts host protein synthesis. Infection was begun in the presence of puromycin, an inhibitor of protein synthesis. After 1 hour, when host protein synthesis would normally have been greatly inhibited by the virus, the drug was removed. For the first 15 to 30 minutes the infected cells synthesized protein normally, but after this time virus-induced inhibition of cell protein synthesis set in (Penman and Summers, 1965); again, this is evidence that a protein product from the incoming RNA strand is responsible for the interruption of host protein synthesis. Virus mutants have recently been isolated which have a temperature-sensitive effect on host protein synthesis, further implying that a virus protein is responsible for interrupting host protein synthesis (Steiner-Pryor and Cooper, 1973).

Double-stranded RNA, which accumulates during infection and inhibits *in vitro* protein synthesis (Ehrenfeld and Hunt, 1970), is unlikely to be responsible for host cell inhibition because (1) host protein synthesis inhibition is complete in the presence of guanidine where little double-stranded RNA could possibly accumulate and (2) double-stranded RNA inhibits all cell protein synthesis *in vitro* including virus-specific synthesis (Celma and Ehrenfeld, 1975).

3. *Does virus infection destroy the cell mRNA?* One simple way to interrupt host protein synthesis would be to destroy the mRNA. This is apparently *not* the mode of interruption of host protein synthesis. Preexisting mRNA is not reduced in size (Liebowitz and Penman, 1971) and both the 3' terminal poly(A) of cell mRNA (Koshel, 1974) and the 5' terminal methylated cap structures are intact (Hewlett et al., 1976; Nomoto et al., 1976; Fernandez-Muñoz and Darnell, 1976). Thus the cell mRNA is simply not translated in the infected cell.

4. *Is the protein synthesis defect in initiation, elongation, or termination of polypeptide synthesis?* The number of ribosomes in polyribosomes is determined by a balance between the rate of ribosome initiation, the rate of elongation, chain completion and ribosome release. Because protein synthesis in cell-free extracts does not maintain a similar balance as *in vivo*, conclusions about which step is affected by virus infection must initially be based on *in vivo* experiments. If initiation is faulty, polyribosomes should become smaller. If chain elongation and/or termination is faulty, polyribosomes should become larger (Figure 15-3). Experimentally, the surviving host cell polyribosomes in infected cells were found to become gradually smaller, averaging perhaps $\frac{1}{3}$ to $\frac{1}{2}$ as many ribosomes per polysome. Thus, poliovirus infection imposes a defect in initiation of protein synthesis. If the infected cells were treated with cycloheximide, a drug that slows chain elongation but has little effect on chain initiation, the polyribosomes containing cell mRNA returned to normal size, again demonstrating that host protein synthesis initiation is defective in poliovirus-infected cells (Leibowitz and Penman, 1971).

The promotion of translation of various mRNA molecules by a 5' terminal cap structure (Shatkin, 1976) suggests the chemical basis for distinguishing host cell and poliovirus mRNA. All cellular mRNA has such cap structures, while poliovirus virion RNA and polyribosome-associated mRNA lack such caps (Hewlett et al., 1976; Nomoto et al., 1976; Fernandez-Muñoz and Darnell, 1976). The viral mRNA might bind ribosomes and initiate protein synthesis with the aid of different initiation factors or the entering strand might initially bind very slowly. Late in virus infection, poliovirus protein initiation is rapid so a viral product might accumulate that favors noncapped mRNA.

5. *Can the inhibition of chain initiation be demonstrated in cell extracts?* In order to finally purify and characterize the protein(s) necessary for inhibiting host protein synthesis an *in vitro* system must be available that mimics the *in vivo* situation, that is, normal elongation and termination rates and much slower initiation rates. This requirement has recently been partially met with extracts of poliovirus-infected HeLa cells in which initiation of endogenous polyribosomal mRNA exhibited decreased initiation per ribosome. The defect apparently resides in polysome-associated initiation (or reinitiation) factors (Kaufman et al., 1976).

Cell extracts from mengovirus-infected mouse cells that are dependent on exogenous mRNA are stimulated equally well by hemoglobin mRNA and virus mRNA. However, the interruption of host protein synthesis is only 50% in this system compared to over 95% with poliovirus-infected HeLa cells (Lawrence and Thach, 1974). It is dangerous to conclude from *in vitro* studies that initiation is normal because one RNA seems to stimulate protein synthesis equally well as another.

A final purification of the putative virus-induced inhibitory substance for host protein synthesis will require a very detailed examination of *in vitro* protein synthesizing systems so that physiologic initiation rates can be achieved with purified components.

EFFECTS OF OTHER VIRUSES ON HOST PROTEIN SYNTHESIS

The details of inhibition of host protein synthesis for other viruses are less well known, but the inhibition appears to require protein synthesis after infection. The VSV mutants that are unable to transcribe mRNA do not affect host protein synthesis (McAllister and Wagner, 1976), while mutants that allow virus protein synthesis but cannot replicate RNA do stop host cell protein synthesis.

In addition, empty reovirus capsids (Lai et al., 1973) and irradiated herpes virions (Sydiskis and Roizman, 1967) do not cause protein synthesis inhibition although they are attached to and enter cells. The disappearance of host polysomes and appearance of virus-specific polysomes, as occurs in herpes infection, does not necessarily signal initiation defects in host protein synthesis but could be simply excessive virus mRNA production. At any rate, since only poliovirus (and other picornaviruses) lacks a capped 5'-terminus the tentative explanation offered for the discrimination between host and viral mRNA in polio-infected cells would not apply for other viruses.

A few virion proteins themselves appear to be toxic; for example the fibers from adenoviruses halt protein, RNA, and DNA accumulation in growing cultures (Ginsberg et al., 1967). During an adenovirus infection, however, the initial exposure of cells to adenovirus fibers does not stop host protein synthesis. Only 6 hours or so after the virus has entered the cell does host protein synthesis substantially decline. Thus, it is new intracellular virus-specific products, perhaps including fibers that are responsible for interrupting host protein synthesis.

Ultraviolet irradiated vaccinia virions inhibit host protein synthesis, but virion RNA polymerase still makes vaccinia mRNA and new virus proteins are synthesized, so this is probably not a case of direct virion toxicity (Bablanian, 1975).

In summary, cytolytic viruses of all types stimulate the synthesis of protein(s) that may or may not be virus capsid structures which have the capacity to interrupt host protein synthesis.

RNA Synthesis Inhibition

Many studies have shown a decrease in the incorporation of nucleotides into RNA of infected cells. Such studies *do not* necessarily establish a decreased *rate* of RNA synthesis. Rather they measure a decreased accumulation of RNA products, mostly ribosomes. Such decreases in incorporation were first observed in L cells infected with mengovirus (Franklin and Rosner, 1962; Franklin and Baltimore, 1962).

By contrast HeLa cells infected with poliovirus for an hour or so still incorporate "pulse" label into pre-rRNA almost as well as do growing cells (Zimmerman et al., 1963), but pre-rRNA processing is considerably depressed within the first 90 minutes of infection (Darnell et al., 1967). Total ribosome accumulation ceases, leading to a decreased incorporation after longer label times (Figure 15-4).

Such variation in response of cells to picornavirus infection resides with the cell not the virus. HeLa cells infected with mengovirus maintain the same rate of uridine incorporation for about 2 hours, while L cells infected with the same virus are profoundly affected within the first 2 hours. The virus, however, grows equally well in either cell (McCormick and Penman, 1967).

Many other viruses interrupt ribosome manufacture. Herpes viruses cause abortion of rRNA processing and synthesis within several hours of infection (Wagner and Roizman, 1969). Adenovirus causes inhibition of ribosome formation in growing cells, but only after 8 to 10 hours (Raskas et al., 1970)

FIGURE 15-4. Decrease in ribosome formation in poliovirus-infected HeLa cells. 45S ribosomal precursor RNA was labeled with both ^{3}*H uridine and* ^{14}*C-methyl-methionine from 50–90 minutes after infection but ribosomal RNA was not satisfactorily processed and new ribosomes did not appear in the cytoplasm. Transfer RNA synthesis was relatively unaffected. Redrawn from Darnell et al. pp. 375, in "Molecular Biology of Viruses" ed. Colter and Parenchych, Acad. Press, N.Y., (1967).*

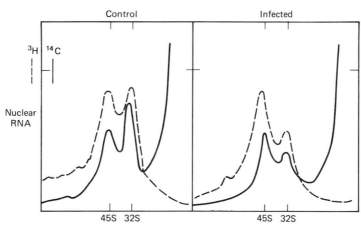

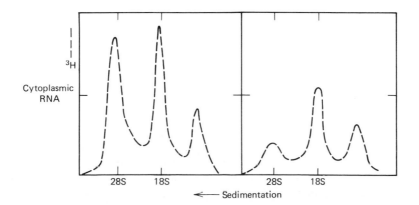

and an early stimulation of ribosome manufacture in resting embryonic kidney cells occurs (Ledinko, 1972). Even very late in adenovirus infection of HeLa cells pre-RNA synthesis and methylation continue, but complete maturation of ribosomes is not achieved (Sommer et al., 1976). Chick fibroblasts infected by mutants of Semliki forest virus, which are temperature-sensitive for viral RNA synthesis, continue host RNA synthesis at high temperature, but stop host RNA synthesis at lower temperatures where virus RNA replicates (Tan et al., 1969).

Virus effects on tRNA metabolism have been studied mainly in the hope of finding virus-specific tRNAs which would have a special role in virus replication. No virus-specific tRNAs have been discovered, and the indications are that cell tRNA formation is normal at least early in most viral infections.

The effects of virus infections on host cell mRNA metabolism is a central area of interest. Because mRNA manufacture in uninfected cells is not yet completely understood, very little is known about how viruses affect host mRNA production. Host mRNA and host hnRNA, the possible reservoir of at least some mRNA precursors, are made early in poliovirus infection, and mRNA enters polyribosomes immediately after infection (Darnell et al., 1967; Leibowitz and Penman, 1971) before inhibition of host protein synthesis has set in. Incorporation of labeled ribonucleoside triphosphates by the total RNA polymerases associated with chromatin is inhibited soon after picornavirus infection (Franklin and Baltimore, 1962; Holland, 1963). The cell DNA is still intact and all three cellular RNA polymerases, partially purified from extracts of infected cells, are as active as the polymerases from normal cells (Apriletti and Penhoet, 1974). Thus the inhibition probably results from some addition to the chromatin proteins of infected cells.

In adenovirus and herpes virus infections (both these viral DNAs enter the cell nucleus), nuclear virus-specific and nuclear cell-specific synthesis of hnRNA go on at least early in infection (Philipson et al., 1974; Wagner and Roizman, 1969). From approximately the midpoints of infection on, however, the only mRNA that enters the polyribosomes is virus-specific. The mechanism for this interesting selectivity is not known. Doubtless some of the secrets of the specificity of transport of particular mRNA molecules lie in the signal sequences at the termini of the mRNA molecules, a topic that is under scrutiny in many laboratories.

DNA Synthesis Inhibition

The general rule with cytolytic viruses is that cell DNA synthesis gradually declines beginning several hours after infection (Tamm, 1975). DNA synthesis occurs during S phase in mammalian cells in more than 10,000 separate replicons, which appear to be activated in clusters at specified times in S phase (Hand and Tamm, 1974). Protein synthesis is required both to activate replicons and to promote maximal rates of chain growth. The role of ongoing RNA synthesis in order to continue DNA synthesis in mammalian cells is not clear but there is evidence that RNA primers serve to initiate DNA segments (Tseng and Goulian, 1975).

Because of the delayed action of viral infections on DNA synthesis and the inhibition by viruses of protein synthesis, it is generally believed that host cell DNA synthesis is secondarily inhibited by picornaviruses, herpesviruses, adenoviruses, and paramyxoviruses. Only reoviruses, among the viruses studied, may cause specific decline in host DNA synthesis which precedes an extensive decline in host protein synthesis. Even here, however, some decline in protein synthesis occurs before DNA synthesis falls (Tamm, 1975). The decline in DNA synthesis is due at least in some infections to a decrease in active replicons with a normal chain growth rate (Hand and Tamm, 1974). Virus specific DNAses are induced by vaccinia virus and DNAse activity increases after herpes infections, but the host cell DNA is not degraded to very small size in either case, nor has wholesale degradation been reported in other virus infections (Joklik and Becker 1964, O'Neill and Rand, 1971). This contrasts with cytolytic bacteriophage infection where host DNA is degraded to deoxyribonucleotides (see Chapters 6–8).

Even though DNA synthesis may proceed at least partially during growth of cytolytic viruses, the general rule is that no cell division occurs longer than a few hours after virus infection. There is no absolute block of mitosis, however, because early in infection with vaccinia, NDV, and other viruses, some cells in mitosis stain with fluorescent antibodies to virus protein (Wheelock and Tamm, 1959). Also, synchronized cells infected with adenovirus in S phase or G2 all divide once normally (Hodge and Scharff, 1969).

Inductive Effects

A large part of the study of virology concerns the "induction" of synthesis of virus-specific products. These products are enzymes necessary for viral nucleic acid synthesis, the replicated nucleic acid itself, proteins for the capsids and envelopes of the virion, probably some proteins necessary for assembly of virions, and, perhaps, proteins that serve to diminish host cell macromolecular synthesis. Viruses can also cause cells to increase the synthesis of certain proteins that are normally made by the cell. In addition at least one class of proteins is induced which is not made normally and which may function as a cellular defense mechanism against viruses. Finally, virus envelope proteins appear in the cell membrane and cause changes in the "social behaviour" of cells, that is, interaction between cells. This section considers these *virus-induced* cell changes, which are not obviously necessary for viral replication.

Interferon

A discussion of interferon that only considered it as a cell protein formed in response to virus infection which then causes other cells to become refractory to infection would ignore the history of how interferon was discovered and the relation of this discovery to the long-recognized phenomenon of viral interference (Schlesinger, 1959; see next section).

For many years it was known that animals are often protected from the virulent action of one virus by simultaneous or previous infection either with unrelated viruses or with less virulent strains of the same virus. The suppressive effect on multiplication of a neurotropic strain of influenza virus by nonneurotropic strains was the first case to be subjected to quantitative analysis (Andrews, 1942). This interference did not require live virus, since ultraviolet-irradiated influenza virus injected into embryonated eggs inhibits production of infectious virus (Henle and Henle, 1947).

Isaacs and Lindeman (1957) then found that the allantoic fluid of eggs that had been exposed to irradiated virus possesses an interfering capacity distinct from the virus itself. The substance responsible for this interference was named *interferon*. It blocks the multiplication of many different RNA- and DNA-containing viruses both in eggs and in cell cultures. Many different animals produce interferon; cultured cells of many types, both normal and malignant, produce interferon, although the yields differ widely. Two especially good cells for the production of interferon are mouse L cells and a specially selected strain of human fibroblasts (Havel and Vilcek, 1972). Circulating white blood cells, probably especially leukocytes, also produce large quantities of interferon (Morganson and Cantell, 1974). Finally, some tissues must store interferon because various nonspecific toxic stimuli, such as bacterial endotoxin, induce a large increase in serum levels of a preformed virus-interfering material, presumably interferon (Younger and Hallum, 1968).

At one time it was thought that interferons were strictly species specific (Merigan, 1964) but this is not true. For example, human and monkey interferons elicit cross-species responses. More recently, cross-reaction of interferons between more distant species, for example, humans and rodents, was recognized (Stewart and Lockart, 1970; Stewart, 1974). These cross-reactions usually vary in the intensity of the antiviral response they elicit.

The level of protection offered to a particular challenge virus is characteristic of the cell, not of the interferon. Human cells exposed to human interferon become more resistant to vesicular stomatitis virus (VSV) than to Semliki forest virus (SFV), and monkey interferon causes the same relative protection in human cells. Monkey cells, in contrast, become more resistant to SFV than to VSV regardless of which interferon is used to protect.

Early tests for interferon activity, such as the protection of whole animals or embryonated chick eggs, have now been largely replaced by end-point dilution assays for the protection of monolayers of cultured cells or by the quantitative methods of *plaque reduction* (Wagner, 1963) or *virus-yield reduction* (Sreevalsan and Lockart, 1962). A cell culture is pretreated with dilutions of an interferon solution and then challenged with an amount of virus known either to produce cytopathic effects within a given time or to produce a known number of plaques or a known virus yield. The delay in cytopathic effect, the reduction in plaque number, or the reduction of virus yield allows a quantitative estimate of the relative content of interferon in different preparations.

With the availability of quantitative assays for interferon, attempts at purification began. Interferon was demonstrated to be sensitive to proteolysis and thus assumed from early studies to be a protein. Even without extensive purification a relatively uniform size of 20,000 to 40,000 daltons was established for the protein active in inducing the antiviral state (Ng and Vilcek, 1972; Merigan, 1974). Purification of interferon to apparent homogeneity was difficult because of the extremely low concentrations present in culture medium. By starting with the output from 10^{10} to 10^{11} cells, Knight (1975 and 1976) succeeded in purifying a human interferon from human fibroblasts and two mouse interferons from mouse L cells. The human interferon is a single homogeneous glycoprotein of 20,000 daltons. In contrast, interferon produced by cultured mouse L-cells consisted of two main species of 22,000 and 38,000 daltons with a variety of active minor glycoprotein species between the two major bands. One possibility for this variability in size is a variability in glycosylation. The two major species appear unrelated because no reduction of the larger material to smaller subunits could be achieved, and both species remained active after strong denaturing treatment. Interferon is a highly active protein; at a concentration of 10^{-11} M, human interferon prevents VSV multiplication in human fibroblasts (Knight, 1976). Polypeptide hormones such as insulin (Rodbell, 1964), glucagon (Cornblath et al., 1963), ACTH (Buonassisi et al., 1962), and others exert their physiological effects in the range of 5×10^{-10} to 1×10^{-8} M.

More than one molecular form of interferon can be demonstrated without complete purification (for example, see Stewart and Desmyter, 1975). Interferons produced by one species can be active in cells from quite distantly related species, for example, humans and rabbits. Stewart and Desmyter assayed the size of impure human leukocyte interferon that protected human cells or rabbit cells. Two active molecular sizes of about 21,000 and 15,000 were found; the lower molecular weight species has a 20-fold greater activity in protecting human than rabbit cells whereas the higher molecular weight species is equal in protecting human and rabbit cells. In addition, the lower molecular weight form (15,000 daltons) is totally destroyed by –S–S– group reduction with β mercaptoethanol while the higher molecular weight species (21,000 daltons) is stable under these conditions. Thus a single molecular species can act as interferon, but many if not most cells produce at least two types of polypeptide with interferon activity.

The induction of interferon production and the induction by interferon of the antiviral state are two closely related but probably separate phenomena. Cells entering the antiviral state may produce interferon. However, interferon is almost certainly not the protein responsible for the antiviral state, for cells require hours to achieve the fully antiviral state and then may not produce detectable interferon. Nevertheless, after challenge by virus infection, interferon-protected cells may produce more interferon. We shall consider two separate topics: the induction of interferon, and the induction and molecular basis for the antiviral state.

INDUCTION OF INTERFERON

In addition to the induction of interferon by live viruses, interferon production also follows treatment of cells with ultraviolet-killed influenza virus (Isaacs and Lindemann, 1957), an observation that has been repeated with many other viruses. Attempts to discover the chemical component of viruses responsible for induction led Isaacs (1963) to conclude that any RNA was capable of inducing interferon. Later, double-stranded RNA (Lampson et al., 1967), single-stranded RNA with stable base-paired loops (DeClerq and Merigan, 1969), and the synthetic double-stranded homopolymer poly(I)·poly(C) (polyinosinic acid·polycytidylic acid) were shown to be active inducers of interferon formation (Field et al., 1968). Only ribohomopolymers (Colby and Chamberlain, 1969) were capable of induction; both deoxyribopolynucleotides and hybrids of ribo- and deoxyribopolynucleotides were inactive. An obligatory role of double-stranded RNA for induction of interferon during virus infection is unsettled. For example, interferon can be induced by vaccinia virus, a DNA virus which forms single-stranded mRNA in the cell cytoplasm (Colby and Chamberlain, 1969). In vaccinia-infected cells small amounts of virus-specific RNAse-resistant RNA can be detected which might be responsible for the capacity of vaccinia to induce interferon. However, certain Sindbis virus *ts* mutants that fail to produce any detectable new viral RNA, single- or double-stranded, can induce interferon (Lockart et al., 1968). Other Sindbis virus *ts* strains, so-called RNA+ strains, produce RNA, single- and double-stranded, but are defective for virion production at high temperature; these mutants induce less than 5% of the normal interferon yield. Therefore, the nature of the inducing signal for interferon is not clear. Double-stranded RNA added to cultured cells has a direct cytotoxic action (Cordell-Stewart and Taylor, 1973), and Lockart (1970) proposed that any of a variety of cellular injuries might induce interferon production.

Whatever the signal for interferon production, new cellular "gene activation" is necessary to form the protein because actinomycin D, which inhibits cellular RNA synthesis, and puromycin or cycloheximide, which inhibit protein synthesis, block the appearance of interferon when added early in course of interferon induction (Lockart, 1970; Ng and Vilcek, 1972).

Studies of the induction of interferon using various drug treatment protocols suggest a complex genetic regulatory apparatus for control of interferon production. The induction of interferon by poly(I)·poly(C) is greatly enhanced if cells are treated during the inducing phase with cycloheximide and then released to form protein normally. In addition, the normal deinduction (that is, decrease in rate of interferon synthesis) that occurs after the removal of poly(I)·poly(C) is blocked by actinomycin D or the RNA synthesis inhibitor, 5,6 dichloro-1-β-D-ribofuranosylbenzimidazole (DRB) (Sehgal et al., 1976a). With the appropriate use of both drugs—add cycloheximide, remove cycloheximide, add actinomycin D—large increases (100-fold or more) in interferon output can be achieved (Figure 15-5; Tan et al., 1970; Vilcek and Havel, 1973; Sehgal et al., 1976b). Cellular regulatory elements requiring intact protein and RNA synthesis appear responsible, first, for the induction of interferon production and, second, for the decrease in production. This necessity for direct action to stop the ongoing synthesis of a protein was first described in mammalian cells for thymidine kinase during vaccinia replication (McAuslan, 1963).

Through the use of interspecies cell hybridization, two elements of the interferon-antiviral system have been mapped on human chromosomes; only those hybrid cell lines carrying human chromosomes 2 and 5 can produce human interferon (Tan and Ruddle, 1974).

The conclusion that cellular nucleic acid and protein synthesis are required for interferon induction, finds support in the fact that the most virulent and fast-acting cytolytic viruses, poliovirus, for example, are notably poor inducers of interferon. Perhaps the cell genome is disarmed by the sudden

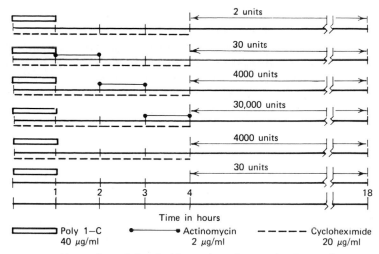

FIGURE 15-5. *Effect of cycloheximide and actinomycin D on the accentuation of interferon production stimulated by poly I-poly C. Five sets of cultures were exposed to cycloheximide (top five lines). A control set (line 6) received no inhibitors. The amount of interferon produced is given above each line. Tan et al. (1970), Proc. Nat'l. Acad. Sci. USA, **67**, 464).*

cytolytic events that occur after virus infection. Such a tug-of-war apparently goes on in cells infected with Chikungunya virus, a togavirus. Such infected cells normally elaborate both large amounts of interferon and several hundred plaque-forming units of virus per cell; if treated with actinomycin, no interferon is produced and a large increase in virus production occurs. It appears that a cellular protective response involving interferon production normally occurs when cells become infected and this response is blocked by actinomycin (Heller, 1963).

THE INDUCTION OF THE ANTIVIRAL STATE BY INTERFERON

Cultured cells induced to produce interferon by either killed virus or by polynucleotides also become virus-resistant. Likewise, many cells treated with interferon make a very large amount of additional interferon after challenge with a virus. However, certain monkey cells (Strain Vero) can become virus-resistant after exposure to monkey interferon (Desmyter et al., 1968) but cannot make detectable amounts of interferon nor be induced to the antiviral state after stimulation with poly(I)·poly(C) or other double-stranded RNAs (Schaefer and Lockart, 1970). In addition, these cells, unlike most monkey kidney cells, fail to become resistant to a variety of secondary viruses during rubella virus infection. Since interferon is a protein, antibodies to it can be produced. Cells treated with the interferon inducer poly(I)·poly(C) in the presence of interferon antibody do not become virus resistant (Vengris et al., 1975).

These results suggest that to trigger the antiviral state the presence of small amounts of interferon at the surface of the cell may be necessary. The pathway of induction of the antiviral state with poly(I)·poly(C) may involve the initial production of interferon followed by the secondary induction of the antiviral state by the interferon itself. After cells are fully induced to the antiviral state, however, further interferon production cannot be detected, and without further exposure to interferon the antiviral state disappears. A number of other experiments support the proposal that interferon induces the antiviral state by interacting with the cell membrane. Purified interferon linked covalently to Sepharose beads that are too large to be ingested into cells still stimulate the cells to assume the antiviral state (Knight, 1974). The activity of interferon-bead complexes can be titrated just as interferon can, so a minimum dose to induce resistance in cells can be determined. The bead-interferon complex can be recovered and used to stimulate additional cells through six cycles of production without losing activity. This would be impossible if the cell protection were due to free interferon release. A surface interaction of interferon with cells appears, therefore, to suffice for induction of the antiviral state. The structural gene for the surface receptor for human interferon appears to be carried on chromosome 21, because antibodies to cell surface proteins made against hybrid cells bearing only this chromosome prevent the action of interferon in inducing the antiviral state (Revel et al., 1976).

THE MOLECULAR BASIS OF THE ANTIVIRAL STATE

The nature of the antiviral state induced by interferon has been the subject of intense activity and controversy. Full protection requires several hours at least and cell RNA and protein synthesis are both required for the development of the antiviral state (Lockart, 1964), suggesting that new host cell proteins must be formed.

The cell nucleus is necessary for induction of the antiviral state, but not its maintenance. "Cytoplasts," the name given to the remaining cytoplasm after enucleation of cells by treatment with cytochalasin B, will allow replication of VSV but cannot be induced to the antiviral state by interferon. Cytoplasts from cells that have already achieved the antiviral state will not replicate the virus (Young et al., 1975).

Although the antiviral state induced by interferon affects a wide variety of RNA and DNA viruses, treated cells are certainly not equally "antiviral" for all viruses. In addition, different amounts of interferon are required to provide comparable levels of protection against different viruses in a given cell culture. Roughly 30 times more interferon is needed in chick fibroblasts to cause equivalent interference with NDV as with a togavirus (Ruiz-Gomez and Isaacs, 1963). Myxoviruses, togaviruses, and vaccinia virus—viruses with lipid-containing envelopes—are more susceptible to interferon than adeno- and enteroviruses, but some of the enveloped viruses, including herpes and NDV, are more resistant than others. The small icosahedral RNA viruses are among the most resistant to interferon action (Friedman, 1970). The meaning of this variable induction of antiviral activity is unknown.

The block in virus infection induced by interferon is past the early events of adsorption and penetration of viruses (Friedman, 1970; Wiebe and Joklik, 1975). Because interferon can prevent both RNA- and DNA-containing viruses from replicating, viral mRNA translation by host ribosomes, a common event for all viruses, would be a logical locus of inhibition (Joklik and Merigan, 1966). Such a situation could be achieved by an antiviral protein that could distinguish cell from virus mRNA. *In vitro* protein synthesis with extracts of interferon treated cells has not produced consistent evidence of failure to translate viral mRNA while translating cell mRNA normally.

Thus, in spite of the attractiveness of the simplifying hypothesis of translational discrimination as the basis of interferon action, no single simple mechanism explains all the facts known about the antiviral state.

In interferon treated vaccinia-infected cells, transcription of early mRNA by the virion DNA-dependent RNA polymerase is not inhibited but the mRNA is not translated; no early proteins are formed (Jungwirth et al., 1972; Bodo et al., 1972). Likewise in reovirus-infected cells, high doses of interferon decrease mRNA formation only slightly compared to the decrease in translation (Wiebe and Joklik, 1975). However, it has not been established in either case that the viral RNA is appropriately modified with a 5' blocked methylated cap structure or in the case of vaccinia, a 3'-poly(A) terminus. Thus the basis of the antiviral state may be that defective mRNA is formed and there is no change in the translation apparatus.

In fact, ineffective methylation of mRNA in interferon treated cells has been observed in an admittedly artificial situation. Reovirus RNA produced by virions is normally methylated but by synthesizing the mRNA in the absence of S-adenosylmethionine, unmethylated mRNA can be produced. This mRNA, upon addition to uninfected normal cell extracts, becomes methylated but this methylation cannot be performed by extracts from interferon treated cells (Sen et al., 1975). Whether this result signals an important *in vivo* change is unknown.

In cells infected with VSV (Vesicular stomatitis virus), a virus notable for its sensitivity to the antiviral state, no virus-specific RNA formation can be detected in infected cells (Marcus et al., 1971). At first this effect was thought to be the result of inhibition of the virion polymerase responsible for VSV mRNA formation. A recent suggestion, however, is that a nuclease activity is established in interferon treated cells which may specifically destroy the viral mRNA (Marcus et al., 1975). Enhanced nuclease activity that inactivates specific tRNA molecules has been detected with extracts of interferon treated, but not normal, cells (Sen et al., 1976). There is no evidence that these activities play a role in the antiviral state.

In poly(I)·poly(C)-treated, SV40-infected cells much less T-antigen (an early virus product) and much less early virus mRNA is present than in normally infected cells. This result was thought to be due to decreased transcription of SV40 DNA (Oxman and Levine, 1971). However, SV40 viral mRNA originates in the cell nucleus and is transported to the cytoplasm after a number of posttranscriptional modifications; the better test for transcription is to label for a short time after which most of the newly transcribed RNA should be nuclear. After a 30 minute label period, interferon treated cells contained about one-half as much nuclear viral-specific RNA as controls (Brandner and Mueller, 1974).

Graessmann (1974) prepared SV40 specific RNA by transcribing SV40 DNA with *E. coli* polymerase; such RNA clearly requires modification (for example, cap and poly(A) addition) to become effective mRNA. Microinjection of the RNA into normal cells induced T-antigen synthesis but in interferon treated cells no T-antigen was formed. The basis for the antiviral effect in this case clearly lies after transcription of the RNA. At least four possible mechanisms for the antiviral state must remain open to consideration in various virus infections: (1) failure of viral mRNA to initiate protein synthesis (2) failure of transcription of viral mRNA (3) failure of proper posttranscriptional modification of viral mRNA (4) destruction of viral mRNA perhaps secondary to (1) or (3).

VIRAL INTERFERENCE NOT MEDIATED BY INTERFERON

Some virus infections foreclose the possibility of second infection by other unrelated viruses or, in some instances, related viruses. This phenomenon, called *viral interference,* unlike interferon, is not based on a response of the host cell genome. It depends on the production of viral products by an initial virus that renders the infected cell incapable of supporting the replication of a second virus (Schlesinger, 1959). Many, many different pairwise combinations of viruses have been tested and it is likely that most cases of interference operate either at the level of preventing mRNA translation of the second virus (Marcus and Carver, 1967) or in a few cases preventing proper entry through the cell membrane.

Two examples of interference will illustrate these conclusions. Sindbis virus mutants, which are temperature-sensitive for RNA replication, establish an interference for VSV replication that is maintained even at the nonpermissive temperature (37°). VSV enters the cell and VSV-specific mRNA is formed by the virion associated transcriptase but no infectious virus is formed. All of this occurs in the presence of actinomycin D which blocks interferon production (Hunt and Marcus, 1974). The mechanism by which VSV mRNA is prevented from functioning is not known and in many other cases of interference it is not known at what intracellular stage inhibition is imposed on the second virus.

A second type of interference, which probably operates at the level of virus entry, occurs between related retroviruses. Cells that are susceptible to transformation by either of two viruses having identical envelopes but different nucleocapsids, one tumorgenic and one not, become refractory to the tumorgenic virus when already producing the nontumorgenic virus. Adsorption is not defective because the tumorgenic virus is taken up by the cells from the medium but it is held for an abnormally long time at the cell surface where it can be neutralized with antiserum (Steck and Rubin, 1966).

These cases of *heterologous* interference should be distinguished from cases of *homologous interference* with subviral particles, *defective interfering particles,* or *DI* particles which exhibit autointerferences, in the case of VSV at least, by preventing normal viral RNA replication (see Chapter 13). Finally, it should be mentioned that interference is not always the rule when two different viruses infect cells. For example, poliovirus, which destroys host protein synthesis completely and interferes with many other viruses, allows the coincident replication of coxsackie and mengoviruses, both of which, like poliovirus, are picornaviruses.

Membrane Changes in Virus-Infected Cells

Many of the changes in the membranes of infected cells may be related to virion formation. For example, the nucleocapsid of enveloped virions exit through the cell membrane in regions where virus-specific glycoproteins are inserted. However, some of the changes that the virus-specific surface proteins impose on cells do not have any apparent connection to replication but are so dramatic they deserve mention. Herpesvirus, measles virus, and respiratory syncytial viruses of humans, prominently among other viruses, cause the phenomenon called *polykaryocytosis* (Figure 15-1) (Poste, 1970), the recruitment of perhaps 50 to 100 cells into one giant cell that contains 50 to 100 nuclei within one cytoplasm. This fusion can go on in the absence of effective virion production and is seen in the tissues of affected patients as well as in cell cultures.

Other membrane changes may be related to virion production in a manner too subtle to have been solved as yet. Or the changes may represent a cellular response to virus infection such as exaggeration of ongoing synthesis or perhaps the induction of new cell products. For example, poliovirus stimulates the synthesis of new cellular membranes as evidenced by an uptake of choline and of glycerol (Penman, 1965; Mosser et al., 1972). The glycerol initially enters rough membranes (those containing polyribosomes) and then shifts to smooth membranes where most of the choline label is also found. This stimulus occurs to a reduced extent even if viral RNA replication is prevented by treating the cells with guanidine. If RNA replication is allowed, the virus RNA synthesis is concentrated in the smooth membranes.

Herpesviruses, which have a prominent nuclear phase in their replication cycle, cause a massive stimulation of nuclear envelope synthesis resulting in redundant folds of nuclear envelope (Figure 15-1; Roizman and Heine, 1972). Arbovirus formation stimulates a burst of perinuclear membrane synthesis although most virion replication is thought to be randomly scattered throughout the cytoplasm (Pfefferkorn and Shapiro, 1974).

How much of these extensive membrane changes is cell response and how much events in viral growth is not known at present. They are mentioned as interesting phenomena that might allow deeper understanding of cellular membrane biogenesis.

Stimulation of Synthesis of Cellular Proteins in Infected Cells

Perhaps the best studied case of virus stimulation of normal cell functions are the events that occur after small DNA tumor viruses infect resting cells. Enzymes that normally increase in concentration before S-phase (thymidine kinase, deoxycytidylate deaminase, and DNA polymerase), all increase significantly. New protein synthesis is required for the stimulation as indicated by lack of increase if cycloheximide is present (Green, 1970).

The enzymes are presumably encoded in the cell genome because mutant cells that lack thymidine kinase are not stimulated to produce that enzyme by SV40 infection (Kit, 1968). In addition, the DNA of polyoma and SV40 is only about 5000 bases long, probably too small to encode all the necessary virus functions as well as the enzymes related to DNA synthesis. The new DNA synthesis stimulated by virus infection is host DNA as indicated both by inclusion of prior DNA label in the replicated DNA as well as progression of some of the stimulated cells into mitosis (Weil et al., 1965). In addition to the host DNA synthesis, histone synthesis is stimulated so that the cell apparently goes through a normal S phase (Winocur and Robbins, 1970).

The *ts* mutants of polyoma virus that do not stimulate cell DNA also cannot transform cells so that the stimulation of host DNA may play an important role in transformation (Benjamin, 1972).

How often particular enzymes of diverse other sorts might be stimulated by the wide variety of existing animal viruses is not known. One such apparently random instance of a virus stimulation of the production of excess amounts of enzymes already being synthesized occurs during togavirus infection. The synthesis of dehydrogenases for glucose-6-phosphate, isocitrate, malate, and lactate are stimulated three- to four-fold and the stimulation is actinomycin D sensitive, suggesting the need for cell RNA synthesis in the response (Cassells and Brooke, 1973).

Many additional instances of cellular responses to invading viruses may come to light as molecular details are learned about virus replication and virus damage.

Animal Virus Diseases

It is not within the scope of this book to discuss virus diseases in detail. It is appropriate, however, to outline briefly how fundamental virology is related to some major problems of clinical virology.

RANGE OF AGENTS AND DISEASES

Many major illnesses—smallpox, poliomyelitis, yellow fever, influenza—as well as the usually mild common infections of childhood—measles, mumps, and chicken pox—are caused by infections with specific viruses. Also, it has long been suspected that a great number of mild illnesses that appear in humans, either as sporadic cases or in near-epidemic form, may be caused by viruses. There is now a growing feeling that some chronic slowly progressive neurologic diseases may also be of viral origin. The viral etiology of human illnesses is difficult to prove by animal inoculation since the only easily detectable results in animals are dramatic effects such as death or paralysis. With the advent of modern tissue culture methods for the detection of cytopathic effects of viruses, the practice of clinical virology has been greatly changed. Not only can virologists isolate quickly and efficiently already known viruses, but they can use the new tools to search for agents responsible for illnesses of unknown origin. This has produced a flood of discoveries revealing hundreds of new agents that can infect human beings (Huebner et al., 1956; Jackson and Muldoon, 1975) and many additional viruses of importance in animal husbandry.

No one method of grouping the disease-causing viruses leads to a simple classification scheme. No one clinical disease state is caused by one virus type nor does one virus group affect only a specific tissue. For example, mild upper respiratory disease can be caused by picornaviruses (the "common cold" viruses are rhinoviruses), adenoviruses, myxoviruses (influenza), and para-myxoviruses (respiratory syncytical virus), and probably others such as reoviruses and the newly discovered enveloped viruses, the coronaviruses. Liver cell infections are caused by togaviruses (yellow fever) and hepatitis virus which is possibly a lipid-containing DNA virus, and central nervous system infections leading to paralysis and death can be caused by togaviruses (dozens of different encephalitis viruses), picornaviruses (polioviruses), and rhabdoviruses (rabies) among others. Systemic virus diseases which cause prominent skin eruptions include perhaps the most dreaded of all virus infections, smallpox, as well as some of the mildest and most common infections—measles, chicken pox, and rubella. The smallpox virus, variola, which until recently killed many people each year in underdeveloped countries, is a typical member of the pox group. Measles virus, which produces a quick debilitating, self-terminating illness with occasional central nervous system involvement, is a paramyxovirus; rubella, which usually is a quite mild disease with a rash as the only major symptom, is a togavirus. The clinical disease called "chicken pox" results not from a pox virus at all but a herpesvirus. This is a highly infectious agent which almost invariably leads to the overt clinical disease.

ACUTE DISEASES AND SUBCLINICAL OR INAPPARENT INFECTIONS

There is a wide spectrum of severity in the course of human infections with any particular virus. For example, only about 10% of people infected with poliovirus show any involvement of the central nervous system leading to even temporary paralysis (Bodian, 1955). Mumps is another widely distributed agent, affecting over 90% of the population, as judged by immunologic evidence of past infection. However, *subclinical* or *inapparent* infection is probably frequent because only about one-half of the adults with mumps antibodies in their blood recall having had the disease. Many of the newly discovered viruses are so frequently recovered from people without symptoms that establishment of disease causation is difficult; inapparent infection is the rule. Measles virus infections, on the other hand, almost invariably produce the classic symptoms of measles in a nonimmune person. In fact, epidemics of measles infections in nonimmune populations can produce catastrophic results. Many American Indians were probably killed by measles virus brought by European settlers and the natives of the Faroe Islands were almost all afflicted with measles resulting in many deaths when the virus was accidentally introduced after many years of absence. The disease still remains an important killer in developing countries (Naficy and Nategh, 1972).

PERSISTENT VIRUS INFECTIONS

Most of the virus infections mentioned above cause whatever symptoms they are going to cause within a few days or two or three weeks at most, and the disease is *acute*, that is, of rather sudden onset and fixed, short duration. A wide and increasing variety of additional long-term interactions between viruses and animals, including humans, have been recognized. (1) *Latent* infections are characterized by intermittent frank lesions bearing virus which spontaneously disappear; the virus becomes "latent"; that is, it can no longer be isolated. (2) *Chronic* infections in which the virus is always present. Symptoms may be totally lacking, or disease may eventually result from complexes between antibody and virus or perhaps antiviral antibody which reacts with infected cells, probably at their membranes (3). *Slow* infections which are progressive, transmissible diseases with extremely long latent periods.

LATENT INFECTIONS.   The classic latent infection, with which too many of us are all too familiar, is herpesvirus infection of the lip. These chronic viral infections flare up in response to a variety of stimuli including excess sunshine (ultraviolet irradiation presumably) and fever, hence, the designation "fever blister." The infection tends not to spread to other parts of the body; in fact, circulating antibodies exist, yet the infected individual does not get rid of the virus (Roizman, 1965).

Latent herpes infections are responsible for another fairly common clinical picture. Chicken pox infections (varicella) are caused by the same virus that can be recovered many years later from the lesion of "shingles" or herpes zoster (Greek, meaning girdle), a disease whose very painful vesicles of the skin are distributed along the tract of a peripheral nerve, frequently an intercostal nerve. Apparently, the varicella virus remains dormant in the ganglion of the affected area and is activated from this "latent" state by a variety of physical or pharmacological insults. As in the case of individuals with fever blisters, patients with shingles have circulating antibodies against the offending latent virus (Hope-Simpson, 1965).

CHRONIC INFECTIONS.   Since modern methods of virus detection in cell culture became available, humans have been found to harbor many viruses for long periods of time without disease symptoms. EB virus (a herpes virus), which can cause infectious mononucleosis and may also be associated with human cancers (Burkitt's lymphoma and nasopharyngeal carcinoma; see Chapter 16), is carried in a chronic asymptomatic form by many humans (Henle and Henle, 1972). Cytomegalovirus, another herpesvirus, can be acquired congenitally and carried without incident even though individuals possess antibodies (Rifkind et al., 1967). Chronic adenovirus infections without disease also occur. When tonsils and adenoids are cultured after removal from asymptomatic children, adenovirus is often recovered, whereas it is much less frequently recovered at the time the tissue is removed. The actual state of the virus in the adenoid and tonsil tissue is unknown (Schlessinger, 1959).

The virus of serum hepatitis (hepatitis B) is one of the most clinically important of the potentially chronic infections (Lebouvier and McCollum, 1970; Dienhardt, 1976). This virus is passed mainly from a chronic asymptomatic carrier to another person through blood transfusions or injections with contaminated needles, although it can also be spread by close contact. By no means are all chronically infected people symptom-free; chronic liver failure can result from infection with this agent. The diagnosis of this disease is now made by the presence in the serum of an antigen, hepatitis B antigen, probably a surface antigen of the virus. Since this antigen can be carried for life, signifying prolonged infection, irradiation of serum products and careful selection of blood donors is most important to prevent spread of this agent.

Lymphocytic choriomenigitis (LCM) of mice is perhaps the most well-studied chronic viral infection of animals. The outcome of infection with this virus is age-dependent; infection of a fetus or newborn by inoculation or vertical infection from an infected mother to offspring results in apparent tolerance to the virus. Throughout the animal's life it produces large amounts of virus without signs of disease. However, antibodies are produced that combine with the excess viruses and therefore are not detectable in the circulation (Oldstone and Dixon, 1970). A chronic kidney disease develops, however, perhaps because of glomerular damage by excessive virus-antibody complex. If adults are infected with LCM they develop an encephalitis in about 10 days to two weeks. If they are irradiated to suppress antibody formation then the encephalitis is prevented. The encephalitis appears because of immune cells (lymphocytes) that attack virus infected brain cells which present the viral antigen on their surface. The role of cell-mediated immunity can be proved by injecting an LCM-infected mouse with isologous lymphocytes from an immune donor (an animal exposed to dead virus). These immune lymphocytes then cause in the infected mouse lesions in the liver, brain, spleen, and elsewhere (Volkert and Lundstedt, 1975).

No exact counterpart to this situation is known to exist in humans but a condition known as subacute-sclerosing panencephalitis (SSPE), which occurs secondary to measles virus infection and results in progressive degeneration especially in the cortex of the brain, is characterized by extensive lymphocytic infiltration. Prominent virus surface antigens as well as cytoplasmic nucleocapsids are present. It could be that the damage in SSPE is mediated by cellular immunity, although a chronic delayed virion infection of brain cells could also be responsible.

Two laboratory findings shed some light on latent infections and/or chronic infections. Infected cell cultures in which a small amount of neutralizing viral antibody is added can form a balanced state of chronic infection. This occurs because a few cells in the culture are infected while the majority remain uninfected and grow as so-called "carrier cultures" (Walker, 1968). Some latent infections might arise in this way. Certain other viruses that tend to produce chronic infection might conceivably integrate into host DNA. For example, herpes simplex virus has been found capable of partial integration into host DNA and may, in fact, be a human tumor virus (Henle and Henle, 1972 and see Chapter 16). The recurrent "fever blister" or herpes zoster could be a reactivation of the viral genome from an integrated state although it is more likely a release of "carrier virus" from antibody restraint.

SLOW VIRUS INFECTIONS.   This group of infections is a fascinating collection of human and animal conditions where after exposure to a virus agent the person or animal becomes ill gradually over many years. One slow virus, visna virus, has been isolated from sheep and identified as the cause of a slow paralyzing neurological disease. The virus is similar to the RNA tumor viruses in possessing a reverse transcriptase and can transform cells in culture.

Slow viruses may therefore integrate with host neurologic cells and produce virus surface antigens. Immunologic reaction to the virus surface antigens may underlie the disease (Thormar and Palsson, 1967). The most dramatic human ailment presumed to be a slow virus was discovered among the headhunters of New Guinea where ritual cannibalism is practiced. All the victims of a deadly neurological disease called *kuru* were shown to have eaten probably contaminated human tissue from several years to as long as 20 years previously. Gadjusek and colleagues (1969) have succeeded in passing and repassing (in chimpanzees) a "filterable" agent from infected human brain tissue. The nature of the virus is unknown.

A similarly uncharacterized filterable agent, scrapie, has been isolated from a disease of sheep and is thought to be mainly vertically transmitted from ewe to lamb. The disease has a latent period of years, but scrapie can be transmitted to mice where encephalopathy is evident in 4 to 6 months. Nevertheless, the disease-causing agent has not been clearly identified and characterized. These agents (scrapie and kuru) share two unusual features. No immunity to them has been detected and protein denaturing agents have little effect on the infectivity which can, of course, only be crudely assayed. The infectivity is also highly resistant to ultraviolet irradiation. It has been suggested that a small RNA intimately associated with membranes may be responsible for these strange diseases (Diener, 1972).

Diagnosis of Virus Diseases

In a modern virology laboratory the identification and classification of a disease-causing virus usually passes through the following pattern: a sample of material such as blood, urine, feces, or throat or eye washings is collected, treated with antibiotics, and either filtered or centrifuged in order to decrease the likelihood of bacterial contamination. The sample is then added to tissue culture cells or to embryonated eggs, again in the presence of antibiotics.

The eggs or cultures are examined at intervals and, if a cytopathic effect is observed, their fluids are tested to see if the effect can be propagated. If a cytopathic agent is recovered, then a classification is attempted by morphological, biochemical, and serological tests. Some information can be gained from the type of cytopathic effect or from the host range of the agent in different cultures, but cultural properties alone do not allow the identification of the virus, as they would with bacteria. Metabolic inhibitors such as actinomycin D or fluorodeoxyuridine may be useful in classifying a virus isolate as a DNA- or RNA-containing virus.

Final identification rests on tests of immunologic relatedness to known viruses. Samples of the newly isolated virus are mixed with antisera against known viruses and the mixtures are tested for cytopathic effect. Protection against such action identifies the serum that is specific for the virus in question. Exhaustive tests with antisera against many viruses are needed before a virus can be accepted as a new agent.

Even when a virus is isolated from a diseased patient, whether it is an already recognized agent or a new one, its causative role in the production of illness is not yet established. Additional information must be sought, the most useful being that provided by serological tests on the patient. Samples of a patient's serum taken at the onset of an illness (*preimmune* serum) would not be expected to contain antibodies to the causative virus, whereas samples taken later (*convalescent* serum) should contain such antibodies. The most commonly used test for antibody is complement fixation (CF), but if the isolated virus has the capacity to hemagglutinate red blood cells then the hemagglutination inhibition (HI) test can conveniently be used. In this test a hemagglutinating virus is mixed with the patient's serum, and red blood cells are then added to the mixture. If the serum has an antibody that reacts with the virus, the ability of the virus to cause hemagglutination is inhibited.

A rise in the CF, HI, or neutralizing antibody content in a patient's serum subsequent to infection is strong evidence of a causal relationship between the virus and the disease. When a particular virus has repeatedly shown this relationship in a particular disease syndrome, it becomes accepted as the cause of that disease. Many new disease-causing agents have passed this test in the last 10 to 15 years, but a large number of the newly discovered viruses have not fulfilled the criteria needed to incriminate them as agents of specific diseases.

It is likely, however, that the major classes of viruses that produce acute disease in humans have by now been identified. Long-term studies on entire families (see, for example, Moffet and Cramblett, 1962 and Jackson and Muldoon, 1975) have been carried out on the prevalence and variety of viruses and virus diseases and have continued to turn up the same groups of viruses. This situation suggests that embryonated chick eggs and the cell strains available for virus isolation—monkey kidney cells, HeLa and other tumor cell lines, and continuous cell lines from normal embryonic and adult tissues—may, by now, have detected all the viruses they can detect.

Transmission of Virus Diseases

MECHANICAL TRANSMISSION

To cause a disease, a virus must first obtain a *portal of entry* which provides access to the cells in which it can grow.

Mechanical transmission by direct contact with contaminated material or by ingestion of contaminated food or water is the common means of natural infection. Some viruses are easily brought into contact with susceptible cells; inhalation of virus in minute droplets of nasal secretions from infected individuals is the main route of transmission of upper respiratory diseases. Viruses producing conjunctivitis also have an exposed body surface for entry. Other viruses, such as measles or smallpox, may easily come in contact with susceptible upper respiratory tissues or skin, but in order to cause full-blown disease they must multiply, invade the bloodstream, and spread throughout the body. The classic study of mouse pox by Fenner (1948) furnishes an example of the time course of a generalized virus infection which documents the route of spread (Figure 15-6).

Other mechanically transmitted viruses must survive passage in the gastrointestinal tract to enter the blood and lymphatics and spread to various

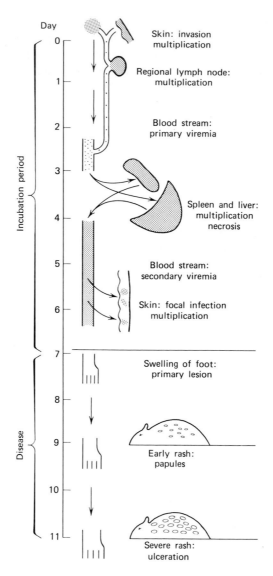

Day
0 — Skin: invasion
multiplication

1 — Regional lymph node:
multiplication

2 — Blood stream:
primary viremia

3 —

4 — Spleen and liver:
multiplication
necrosis

5 — Blood stream:
secondary viremia

6 — Skin: focal infection
multiplication

7 — Swelling of foot:
primary lesion

8 —

9 — Early rash:
papules

10 —

11 — Severe rash:
ulceration

Incubation period

Disease

FIGURE 15-6. *Scheme illustrating the possible sequence of events during the incubation and development of signs of disease in mousepox. From Fenner, (1948, Lancet 2, 915).*

organs. Poliovirus enters the body in this manner and may multiply in the throat, but more importantly survives passage through the stomach because of acid stability and multiplies in the lymphatic tissue of the intestine to finally reach and enter the central nervous system. The mechanism of such entry is still unsettled; the two possibilities are direct invasion from the blood, and invasion of the regional nerve ganglia followed by transport to the brain and spinal cord (Bodian, 1955).

The need for a secondary spread in the development of certain virus diseases explains why circulating antibody can prevent these diseases by preventing the establishment of infection remote from the point of entry.

VECTOR TRANSMISSION

Viral diseases resulting from vector-carried viruses are considerably less common in the Western world today than mechanically transmitted viral disease. However, control of vector-transmitted viruses is an important part of the exciting history of medical microbiology. After the dramatic proof by Walter Reed and collaborators (1902; 1911) of the role of mosquitos in transmitting yellow fever, this disease was virtually eradicated in North America. Pasteur's discovery of the infectious cause of rabies and the vaccination of dogs (Pasteur, 1885) resulted in eliminating dogs as vectors, at least in most advanced countries. Pasteur, incidentally, did not recognize the viral nature of rabies. These classic examples illustrate the important epidemiologic fact that eradication of a vector-spread virus is easier to accomplish than the control of virus diseases spread by mechanical means.

The most important group of vector-transmitted viruses are the arthropod-borne viruses, most of which are togaviruses, that cause various syndromes in humans, encephalitis being the most important one in the United States. These viruses also cause yellow fever, dengue (severe muscle and bone aches due to muscle infection), hemorrhagic diseases, and undifferentiated fevers. There are two major groups of these viruses, totaling over 200 members, at least 50 of which cause human diseases (Casals, 1957; Kisilius, 1960). Epidemics caused by these viruses continue to occur, and, because there are so many serologically distinct agents, vector control seems to be the only practical public health measure against them.

In most cases the arboviruses are maintained in nature by a vector-host cycle which does not include humans. Four families of blood-sucking insects (Culicidae, Ceratopogonidae, Psychodidae, and Ixodidae) are the vectors for these viruses; various vertebrates, generally wild mammals and birds, serve as the natural hosts and reservoirs. The virus is maintained only if balanced populations of host and vectors exist, and if contact between them is assured.

One of the most interesting properties of the arboviruses (togaviruses) is that they multiply both in a vertebrate host at temperatures up to 39° or in insects and insect cell cultures at lower temperatures (Hurlbut and Thomas, 1960; Pfefferkorn and Shapiro, 1975). The insect does not seem to be adversely affected by the virus and in cultured cells chronic infections can result.

TRANSPLACENTAL TRANSMISSION

In the past 30 years mild virus infections in pregnant women have been recognized to cause various serious effects on the fetus. This unusual route of transmission of viruses, which presumably requires viremia in the mother and infection of some placental cells, emphasizes the difference in receptivity to viruses by tissues at different ages. In humans rubella infection of children and adults produces a mild rash and little else, yet during the first trimester of pregnancy this virus frequently causes fetal death and congenital abnormalities (Rawls, 1968). In addition, other relatively mild viruses such as herpes simplex and cytomegalovirus (another herpesvirus) can cause microcephaly and other severe central nervous system defects in the fetus. Coxsackie viruses, group B, which are picornaviruses, have receptors in both heart and skeletal muscle in adults and can cause congenital heart lesions in fetuses. Careful avoidance by pregnant women of exposure to virus infections is obviously important, and passive immunity by gamma globulin injections is frequently recommended if exposure has occurred.

Termination of Viral Infection: Nonspecific Inflammatory Responses

The cause of termination of a virus disease is an unsolved problem. The question arises first at the cellular level. Why does the production of most animal viruses stop when only a small proportion (0.1 to 2%) of the mass of the infected cell is virus, whereas plant and bacterial cells can make virus up to 20% or more of their dry weight? In dealing with the whole animal the question becomes: why does virus multiplication and transfer to new susceptible cells not always proceed until every susceptible cell is infected?

Lwoff (1959) summarized the effects of certain variables that influence the propagation of poliovirus in a homogenous cell population *in vitro*. Improper temperature, either too high or too low, inhibits viral multiplication; the pH of the medium optimal for virus proliferation has relatively narrow limits, and with certain cultured cells the amount of oxygenation determines the final yield of virus. All these variables are known to change in areas of the body where an infection exists. The body temperature rises, the pH of the extracellular fluid in the affected area falls, and oxygenation may be somewhat impaired. The extent to which these nonspecific changes, which accompany any inflammatory reaction, affect the outcome of viral infection is uncertain, but the magnitude of the changes is potentially sufficient to play some role in controlling virus growth. For example, when mice infected with certain heat-sensitive strains of encephalomyocarditis virus are maintained with elevated body temperatures, they are spared the more deadly effect of the infection (Perol-Vauchez et al., 1961). It is now common practice in clinical medicine to allow fevers early during virus infections to remain fairly high in an attempt to retard virus growth.

IMMUNOLOGICAL RESPONSES

The most specific reaction to viral infection is, of course, the formation of antibodies. Circulating antibodies would certainly seem to be important in preventing certain viral infections; for example, long-lasting immunity with detectable serum antibody levels occurs both after vaccination and natural infection with many viruses. In addition, the barrier to the systemic spread of many viruses is probably circulating antibody (IgM and IgG type). This is suggested by the ability of injected γ-globulin to block the development of mumps and measles even if administered to an individual who is already in the early stages of infection (Deinhardt and Shranek, 1969; Krugman, 1971). Furthermore, the spread of a primary infection to a secondary site might also be stopped in natural infections by rapidly developing circulating antibodies. Poliovirus injection into rabbits raises detectable circulating neutralizing antibodies (IgM) within 24 hours provided sensitive enough assays are used (Svehag and Mandel, 1964). It is possible therefore in humans that polio does not spread more frequently from the gut or throat because of these early low levels of antibody. Likewise, rabies vaccination immediately after the bite of an infected animal is thought to protect the individual from central nervous system damage.

Before circulating antibodies are given all the credit for containing virus infections, however, several other observations must be considered. Diseases that affect surface tissues at a portal of entry such as influenza virus (and respiratory infections in general) are probably little affected by circulating antibodies. IgA antibodies, which are produced by exocrine cells (such as mucous glands, salivary glands, and breast tissue), however, are probably important in recovery from these diseases (Tomasi and Bienenstock, 1968). Passive transfer of such antibodies to breast-fed infants may also be important early in life.

In addition, recovery from a number of viral infections, may depend less on secreted antibodies than on the so-called cell-mediated immunity (Allison, 1972). Heritable human dysgammaglobulinemias exist in which circulating antibodies are grossly deficient, but these individuals recover from many virus diseases normally, although they often succumb to bacterial infections. By contrast, individuals with congenitally defective "delayed hypersensitivity" or ineffective cell-mediated immune responses do not recover from virus infections well. For example, vaccinia innoculation in such people can cause generalized extensive infection rather than the usual simple localized skin lesion (Fulginiti et al., 1968).

INTERFERON

Another factor that has been proposed as a major barrier to an unlimited course for viral diseases is interferon (Isaacs, 1963). Production of interferon in the lungs of mice infected with sublethal doses of influenza parallels the decline of disease symptoms. Also, vaccinia infection of the rabbit skin and guinea pig skin can be prevented by prior application of interferon (but not blocked by application after infection). In the chick embryo there is a concomitant increase in both capacity to survive infection and ability to produce interferon. It seems likely, therefore, that interferon plays some role in regulating virus infections in animals. Mice infected with lymphocytic choriomeningitis virus, however, are resistant to a number of other unrelated viruses, although no interferon can be detected, an indication that other humoral factors exist that play a role in controlling virus disease (Wagner, 1963). Attempts to use interferon in the treatment of human disease have foundered mainly because of lack of enough material. In addition, continual administration of proteins that are impure subjects the recipient to the risk of allergic reactions. Purified interferon can be produced to determine whether it effectively prevents human disease, but whether it could ever be administered to enough people to be practical is questionable. Therefore the finding of interferon induction by polyI:C caused a great deal of excitement in clinical virology. Clinical trials with poly I:C in humans, however, have produced only modest interferon responses. Work is now focused on finding better interferon inducers, and/or better means of delivery of inducer (Hill, 1973).

DI PARTICLES

A final means by which viral replication may be limited is built into the viral replication cycle itself. In cell culture at high multiplicity of infection defective interfering, DI, particles accumulate with many types of viruses. If these particles accumulate sufficiently in the body they might tend to make infection self-limiting.

Control of Virus Infection

One of the ultimate aims of the virologist is to gain sufficient information about the epidemiology and molecular biology of viruses and viral illnesses to combat these diseases effectively. In addition to the previously mentioned eradication of vectors, there are two major approaches to the control of virus infections. One, *vaccination,* is proven and has been widely used for many years. The second, *chemotherapy,* is relatively new and has thus far been useful in specialized cases.

IMMUNIZATION: LIVE AND KILLED VACCINES

Successful immunization against smallpox by Edward Jenner (1798) marked the first success against any infectious agent, viral or bacterial. Jenner's triumph was based on the use of cowpox, a virus immunologically related to smallpox, but sufficiently different so that no widespread disease occurred in the vaccinated patient. For many years, it was believed that for long-lasting and really effective immunity against a viral disease, Jenner's approach must be followed. For example, the original Pasteur strains of rabies were attenuated by passage in rabbit brain; the yellow fever vaccine was also an *attenuated strain.*

Another approach, the use of *killed virus vaccines,* received an important boost from the development by Salk (1953) of a trivalent poliovirus vaccine. This achievement was a triumph not only for clinical virology, but also for the techniques of animal cell culture. Preparation of an effective killed poliovirus vaccine was made possible because of the development (Enders et al., 1949) of a satisfactory means of growing large amounts of virus reproducibly. The only treatment that the infected tissue culture fluid needed was formalin inactivation of the virus.

In one of the largest experiments ever conducted using human subjects it was shown that Salk's vaccine could provide effective immunity to poliovirus, thus proving that if sufficient virus antigen is available, an effective killed vaccine can be prepared (Francis et al., 1957). The widespread use of this vaccine began in the spring of 1954. By 1957 a marked reduction in the attack

TABLE 15-2 HUMAN VACCINES CURRENTLY AVAILABLE

Live vaccines, attenuated	Killed virus vaccines
Smallpox	
Poliovirus	Poliovirus
Measles	Measles
Rubella	
Yellow fever	
Rabies*	Rabies
Adenovirus, types 3, 4, 7**	Adenovirus 3, 4, 7
Mumps	Mumps
	Influenza
	Respiratory syncytial virus

* Original Pasteur vaccine strain was passed in rabbit brain to attenuate and tissue was dried to produce vaccine probably killing majority of virus.

** Seldom used except in epidemic situations because of feared, but undemonstrated human oncogenic potential

rate of polio had occurred in the United States. Live attenuated polio vaccines were also developed (Koprowski et al., 1952; Sabin, 1957) and proved equally or more effective in reducing the incidence of poliomyelitis. Live attenuated vaccines possess substantial practical and theoretical advantages. First, they are simple to administer and enter through the normal route. Perhaps more important, they induce higher levels of circulating antibody as well as the production of IgA antibody from local exocrine cells. Finally, since the attenuated strains of polio are multistep mutants, it is a rare event for them to regain virulence (Melnick, 1971).

Both killed and attenuated vaccines against a number of virus diseases have now been prepared and are listed in Table 15-2.

DEFICIENCIES OF VACCINES

On the basis of the successes just mentioned with smallpox, yellow fever, poliomyelitis, and measles, it seems reasonable to expect that any virus disease caused by one or a very small number of viral agents can be effectively suppressed by adequate vaccination programs. Many difficulties, however, will have to be overcome before all virus diseases can effectively be controlled with either killed or live vaccine. For example, vaccines of formalinized influenza virus are effective in reducing the influenza attack rate (Davenport, 1971), but this immunity is not lasting and the vaccine is quite toxic, especially to infants. Circulating antibody declines fairly quickly after vaccination and resistance to influenza virus challenge, particularly with Type A strains, is substantially reduced within six months after vaccination (Francis, 1953).

The problems of short-lived immunity might be overcome by yearly vaccinations, but the variability of naturally occurring influenza virus strains renders even the vaccinated individual subject to attack. This antigenic variability has made its impact felt in the cyclic worldwide influenza epidemics that have occurred in 1889, 1918, 1957, 1968, and threatens again in the late

70's. The severity of such epidemics can be greatly limited by prompt production and administration of vaccine against the current particular strain of influenza before its wide occurrence is established. If the capacity of influenza viruses for antigenic variation should prove not to be limitless, but rather cover a limited range of variations, it may eventually be possible to include all major groups in a polyvalent vaccine which, if given frequently, might prevent influenza.

When general programs of vaccination against other respiratory viruses are considered, the problems are equally complex. No single virus causes any large proportion of the respiratory illnesses in humans (Jackson and Muldoon, 1975) so that effective vaccination would have to include scores of different viruses. For live vaccines, attenuated strains of each virus would have to be developed which should then be given sequentially, since interference would certainly prevent immunity to some if they were given simultaneously.

Even if a single virus type causes a clinically significant disease there are problems in developing completely successful vaccines. Measles, for example, is caused by a single virus and both killed and attenuated vaccines exist. The first attenuated vaccines still caused a fairly severe systemic disease with high fever in about 25% of children inoculated. A less virulent strain, the Edmonston strain, gained widespread use in the United States and measles incidence declined to perhaps only 10% of the former levels. Even the Edmonston strain caused fevers and measles in 10 to 15% of children, but did not cause the encephalitis which can occur with wild type measles virus. The currently used Schwarz strain produces even fewer symptoms but an adequate immunity (Krugman, 1971). A killed virus vaccine has also been produced but in addition to inducing rather poor levels of antibodies, some children have a very severe reaction, probably allergic in nature, upon becoming exposed to the wild type virus.

A most critical potential problem in the use of live virus vaccines for the control of relatively mild illnesses is the possibility of contamination with viruses capable of causing tumors.

As is discussed in Chapter 16, many viral agents that incite tumor production in animals are known. Some of these viruses are found in chicken eggs and monkey kidney cultures that are used for vaccine production. Even though no evidence exists at present that any person receiving such vaccines has had a tumor caused by the vaccine, the potential risk is still unknown and it seems unwise at present to hope that indiscriminate use of live virus preparations would not produce any tumors. This is especially true in young children who, of course, would benefit most from the administration of vaccines.

It is widely believed that if the problem of vaccination against the less common or less serious virus infections has a solution at all, the solution lies in the production of purified virus vaccines (Bachrach and Breese, 1968). Since knowledge of the nature of viruses and means for inactivation of their nucleic acid without harming the protein have advanced so much in recent years, killed virus vaccine from pooled samples of many purified viruses might be made and might have sufficient potency and reasonable safety with respect to possible contamination by tumor viruses.

CHEMOTHERAPY OF VIRUS INFECTIONS

The rapid increase in understanding of gene expression involving the DNA-dependent transcription and subsequent translation of messenger RNA (see Chapter 4) has been paralleled and aided particularly in mammalian cells by the understanding of the mode of action of inhibitory drugs such as puromycin, actinomycin, cycloheximide, and mitomycin. These drugs, however, show no preferential action on virus-controlled synthesis compared to host cell syntheses, hence they are useless as antiviral chemotherapeutic agents. Indeed, the interrelation of host cell and viral synthetic activities is so intimate that some workers have despaired of finding drugs that specifically interrupt viral synthesis and leave host cell synthesis undisturbed. Yet compounds have been found that are active against a limited number of virus infections, and work continues with the ultimate hope of finding antiviral agents of practical value (Carter, 1973).

There are three major possible points of interruption of viral multiplication (1) attachment and/or penetration before the virus successfully begins to dictate intracellular synthesis, or (2) interruption of virus nucleic acid and/or protein function and (3) interruption of virus maturation and/or exit from the cell.

Perhaps because more is known at the fundamental level about virus nucleic acid and protein synthesis, more drugs that are successes or partial successes have been reported in this area. However, at least one effective drug in each of the other categories is also known.

INHIBITORS OF VIRAL ENTRY.   If neuraminidase could be safely administered by inhalation on a wide scale just before contact with influenza virus, it would obviously protect against influenza infection by destroying the receptors for the virus. In fact, mice can be protected by such inhalation treatment (Stone, 1948). Antibodies against purified viral neuraminidase also protect mice from influenza (Schulman et al., 1968). Thus entry of influenza virus may be a particularly vulnerable step in virus replication.

Considerable interest therefore has attended the discovery of an antiviral agent, 1-adamantanamine hydrochloride, which appears to stop myxoviruses from fusing with the cell membrane (Hoffman, 1973). The compound has preventive value and possibly some therapeutic value particularly if used early in influenza infection. As more is learned about the chemical nature of virus receptors on the cell surface and how virus protein-cell surface protein interactions lead to virus entry, additional clues should be uncovered as to how virus infections can be safely stopped at this level.

INHIBITORS OF VIRAL NUCLEIC ACID OR VIRAL PROTEIN FUNCTION.   A number of agents have been described which act by being incorporated into macromolecules, thus preventing virus nucleic acid synthesis or disturbing viral nucleic acid function.

Two agents, 2(α-hydroxybenzyl)-benzimidazole (HBB) and guanidine, block the development of some of the enteroviruses (polio, coxsackie, and echo viruses, Caliguiri and Tamm, 1973). Cells treated with either of these compounds adsorb virus, but do not make viral RNA, viral RNA polymerase, or viral capsid protein in detectable amounts. If the drugs are added after virus synthesis is underway, virus proteins continue to be made but single-stranded viral RNA output is blocked. However, guanidine does not block the synthesis of the 3' terminal poly(A), the last portion of the viral molecule to be replicated (Spector and Baltimore, 1975) and it has no effect on chain elongation of nascent chains *in vitro*. It therefore remains a mystery how this drug acts.

Although the spectrum of activity of HBB and guanidine is generally similar, there are some significant differences. Table 15-3 shows that HBB fails to inhibit coxsackie A and certain rhinoviruses that are inhibited by guanidine. Also mutant virus strains resistant to one drug are only partially resistant to the other. These drugs were earlier thought to be impractical as therapeutic agents, however, because of the high rate of virus mutations to resistance. Recent work has rekindled interest in the possible effect of using the two drugs in combination. Because of an apparent synergistic action, mice infected with Echo 9 and Coxsackie type A9 viruses were protected from paralysis and death with HBB plus guanidine (Eggers, 1976).

Another class of compounds, the thiosemicarbazones, especially isatin β-thiosemicarbazone or IBT, has a protective effect against poxvirus infections in mice (Thompson et al., 1953). Electron micrographs of vaccinia-infected cells treated with IBT show what appear to be immature virus forms containing both viral protein and DNA, but they apparently are not infectious (Easterbrook, 1962). The early events in vaccinia virus synthesis seem to proceed normally, but vaccinia mRNA is not properly utilized and synthesis of whole infectious virus is depressed (Woodson and Joklik, 1965).

TABLE 15-3　INHIBITION OF PICORNAVIRUSES BY GUANIDINE AND 2-(a-HYDROXYBENZYL)-BENZIMIDAZOLE (HBB)

Virus	Inhibition by	
	Guanidine*	HBB†
Polio 1, 2, 3	+	+
Coxsackie A9	+	+
Coxsackie A7, A11, A13, A16, A18	+	−
Coxsackie B1, B3, B5	+	+
Echo 1, 5, 6, 9, 12	+	+
Echo 22, 23	−	−
Rhino 1B	−	−
Rhino 2	+	−
Bovine enteroviruses, G-UP, 51/60, 100/60, 328/60	+	+
Foot-and-mouth disease‡	−	−

* The concentration of guanidine ranged from 0.47 mM to 6 mM, with 1 mM as the common concentration.
† The concentration of HBB ranged from 0.098 mM to 0.49 mM, with 0.22 mM as the common concentration.
‡ Wild type strains are inhibited by guanidine in Earle's salt solution + serum, but not in the commonly used Eagles's minimum essential medium, whereas attentuated strains are inhibited in both media. (From Calaguiri and Tamm, 1974).

Prevention of smallpox has been clearly demonstrated with N-methylisatin-β-thiosemicarbazone. In Madras, India, a large group of unvaccinated people were exposed to smallpox. Only 3 out of 1100 people treated with the thiosemicarbazone derivative got smallpox, whereas 78 out of about 1100 untreated people contracted the disease (Bauer et al., 1969). This was an important breakthrough in viral chemotherapy because until several years ago smallpox still killed large numbers of people in areas of the world where vaccination is not regularly practiced. The disease has, however, been virtually wiped out at the present time.

Another instance of successful viral chemotherapy is the use of bromo- and iododeoxyuridine, BUDR and IUDR, against herpes infections of the eye (Kaufman, 1962). This fortunate result was somewhat surprising because fluorodeoxyuridine (FUDR) has a marked effect on both virus synthesis and cell growth in tissue culture, but no selective action on the virus. The reason for the difference is that FUDR acts through an inhibition of DNA synthesis by blocking the synthesis of thymidylic acid, whereas BUDR and IUDR do not stop DNA synthesis, but are incorporated into DNA in the place of thymidine, producing defective DNA molecules. Cell growth can continue in the presence of low concentrations of these drugs for a number of generations, whereas synthesis of infectious virus, either vaccinia or herpes, is quickly inhibited and noninfective virus particles are produced. This slight selectivity is apparently sufficient to make BUDR and especially IUDR effective in treating herpes keratitis (corneal infection), as well as vaccinial keratitis. The drug is probably effective only in the eye because here the infected cells are readily accessible and cellular multiplication with attendant incorporation of IUDR into cellular DNA is so slow that no damage to the host cells is produced. Generalized herpes infections cannot be treated with IUDR because of toxicity. Strains of herpesvirus resistant to IUDR arise and recently improved results in herpes keratitis have been reported with another halogenated pyrimidine, trifluorothymidine (F3T). The basis of action of F3T is presumably the same—incorporation into viral DNA rendering the DNA inactive (Sugar and Kaufman, 1973).

Specific inhibitors of viral RNA polymerases would seem to be a possible target for successful chemotherapy, especially those virions that carry their own polymerases. Several rifampicin derivatives are at least partially effective against the *in vitro* action of reverse transcriptase of RNA tumor viruses and this is presumably the basis for prevention of focus-formation by these viruses (Moss, 1973).

One additional vulnerable site for viral replication might seem to be the double-stranded RNA molecules that exist during picornavirus, togavirus, and reovirus replication. However, it may be that inside the cell during picornavirus infection, relatively little of these potential double-stranded structures are in fact in a double-stranded form (Oberg and Philipson, 1971). However, given a drug that specifically bound double-stranded RNA, it would certainly be important to test its usefulness against the single-stranded as well as double-stranded RNA viruses.

INHIBITION OF VIRUS ASSEMBLY. Selective chemotherapy by blocking cleavage of viral precursor protein molecules to effective capsid proteins is

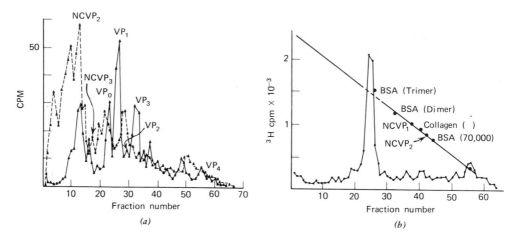

FIGURE 15-7. *Block of poliovirus protein cleavage by protease inhibitor. HeLa cells infected with poliovirus produce poliovirus-specific polypeptides labeled NCVP— and VP- (detected by electrophoresis of amino acid labeled polypeptides). This pattern of labeled proteins does not develop if cells are treated with tosyl-lysyl-chloromethyl ketone (TLCK; ---▲---). The largest polypeptide was recovered (from slice 4) after electrophoresis and compared in size to various other proteins in part (b). The size is approximately 200,000 daltons, larger than the trimer of bovine serum albumin. Redrawn from Korant, (1972, Virology* **10,** *751).*

another logical avenue of attack. It is clear that a number of agents that block proteolytic enzymes (TPCK and TMCK, chlorinated ketones, iodacetamide, high concentrations of Zn^{++}) can halt poliovirus and rhinovirus infections in cell culture by blocking viral protein cleavages (Korant, 1972; Korant et al., 1974; see Figure 15-7). These compounds are, however, too toxic for use in infections. However, recent evidence indicates that in poliovirus infection a new virus-specific proteolytic activity is developed within cells, increasing the interest in finding proteolytic blocking agents that would affect virus activity more than cellular activity (Korant, pers. comm.).

In vaccinia infection, inhibition of virus maturation by rifampicin, which in *E. coli* inhibits RNA polymerase, may be due to blocking cleavage of a larger protein (Moss, 1973) thus preventing virus maturation. Selective agents that block virus protein cleavage and virion maturation therefore offer some promise.

Conclusion

The understanding and eventual control of human disease will continue to be intimately connected with the study of virology and immune reactions to virological agents. The whole area of tumor virology has undergone an explosive surge of interest since 1970, not because of greater likelihood of human tumors being caused by viruses now than 30 to 40 years ago, but because basic studies have made possible definitive probes to search for human tumor viruses. The role of infectious agents, including viruses in a whole array of chronic human diseases, has been suggested for many years. But it is the new findings on the biology and biochemistry of slow viruses and on the possible autoimmune nature of some virologically induced central nervous system diseases that make it imperative now to reexamine virtually all slowly progressive diseases for a possible originating infectious cause. The more variety is discovered in molecular mechanisms of virus growth, the more variable are the possible ways in which virus disease might be caused.

Tumor
16 Viruses

Since the original demonstration of sarcoma induction by a virus of chickens (Rous, 1911), two distinct classes of cancer-causing viruses have been identified in a range of vertebrate species, DNA viruses and retroviruses. The oncogenic (tumor-inducing) DNA viruses include papovaviruses, adenoviruses, and herpesviruses. Retroviruses are the only RNA viruses able to cause cancer.

The kinds of tumors caused by oncogenic viruses vary enormously. Polyoma virus causes mainly salivary gland tumors but, as its name suggests, it also causes a wide range of other malignancies. The retroviruses cause mainly leukemias and sarcomas but can cause mammary as well as other tumors.

In Chapter 14, the potentially oncogenic viruses were discussed only from the point of view of how they multiply. In this chapter, their ability to stably associate with and change the growth properties of cells is emphasized. Although cancer is a disease definable only in whole animals, an analogue of malignancy called *cellular transformation* provides the *in vitro* model on which almost all work with oncogenic viruses is based (Macpherson, 1970; Ponten, 1971). Transformation provides the basis for quantitating many tumor viruses (Temin and Rubin, 1958) as well as the material for comparing the physiology of normal and tumor cells.

Details about the oncogenic effects of viruses on animals can be found in Gross (1970) while *The Molecular Biology of Tumor Viruses* (Tooze, 1973) reviews attempts to understand how the viruses affect cell growth.

There are two central questions about tumor viruses: What differentiates transformed cells from normal cells? How do viruses maintain cells in the transformed state?

What Is a Transformed Cell?

The operation that generates a transformed cell population is infection of normal cells with an oncogenic virus, such as Rous sarcoma or polyoma virus, followed by isolation of colonies of cells with altered properties. The types of alterations that are found include changed cellular morphology (for example, rounding of cells), piling up of cells that normally appear to grow as a monolayer, and growth of cells in semisolid medium under conditions where normal cells will not grow. Other criteria exist and, in general, selection of cells by one criterion yields cells that are transformed by most criteria (but see Risser and Pollack, 1974). Most DNA tumor viruses and the sarcoma-inducing retroviruses will cause cellular transformation. Leukemia-causing retroviruses usually grow in cells without causing transformation.

The normal cells used for transformation studies are of two types: cells taken from embryos, often of chickens or rodents, and permanent lines of mammalian cells that can be cloned. Use of cloned permanent cell lines assures that the transformed cells differ from normal cells only by transformation-induced characteristics. The deficiency of such cell lines is that they are not really normal. For instance, cultures initiated from an animal have a finite life span, while cell lines are immortal. Many workers therefore prefer to use primary cells. The normal cells used for transformation assays grow adherent to a solid substratum (glass or plastic) and are generally called "fibroblasts" although they may represent a variety of cell types.

Once a transformed cell culture has been established by any one criterion, a wide range of parameters can be used to differentiate it from a normal cell. Table 16-1 describes the variety of transformation-induced changes already recognized in transformed cells. A few of these characteristics are described in more detail below. Two broad categories of changes are known: (1) alterations in growth control and longevity; (2) alterations in properties of the surface (plasma membrane).

ALTERATIONS IN GROWTH PROPERTIES

IMMORTALIZATION.   Primary or secondary cell cultures have a finite life expectancy. Human cell cultures, for instance, die about 50 cell generations after they are established (Hayflick and Moorhead, 1961). Chicken cell cultures have a much shorter life span. Mouse cultures start to become moribund not long after they are prepared from mouse embryos but a few cells usually survive to become cell lines (Todaro and Green, 1963). By contrast, most transformed cells are immortal, they will grow indefinitely. For instance, colonies of rounded cells can be induced by mouse sarcoma viruses in cultures of primary cells from many species. Such cells will grow thereafter for as long as experimenters are willing to keep dividing and refeeding the cultures to prevent the cells from killing each other by crowding and exhausting the medium.

EB virus, the herpesvirus that is associated with Burkitt's lymphoma, can be assayed only by an immortalization response. This virus specifically infects primate or human B-lymphocytes which respond by starting to grow continuously as nonadherent cells in suspension culture (Epstein, 1970).

TABLE 16-1 CHANGED PROPERTIES IN TRANSFORMED CELLS

Morphological and behavioral changes
Become more rounded, have looser attachment to substratum
Mutual orientation more random, lose contact inhibition of movement
Grow on top of each other
Grow in suspension, lose anchorage dependence
Grow to high or indefinite saturation density, kill themselves rather than stop growth
Have decreased serum requirement
Become invasive

Surface alterations
Hyaluronic acid increased
Protein-linked sialic acid decreased
Ganglioside content of lipids decreased
250,000 dalton surface protein disappears
Sugar transport increased
More easily agglutinated by plant lectins
Surface proteins more mobile
Lipid fluidity not changed
Microfilaments (actin) cables disappear but diffuse actin remains
Myosin-like filaments disappear
Microtubules disaggregate
Fetal antigens become evident
Virus-specific transplantation-rejection antigens appear

Nonsurface biochemical changes
Release of proteases
Transcription of fetal genes

INCREASED SATURATION DENSITY. Most normal or transformed cells grow adherent to a substratum of glass or plastic. Normal cells will stop growing before they have exhausted the nutrients in the medium and will then remain attached to the culture dish as quiescent, viable cells. If they are removed from the dish and the culture is reestablished at lower cell density, they will reinitiate growth. At the time they stop growth, the cells appear to have formed a *monolayer* on the surface of the culture dish because their nuclei, the most prominent part of the cell, do not overlap. Actually, the cell cytoplasms often overlap extensively in such limit cultures but nevertheless the term monolayer is used to describe them (Figure 16-1).

In sharp contrast, most transformed cells will grow continuously until they kill themselves—they are unable to go into a quiescent state (Pardee, 1974). This is possibly the single most characteristic property of transformed cells. The continuously growing cells also do not respect each other's borders; they grow chaotically over and under each other forming disorganized, multilayered masses of cells (Figure 16-2). Transformed cells truly look malignant in comparison to their orderly, normal counterparts.

Because of the disordered social organization of transformed cells compared to normal cells, it appears that normal cells can sense each other when

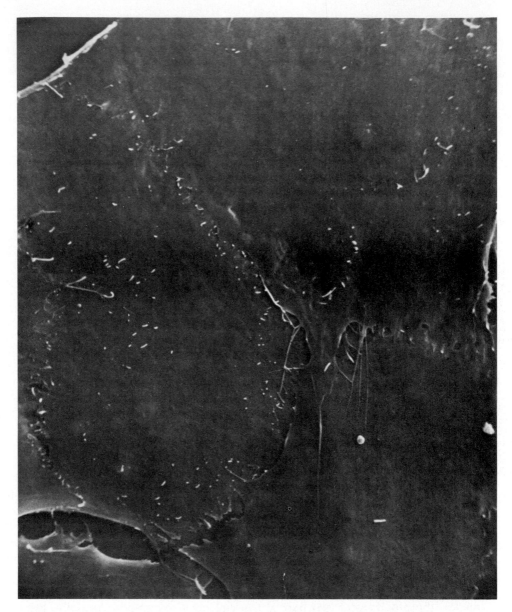

FIGURE 16-1. Scanning electron micrograph of a culture of BALB/c 3T3 cells, clone A31 at confluence. The overlap of the cytoplasms is evident as is the flat, unruffled surface of the cells. Kindly provided by Dr. Keith Porter.

they come into contact and stop growing but that transformed cells lack such sensing mechanisms and so grow on top of each other. Using time-lapse microcinematography, it can be shown that when, in a sparse culture, two normal cells come into contact, one or both will stop moving and then begin moving again in another direction (Abercombie and Heaysman, 1974; Abercombie, 1962). This well-defined phenomenon is known as *contact inhibition of movement* and is lacking in transformed cells. It leads to an orderly pattern of cell growth in normal cultures and is responsible for the apparent monolayer in a limit culture. However, the extrapolation made by numerous investigators that contact inhibition can prevent cell multiplication as well as movement has received less support. It is still not completely clear what causes normal cells to stop growing and become quiescent in a crowded culture dish, but availability of limiting factors in the medium is probably a more crucial element of the response than is contact (Holley and Kiernan, 1968). The responsible factors are thought to be ones found in the serum used to support cell growth; transformed cells require much less serum for optimal growth than do normal cells. The nature of the limiting factors, though, is still obscure.

ANCHORAGE DEPENDENCE: GROWTH IN SEMISOLID MEDIUM. A practically very useful phenomenon is that in order to grow, most normal cells must be attached to a rigid substratum. Transformed cells will grow without such attachment and so will form colonies when suspended in a gelled medium such as one containing agar or methyl cellulose. Exposure of a normal cell population to a transforming virus followed by plating in an agar-based medium allows direct selection of transformed cell colonies (Macpherson and Montagnier, 1964).

The issue often arises whether transformed cells are truly malignant. Of all the characteristics mentioned here, loss of anchorage dependence correlates best with the ability to form a malignant tumor *in vivo* (Shin et al., 1975). For instance, SV40 can transform rat embryo cells to be immortal and to require less serum but only if they are transformed by the criterion of growth in agar will they be tumorigenic. For such tumor-forming assays, *nude* (athymic) mice have been used because no immunologic rejection of tumors occurs in such animals, thus providing a pure test of "tumorigenicity."

The debate about whether cells transformed by one or more criteria are truly tumor cells is ultimately a futile one. The difficulties are best shown by the observation that if a standard monolayer, transformable mouse cell line, called BALB/3T3, is injected into a syngeneic mouse it will not form a tumor, but if the cells are grown on glass beads and these are injected, the cells will initiate tumor formation (Boone, 1975). The anchorage dependence of the cells is satisfied by the glass, allowing them to express their malignant potential. It becomes semantic whether such cells are called normal or malignant but even if one wishes to consider them malignant, their utility for transformation assays is unquestionable because virus infection changes a constellation of cellular characteristics in a reproducible and useful way.

ALTERATIONS IN SURFACE PROPERTIES

The changes in cellular morphology and growth properties produced by transforming viruses are paralleled by profound alterations in biochemical and biophysical aspects of cellular structure and function. Most studies to date have concentrated on the plasma membrane because of the possibility that alterations in its structure and properties might be the cause of transformation, but there are equally important changes in other parts of transformed cells.

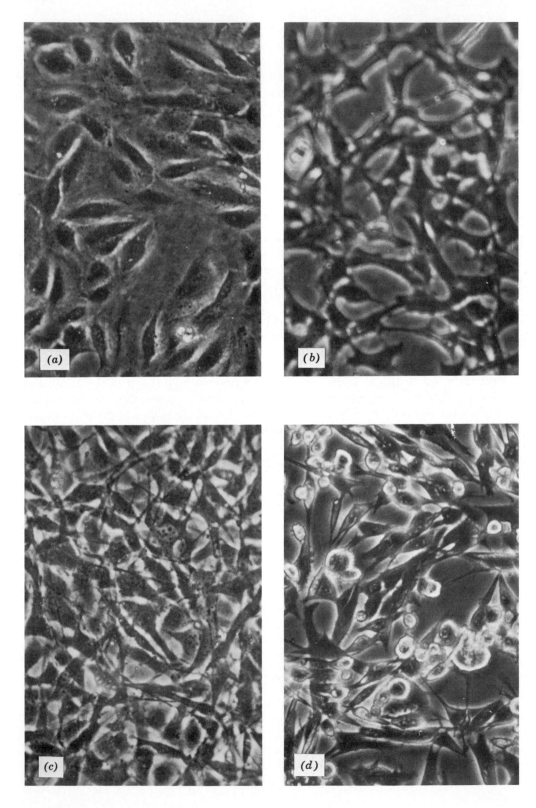

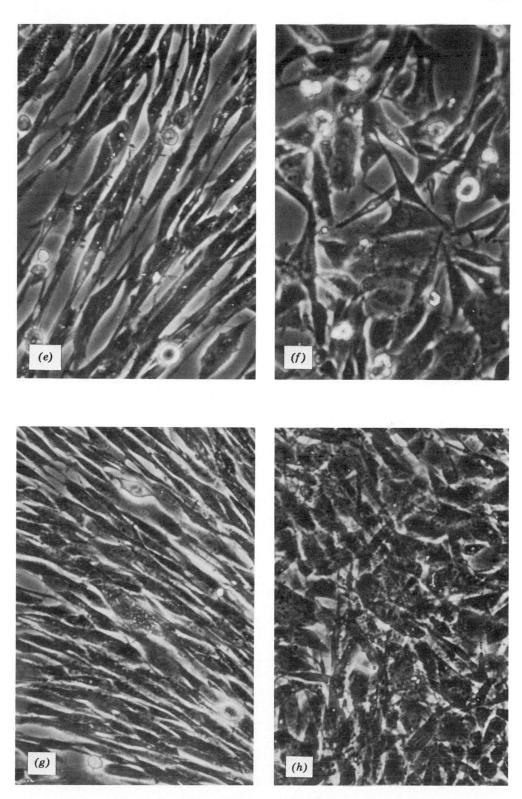

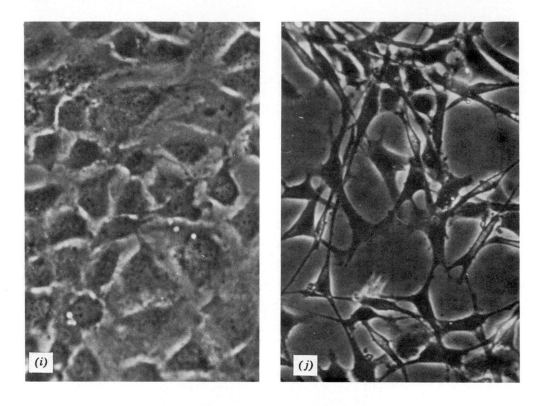

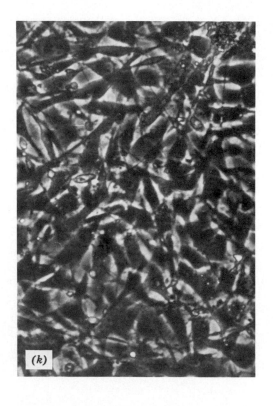

Some of the first changes observed in transformed cells were functional alterations. One was an increased rate of sugar transport after transformation (Hatanaka et al., 1971). Another was an increased ease of agglutination of cells by plant lectins such as concanavalin A (Burger and Goldberg, 1967). Such lectins are natural substances that exhibit multivalent binding to specific carbohydrates. They agglutinate cells by cross-linking the glycoproteins on the cell surface. Almost all tumor cells are more easily agglutinated by lectins than are their normal counterparts (Aub et al., 1965).

There are numerous structural differences between the plasma membranes of normal and transformed cells. Both hyaluronic acid and protein-linked sialic acid are increased in the transformed cell membrane while the ganglioside content of the membrane is decreased. A major 250,000 dalton protein of the normal cell surface, called LETS protein, is absent from the surface of most transformed cells, as are some minor proteins (Hynes, 1974). The surface of tumor cells also shows more ruffling than that of normal cells (Figure 16-3).

Another effect of transformation is to increase the mobility in the plane of the membrane of the proteins that bind concanavalin A. Cross-linking of surface proteins by this plant lectin causes the proteins to coalesce into patches. Transformation greatly increases the amount of patching by a given concentration of concanavalin A, indicating that the proteins can more easily move on the surface of transformed cells than on normal cells (Ash et al., 1976). It has been suggested that the mobility of surface proteins is modulated by their transmembrane attachment to elements of the cell's cytoskeleton (the microtubules and microfilaments along with myosin-like proteins) (Nicholson, 1976). Consistent with this suggestion, in transformed cells the cytoskeleton is depolymerized (Pollack et al., 1975; Edelman and Yahara, 1976; Ash et al., 1976; Figure 16-4). The increased mobility of proteins in transformed cells is not due to a changed fluidity of the lipids in transformed cell membranes (Robbins et al., 1974) and may well be a consequence of changes in the cytoskeleton. Although an altered cytoskeleton may be an important cause of differences between normal and transformed cells, there is no evidence that it is directly affected by viral proteins.

The increased mobility of surface proteins in transformed cells probably explains the ability of lectins to agglutinate tumor cells more easily than normal cells. The patches of aggregated lectin receptors on the transformed cell surface provide foci for tight cross-linking of cells; on normal cells, where lectins aggregate receptors with less efficiency, strong intercellular binding occurs much less readily.

←───

FIGURE 16-2. Growth patterns and morphological changes in normal and transformed established cell lines. (a) BALB-3T3 cells, nearly confluent; (b) BALB-3T3 transformed by polyoma virus at cell density similar to BALB-3T3 in (a); (c) same as (b), 24 hours later; (d) BALB-3T3 transformed by mouse sarcoma virus; (e) BHK cells nearly confluent; (f) BHK cells transformed by polyoma virus; (g) RECL$_3$; (h) RECL$_3$ transformed by polyoma virus; (i) 3T3 (Swiss), confluent; (j) 3T3 transformed by polyoma virus, low density; (k) same as (j), high density. Phase contrast, ×75.

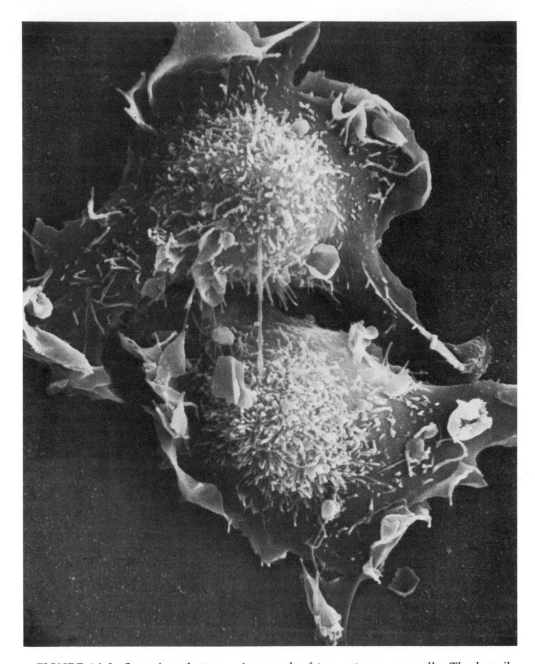

FIGURE 16-3. Scanning electron micrograph of two rat sarcoma cells. The heavily ruffled surface is evident. Kindly provided by Dr. Keith Porter.

Many of the characteristics of transformed cells can be mimicked by treating quiescent normal cells with proteases. Furthermore, many transformed cells release proteases whereas normal cells do not (Unkeless et al., 1973). Although the correlation of protease release and transformation is not perfect, secretion of proteases in some systems correlates very well with loss of

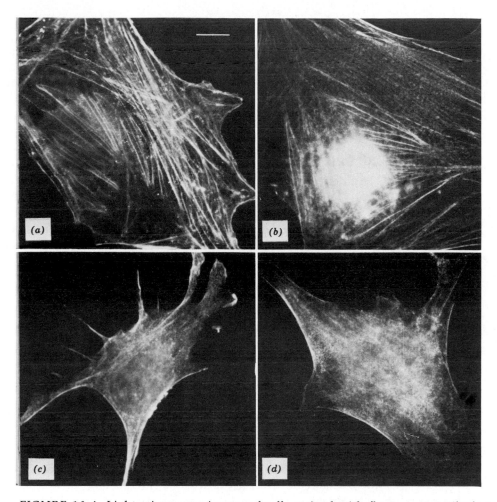

FIGURE 16-4. *Light microscope images of cells stained with fluorescent antibody to actin and myosin. Cells were illuminated with ultraviolet light and the emitted fluorescence was photographed. The obvious cables of actin and myosin are evident in the normal cell (a and b) but have largely disappeared in the transformed cells (c and d). In place of the cables, diffuse actin and myosin can be seen in the transformed cells (c and d). Biochemical studies have shown no decrease in the amount of actin when cells are transformed so transformation causes only a disaggregation of cables. (a). Normal 3T3 cells stained with antiactin antibody. (b). Normal 3T3 cells stained with antimyosin antibody. (c). 3T3 cells transformed by SV40 virus and stained with antiactin antibody. (d). 3T3 cells transformed by SV40 virus and stained with antimyosin antibody. Reprinted from Pollack et al. (1975) with permission.*

anchorage dependence (Pollack et al., 1974). The possibility of viral proteins causing release of extracellular protease as a mechanism for transformation is intriguing but no direct evidence exists that links the events.

One of the most important attributes of tumor cells is their invasiveness. Transformed cells are also invasive, as demonstrated by their ability to penetrate the chorioallantoic membrane of chicken eggs (Scher et al., 1976). Neither primary cells nor established cell lines are invasive when cultured in this fashion. Invasiveness could be a result of virus-induced alterations in the plasma membrane and/or protease release by cells.

Not all the known differences between normal and transformed cells involve the plasma membrane. Rous sarcoma virus infection of chicken embryo cells induces transcription of fetal hemoglobin genes showing that transformation can directly affect gene functions (Groudine and Weintraub, 1975). Possibly the best suggestion for how transformation occurs is that viral gene products could act as general derepressors of extensive regions of the cellular genome. The derepressed proteins would have the ultimate result of maintaining the cell growing under conditions where normal cells cease division.

Clearly, the mechanism by which viruses transform cells has not been determined by studying the differences between transformed and normal cells. Although a multitude of differences has been found, cause and effect relationships have been elusive.

What Induces and Maintains Transformation?

The second avenue of investigation of virus-induced cellular transformation has focused on the linked questions of whether the virus actively maintains transformation, or works by a hit-and-run mechanism, and what viral proteins are responsible for transformation.

Before considering these questions a crucial difference between RNA and DNA oncogenic viruses must be appreciated. Retroviruses, because of their mechanism of multiplication, can productively infect cells without killing them, allowing retroviruses to transform the cells in which they grow. DNA viruses are generally unable to both grow in and transform the same cells because productively infected cells are killed in the process of making new virus. However, situations have been found in the laboratory which allow papovaviruses, adenoviruses, and herpesviruses to transform cells.

The growth cycles of DNA tumor viruses all have early and late phases. Early gene expression comes from the few input genomes that initiate the infection and neither the amounts of virus-specific synthesis nor the functions of the early gene products will cause death. Once late genes begin to function, and DNA replicates, the virus so completely dominates cellular function that the cell can no longer live. Hence DNA viruses must transform cells under conditions where the switch to late functions is either blocked or markedly suppressed.

Three different conditions have been found that allow DNA viruses to transform cells. One is infection of nonpermissive cells, cells that for unknown reasons do not provide the conditions for the virus to enter the late phase of the infection cycle. These cells are the most common targets in transformation assays. In an assay of polyoma or SV40 on such cells, many of them first become transformed and then rapidly revert to normal; they are said to be "abortively transformed" (Stoker, 1968). The small number of permanently transformed cells then form visible colonies of altered cells. Of course, transformation of nonpermissive cells is a dead end for the virus.

The second instance of cells transformable by DNA tumor viruses is semipermissive cells, ones that make so little virus they are not killed. Human cells, for instance, are semipermissive for SV40.

The third type of interaction leading to transformation is the infection of a permissive cell by a suitably defective virus. For SV40, only cells from primates are permissive; for polyoma, mouse cells are the permissive ones. If polyoma virus is mutated so as to be unable to initiate the late phase of its growth cycle, it can transform mouse cells. In similar ways, other DNA viruses can be made capable of transforming permissive cells. Because few cell types are nonpermissive for herpesviruses, most transformation studies with it have required the use of virus heavily irradiated with ultraviolet light.

In practice, many transformed cells are derived after massive infection of semipermissive cells. The resulting transformed cells, which occur at very low frequency, are found to carry fragments of viral DNA but no complete genomes. Apparently the transformation results from both a deleted viral genome and a natural partial block to a permissive interaction.

One DNA tumor virus that is an exception to these generalities is EB virus. It infects only one known cell type, B-lymphocytes of primates, and establishes an apparently productive infection while simultaneously transforming the cells. Only rare cells in the culture produce virus whereas most cells express the very early viral antigens. Those cells that begin to express later stage antigens, and those that produce virus, are killed. So the virus establishes both a productive and a nonproductive interaction with the same cell population. It is formally a situation similar to that of a temperate bacteriophage which expresses only very early repressing functions in most cells but in occasional cells slips into a productive mode, kills the cell, and releases a burst of virus.

Although retroviruses can both transform and be produced by the same cells, in many cases the transforming retroviruses are defective. Such viruses give rise to nonproducer, transformed cells from which transforming virus can be rescued by superinfection with a helper, nondefective retrovirus.

Viral Information in Transformed Cells

All cells transformed by viruses contain viral genetic material. Except for EB virus, which may maintain itself as a plasmid in transformed lymphocytes (Lindahl et al., 1976), the viral DNA is covalently integrated into host cell DNA. For retroviruses, integration of the provirus is a natural part of the viral life cycle. For DNA tumor viruses there is no evidence that integration plays an obligate role in their lytic cycle, although joined host and viral sequences have been found in cells productively infected by DNA viruses.

The most persuasive evidence that viral DNA is linearly integrated into cellular DNA was obtained by Botchan et al., (1976) who fragmented the DNA of SV40-transformed cells with restriction endonucleases, found one or more whole integrated genomes, and deduced a linear fragment sequence for them.

In each cell line they examined, the circular SV40 gene had been broken at a different point and attached to different regions of cellular DNA. Thus, SV40 DNA integration is not confined to a single site either on the viral DNA or the cellular DNA. It seems most probable from these data that integration is a nonspecific occurrence very different from the precisely positioned integration of bacteriophage lambda.

In most cases, only fragments of viral DNA can be found in transformed cells. For papovaviruses, transformed cells have the early region of viral DNA but can lack some or all of the late region, indicating that the late region is not involved in transformation. For adenoviruses, a small fraction of the left end of the genome is all that is present in many transformed cells. Again, this is a region expressed early in the lytic cycle (Flint et al., 1976).

Cells transformed by DNA viruses contain virus-specific RNA (Benjamin, 1966), and viral antigens can be detected in most transformed cells using fluorescein-conjugated sera from tumor-bearing animals. T-antigen is universally present in cells transformed by papovaviruses—this antigen was already mentioned in Chapter 14 as a protein made early in the lytic cycle. Adenoviruses also specify a T-antigen in transformed cells and all EB virus-transformed cells have an antigen called EBNA (EB nuclear antigen). Transformed cells therefore contain viral DNA, RNA, and proteins, suggesting that virus-encoded products could be responsible for the transformed state.

For retroviruses, the question of viral gene expression in transformed cells is complicated by the fact that many transformed cells are productively infected and therefore must make the RNA and proteins needed for virus maturation. In fact, all retrovirus-transformed cells, whether productive or not, do make viral RNA, and genetic evidence, to be presented below, strongly argues that a viral protein is crucial to the maintenance of the transformed state.

Transforming Protein for Fibroblasts

Studies of Rous sarcoma virus have provided the best evidence for virus-specific proteins that maintain the transformed state (Hanafusa, 1976). Martin (1970) and others have isolated mutants of Rous sarcoma virus that are temperature-sensitive for transformation. The mutant viruses grow equally well at both permissive and nonpermissive temperature but they transform only at permissive temperature. When cells transformed at the permissive temperature are shifted to nonpermissive temperature they revert to normal by virtually every one of the criteria shown in Table 16-1. This process is completely reversible, allowing cells to be shifted back and forth between transformed and normal states at will. For certain of these mutants, merely treating transformed cells at the permissive temperature with an inhibitor of protein synthesis causes them to revert to normal (Ash et al., 1976).

Both the thermolability and the lability to inhibition of protein synthesis argues very strongly that Rous sarcoma virus makes a protein which acts to maintain cells in the transformed state. Genetic studies indicate that only one virus-specified protein is involved and deletion studies suggest that it may have a size of less than 50,000 daltons (Vogt, 1976). The gene for this protein has been designated *src* to denote its ability to direct synthesis of a sarcomagenic protein. It seems remarkable, yet difficult to escape, that a single, small virus-specified protein is able to cause the widely pleiotropic effects called transformation.

How DNA tumor viruses maintain the transformed state is more obscure because the study of mutants has given ambiguous results. For SV40, the *ts*A groups of mutants—ones that map in the early region of the genome and block initiation of viral DNA synthesis during the lytic cycle—generate partially temperature-sensitive transformed cells (Brugge and Butel, 1975). The extent of return to untransformed growth properties under nonpermissive conditions is not as dramatic as with the *src^{ts}* mutants of retroviruses so that the A gene can only tentatively be assigned a transforming function. Cells transformed by polyoma *ts*A mutants do not, in fact, revert to normal under nonpermissive conditions (Fried, 1965).

For polyoma virus, the *hr-t* mutants (host-range transforming), whose mutations map apart from the *ts*A mutations but in the early region, are absolutely transformation negative and are complemented for transformation by *ts*A mutants. No such mutants have been isolated from SV40 but it seems that two separate transforming functions may exist in papovaviruses, which might even be part of the same protein. It could be that the two adenovirus proteins found in transformed cells play the dual roles of a single papovavirus protein.

In summary, we now have a reasonably clear picture of certain necessary events for viruses to transform cells although we completely lack any mechanistic details. The virus integrates its DNA into the cellular DNA and from the integrated genome virus-specific mRNA is formed. This encodes one or two proteins that act to abrogate normal growth control mechanisms and to cause major changes in plasma membrane structure and function. Whether the transforming proteins directly affect membrane-related processes, or cause alterations in transcription or in other central genetic processes, is a major question for the future.

Other Transforming Proteins

Thus far only transformation of adherent, fibroblastic cells has been discussed and the genes identified have affected such cells. But oncogenic viruses also cause cancers of nonadherent cells: the majority of cancers caused by retroviruses, for instance, are leukemias, malignancies of white blood cells. Such cells normally exist as free cells in bone marrow, lymphoid organs, or blood. Retroviruses that lack the *src* gene can cause leukemia, even though they do not transform fibroblasts. Such viruses may have *leuk* genes, but no *leuk* genes have been defined by mutations. Since it is possible to transform fetal liver, spleen, and bone cells *in vitro* with certain leukemia viruses, it may be possible to isolate such mutants in the future (Rosenberg and Baltimore, 1976; Figure 16-5).

Transformation of nonadherent cells can be accomplished by two kinds of

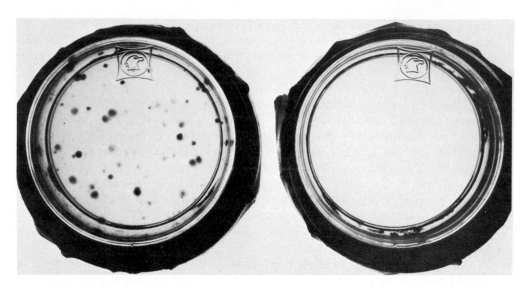

FIGURE 16-5. Transformed colonies of mouse bone marrow cells after exposure to Abelson mouse leukemia virus. A million bone marrow cells either infected (plate on the left) or uninfected (plate on the right) were suspended in semisolid agarose and incubated for 15 days. The development of visible colonies in the infected preparation is evident. No colonies appear in the control, uninfected preparation. Each of the colonies can be used to initiate an immortal culture of transformed lymphoid cells. Reprinted from Rosenberg and Baltimore (1976) by permission.

viruses, retroviruses that cause leukemia and EB virus. Basically only one characteristic distinguishes normal nonadherent cells from their transformed counterparts—immortality. Nonadherent cells usually grow poorly in culture and only for a short time, transformants grow well and indefinitely.

Role of Transformation in Lytic Infection

The ability of a virus to cause a disease is generally considered a by-product of the mechanisms by which the virus multiplies itself. There is no obvious reason why a virus should "want" to cause the symptoms of a specific disease. One rationale for the ability of oncogenic viruses to cause cancer is that the virus multiplies best in growing cells, hence it may induce cells to grow. For the papovaviruses and adenoviruses this explanation has force: they induce DNA synthesis in resting cells as part of their lytic cycle and the genes responsible for induction of DNA synthesis may be the ones responsible for transformation. For these viruses, oncogenicity could be a direct result of their mechanism of replication. Because they only transform under conditions where they do not grow, their ability to generate continuously growing transformed cells can play no role in their multiplication. Furthermore, the variability of the integration sites in both cell and virus suggests that integration could be an accident rather than a process for which the virus has evolved its own mechanisms. In fact, there is no known natural situation in which papovaviruses cause malignant disease although they do cause benign tumors such as human warts and rabbit papillomas.

For the oncogenic retroviruses, the *src* function is entirely gratuitous to the virus' growth cycle. The *src* function can be expressed only after a cell has been productively infected, hence the gene can play no initiating role in the infection cycle comparable to the induction of cell DNA synthesis by DNA tumor viruses. Furthermore, only rare isolates of retroviruses have a *src* function and its deletion is not deleterious to virus growth. If *leuk* functions exist they may be more central to virus growth.

Origin of Transforming Genes in Retroviruses

Because the *src* gene is not crucial to growth of retroviruses it may not have even evolved as a viral gene. Retroviruses might be able to incorporate cellular genes into their viral genome and *src* could be such a cellular gene. This gene, which might play a constructive role in the growth of special cell types, would be oncogenic as a consequence of its inappropriate expression in virus-infected cells.

This possibility has received strong support from the identification of *src* information in normal cells. Because *src* is gratuitous to viral growth, identical viruses with and without *src* can be constructed. Using them, it has proved possible to purify nucleic acid sequences of *src* (Stehelin et al., 1976a) and, by using nucleic acid hybridization methods, to probe whether *src* exists in uninfected cells. Surprisingly, *src*-related sequences occur in the DNA of most vertebrate species even though among the viruses these *src* sequences are restricted to avian sarcoma viruses (Stehelin et al., 1976b). Other evidence links transforming genes to normal cellular processes and the hypothesis of a cellular origin of viral oncogenic information, at least for retroviruses, is gaining support.

Tumor Induction

Many factors determine whether a virus can cause a tumor in an animal. Properties of the individual virus can be crucial, as in the selective induction of mammary tumors by one class of murine retroviruses. Properties of the target cell are relevant, such as the existence of appropriate, unblocked receptors or the occurrence of intracellular restricting factors. Properties of nontarget tissues may be important, such as the animal's ability to mount an immune response to the virus or to the infected cells. Tumorigenesis by a virus requires that all of the relevant factors be permissive, allowing many physiological and genetic factors to block tumor induction or progression.

The first prerequisite for viral tumorigenesis is that sensitive cells exist in the animal's body. For viruses that have a very limited range of cell targets—such as Friend virus which specifically acts on immature mouse erythroid cells—there may be a lack of target cells in an animal because of its age or physiological state. Also, alleles of specific genes of the mouse that affect development of the hematopoietic system, such as *W* (dominant spotting) and *Sl* (steel), can abolish susceptibility to Friend virus.

Target cells might exist in the animal but their virus receptors could be blocked. For instance, endogenous retrovirus-related genes in the cell can produce the viral glycoprotein and so block the cell's receptors for an added virus.

In mice a number of genetic loci have alleles able to directly prevent retrovirus infection or tumorigenesis (Lilly and Pincus, 1973). The best-studied is the Fv-1 gene, which makes a product that can reduce the susceptibility of an animal's cells to certain retroviruses by 100-fold. The Fv-1 product apparently acts on some step in the viral life cycle after reverse transcription but before integration of the provirus (Jolicoeur and Baltimore, 1976).

The Fv-1 alleles are able to prevent certain murine retroviruses from inducing leukemia even if the viruses are inherited. Because Fv-1 does not appear to affect induction of viruses, it must act by preventing spread of the virus from cell to cell. Thus, tumorigenesis by an inherited virus does not occur simply by induction of the latent genome in a target cell. Rather, the virus must grow efficiently in the animal's cells, spreading from some initial site of induction to the target tissue. Direct evidence that external infection has occurred in the tumor cells of animals with inherited leukemia has come from the demonstration of an increased number of viral genes in tumor tissues (Berns and Jaenisch, 1976).

Another type of mouse gene that can modify susceptibility to many leukemia viruses are genes closely linked to those governing the major histocompatability system of the animal, H-2. The H-2 linked genes include Rgv-1, a gene with alleles that can make an animal resistant to leukemogenesis by the Gross murine leukemia retrovirus. Such genes probably function by altering the animal's immune response to the tumor cells.

The immune system of animals can respond to tumor cells if the cells have new surface antigens. Animals preimmunized with a tumor induced by a papovavirus can reject implantation of another tumor induced by the same virus, showing that these viruses induce production of TSTA (tumor-specific transplantation antigen) on transformed cells (Habel, 1961; Sjogren et al., 1961). This antigen is virus-specific—polyoma and SV40 induce different ones—but it is not certain whether it is encoded by the viral nucleic acid. Retroviruses may also induce transplantation antigens but the only ones characterized thus far have been structural proteins of the virions. Such proteins can be excellent targets for tumor rejection (Pasternak, 1969).

Viruses and Human Cancer

Although viruses are clearly established as causes of cancer in numerous animal species, it has not been possible to prove unequivocally that any human cancer is caused by a virus. The best case is the virtually perfect association of EB virus with Burkitt's lymphoma, a tumor that occurs mainly in a restricted portion of Africa, and with nasopharyngeal carcinoma, an epithelial tumor occurring in certain Chinese populations (Klein, 1972). In both cases, the virus can be recovered from tumor cells of almost every patient and all tumor cells are positive for EBNA, the EB-specific antigen.

Although EB virus will cause tumors in marmosets, proving that it causes Burkitt's lymphoma or nasopharyngeal carcinoma has been difficult because of the impossibility of doing controlled experiments in humans and because the virus is very widespread within human populations. If the virus is a cause of the disease, there must be other critical factors or everyone should have the tumor. The ultimate test of causation may have to be a vaccination project.

For other human cancers the association with a virus is more tenuous. Herpes simplex, type 2, has been found in conjunction with cervical cancer often enough to be suspect. Antigens, enzymes, and nucleic acid sequences of retroviruses have often been reported to be associated with various human tumors. In no case is there even a suggestion of a causal relationship but extensive investigations of possibilities are in progress.

It was suggested a number of years ago that human as well as animal cancer might result from activation of inherited genomes of retroviruses (Huebner and Todaro, 1969). In many animal species, including certain primates, inherited retroviruses are present as part of the genome (see Chapter 14) and inherited mouse viruses, but not those from most other species, are oncogenic. In many species lacking demonstrable inherited viruses, such as humans, nucleic acid sequences related to viruses of other species can be demonstrated. Whether these are evolutionary vestiges or intact viral genomes, the inducers for which have yet to be found, is an open question. Thus the situation exists whereby inherited viruses could play a role in human cancer but there is no evidence at present to suggest such a role.

One type of evidence has been considered an argument against a crucial viral involvement in much human cancer: cancer is almost never a communicable disease whereas if viruses were involved person-to-person transmission might be expected. If activation of inherited viruses by environmental factors played a role, however, then evidence of inheritance of a tendency to develop cancer might be expected. In fact, a genetic propensity to develop certain cancers has been found (Heston, 1976), although it could have a variety of explanations.

In spite of 10 years of intensive work under special government programs, the link between viruses and human cancer is still problematic. It seems odd that oncogenic viruses can play such obvious roles in other species but not affect humans.

Interaction of Plant Viruses with Their Hosts

17

Experimental Systems

Until recently most of what was learned about virus multiplication and other aspects of virus-host interaction had to be learned by combinations of indirect approaches because no experimental system was available for quantitative studies of virus infection in isolated cultured cells. In the last decade progress has been made in two directions: the preparation of plant cell protoplasts, which proved infectable *in vitro* (Zaitlin and Beachy, 1974), and the development of monolayer cultures of susceptible cells from insects that are vectors of certain plant virus diseases (Chiu and Black, 1967). Both methods promise a variety of useful applications to virus titration and to the study of viral biosynthesis in single cells. It has already become clear that the overall course of virus-host interaction at the cell level is congruent with what is known about comparable animal viruses.

All but one group of the currently known plant viruses (see Table 1-2) have an RNA genome, the exception being a group of DNA viruses typified by cauliflower mosaic virus (Shepherd, 1976). Certain well-studied viruses are convenient prototypes for the groups to which they belong: tobacco mosaic virus (TMV) for the rod-shaped viruses with single-stranded RNA; potato yellow dwarf virus (PYDV) for a group of rhabdoviruses resembling vesicular stomatitis virus; turnip yellow mosaic virus (TYMV) for small icosahedral viruses; and wound tumor virus (WTV) for arthropod-transmitted viruses whose icosahedral virions contain two-stranded DNA.

A peculiar situation has arisen in recent years following the discovery in plants of parasites such as mycoplasmas and related organisms (Doi et al., 1969) and of small gram positive or gram negative bacteria (Davis and Whitcomb, 1971). A number of plant diseases such as aster yellows and clover club leaf, whose etiology had been thought to be viral but whose agents had

455

not been characterized, are now suspected to be caused by members of the newly discovered groups of pathogen. This state of affairs is, of course, a reflection of the fact that some plant diseases had been (and still are) considered as of viral etiology on the basis only of symptoms and of transmissibility, without precise characterization of the causal agents.

The virions of some well-characterized viruses are easily detected by electron microscopy of sections or extracts of infected plants, in which the virions accumulate in enormous amounts. A single hair cell of an infected tobacco plant has been estimated to contain over 10^7 TMV virions. As much as 10% of the dry weight of infected leaves may be TMV. Not all plant viruses are present in such large amounts, however, and the search for the pathogenic organism can be a demanding enterprise.

Isolated protoplasts from mesophyll cells of tobacco can be infected with TMV directly (Takebe, 1975). A single cycle of infection can then be studied. There is at first a rapid uncoating or decapsidation of the RNA from the entering virions. As a consequence, infected cell extracts exhibit reduced infectivity, corresponding to an eclipse period. Decapsidation of TMV occurs by removal of subunits from the helical capsids, whereas TYMV, an icosahedral virus, apparently releases RNA without disintegration of its protein shell (Schneider et al., 1972). Viral RNA released from the virions is converted into reproductive forms and reproductive intermediates, presumably by mechanisms similar to those described for RNA phages and picornaviruses (see Chapters 8 and 13). The total amount of viral RNA increases exponentially at first; later, plus strands are made preferentially and at a linear rate. Maturation starts after 4 to 5 hours and thereafter any newly formed viral RNA is rapidly encapsidated. Little virus gets released, however; it is characteristic of most plant virus diseases that infected cells can continue to manufacture virus without either lysing or even dying. This explains the enormous virus concentrations that can accumulate in plant cells infected by viruses such as TMV. Transfer of infection from cell to cell occurs mainly by direct transfer of virions through intercellular bridges (plasmodesmata).

Cycles of virus growth can be studied accurately for viruses that multiply in insect cell monolayers (Hsu and Black, 1974). The rhabdovirus PYDV (var. *sanguinolenta*) in leafhopper cell culture has a growth curve (Figure 17-1) with an eclipse period of 9 hours, a growth period lasting 20 hours, and a yield of over 10,000 per cell. Such precise analysis of the growth curve was possible because 100% of the cells could be synchronously infected, as shown by fluorescent antibody, and the new virus produced could be titrated efficiently on insect cell monolayers. In infection of intact plants it is difficult to analyse the first cycle of infection because only few cells are infected at first and subsequent cycles overlap before enough virus or viral products are available for measurement. If the PYDV virus occurs in large excess (several hundred infective units per cell), it promotes fusion of insect cells in culture (Hsu, 1975) by a mechanism that may be similar to cell fusion promoted by paramyxoviruses.

The RNA of plant viruses is presumably replicated by specific RNA replicases (see Hamilton, 1974). From TMV-infected leaves a replicase has been isolated in two forms (Zaitlin et al., 1973): an insoluble form, bound to TMV RNA and capable of synthesizing specific RNA using riboside triphos-

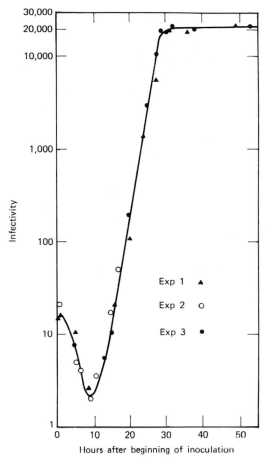

FIGURE 17-1. *Growth of yellow dwarf virus purified from plants on AS2 cells from the vector* Aceratagallia sanguinolenta *in monolayer culture. Infectivity was measured by infection of AS2 cells in monolayers. Redrawn from Hsu and Black,* Virology **59,** *331 (1975).*

phates; and a soluble form, about 150,000 daltons, requiring a template and capable of replicating nonspecifically any one of several RNAs. The bound form extracted late in infection synthesizes preferentially plus (viral) strands.

TMV RNA, a single strand 2.3×10^6 daltons in weight, has a coding capacity of about 2.4×10^5 daltons of protein. *In vivo* it directs the synthesis of several proteins, including the coat protein (17,500 daltons) and two large peptides, about 160,000 and 140,000 daltons, respectively (Paterson and Knight, 1975). Altogether, this amount of protein is too much to be coded independently by the TMV plus strand; either there are overlapping peptides, or some peptide is first made and then split into smaller ones, or some peptides are translated off the minus strand. Cell-free translation systems from wheat germ synthesize, from a TMV virion RNA template, mainly the 140,000 dalton peptide and a smaller amount of the 160,000 dalton species (Roberts et al., 1973). One of these (or both) may represent the replicase. Coat protein is not made *in vitro* from a RNA template, either as such or as part of the longer peptides, as shown by fingerprint analysis (Knowland et al., 1976; Figure 17-2). The 140,000 dalton polypeptide (but not the coat protein) is synthesized when TMV RNA is injected and translated in oocytes of the amphibian *Xenopus* (Knowland, 1974).

Extraction of TMV-infected leaves yields a TMV-specific RNA, 250,000 daltons, which is believed to be the messenger for the coat protein and does in fact code for coat protein in a wheat-germ cell-free system (Hunter et al., 1976). This messenger RNA appears to correspond to the 3' end of the *plus* strand, that is, the virion strand, which may acquire the ability to be translated only after some modification. In the *plus* strand from the virions the interaction of the coat messenger with ribosomes may be blocked by secondary structure, as for the A gene in the RNA phages (see Chapter 8), or the RNA may have to be split endonucleolytically before its 5'-end is appropriately modified to serve as template.

An intriguing evidence of a specific tri-dimensional structure in plant virus RNAs is that TMV RNA and TYMV RNA, as well as other plant virus RNAs, can function as amino acid acceptors in the presence of ATP and an appropriate activating enzyme: TMV for histidine (Öberg and Philipson, 1972) and TYMV for valine (Litvac et al., 1973). The 3'—OH terminal sequence in both these viral RNAs is the same as in tRNAs, —C·C·C·A, the terminal adenylate being added not by complementary strand coding but enzymatically. The specific amino acid acceptor activity is indicative of a tertiary structure resembling that of the corresponding tRNAs. Whether this is accidental or in any way related to the template function of the viral RNAs remains to be seen. The histidine tRNA portion of TMV RNA can be methylated by an *E. coli* methylase using s-adenosylmethionine as a methyl donor (Marcu and Dudock, 1976).

The 5'-terminus of the TMV·RNA is not free but, like that of other messengers from eucaryotic cells (Furuichi et al., 1975) is blocked by a special group, which in TMV RNA has the sequence 7-methyl G-5'-ppp-5'-Gp . . . (Zimmern, 1975). This is similar but not identical to the 5' terminal sequence of certain animal cell messengers. The blocking probably plays some necessary role in translation, but the nature of that role is still unknown.

FIGURE 17.2. The possible relation of known TMV peptides (broken lines) to the viral RNA. The arrows indicate the initiation sites. The one for coat protein may be masked in the RNA from virions. After Knowland et al., Proc. INSERM-EMBO Colloquium **47,** *211 (1976).*

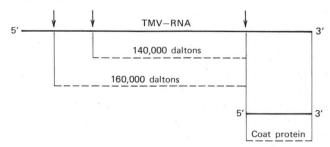

ASSEMBLY OF VIRIONS

Since the TMV virions are among the best characterized macromolecular structures it is not surprising that the process of their assembly should have been analysed in great detail. The problem can be summarized as follows: How do 2130 identical molecules of coat protein plus one molecule of RNA come together to form the helically organized final structure? There must be a specific assembly-initiating process, an elongation mechanism, and a termination signal, which can be provided by the RNA itself. A major advance was the recognition (see Klug, 1972) that both *in vivo* and *in vitro* the TMV coat protein tends to aggregate by itself in either of two forms depending on the pH: a helical, virion-like form (see Figure 3-5) and a cylindrical form (Figure 17-3). As already discussed in Chapter 3, at physiological pH and low ionic strength the primary subassembly is a disk (Markham et al., 1964) consisting of 34 subunits, two layers of 17 subunits each (Champness et al., 1976). A disk can, by a slight conformational change, turn into a "lock-washer" form, which is the constituent of the helical structure (Figure 17-4). An important feature distinguishing disks from helical forms reflects the interaction between two carboxyl groups in the coat protein molecule. At neutral pH these groups have normal pK's and, being ionized, are mutually repulsive. In the helical structure they are pushed together so that one proton is firmly bound. The transition between the disk and the helical form can occur either upon lowering of pH or, presumably, by association with the viral RNA. The pairs of carboxyl groups would act as electrostatic switches between forms.

The surprising conclusion reached by TMV workers (Butler, 1976) is that the assembly of virions begins preferentially with the two-layer disks rather

FIGURE 17-3. (a) *Electron micrographs of TMV protein disks. Insert: rotationally filtered image of a disk.* (b) *Electron micrograph of a stacked disk rod. Insert: tridimensional image reconstruction. From Klug,* Feder. Proc. **31,** 30 (1972). *Courtesy Dr. A. Klug.*

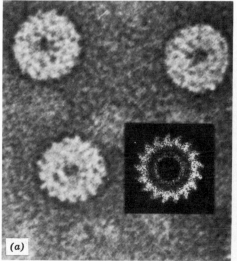
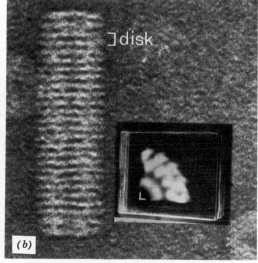

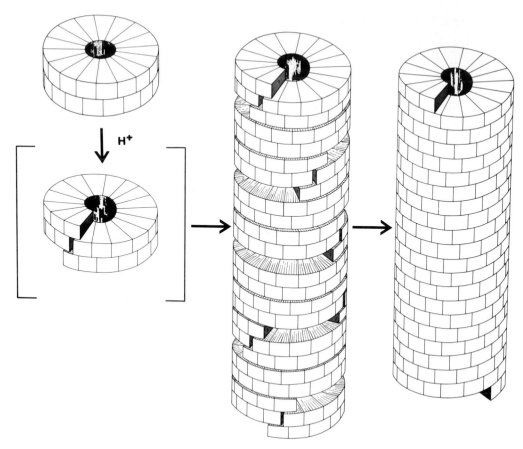

FIGURE 17-4. Diagram of stages in the disk-to-helix transformation of TMV protein. From Klug, Feder. Proc. **31,** 30 (1972) Courtesy Dr. A. Klug.

than with the coat protein subunits or its small aggregates. Reconstitution of TMV *in vitro,* which was one of the landmarks in the history of molecular biology (Fraenkel-Conrat and Williams, 1955) is much more efficient with disks than with the subunits themselves. There is still controversy as to whether the growth of the TMV virion, as distinct from initiation of assembly, actually occurs by the addition of small aggregates (Richards and Williams, 1973) or of disks of coat proteins (Butler, 1976). Figure 17-4 should make the reader aware of the added complexities arising if two-tier disks rather than small aggregates of subunits are used for elongation. Specifically, a disk being converted into two turns of helix must open up in such a way as to accommodate the RNA. Despite these complexities the growth-by-disks idea is currently favored for TMV. Disassembly of TMV virions entering a host cell might be facilitated if it gave rise to disaggregated protein rather than to disks, so that reassembly by protein coating of the RNA would not occur prematurely.

It was believed that assembly of TMV virions started at the 5' terminus of the RNA. Recent evidence, based on analysis of the nucleotide sequences that are protected against nuclease action in partially assembled structures (Zimmern, 1976) indicates that the protected portion of the RNA includes neither the 5' nor the 3' terminus. It must, therefore, be located at some inner position and assembly must proceed bidirectionally. The reader may notice a similarity between this process and the assembly of RNA phages, in which coat protein has a specific affinity for an internally located sequence at the start of the replicase gene and may nucleate virion assembly from that position (see Chapter 8).

A number of plant viruses are multipartite; that is, their genome is packed severally into more than one virion, often in unequal size portions (Lister, 1969; Jaspars, 1974). The assembly of multipartite viruses such as cowpea mosaic, tobacco rattle, and alfalfa mosaic has not yet been studied in any such detail as that of TMV. Apparently the coat proteins of each of these viruses can assemble around the various portions of which the viral genome consists (Figure 17-5). The coat of each virus type may consist of a single type of protein subunit, as in TMV, or may include several proteins. Multipartite viruses have, of course, an anomalous titration curve: one infectious unit results from the coincidental entry of all the needed viral portions into one cell.

Wound tumor virus (Black, 1972) is not a multipartite virus, but has in common with the reoviruses (Shatkin, 1971) and with several other plant viruses the problem of a genome that consists of several double-stranded

FIGURE 17-5. *Requirements for infectivity of alfalfa mosaic virus, a multipartite plant virus. Infectivity can be conferred to the RNA mixture from B, M, Tb particles by addition of some virion protein or of Ta RNA, which may generate the needed protein* in vivo. *Modified from Jaspars,* Adv. Virus Research **19**, *37 (1974).*

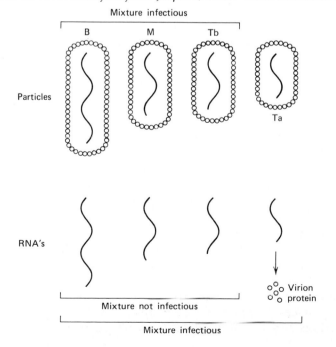

elements, all of which have to be packed into one virion. How this is done remains to be discovered. In its virions WTV carries a protein with RNA-dependent RNA polymerase activity (Black and Knight, 1970) which may be its specific replicase.

Tobacco necrosis virus (TNV; Rothamsted strain) has icosahedral virions. In infected plants these are often accompanied by the smaller icosahedral virions of another, serologically unrelated virus called TNV-satellite. The satellite virus is a defective virus that can multiply only in cells infected with TNV. The RNA genome of the satellite, 1200 nucleotides long, has just about enough information to code for the coat protein with its 372 amino acids (Reichman, 1964). It can actually direct the synthesis of that protein in a cell-free translation system. Evidently, TNV provides the machinery for satellite replication. Only TNV can act as helper virus for the satellite; other viruses are inactive. Presumably TNV provides a replicase and any other proteins needed for the satellite virus replication.

Equally interesting, although different, is the satellite of tobacco ringspot virus (Schneider et al., 1972). It has an RNA "genome" of only about 300 nucleotides, like the viroids described below. It cannot replicate or mature except in the presence of the ringspot virus, but in mixed infection the satellite RNA enters into virions made of ringspot coat protein. A single virion may incorporate anywhere between one and 12 satellite RNAs.

Virion assembly is not always specific for viral RNA. Some coat proteins can incorporate host RNA and even (in vitro) synthetic polynucleotides. The elements with host RNA are described as pseudovirions and may contain either chloroplast or nuclear RNA (see Hamilton, 1974).

VIROIDS

There is a class of infectious plant diseases that were considered to be viral and are now reported to be caused by a novel class of pathogenic agents celled *viroids* (Diener, 1972). Viroids are not simply maturation defective viruses. They consist of RNA, have no protein coat, and their "genomes" are quite small, about 300 nucleotides for the potato spindle tuber viroid and about 400 for the exocortis viroid (Semancík et al., 1975). These RNAs have a tertiary structure, no amino acid acceptor function (Hall et al., 1974), and consist of covalently closed circles (Sanger et al., 1974). Their synthesis is inhibited by actinomycin (Diener and Smith, 1975). Hybridization studies with host DNA indicate that these transmissible agents are produced by host cell enzymes in the nucleus of infected cells (Takahashi and Diener, 1975). The small size of viroid RNA is more compatible with some kind of regulatory action than with messenger function. If in fact the viroids are infectious gene products, one must assume that in the infected cells they cause the production of more viroid copies by stimulating transcription of the corresponding genes. Altogether the viroids as currently interpreted would represent a unique class of biological elements.

An interesting and important analogy (Diener, 1972b) is with the agent of the puzzling disease of sheep called *scrapie*, which also appears to be caused by an infectious entity of very small size without a protein coat. It is possible that other agents of this class remain to be discovered as causative agents of diseases, including some human diseases.

CYTOLOGICAL FINDINGS IN VIRUS INFECTIONS

It has long been believed that TMV replication occurs only in the cytoplasm. Some cytological and cytochemical evidence, however, suggests that the primary site of replication of viral RNA may be the nucleus (Bald, 1964; Smith and Schlegel, 1965). Ferritin-conjugated TMV antibody shows the TMV protein first in the cytoplasm, mainly around the nuclear membrane, where it can be seen a few days after infection of leaf cells. Intranuclear TMV protein is also found, but complete virions are seen by electron microscopy only in the cytoplasm, not in the nucleus. The virions of some other viruses, such as barley stripe virus, are actually found within the nucleoplasm.

The virions of potato yellow dwarf virus, an enveloped virus whose morphology is reminiscent of that of rhabdoviruses (Francki, 1973) are seen in sections of infected cells closely associated with the nuclear membrane (MacLeod et al., 1966). It is possible that this membrane contributes some components to the viral envelope. The virions are not seen inside the nucleus.

A frequent characteristic effect of plant virus infection is the production of microscopically recognizable *intracellular inclusions*. These consist, in most instances, of agglomerations of virus particles, associated or not with cellular constituents. The morphology and location of inclusions vary depending on the virus infection. Most viruses give only intracytoplasmic inclusions, but tobacco etch virus and some other viruses produce crystalline inclusions in the cell nuclei. Whether these inclusions are crystals of virions is not certain. When a cell with etch virus crystals in its nucleus divides, the crystals are extruded into the cytoplasm and new ones are formed in the new nuclei (see Bawden, 1964). Sugar beet yellow virus causes the production of crystalline inclusions in the chloroplasts.

Two types of cytoplasmic inclusions are found in plant cells infected with TMV. One type, the ameboid, amorphous, or X-bodies, contain TMV virions and also some cytoplasmic components. The other type consists of hexagonal crystals (Figure 17-6) made up almost entirely of virions. Treatment with dilute acids converts the crystals to masses of TMV paracrystals. In the hexagonal crystals the particles are oriented more regularly than in the paracrystals, forming successive layers aligned parallel to one another (Wilkins et al., 1950; Figure 17-7). The conditions that determine the type of inclusion that is made in a given cell are still unknown. Some TMV mutant strains do not produce any crystalline inclusions, but only long fibers of viral materials.

In TMV-infected leaf tissue the chloroplasts become distorted and often degenerate. Formation of new chloroplasts is inhibited. These changes are responsible for the chlorosis observed in infected plants. TMV virions, however, are generally not found within chloroplasts of infected cells until the late stages of infection; they probably enter chloroplasts only after these have become damaged. Accumulation of viral protein around the chloroplasts, however, is seen very early, suggesting that virus biosynthesis is associated with sites of high ATP concentration. In TYMV infection the chloroplasts are affected early and extensively (Reid and Matthews, 1966).

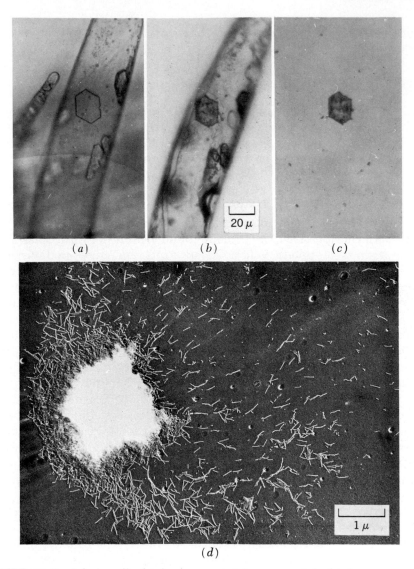

FIGURE 17-6. *A hair cell of tobacco, containing a crystal of tobacco mosaic virus, is photographed* (a) *before and* (b) *after freeze-drying. The crystal is extracted by micromanipulation in the frozen-dry state* (c). *Upon partial dissolution in water such crystals release characteristic virions* (d). *From Steere and Williams, Am. J. Bot.* **40,** *81 (1953). Courtesy Dr. R. C. Williams.*

Inclusions formed by viruses other than TMV may also be crystalline or ameboid. The cytoplasmic spherules or *viroplasms* found in the cells of tumors produced by wound tumor virus consist of microcrystals of virions (see Figure 17-8).

VARIATION IN PLANT VIRUSES

Naturally occurring variant strains of many plant viruses are well known. Common origin is inferred by similarity in virion morphology and in

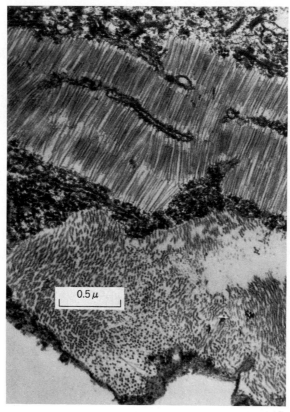

FIGURE 17-7. Electron micrograph of part of a cell in the periphery of a 10-day-old TMV lesion in a leaf of Chenopodium amaranticolor. TMV aggregates are seen in cross section and in profile. From Milne, Virology **28,** 520 (1966). Courtesy Dr. R. G. Milne.

serological properties. Mutations can alter virion properties so radically that two related viruses, after purification, may even crystallize in different shapes (Bawden, 1964). Variants sometimes appear as patches of tissue with unusual symptoms, for example, yellow spots in green mosaic-infected leaves.

Mutants can readily be obtained by mutagenic treatments either of the intact virus or, even better, of infectious RNA (see Mundry, 1965). An early application has been the use of TMV mutants in the analysis of the genetic code. Since the TMV protein monomer was one of the few proteins whose amino acid sequence was fully known analysis of the amino acid changes produced by mutagens contributed to the elucidation of the codons corresponding to various amino acids, as discussed in Chapter 4. Some TMV mutants derived by *in vitro* mutagenesis produce different symptoms in infected plants but have the normal coat protein, with an unchanged amino acid sequence. These mutations must affect some of the virus-coded proteins other than the capsid monomer. The position of a gene locus responsible for production of local necrotic lesions instead of mosaic on tobacco leaves has been localized in the RNA molecule of TMV by the ingenious expedient of

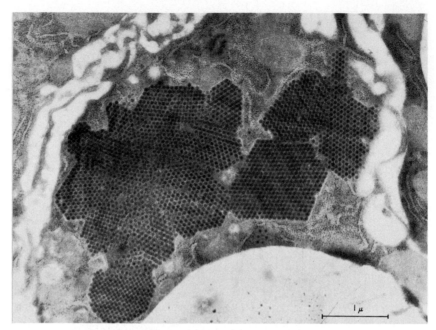

FIGURE 17-8. Microcrystals of wound tumor virus in the cytoplasm of a fat body cell of the leafhopper vector Agallia constricta. From Shikata and Maramorosch, Nature **208**, 507 (1965). Courtesy Dr. K. Maramorosch.

polar stripping of protein subunits from the virus particles, followed by mutagenesis on the exposed RNA (Kado and Knight, 1966).

Some TMV mutants are defective in the process of viral assembly or maturation; symptoms of disease are observed and infective RNA is produced, but few or no complete virions are made (Siegel et al., 1962). Some capsid protein can be demonstrated by means of fluorescent antibody even though it is so altered that it cannot assemble with RNA to form normal virions.

MIXED INFECTION AND RECOMBINATION

Evidence of genetic recombination between plant viruses has long been sought in plants mixedly infected with two related viruses. It is certain that two unrelated viruses can often grow in the same cell. Inclusions of two types characteristic of two different viruses are observed in doubly infected cells, and mixed lesions containing two strains of TMV have been produced by mechanical introduction of a mixture of strains into a single cell (Benda, 1958).

When two related viruses are present in the same plant, new strains with apparently hybrid properties are occasionally found among the viral progeny. This has been observed with strains of tomato spotted wilt virus (Best and Gallus, 1955) and with potato virus X as well as other potato viruses (Thompson, 1961). The present evidence is suggestive but insufficient to prove the occurrence of recombination among virus with unipartite genome.

True viral hybrids have been demonstrated in mixed infection with two multipartite viruses (Sänger, 1968; Jaspars, 1974).

SPREAD OF VIRUSES IN PLANTS

Mechanically inoculated viruses spread slowly through nonvascular tissue from the initially infected cells to neighboring cells; the spreading rate for TMV is 1 mm per day or less. Presumably, multiplication precedes the spread to new cells, which probably takes place through intercellular channels or plasmodesmata. Several hours may be required before TMV moves from the initially infected cell to its neighbors. Both intact virions and viral RNA can pass from cell to cell; virions have been photographed within plasmodesmata, and nonmaturing virus variants can spread in plant tissues.

When a virus reaches the vascular tissue, either from the parenchyma or directly from an insect vector, it moves rapidly down the veins, the leaf petiole, and into the stem. It may then either spread to all parts of the plants or not, depending on the properties of host and virus. Most mechanically transmissible viruses are histologically "nonlimited"; that is, they can invade most tissues of the infected plant (Esau, 1956). Actively growing tissues and roots are generally attacked first. The transfer of virus to distant points occurs mainly in the phloem, accompanying food transport, although movement through the water transport system in the xylem can also take place. Viruses can pass with liquid food through the sieve-tube cells of the phloem without necessarily infecting them, but phloem necrosis is not infrequent in viral diseases, witness the phloem necrosis of potato leaf roll.

Transmission of virus by seed is infrequent and pollen transmission is even rarer. The morphogenetic processes of flower development tend to prevent virus invasion of the gametes. More generally, the apical meristems of virus-infected plants tend to have low concentrations of virus and are occasionally virus-free.

The spread of viruses in infected plants depends on the reaction of the infected cells, a reaction that may be extremely variable. At one end of the virulence scale are cases of necrotic lesions; the cells die so fast that they often fail to transfer the virus to their neighbors. An increase in polyphenol oxidase has been detected in cells infected with necrotizing viruses and the highly toxic products of polyphenol oxidation are considered partly responsible for local necrosis (Farkas et al., 1960). There is no evidence that the oxidases are enzymes specifically determined by the viruses.

At the other extreme of virulence are almost silent infections, revealed only by mild symptoms or by the discovery, in apparently healthy plants, of virus infective for other plants. The infected cells are not seriously damaged and can still divide. Extensive cellular proliferation and even tumoral transformation are the typical responses to some viruses. The mechanism of the tumorigenic action of wound tumor virus is still unknown and the role of wounds in its expression is not yet understood (Black, 1972). Plant hormones can act synergistically with the virus in initiation and stimulation of tumor growth.

No generalization appears justified concerning the metabolic effects of infection with plant viruses. Most of the changes observed are probably indirect effects of interference with various metabolic processes such as photosynthesis, respiration, growth regulation, and with transport of water, nutrients, and other substances. Disturbances of growth regulation result in morphogenetic abnormalities ranging from mosaic of leaves and flowers through necrotic spots and streaks to leaf enations and tumors. A number of substances such as scopoletin, a fluorescent aromatic compound, often accumulate in necrotic lesions, but there is no evidence of a specific role of viruses in their biosynthesis.

RECOVERY, ACQUIRED IMMUNITY, AND CROSS-PROTECTION

Virus-infected plants, if they do not die, often show decreasing symptoms of disease and even apparent recovery; the new growth may be completely free of symptoms. Yet, the plant is not free of virus. This is shown better by tests of *acquired immunity* than by recovery of virus from extracted leaves because the amount of virus present in recovered plants is often much reduced. Acquired immunity is manifested when plants that have apparently recovered are markedly refractory to local or systemic infection if reinoculated with the same or related viruses (Price, 1940).

Recovery is not a prerequisite for acquired immunity; leaves that have been systemically infected with a nonnecrotizing virus such as TMV fail to give necrotic lesions when inoculated a few days later with a necrotizing strain of the same virus (Figure 17-9). The excluded virus strain is actually prevented from multiplying. The mechanism of exclusion remains unknown. It is possible that phenomena similar to immunity in lysogenic bacteria may play a role, repressing the function of essential genes of a virus that superinfects an already infected cell. Alternatively, a virus may initiate sequences of regulatory processes, as observed in phage and poxvirus infec-

FIGURE 17-9. *Leaves of* Nicotiana sylvestris *partly protected against aucuba mosaic by direct inoculation with TMV. The areas free of the necrotic aucuba lesions were rubbed with TMV five days before inoculation with the aucuba virus. Courtesy Dr. L. O. Kunkel.*

tions, and these may interfere with the development of other viruses. Both these possibilities . would be compatible with the observation that cross-immunity is most frequent among related viruses.

Less specific mechanisms, however, are probably also involved. For example, leaves bearing local necrotic lesions of one virus on a half-leaf may prove much more resistant to other necrotizing viruses inoculated on the other half-leaf. Such effects can also spread from lower to higher leaves, suggesting that the necrotic tissue may release inhibitory substances.

Cross-protection is a form of immunity that occurs among viruses that are known, by other criteria, to be related. It has been widely used as a test of relatedness (see Price, 1940). The lack of evidence concerning the mechanisms of interference among plant viruses makes it impossible to provide any rationale for cross-protection tests as criteria of relatedness.

Transmission Mechanisms

With a few exceptions, for example, TMV, mechanical transmission of viruses onto leaf surfaces as done experimentally is unlikely to be a major source of virus spread in nature. Transmission by grafting, which is possible with almost all viruses, is mainly an experimental procedure that, however, may play a significant role in the propagation of diseases of fruit trees and ornamental plants. Parasitic dodder, which sends feeding haustoria into the stems of the host plants to make a living connection between the vascular systems of host and parasite, is a useful research tool for transmission of viruses to new hosts (Johnson, 1941) but is evidently not an important mechanism of transmission in nature.

Some virus diseases can be transmitted from plant to plant only by grafting. Some of these diseases are caused by viruses that have lost by mutation their transmissibility by vectors. Paracrinkle virus (Bawden et al., 1950) may have originated in this way. The possibility that the pathogenic agent may sometimes be nonviral must be considered.

TRANSMISSION BY ARTHROPOD VECTORS

The most important natural means of plant virus transfer is provided by animals that feed on the plants. Sometimes the vector-mediated transfer is a purely mechanical process, but in most cases the transfer by animal vectors is a specific process that reflects special relationships among virus, vector, and plant (Sinha, 1968). Arthropods, especially insects but also mites, are the most important vectors of plant viruses. There are numerous similarities between the relation of some plant viruses to their arthropod vectors and that of the *arbo* (arthropod borne) animal-pathogenic viruses to their vectors.

Special techniques have been developed for serological detection of virus in single insects weighing about 1 mg, for injecting extracts into these minute vectors, and for inducing insects to feed on droplets of virus-containing fluids (Bennett, 1935). More recently tissue cultures of insect vectors have proved exceedingly useful experimental materials.

The sucking arthropod vectors are efficient transmitters of plant viruses because of their ability to introduce viruses into the relatively deep plant tissue, the phloem (Figure 17-10). A few viruses are xylem-limited; many more are phloem-limited. The most common vectors are aphids such as *Myzus persicae,* a green louse that can transmit more than 50 different viruses of potato, bean, and other plants. Transmission by insects can be of different types:

1. *External, nonpersistent, or stylet-borne transmission.* The vector takes up virus on the tip of its stylets while feeding on one plant and can immediately transmit it to another plant. The transmitting ability may be lost quickly or may last for days but is not carried through molting. Some plant products may be involved in facilitating virus transmission (Kassanis and Glover, 1971).

2. *Regurgitative transmission.* Aphids and beetles can store virus in the foregut and transmit it to healthy plants for relatively long periods of time.

3. *Circulative transmission.* Here the ability to transmit virus manifests itself only after a latent period of several hours or days from the feeding time and lasts much longer than regurgitative transmission. Virus is found circulating in the insect's tissues. The transmitting ability is not lost upon molting.

4. *Propagative transmission.* The virus actually multiplies in the insect tissues before reaching the mouth parts and the latent period is the time

FIGURE 17-10. *Section through a potato leaf and an aphid* (Myzus pseudosolani) *showing the path of the insect's stylet reaching into the phloem. From Dykstra and Whitaker,* J. Agr. Res. **57,** *319 (1938).*

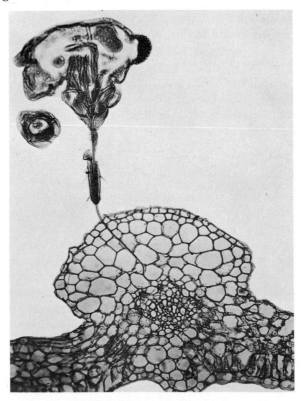

required for multiplication. The virus eventually reaches the salivary glands and is inoculated into plants. Propagative transmission is almost the rule with a large class of virus vectors, the leafhoppers (*Cicadellidae*), which include the vector of wound tumor virus, *Agallia constricta*, and many others. It has been proved beyond doubt (Brakke et al., 1954) that some viruses can multiply indefinitely in the leafhopper tissues. Transovarial transmission also does occur in some of these vectors.

Viruses multiplying in insect tissues can be traced and identified by fluorescent antibodies prepared against virus extracted either from the plants or the insect vectors (Sinha and Black, 1962); the same viral proteins are evidently made in both hosts. Examination of the organs of leafhoppers at intervals after feeding on plants infected with wound tumor virus reveals a sequential spreading of virus, which multiplies first in the filter chamber of the intestine and then after about two weeks appears in the hemolymph and in various organs.

The ability of insects to transmit a given plant virus is genetically controlled. A single gene difference may determine transmission, presumably by controlling the permeability of the gut to the virus. Puncturing the abdomen can convert a nontransmitter into a transmitter, as is also true of insect vectors of some animal viruses.

The ability of some plant viruses to multiply in insects raises intriguing problems, not only of semantics—are these plant viruses or animal viruses?—but also, and more important, of viral evolution. Given the intimate feeding relations of insects with plants, it is conceivable that viruses have spread from one class of hosts to the other in the course of evolution and have become selectively adapted to a "double life." The essential act of infection is the invasion of a cell by a viral nucleic acid whose genetic potentialities, jointly with those of the cell, permit virus replication. How the penetration into the cell is accomplished may be less important than the metabolic and biosynthetic conditions the viral genome finds there.

As mentioned earlier, some insect-transmitted plant viruses resemble animal viruses in some characteristics. The virions of tomato spotted wilt have envelopes (Best and Palk, 1964) and those of potato yellow dwarf resemble rhabdoviruses (Francki, 1973).

Wound tumor virus, a typical insect-transmitted virus, has many properties in common with the reoviruses. The virions are morphologically similar and contain double-stranded RNA of similar base compositions (Gomatos and Tamm, 1963). If some plant and vertebrate viruses are indeed close relatives, what evolutionary history has made some of them human pathogens, others plant pathogens?

TRANSMISSION OF PLANT VIRUSES BY NEMATODES AND FUNGI

Soil-borne infection of plants has long been known. The role of nematodes as vectors of several viruses, including tobacco ringspot and tobacco rattle, has been ascertained (Hewitt et al., 1958). These worms acquire and transmit virus while feeding on roots and can harbor virus for months, but do not transmit it to their progeny through the eggs.

Tobacco necrosis virus and at least two other viruses are transmitted by a parasitic primitive phycomycete, *Olpidium brassicae* (Kassanis and McFarlane, 1965). The transmission is specific both with respect to virus strains and to vector strains and the virus is probably carried on the fungal zoospores. Other fungi can probably act as vectors of soil-borne viruses. Virus transmission by various types of organisms has been reviewed by Kado and Agrawal (1972).

Insect-Pathogenic
18 Viruses

Inclusions and Their Relations to the Virus

Insects are subject to a variety of virus diseases. Some of these affect useful insects like the silkworm, others affect insect pests and may be important in their control.

A fairly common feature of many virus diseases of insects in the presence of polyhedral inclusions in the cells of the infected animals; hence these diseases are called *polyhedroses* (see Bergold, 1958; Smith, 1959). The polyhedra are specific products of the infecting viruses (Figure 18-1). In some infections they are present in the nucleus, in others in the cytoplasm of infected cells. In some insect virus diseases, called *granuloses*, there are no polyhedra, but granular inclusions or capsules (see Figure 3-24). Some insect-pathogenic viruses cause no intracellular inclusions at all.

Research on the biology of the insect viruses has been restricted by the inadequacy of techniques to study viral growth in controllable systems. Apart from studies on the pathology of infected animals, most work was concerned with the chemistry and morphology of viral particles and the inclusion bodies. Whatever was known of viral multiplication was deduced from electron microscopic studies. Patient efforts to develop cultures of insect tissues suitable for research on these viruses have been rewarded with some success (Grace, 1969; Vago and Bergoin, 1968).

NUCLEAR POLYHEDROSES

Nuclear polyhedroses have been described in lepidoptera, hymenoptera, and diptera. A typical instance is the polyhedrosis of the silkworm, *Bombix mori* (Figure 18-2). A few days after larvae become infected by injection or feeding, small inclusions appear within the nuclei of most body tissues. The

473

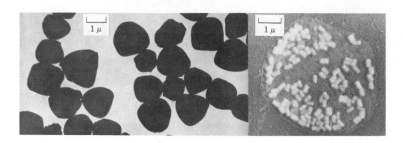

FIGURE 18-1. (a) *Polyhedra from* Choristoneura fumiferana *as seen in the light microscope. From Bergold,* Can. J. Zool. **29,** *17 (1951). Courtesy Dr. G. H. Bergold.* (b) *A polyhedron from* Prodenia praefica *partly dissolved in alkali, showing polyhedral envelope and bundles of virions. From Hughes,* J. Bacter. **59,** *189 (1950). Courtesy Dr. K. M. Hughes.*

inclusions increase in size and numbers, reaching up to 10 or 15 μ in size. As many as 100 polyhedra may be present in a nucleus. The nuclear chromatin disappears, the cells finally die, and free polyhedra appear in the hemolymph.

The silkworm polyhedra are crystals of a high molecular weight protein, around 300,000 daltons, consisting of subunits of about 20,000 daltons. The polyhedral protein is highly resistant to proteolytic enzymes, has an isoelectric point around pH 5.6, and is soluble only in alkali—properties that may account for its crystallization in the nuclear sap. The polyhedral proteins are generally good antigens; serological cross-reactions are found between polyhedral proteins from virus diseases of different insects, but not between host proteins and polyhedral proteins (Krywienczyk and Bergold, 1960). It is probable that the polyhedral proteins are determined by structural genes in the viral genome. Some viruses have been transmitted experimentally to new insect hosts, although the frequent presence of latent viruses complicates this transmission. The protein of polyhedra supposedly produced by a given virus in different hosts is serologically unchanged.

Embedded within polyhedra are virus particles, either singly or in bundles (see Figure 18-1). The virions are readily released by mild alkaline hydrolysis of the polyhedra and consist mainly of rod-shaped capsids, with a core of double-stranded DNA surrounded by two membranes, an *intimate* membrane, probably a single protein layer corresponding to a capsid, and an envelope that surrounds the viral bundles and contains protein and lipids (Figure 18-2).

The DNA of each insect virus has characteristic base composition. Its size has been estimated at about 10^8 daltons per virion. The virions have sizes reported as about 30 to 50 by 200 to 300 nm for different viruses and are often accompanied by smaller, spherical particles. The suggestion has been made that the spherical elements are developmental forms and a developmental cycle has been described, leading from the spherical particles through immature rods to mature rod-shaped virions (see Bergold, 1958). The evidence for this cycle is only morphological, however, and the small particles may be incomplete or degraded virions rather than developmental forms.

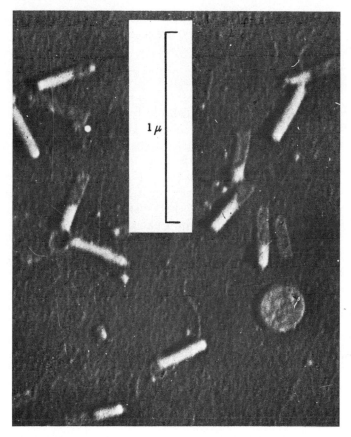

FIGURE 18-2. Virions from polyhedra of Bombix mori (silkworm), some intact, others showing empty "intimate membranes." Courtesy Dr. G. H. Bergold.

CYTOPLASMIC POLYHEDROSES

These are found in lepidoptera and diptera and are characterized by formation of masses of small polyhedra within membranes. Any given virus apparently produces either cytoplasmic or nuclear inclusions, not both. The virus particles in the cytoplasmic polyhedra are generally near-spherical and contain double-stranded RNA in 10 separate segments, similar to that of reoviruses (Lewandowsky and Millward, 1971). Like reoviruses, a cytoplasmic polyhedrosis virus introduces into the host cell a preformed RNA polymerase.

GRANULOSES

Granuloses are diseases of lepidoptera. The inclusions are oval grains or capsules of protein, 200×500 mμ, each containing one or a few rod-shaped, DNA-containing virions, resembling those of nuclear polyhedroses (see Figure 3-24). The capsules are found mainly, but not exclusively, in the cytoplasm of the infected cells and are released when the cells disintegrate. Hughes (1952) has described a process of formation of capsules around virus particles proceeding from one end of the virion to the other. Long branched filaments, possibly abnormal virion derivatives, are also found in cells of infected larvae (Smith and Brown, 1965).

NONINCLUSION DISEASES

Noninclusion diseases are exemplified by the iridescent virus diseases of the crane fly *Tipula paludosa* (Xeros, 1954) and of the beetle *Sericesthis pruinosa* (Steinhaus and Leutenegger, 1963). The sacbrood disease of the honey bee, in which no inclusions have been observed, is also attributed to a virus. The capsids of the iridescent viruses are perfect icosahedra, 120 to 150 mμ across (see Figure 3-1). These virions form crystals within the fat bodies of the infected larvae and the presence of the crystals gives the infected larvae the typical iridescent coloration, also apparent in pellets of virions collected by centrifugation. The iridescent viruses can be titrated more efficiently than the inclusion viruses and lend themselves better to growth studies in tissue culture (Bellett, 1968).

VIRUS MULTIPLICATION

Descriptions of developmental cycles of polyhedral viruses based on morphological appearances are at best suggestive of the processes of viral multiplication. The assay of virus infectivity in larvae is complex and inefficient and quantitative growth experiments are practically impossible. The

FIGURE 18-3. Growth curve of Sericesthis iridescent *virus in tissue culture of* Antherea eucalypti. ● *Cells stained by antiviral fluorescent antibody.* ○--○ *Titer of cell-associated virus.* ○—○ *Titer of released virus. From Bellett,* Virology **26,** *132 (1965).*

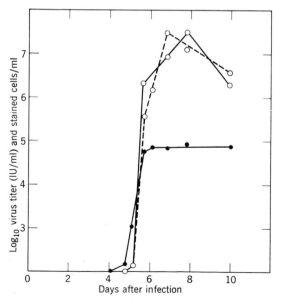

situation has been improved greatly by the development of methods for continuous cultivation of ovarian cells from insect larvae, for example, those of *Antherea eucalypti*. These cell lines can be infected with polyhedral viruses and develop polyhedra. They have been used even more successfully as hosts for the *Sericesthis* iridescent virus (Bellett, 1965).

Titration is done using fluorescent antibody, which reveals intracytoplasmic foci of infection within a few days after addition of virus. Tritiated thymidine and acridine orange detect the presence of DNA in the foci of virus growth. In multiple infection, several foci can form in a single cell, where they develop in synchrony. The infectivity ratio of about one focus per 80 virions adsorbed is about as high as that found with the same virus titrated in intact animals. The intracellular foci contain virus particles, and additional virions are found free in the cytoplasm and in buds on the cell surface. As the particles are released from these buds they acquire an external membrane, probably of cellular origin, around their icosahedral capsids. This process resembles the release of myxoviruses and other animal viruses with lipid envelopes.

A cell infected with *Sericesthis* virus may contain as many as 500 infectious units; that is, 40,000 or more virions, most of which are released by the end of a single growth cycle lasting about a week (Figure 18-3). Despite this extensive viral production in the cytoplasm, the cells continue to make their own DNA.

VIRAL LATENCY

A remarkable feature of insect viruses is their ability to persist in a latent state through many generations of the insect host. Transovarial transmission almost certainly occurs, although reinfection of larvae by feeding has not been excluded. In the latent state the viruses produce no detectable symptoms, but infectious virus and symptoms of disease appear suddenly as though activated by some internal or external factor. A thermal shock or a food change, for example, a new source of mulberry leaves, can be the inducing factor. Similar effects are elicited by X-rays or chemicals. Yamafuji (1964) reported that formaldehyde, hydroxylamine, peroxides, oxines, and nitrite can serve as stimuli in provoking silkworm polyhedrosis.

Even before the mutagenic action of many of these chemicals was known, these observations led Yamafuji to develop a theory of the origin of polyhedral viruses from host genetic materials in response to mutagenic treatments that affected the chromosomal DNA. This theory has not gained much support because it is known that deliberate inoculation of silkworm larvae with polyhedrosis virus can often lead to inapparent, latent infection rather than disease (Vago, 1951). Hence the silkworms in which virus appears after chemical treatment may have been virus carriers. A provirus theory of latency of polyhedrosis viruses seems preferable to one of *de novo* conversion of host genetic elements into viral genomes.

THE CO_2 SENSITIVITY VIRUS OF *DROSOPHILA*

An important series of studies has been devoted to a transmissible agent that controls CO_2 sensitivity in certain races of *Drosophila*; the sensitive flies are paralyzed and killed by an exposure to CO_2 that would produce only a reversible narcosis in resistant flies. The genetic factor, called *sigma*, is similar to viruses in being transmissible and mutable and in its size and host specificity. In some of its behavior it resembles temperate bacteriophages. The original findings can be summarized as follows (L'Héritier, 1951, 1958).

The CO_2-sensitive flies contain a virus that can be extracted and injected into resistant flies and renders them sensitive to CO_2. Sensitivity appears after several days, the length of this incubation period depending on the size of the inoculum. In the newly inoculated flies, the infectious titer as measured on extracts first decreases to a low level, then rises rapidly until a maximum is reached at the time when CO_2 sensitivity appears.

Sensitivity to CO_2 passes from a sensitive female to its progeny in one of two ways. A recently infected fly will transmit sensitivity only irregularly to some of its offspring, probably by infection of oocytes as they develop in the ovary. Some of those offspring that have been infected in the egg stage become "stabilized sensitives," which remain sensitive for life and transmit sensitivity to their offspring much more effectively. In fact, stabilized-sensitive females transmit sensitivity to all their progeny, whereas stabilized-sensitive males, when mated to a nonsensitive female, give a variable proportion of sensitive offspring. These father-infected flies behave like newly infected, unstabilized sensitive flies. In the stabilized sensitive flies the level of infectious *sigma* virus is significantly lower than in flies made sensitive by recent infection. *Sigma* has some pathogenic effect on the host in that it reduces viability of the eggs (Seecof, 1964).

The *sigma* virus is mutable. Some of its mutants are defective; they do not produce infectious viral progeny, but confer immunity against exogenous virus. Reverse mutations can restore the normal efficiency. Host genes controlling refractoriness to *sigma* are known, as well as host range mutants of *sigma* that become able to infect refractory hosts. Some mutations of *sigma* are listed in Table 18-1

These and other facts fit the following interpretation (L'Héritier, 1958). Upon infecting CO_2-resistant, virus-receptive flies, the virus, after an apparent eclipse, multiplies vegetatively and matures, producing a maximal crop of

TABLE 18-1 MUTATIONS OF THE VIRUS *SIGMA* OF *DROSOPHILA*

Gene	Allele	Phenotype
g	g^+	Hereditary transmission present after infection
	g^-	Hereditary transmission absent after infection
pr	pr^+	Prolonged hereditary transmission in pr^+g^+ flies
	pr^-	Shorter hereditary transmission in pr^-g^- flies
P	P^+	Can multiply in refractory flies
	P^-	Cannot multiply in refractory flies
v	v^+	High frequency of transmission from stabilized male flies
	v^-	Low frequency of transmission from stabilized male flies
Tr	*Tr*	Replicates at 30°
	Ts	Does not replicate at 30°

Data from Ohanessian-Guillemain, *Ann. Génétique* 1: 59 (1959).

progeny. CO_2 sensitivity appears when the amount of vegetative virus has reached a threshold value. The stabilized condition may represent a more intimate integration of the virus in a noninfectious form which only occasionally gives rise to mature virus. The stabilized form can be transmitted by a female to all its offspring, female or male. When carried by a male gamete into a zygote it is changed to the unstabilized, vegetative form.

A surprising and important development has been the identification of *sigma* virions as rhabdovirus-like in morphology (Printz, 1973). Moreover, the rhabdovirus vesicular stomatitis virus, injected into *Drosophila*, can persist and produce a stable infection and persistent CO_2 sensitivity. Thus one more link is established among viruses discovered in very distant types of hosts.

Origin and Nature of Viruses

Virus as Independent Genetic Systems

Where do viruses belong in the biological world? What are their origins and closest relatives? The knowledge of viruses outlined in the preceding chapters has illustrated the thesis, put forward at the outset of this book, that viruses are not small cells. They are elements of genetic material that possess an evolutionary history of their own because they have the intrinsic ability to mediate their own transfer from one host to another.

In this sense, viruses are independent genetic systems. They are not accidentally separated fragments of a cell genome. They are endowed with genetic continuity and mutability and contain sets of genes working in concert to make more virus. They have their own evolution, which is independent, to some extent at least, of the evolution of the organisms in which they reproduce.

Yet viruses are not strangers to the evolutionary history of cells and organisms. Their genetic material is chemically akin to that of all cells, even though in many viruses it consists of RNA, a coded polymer that in the evolution of cells has been relegated to a subordinate position, serving as a delegate bearer rather than a primary repository of genetic messages. If DNA is the sun, cellular RNA resembles the planets, shining with reflected light; but in RNA viruses we discover that some planets glow with a radiance of their own.

The independence of the viruses as genetic systems is itself subject to evolutionary changes. The genome of a temperate bacteriophage may become physically and functionally integrated into that of a bacterium; it has two ways of existence, that of a virus and that of a set of chromosomal genes. When a

481

temperate phage loses by mutation its ability to become prophage it loses one of its existences; it becomes "more virus" and less of a set of cellular genes. In the opposite direction, when a prophage mutates to defectiveness, that is, to inability to carry out all the functions necessary for its own reproduction and for transfer to other hosts, it becomes "less virus," dependent for continued existence on the persistence of its host cell line or the helper function of other, nondefective viruses.

The physical integration of viruses into the host's chromosome has as yet been analyzed in detail only for phages, but many tumor viruses are also capable of chromosomal integration. Examples of defective and helper-dependent viruses are known in all groups and include not only rare accidents of experimentation, but natural cases of undoubted evolutionary significance, such as the satellite virus of tobacco necrosis and the Rous sarcoma virus.

The conversion of a virus to a defective virus residing in the genome of a host cell may formally be considered as a conversion of a set of viral genes into a subset of cellular genes. Conversely, groups of cellular genes might become viral genomes, a possibility that may not be limited to genes that had been brought in by viruses. Thus viral genomes can conceivably evolve from nonviral genetic elements. The question to which we must address ourselves is: Which events play significant evolutionary roles in the origin of viruses as organisms and in the natural history of their genetic material?

Theories of Virus Origin

To guide us in the search we have our knowledge of viruses as genetic elements as well as our knowledge of cells, their evolution, their genetic components, and their functional organization.

Cellular heredity (as distinct from developmental specification) can in principle be embodied in informational patterns of two kinds: genes and primers. Genes are coded macromolecules of nucleic acids, which provide templates for the synthesis of specific macromolecules. Primers, a less well-understood group, consist of structures that supposedly direct the specific extension of existing molecular patterns (Sonneborn, 1967). Priming mechanisms are displayed in the role of preexistent glycogen molecules in the enzymatic growth of glycogen chains (Robbins et al., 1966) and in the role of the bacterial murein cell wall layer (Weidel and Pelzer, 1964). In these cases the existing molecules are needed as acceptors of newly made elements. Notice that this primer action differs from the intrinsic assembly-promoting qualities of structural proteins such as actin, myosin, or viral capsid subunits.

Informational patterns underlying viral heredity are not ambiguous. Viral genomes are composed of nucleic acid genes. If we search for relatives of viruses we must look for nucleic acid containing structures. Specifically, we must ask whether viruses may have evolved from entire cells, or individual cellular components, or precellular or acellular forms of genetic materials. In this search we must keep in mind that there is no *a priori* reason to assume that all viruses have originated in the same way. In fact, the existence of RNA and DNA viruses suggests that at least two avenues of evolution have been followed.

The Regressive Theory of Virus Origin

The complexity of some of the largest DNA viruses, with hundreds of genes and elaborate developmental processes, lends some plausibility to a theory of virus origin according to which virus represent the product of a regressive evolution from free-living cells.

Studies on comparative biochemistry have shown that all cells, from bacteria to mammalian cells, share certain basic synthetic and metabolic processes and that similar essential metabolites are needed by all organisms as building blocks. The earliest self-reproducing units, the primitive life forms, may have been nucleic acid molecules that came into existence by chemical reactions utilizing energy of ultraviolet light and electrical discharges, with hydrogen cyanide as a probable intermediate (Miller and Orgel, 1974). Such nucleic acid molecules, using the available supply of organic compounds also derived by prebiotic chemical reactions, may have succeeded in directing the synthesis of primitive enzymes to catalyze their own reproduction. Cellular organization was probably a relatively early event leading to more efficient use of substrates. The first cells may then have acquired more and more synthetic power by mutation (Horowitz, 1945) and the synthetically more efficient forms may have selectively been favored when the external supply of organic substances dwindled, leading to the establishment of autotrophic forms.

In the course of subsequent evolution some of the synthetic powers were often lost again. Today most animals and microorganisms, as well as some plants, depend on organic substrates. This regressive biochemical evolution (Lwoff, 1943) has made many classes of organisms dependent on others for their nutrition. Plants, photosynthetic bacteria, and chemo-autotrophic bacteria provide the entry point for synthetic energy into the biological world and manufacture the primary foods for all other living things. In extreme cases, nutritional dependence has evolved into true parasitism.

VIRUSES AS REGRESSED PARASITES

A case has been made for an origin of viruses by an extreme process of regressive evolution (Green, 1935; Burnet, 1945). Free-living microorganisms requiring a number of preformed metabolites would become parasitic on other organisms that supply the needed substances. If the requirements involved substances that cannot pass through cell membranes, intracellular parasitism would become the only way to survival. Parasitism, and especially intracellular parasitism, could then favor further loss of synthetic abilities by providing an environment where most needed metabolites are present, thus placing little or no handicap on synthetically deficient mutants. This process could continue to such an extent as to cause the loss of most of the enzymatic machinery and genetic material of the parasite, with accompanying reduction in size and transformation into a virus. A virus could ultimately result, corresponding to a small portion of the genome of a formerly free-living organism.

Today this theory has little to recommend it, at least in its original form. Although there is some parallelism between the degree of parasitism and the nutritional requirements of facultative parasites—bacteria, fungi, and protozoa—hardly any transitional stages are known between viruses and obligately parasitic protozoa such as the malarial plasmodia, or between viruses and obligately parasitic bacteria such as the leprosy bacillus.

Rickettsiae and the Chlamydozoaceae are intracellular parasites that have sometimes been considered as intermediate between viruses and bacteria. We now know that rickettsiae are small gram-negative bacteria which multiply by growth and fission, possess a bacterial-type cell wall, including a rigid mucopeptide layer, and have an autonomous energy-producing system (Moulder, 1962). The exact nature of the synthetic and enzymatic deficiencies that constrain the rickettsiae to intracellular parasitism is still unknown. The pathogenic rickettsiae are primarily parasites of arthropods, only occasionally pathogenic for vertebrates. They may be related to a group of bacteria-like symbionts found in the cells of many arthropods. These, as well as yeast-like symbionts, are carried in the cells of special organs called *mycetomes* and are apparently transmitted from the female to its brood by infection of the egg. Some of these parasites perform functions needed by the arthropod host since, if they are removed with antibiotics, the host becomes sterile or dies (Steinhaus, 1946).

The Chlamydozoaceae (Moulder, 1964) are also procaryotic parasites, only further regressed. For a long time they were considered as viruses (the lymphogranuloma-psittacosis-trachoma group). They have no complete rigid cell wall but possess some components of bacterial mucopeptide. They contain both DNA and RNA and multiply by binary fission. There is no evidence that these organisms ever replicate in the form of free nucleic acid; they retain a cellular structure, resembling in this respect the mycoplasma (or *pleuropneumonia-like organisms*) and the L forms of bacteria (Smith, 1971) which, however, are not committed to an intracellular parasitic life.

The strongest argument against the regressive origin of viruses from cellular parasites is their noncellular organization. The viral capsids are morphogenetically analogous to cellular organelles made up of protein subunits, such as bacterial flagella, actin filaments, and the like, and not to cell membranes. The outer envelopes, when present, either have no membrane-like structure or, if they possess a true membrane, its phospholipid bilayer is of cellular origin, although the membrane can incorporate viral enzymes and other proteins.

Some components of virions are functionally active, including RNA polymerases, replicases, and reverse transcriptases described in preceding chapters. The picture of virions as purely passive packaging devices is oversimplified: precise assembly mechanisms preside over the orderly incorporation of functional components (Casjens and King, 1975) comparable, for example, to the assembly of functional proteins in the contractile apparatus of muscle fibers. But even the most complex virions like those of pox viruses, which contain the means for establishing their own viral "factories" in the host cells, exhibit none of the features of cellular organization.

No known virus packages into its virions the machinery needed for such basic cellular functions as energy generation or protein synthesis. All viruses depend for these functions on their ability to coopt host enzymes and ribosomes for their own ends. Nor does any virus undergo anything resembling binary fission. Even where viral nucleic acid fails to dissociate completely from virion proteins in infection, the virion structure as such is dissolved before the nucleic acid is replicated and later reassembled into new virions.

There is thus a profound discontinuity between cellular parasites like the Chlamydozoaceae and the viruses. If regression from parasites were the origin of viruses, we should postulate intermediate stages involving the loss of most of the genome and evolution of capsid and envelopes. Although such a sequence of events cannot be excluded, it must be realized that these steps are in no way bound to parasitism. That is, evolution of replicative autonomy and of virion components could have involved just as well some portions of the genome of independent cells as of the genome of parasitic cells.

Thus once we conceive of viruses as genetic elements less complete than a cell genome, the distinction becomes blurred between the theory of regressive origin and the next group of theories, which consider viruses to be subcellular elements related to cellular constituents.

Viruses and the Genetic Elements of Cells

We can distinguish three classes of genetic elements in cells: (1) chromosomal DNA, organized somewhat differently in procaryotes and eucaryotes; (2) organellar DNA as in mitochondria and chloroplasts; and (3) the DNA of bacterial plasmids (Schlessinger, 1975). No evidence exists today for the presence in cells of genetic RNA.

VIRUSES AND EUCARYOTIC ORGANELLES

Mitochondrial and chloroplast DNA are not particularly suggestive of analogies to viral genomes, even though their size—between 10^7 and 3×10^7 for mitochondria (Borst, 1972) and about 10^8 for chloroplasts (Gibor, 1965)— falls in the range of viral DNAs. These genetic elements code for the RNA of the organellar ribosomes and probably also for a few organellar proteins, but most of the functional ones, such as the cytochromes and the enzymes of the Krebs cycle, are of chromosomal origin.

A number of biochemical arguments have recently been collected (Margulis, 1970) in support of the long-standing suggestion that mitochondria and chloroplasts may be regressed parasites of procaryotic origin. The protein synthesizing machinery of these organelles resembles the bacterial machinery and differs from that of the eucaryotic cytoplasm. Whereas this fact has sometimes been taken to suggest that organelles are regressed parasites, it must be appreciated that all the genes in eucaryotes, both nuclear and organellar, must ultimately have been derived from procaryotic ancestors. The point at issue is thus not whether organellar genes are related to those of procaryotes, but whether they came from a procaryotic parasite, rather than from the procaryote-derived genes of early eucaryotes. The biochemical similarities between organelles and present-day procaryotes might indicate simply that the rate of gene evolution has been more rapid in the eucaryotic nucleus than in organelles or free-living procaryotes.

If organelles are in fact regressed, highly specialized parasites, their evolution, presumably restricted within a cell lineage, would probably have been very different from the evolution of viruses, which must involve unique selective pressures in the course of host-to-host transfers. Yet, the evolution of mitochondria and chloroplasts from procaryotic parasites, if confirmed and further elucidated, might shed light on some aspects of the merging of genetic elements into a functionally integrated cellular system and might help interpret other instances of genetic merging, such as those observed in viral infections.

Other components of eucaryotic cells, such as centrioles and kinetosomes, have been reported to contain DNA or RNA (Randall and Disbrey, 1965). The origin and independence of this nucleic acid is not established, however. Amplification of specific groups of chromosomal genes, such as occurs for example for ribosomal RNA genes in the maturation of oocytes, might be its source.

A central point needs emphasis. There is nothing in the normal eucaryotic cell that resembles viral capsids, that is, protein shells specifically determined by the genetic elements that they will enclose. Whether or not all naturally occurring virus-like objects are related to infectious agents, they all seem to be programmed for export from the cells in which they were assembled, rather than being endowed with intracellular genetic continuity. If viral genomes originate from cellular genetic elements the invention of the capsid must have been a novelty, utilizing the self-assembly propensity of certain proteins for a new experiment in evolution.

BACTERIOPHAGES AND THE GENETIC MATERIAL OF BACTERIA

The relation between bacteriophages and the bacterial DNA is better understood and, therefore, more illuminating.

Temperate phages can enter and remain in physical association with the host chromosome as prophages. When a prophage is induced to undergo unrestricted vegetative multiplication of the phage genome, it is as though a chromosomal fragment multiplied out of context. In fact, specialized transduction entails just such multiplication of bacterial genes. It is tempting, therefore, to consider phages or at least temperate phages as fragments of bacterial genome that have acquired or regained the property of independent replication and a capacity to construct virions. Replication would then be a kind of gene amplification.

Bacteria often contain other accessory genetic factors besides temperate phages (Hayes, 1968). Some of these factors or plasmids are strictly extrachromosomal, multiplying autonomously as separate *replicons* (Jacob et al., 1963). This is true, for example, of the colicinogenic (Col) factors (Hardy, 1975), which determine production of bacteriocins. These plasmids replicate in more or less strict unison with the chromosome (stringent or relaxed control; see Chapter 10). Col factors, however, also undergo unrestricted replication, resembling that of a phage genome. Unrestricted replication of Col factors is usually associated with the production of colicin protein, but is not bound to it since a plasmid-like ColE1 will produce thousands of copies per cell when protein synthesis is inhibited (Schlessinger, 1975).

Another class of bacterial plasmids, often referred to as *episomes* (Jacob and Wollman, 1961), resemble temperate phage genomes in their ability to integrate into the bacterial chromosome. The best-known instance is that of the F (fertility) factors that determine mating behavior in bacteria. An F factor can persist as an independent replicon or become inserted into the bacterial chromosome at a variety of locations. The DNA of the F factors as well as of other plasmids is circular (Schlessinger, 1975), a fact that places the integration event on a par, topologically speaking, with the insertion of a circular λ genome. The locations of insertion depend on the presence in both F factors and host chromosome of specific "insertion sequences" about 1000 nucleotides in length (see Table 10-2). The insertion event itself can proceed by precise recombination between homologous insertion sequences in F factor and chromosome. Insertion can also take place, at a somewhat lower rate, by mechanisms specific to the insertion sequences, perhaps involving specific integrase enzymes like that of phage λ.

Recombinational integration of plasmids and its reversal can lead to a variety of suggestive situations. Two plasmids may fuse, for example, and replicate under the control of either replicon. Or a plasmid that has entered the bacterial chromosome may later emerge carrying with it a segment of chromosomal material that was located next to it and thereby becomes part of the plasmid—in analogy with temperate phage genomes.

Plasmids share with phages the property of being replicons. It should be made clear that ability to replicate autonomously is not a property of any fragment of DNA that enters a bacterial cell. Fragments entering either through the cell wall or from transducing phage particles fail to multiply unless they become integrated into the cellular genome.

Another common attribute of phages and certain bacterial plasmids is that both possess genes that code for structures specialized for cell-to-cell transfer of genetic materials. We may suppose that genetic transfer elements of all sorts have evolved in procaryotic cells, some more, some less successful in evolution. Phages could represent one group of such elements. By transfer to new hosts they may have acquired parasitic, destructive habits and may then have undergone much evolution of their own. It is not surprising, therefore, that the DNA of some phages contains odd bases—uracil, hydroxymethyl uracil, or hydroxymethyl cytosine—replacing the usual ones. These changes may have served the adaptation to new hosts by allowing more successful competition. It is not even excluded that the odd phage DNAs may represent remnants of early, less successful experiments in the evolution of genetic material.

RNA Viruses and Cellular RNA

The RNA viruses present formidable problems. There are no satisfactory analogies, in bacteria or other organisms, for genetic elements replicating as RNA. A major reservation is suggested by the finding that different RNA phages of the same group have different, specific replicating mechanisms (Haruna and Spiegelman, 1965). If that is so, the existence of unrecognized, cellular RNA-replicating systems must be considered as a possibility and has in fact been postulated, mostly on inadequate evidence.

Otherwise, we must fall back on one of several choices. For example, we might consider the RNA viruses as a unique group of genetic elements forming a separate line of evolution. A second alternative is to think of RNA viruses as derived from DNA viruses whose messenger RNA has become capable of direct replication, so that transcription from DNA is unnecessary. But if such an event were possible, there might be no reason to restrict the possible origin of RNA viruses to viral mRNA. Just as we must give serious consideration to an origin of DNA viruses from cellular DNA, so must we also consider a possible origin of RNA viruses from cellular messengers. The required steps would be acquisition of a replicating mechanism and formation of a virion.

The existence of viruses that code for a reverse transcriptase, which transcribes the RNA of the virion into DNA, offers another possibility: some such viruses may have acquired an RNA replicase and thereby succeeded in bypassing reverse transcription, and, incidentally, prevented integration into the cellular chromosomes.

These various possibilities, taken as a whole, illustrate a series of processes by which segments of genetic material may have shifted from chromosomal to viral status (and vice versa) through mechanisms involving RNA replication. There are, unfortunately, too many gaps in our knowledge to justify formulation of a useful model.

Another puzzle is the existence of a class of viruses with multipartite double-stranded RNA genomes, which includes parasites of all kinds of organisms, including: bacteria, fungi, plants, insects, and vertebrates. Do all these viruses stem from a common ancestor? or did the multipartite, double-stranded RNA genome arise repeatedly in several lines of descent because of some intrinsic selective value?

Virus Origin and the Origin of the Cell

The discussion of virus origin is essentially one of the reproductive and evolutionary independence of genetic elements. The critical questions are, how much independent evolution viruses may have undergone, and at which point their evolution started to diverge from that of the genetic elements now found in cells.

A virus may enter a cell and remain in it for only a fraction of a cell generation; or it may enter and remain for many cell generations. Some viruses can be transmitted through the gametes to successive generations of a sexually reproducing organism. A long-persisting virus is practically indistinguishable from a cell component. It might be considered a virus, or a plasmid, or a gene, depending on the type of effect by which it happens to be detected.

The problem of virus origin involves (1) the relation of viruses to cellular components; (2) the origin of cellular components; and (3) the relations among viral genes. It has been a rather widely held opinion that the cell is monophyletic in origin; that is, the genetic complement of a cell, its genome, has originated by differentiation of an original self-reproducing single element, copies of which have occasionally remained together and by mutation have assumed different forms and functions. Such gene groups would then have developed into chromosomes because of strong advantages that an organized mechanism for equipartition of the genetic material offers in preserving favorable gene combinations. Later in evolution, sexual phenomena brought about complications, but in presexual organisms all genes would have originated within one single-cell lineage. The cytoplasm, in the simplest hypothesis, would be entirely derived from the activity of the genes. As a result, any genetic component of a cell should be a close relative of all others within one genetic line. The transmission of such a component—gene or chromosome—to another cell would represent a *merging* of a portion of a genetic line into another line.

The normal cell might, on the other hand, be polyphyletic in origin. Several primitive self-replicating molecules may have come together into a successful combination and developed into a cell genome. Alternatively, some such elements may have entered an already formed cell. The merging of genetic lines might have occurred relatively early in the evolution of the cell, and the case of genes, chromosomes, or plasmids becoming transmissible from cell to cell would represent a reacquisition of the original independence and a repetition of the original merging process.

Thus all speculations about origin of viruses lead to consideration of possible modes of merging two or more genetic elements to produce a functioning genetic system. Such merging may not be obvious in the case of rapidly destructive viruses, and its fundamental importance escaped early virologists who inevitably thought of viruses in a cell as of bacteria in a culture. But even a cell infected with a virulent virus and doomed to early destruction is a functional system whose ultimate fate—death and disintegration—is incidental to the primary event of genetic and biochemical integration of viral and cellular machinery.

Merging leads to a lasting integration with viruses that are carried for several cell generations or even through the sexual process. Integration may become almost permanent for prophages and possibly some tumor viruses; some plasmids and even chromosomal segments may have evolved in this way. On the other hand, the evolution of mechanisms for genetic transfer may convert genes and groups of genes into plasmids and viruses.

It has been suggested (Hershey and Dove, 1971; Campbell, 1976) that groups of related viruses may have evolved not by mutational events from a common viral progenitor but by the coming together of a number of "modules," that is, segments of genetic material coding for coordinate functions such as capsid assembly or DNA replication. Different viruses could have received one or a few modules common to other viruses and recombination could then generate greater variety. Merging of genetic elements would then be of the essence, not only in the entry of viral genomes into chromosomes, but also in the very invention of viruses.

Such merging is in fact implicit in some of our previous discussions, though couched in different terms. When we speak, for example, of the possibility that a fragment of host DNA might acquire the ability to replicate, we do not mean to imply that a new polymerase gene and replication origin have arisen from base sequences within that fragment. Although such a process might sometimes happen, a more frequent route is likely to be the addition to the fragment of a DNA segment, either viral or host in origin, that already contained replication genes. In that event, it is important to note that the new element thus formed has a dual ancestry. We may elect to focus our attention on the original host fragment, and describe the event as the acquisition by that fragment of ability to replicate; but another observer recording the same facts might say that the replication segment had acquired the genes of the fragment.

Of all living things viruses are perhaps least likely to have a monophyletic origin, because they always replicate in the presence of large amounts of nonviral nucleic acids, which can become incorporated into the viral genome.

Ultimately, the definition of a certain element—gene, plasmid, or virus— will depend on the duration of its joint evolutionary history with other components of the genome. The ability to retain independence may be a function, not only of intrinsic mutability, but also of the duration of the partnership, which may force an ever-increasing interdependence among the various cell components. An exogenous element, placed within a cell lineage, would probably undergo as much evolutionary change as any other genetic components of the cell and would not resemble its primitive ancestor any more than these components resemble theirs.

Thus the basic similarity of viruses to other genetic elements has led us to a somewhat unified viewpoint by showing that different theories of virus origin differ only in their interpretation of the relative duration of the companionship between genetic elements. Merging of genetic lines and integration at the cellular level are the common denominator of all theories. Viruses could derive from one (or several) of the genetic components of cells, either by regressive evolution, or by acquisition of a viral transfer mechanism. Once a transfer mechanism has been acquired, further evolution can then proceed in other hosts.

If a transferred element proved to be rapidly destructive for the new cell system it would generally be lost and could only maintain itself through the availability of innumerable suitable hosts. Often, however, the merging might last. The meaningful question remains—and cannot be answered at present— whether such merging is a novel and exceptional feature, leading mainly to the formation of abnormal complexes of low evolutionary value, that is, of diseased cells, or if it is an example of a process that has played and is still playing an important role in evolution and possibly in development.

A virus may be both a regressed parasite and a partial cell genome that has become infectious, depending on which phase of the evolutionary history of its genetic material we are observing. It may have been both things at different times.

We see then that, just as the study of virus structure and multiplication always leads us back to the cell as the system in which the phenomenon of life takes place, so the problem of virus origin has led us back to the origin of the cell as an integrated whole. A virus is essentially part of a cell. We observe and recognize as viruses those parts independent enough to pass from cell to cell and we compare them with other parts that are more tightly tied up with the whole system. It is indeed this aspect of viruses that makes them invaluable to the biologist, whom they present with the unique opportunity to observe in relative isolation the active determinants of biological specificity, which are truly the stuff of which all life is made.

Bibliography

Abercrombie, M. (1962). Contact-dependent behavior of normal cells and the possible significance of surface changes in virus-induced transformation. *Cold Spring Harbor Symposia Quant. Biol.* **27**, 427–431. [438]

Abercrombie, M., and Heaysman, J. E. M. (1954). Observations on the social behavior of cells in tissue culture. *Exp. Cell Res.* **6**, 293–306. [438]

Abraham, G., Colonno, R. J., and Banerjee, A. K. (1976). Evidence for the synthesis of a "leader" RNA segment followed by the sequential transcription of the genes of vesicular stomatitis virus. In: *Animal Viruses* (ed. D. Baltimore et al.). Academic Press, New York, pp. 439–456. [330]

Abraham, G., and Cooper, P. D. (1975). Relations between poliovirus polypeptides as shown by tryptic peptide analysis. *J. Gen. Virol.* **29**, 215–227. [319]

Acheson, N. H. (1976). Transcription during productive infection with polyoma virus and simian virus 40. *Cell.* **8**, 1–12. [357]

Acheson, N. H., Buetti, E., Scherrer, K., and Weil, R. (1971). Transcription of the polyoma virus genome: synthesis and cleavage of giant late polyoma-specific RNA. *Proc. Natl. Acad. Sci. U.S.* **68**, 2231–2235. [357]

Acs, G., Klett, H., Schonberg, M., Christman, J., Levin, D. H., and Silverstein, S. C. (1971). Mechanism of reovirus double-stranded ribonucleic acid synthesis *in vitro* and *in vivo*. *J. Virol.* **8**, 684–689. [341]

Adams, M. H., and Park, B. H. (1956). An enzyme produced by a phage-host cell system. II. The properties of the polysaccharide depolymerase. *Virology* **2**, 719–736. [17]

Agar, A. W., Alderson, R. H., and Chessoe, D. (1974). Principles and practice of electron microscope operation. North-Holland/American Elsevier, New York. [92]

Alberts, B. M., and Frey, L. (1970. T4 bacteriophage gene 32: a structural protein in the replication and recombination of DNA. *Nature* **227**, 1313–1318. [167]

Albertson, P. A. (1967). Two phase separation of viruses. In: *Methods in Virology*, Vol. II (eds. K. Maramorosch and H. Koprowski). Academic Press, New York, pp. 303–322. [105]

Alexander, H. E., Koch, G., Mountain, I. M., Sprunt, K., and Van Damme, O. (1958). Infectivity of ribonucleic acid of poliovirus on HeLa cell monolayers. *Virology* **5**, 172–173. [309]

Allison, A. C. (1972). Immunity against viruses. In: *The Scientific Basis of Medicine; Annual Review*. Athlone Press, London. [425]

Aloni, Y. (1974). Biogenesis and characterization of SV40 and polyoma RNA's in productively infected cells. *Cold Spring Harbor Symposia Quant. Biol.* **39**, 165–178. [357]

Aloni, Y., Shari, M., and Reveri, Y. (1975). RNA's of simian virus 40 in productively infected monkey cells: kinetics of formation and decay in enucleate cells. *Proc. Natl. Acad. Sci. U.S.* **72**, 2587–2591. [357]

Alwine, J. C., Reed, S. I., Ferguson, J., and Stark, G. R. (1975). Properties of T antigens induced by wild-type SV40 and tsA mutants in lytic infection. *Cell* **6**, 529–533. [354]

Amako, K., and Dales, S. (1967). Cytopathology of Mengovirus Infection. 1. Relationship between cellular disintegration and virulence. *Virology* **32**, 184–200. [306]

Anderson, C. W., Baum, P. R., and Gesteland, R. F. (1973). Processing of adenovirus 2-induced proteins. *J. Virol.* **12**, 241–252.

Anderson, T. F. (1952). Stereoscopic studies of cells and viruses in the electron microscope. *Am. Naturalist* **86**, 91–100. [35]

Andrewes, C. H. (1942). Interference by one virus with the growth of another in tissue culture. *Brit. J. Exp. Path.* **23**, 214–220. [408]

Apriletti, J. W., and Penhoet, E. E. (1974). Recovery of DNA-dependent RNA polymerase activities from L cells after mengovirus infection. *Virology* **61**, 597–601. [406]

Arber, W. (1965). Host-controlled modification of bacteriophage. *Ann. Rev. Microbiol.* **19**, 365–378. [155]

Arber, W. (1974). DNA modification and restriction. *Prog. Nucleic Acid Res.* **14**, 1–38. [75, 81, 155]

Arber, W. (1974). An *Escherichia coli* mutant which inhibits the injection of phage λ DNA. *Virology* **58**, 504–513. [149]

Arber, W., and Linn, S. (1969). DNA modification and restriction. *Ann. Rev. Biochem.* **38**, 467–500. [156]

Armstrong, J. A., Edmonds, M., Nakazato, H., Phillips, B. A., and Vaughn, M. (1972). Polyadenylic acid sequences in the virion RNA of poliovirus and eastern equine encephalitis virus. *Science* **176**, 576–578. [309]

Arscott, J. T., and Goldberg, E. B. (1976). Cooperative action of the T4 tail fibers and baseplate in triggering conformational change and in determining host range. *Virology* **69**, 15–22. [161]

Arstila, P., Halonen, P. E., and Salmi, A. (1969). Hemagglutinin of vesicular stomatitis virus. *Arch. Fur Ges. Virusforsch.* **27**, 198–205. [284]

Ash, J. F., Vogt, P. K., and Singer, S. J. (1976). The reversion from transformed to normal phenotype by inhibition of protein synthesis in rat kidney cells infected with a temperature-sensitive Rous sarcoma virus mutant. *Proc. Natl. Acad. Sci. U.S.*, in press. [443, 448]

Atchison, R. W. (1970). The role of herpeviruses in adeno-associated virus replication *in vitro. Virology* **42**, 155–162. [344]

Aub, J. C., Sanford, B. H., and Cote, M. N. (1965). Studies on reactivity of tumor and normal cells to wheat germ agglutinin. *Proc. Natl. Acad. Sci. U.S.* **54**, 396–399. [443]

Avery, O. T., Macleod, C. M., and McCarty, M. (1944). Studies on the chemical nature of the substance inducing transformation of pneumococcal types. *J. Exp. Med.* **79**, 137–158. [5]

Axel, R., Melchior, W., Sollner-Webb, B., and Felsenfeld, G. (1974). Specific sites of interaction between histones and DNA in chromatin. *Proc. Natl. Acad. Sci. U.S.* **71**, 4101–4105. [257]

Bablanian, R. (1975). Structural and functional alterations in cultured cells infected with cytocidal viruses. *Prog. Med. Virol.* **19**, 40–83. [404]

Bablanian, R., Eggers, H. J., and Tamm, I. (1965). Studies on the mechanism of poliovirus-induced metabolic and morphological alterations in cultured cells. *Virology* **26**, 100–113. [393]

Bachenheimer, S., and Darnell, J. E. (1975). Adenovirus-Z in RNA is transcribed as part of a high-molecular-weight precursor RNA. *Proc. Natl. Acad. Sci. U.S.* **72**, 4445–4449. [364]

Bachrach, H., and Breese, S. S. (1968). Cell cultures and pure animal virus in quantity. In: *Methods in Virology*, Vol. IV (eds. K. Maramorosch and H. Kaprowski). Academic Press, New York, pp. 351–375. [428]

Bachtold, J. G., Bubel, C. H., and Gebhardt, L. P. (1957). The primary interaction of poliomyelitis virus with host cells on tissue culture origin. *Virology* **4**, 582–589. [279]

Baglioni, C., Vesco, C., and Jacobs-Lorena, M. (1969). The role of ribosomal subunits in mammalian cells. *Cold Spring Harbor Symposia Quant. Biol.* **34**, 555–565. [272]

Bald, J. G. (1937). The use of numbers of infections for comparing the concentration of plant virus suspensions. *Austral. J. Exptl. Biol. Med. Sci.* **15**, 211–220. [27]

Bald, J. G. (1964). Cytological evidence for the production of plant virus ribonucleic acid in the nucleus. *Virology* **22**, 377–387. [463]

Ball, L. A., and White, C. N. (1976). Order of transcription of genes of vesicular stomatitis virus. *Proc. Natl. Acad. Sci. U.S.* **73**, 442–446. [330]

Baltimore, D. (1964). *In vitro* synthesis of viral RNA by the poliovirus RNA polymerase. *Proc. Nat. Acad. Sci. U.S.* **51**, 450–456. [312]

Baltimore, D. (1966). Purification and properties of poliovirus double-stranded ribonucleic acid. *J. Mol. Biol.* **18**, 421–428. [312]

Baltimore, D. (1968). Structure of the poliovirus replicative intermediate RNA. *J. Mol. Biol.* **32**, 359–368. [312, 315, 316]

Baltimore, D. (1969). The replication of picornaviruses. In: *The Biochemistry of Viruses* (ed. H. B. Levy). Marcel Dekker, New York, pp. 101–176. [309, 315, 316, 317]

Baltimore, D. (1970). RNA-dependent DNA polymerase in virions of RNA tumour viruses. *Nature* **226**, 1209–1211. [130, 384, 387]

Baltimore, D. (1971a). Expression of animal virus genomes. *Bacteriol. Rev.* **35**, 235–241. [306]

Baltimore, D. (1971b). Viral genetic systems. *Trans. N.Y. Acad. Sci.* **33,** 327–332. [306]

Baltimore, D. (1974). Tumor viruses. *Cold Spring Harbor Symposia Quant. Biol.* **39,** 1187–1200. [381, 389]

Baltimore, D. (1976). Viruses, polymerases and cancer. *Science* **192,** 632–636. [389]

Baltimore, D., and Franklin, R. M. (1963). A new ribonucleic acid polymerase appearing after mengovirus infection of HeLa cells. *J. Biol. Chem.* **238,** 3395–3400. [130, 312]

Baltimore, D., Franklin, R. M., and Callender, J. (1963). Mengovirus-induced inhibition of host RNA and protein synthesis. *Biochem. Biophys. Acta* **76,** 425–428. [399]

Baltimore, D., and Girard, M. (1966). An intermediate in the synthesis of poliovirus RNA. *Proc. Natl. Acad. Sci. U.S.* **56,** 741–748. [312]

Baltimore, D., Girard, M., and Darnell, J. (1966). Aspects of the synthesis of poliovirus RNA and the formation of virus particles. *Virology* **29,** 179–189. [311, 312]

Baltimore, D., Huang, A. S., and Stampfer, M. (1970). Ribonucleic acid synthesis of vesicular stomatitis virus. II. An RNA polymerase in the virion. *Proc. Natl. Acad. Sci. U.S.* **66,** 572–576. [330]

Baltimore, D., and Smoler, D. (1971). Primer requirement and template specificity of the DNA polymerase of RNA tumor viruses. *Proc. Natl. Acad. Sci. U.S.* **67,** 1507–1511. [387]

Baltimore, D., and Smoler, D. (1972). Association of an endoribonuclease with the avian myeloblastosis virus deoxyribonucleic acid polymerase. *J. Biol. Chem.* **247,** 7282–7287. [80]

Baluda, M. A. (1972). Widespread presence in chickens of DNA complementary to the RNA genome of avian leukosis viruses. *Proc. Natl. Acad. Sci. U.S.* **69,** 576–580. [386]

Baluda, M., and Nayak, D. (1970). DNA complementary to viral RNA in leukemia cells induced by avian myeloblastosis virus. *Proc. Natl. Acad. Sci. U.S.* **66,** 329–336. [384]

Bancroft, J. B. (1970). The self-assembly of spherical plant viruses. *Adv. Virus Research* **16,** 99–134. [122]

Banerjee, A. K., and Rhodes, D. P. (1973). *In vitro* synthesis of RNA that contains polyadenylate by virion-associated RNA polymerase of vesicular stomatitis virus. *Proc. Natl. Acad. Sci. U.S.* **70,** 3566. [330]

Barrell, B. G., Air, G. M., and Hutchinson, C. A. III (1976). Overlapping genes in bacteriophage øX174. *Nature* **264,** 34–41. [204]

Barrett, K., Barclay, S., Calendar, R., Lindqvist, B., and Six, E. (1974). Reciprocal *trans*activation in a two chromosome phage system. In: *Mechanisms of Virus Disease* (eds. W. B. Robinson and C. F. Fox). W. A. Benjamin, Menlo Park, California, pp. 385–402. [246]

Bartlett, N. M., Gillies, S. C., Ballivant, S., and Bellamy, A. R. (1974). Electron microscopy study of reovirus reaction cones. *J. Virol.* **14,** 315–326. [339, 341]

Baserga, R., and Weibel, F. (1969). The cell cycle of mammalian cells. *Int. Rev. Exp. Path.* **7,** 1–50.

Basilico, C., Matsuya, Y., and Green, H. (1969). Origin of the thymidine kinase induced by polyoma virus in productivity infected cells. *J. Virol.* **3**, 140–145. [356]

Bauer, D., St. Vincent, L., Kempe, C. H., Young, P., and Downie, A. (1969). Prophylaxis of smallpox with methisayone. *Am. J. Epidemiol.* **90**, 130–138. [431]

Bauer, W. R., Ressner, E. C., Kates, J., and Patzke, J. V. (1977). A DNA nicking-closing enzyme encapsidated in vaccinia virus: Partial purification and properties. *Proc. Natl. Acad. Sci. U.S.* **74**, 1841–1845. [376]

Baum, S. G., Reich, P. R., Muebner, R. J., Rowe, R. N. P., and Weissman, S. M. (1966). Biophysical evidence for linkage of adenovirus and SV40 DNA's in adenovirus 7-SV40 hybrid particles. *Proc. Natl. Acad. Sci. U.S.* **56**, 1509–1515. [367]

Bautz, E. K. F., Bautz, F. A., and Dunn, J. J. (1969). *E. coli* σ factor: a positive control element in phage T4 development. *Nature* **223**, 1022–1024.

Bautz, E. K. F., Kasai, T., Reilly, E., and Bautz, F. A. (1966). Gene specific mRNA. II. *Proc. Natl. Acad. Sci. U.S.* **55**, 1081–1088. [179]

Bawden, F. C. (1964). *Plant Viruses and Virus Diseases.* 4th ed. Ronald Press, New York. [5, 463, 465]

Bawden, F. C., Kassanis, B., and Nixon, H. L. (1950). The mechanical transmission and some properties of potato paracrinkle virus. *J. Gen. Microbiol.* **4**, 210–219. [469]

Bayer, M. E. (1968). Adsorption of bacteriophages to adhesions between wall and membrane of *Escherichia coli. J. Virol.* **2**, 346–356. [161]

Beckendorf, S. K., Kim, S. J., and Lielausis, I. (1973). Structure of bacteriophage T4 genes 37 and 38. *J. Mol. Biol.* **73**, 17–35. [158]

Beckendorf, S. K., and Wilson, J. H. (1972). A recombination grading in bacteriophage T4 gene 34. *Virology* **50**, 315–321. [180]

Becker, Y., and Joklik, W. K. (1964). Messenger RNA in cells infected with vaccinia virus. *Proc. Natl. Acad. Sci. U.S.* **51**, 577–585. [399]

Beckwith, J., and Zipser, D. (1970) eds. "The Lactose Operon." Cold Spring Harbor Laboratory.

Beemon, K. L., Faras, A. J., Haase, A. T., Duesberg, P. M., and Maisel, J. E. (1976). Genomic complexity of micrine leukemia and sarcoma, reticuloendotheliosis and visna viruses. *J. Virol.* **17**, 525–537. [381]

Bellamy, A. R., and Joklik, W. K. (1967). Studies on reovirus RNA. II. Characterization of reovirus messenger RNA and of the genome RNA segments from which it is transcribed. *J. Mol. Biol.* **29**, 19–26. [339]

Bellett, A. J. D. (1965). The multiplication of *Sericesthis* iridescent virus in cell cultures from *Antheraea eucalypti* Scott. III. Quantitative experiments. *Virology* **26**, 132–141. [477]

Bellett, A. J. D. (1968). The iridescent virus group. *Adv. Virus Research* **13**, 225–246. [476]

Benda, G. T. A. (1958). The introduction of tobacco mosaic virus into single hair cells of *Nicotiana langsdorffii* leaves. I. The concentration of virus. *Virology* **6**, 49–54. [466]

Bender, W., and Davidson, N. (1976). Mapping of poly(A) sequences in the electron microscope reveals unusual structure of type C oncornavirus RNA molecules. *Cell* **7**, 595–607. [381]

Benjamin, T. (1970). Host range mutants of polyoma virus. *Proc. Natl. Acad. Sci. U.S.* **67,** 394–399. [355]

Benjamin, T., and Goldman, M. (1974). Indirect complementation of a non-transforming mutant of polyoma virus. *Cold Spring Harbor Symposia Quant. Biol.* **39,** 41–44. [355]

Benjamin, T. (1972). Physiological and genetic studies of polyoma virus. *Curr. Topics Microbiol. Immunol.* **59,** 107–133. [416]

Benjamin, T. L. (1966). Virus-specific RNA in cells productively infected or transformed by polyoma virus. *J. Mol. Biol.* **16,** 359–373. [448]

Benjamin, T. L. (1974). Methods of cell transformation by tumor viruses. In: *Methods in Cell Biology*, Vol. VIII. Academic Press, New York, pp. 367–437.

Bennett, C. W. (1935). Studies on properties of the curly top virus. *J. Agr. Res.* **50,** 211–241. [469]

Benz, W. C., and Berger, H. (1973). Selective allele loss in mixed infections with T4 bacteriophage. *Genetics* **73,** 1–11. [193]

Benz, W. C., and Goldberg, E. B. (1973). Interactions between modified phage T4 particles and spheroplasts. *Virology* **53,** 225–235. [146, 163]

Benzer, S. (1957). The elementary units of heredity. In: *The Chemical Basis of Heredity* (eds. W. D. McElroy and B. Glass). Johns Hopkins Press, Baltimore. [157, 187, 190]

Benzer, S. (1961). On the topography of the genetic fine structure. *Proc. Natl. Acad. Sci. U.S.* **47,** 403–415. [150, 187]

Berg, D. E. (1974). Genes of phage λ essential for λdv plasmids. *Virology* **62,** 224–233. [235]

Berg, P., and Ofengand, E. J. (1958). An enzymatic mechanism for linking amino acids to RNA. *Proc. Natl. Acad. Sci. U.S.* **44,** 78–83. [67]

Berger, P., and Warner, H. R. (1975). Identification of P48 and P54 as components of bacteriophage T4 baseplate. *J. Virol.* **16,** 1669–1677. [174]

Berget, S. M., Flint, S. J., and Sharp, P. A. (1976). Characterization of viral, single-stranded DNA sequences in adenovirus infected cells. In: *Animal Virology* (eds. D. Baltimore et al.). Academic Press, New York, pp. 81–96. [363]

Berglund, O. (1975). Ribonucleoside diphosphate reductase induced by bacteriophage T4. II. Isolation and characterization of proteins B1 and B2. *J. Biol. Chem.* **250,** 7450–7455. [167]

Bergold, G. H. (1958). Viruses of insects. In: *Handbuch der Virusforschung*, Vol. 4 (eds. C. Hallauer and K. F. Meyer). Springer, Vienna, pp. 60–142. [473, 474]

Bernal, J. D., and Fankuchen, I. (1941). X-ray and crystallographic studies of plant virus preparations. *J. Gen. Physiol.* **25,** 111–165. [37]

Bernhard, W. (1960). The detection and study of tumor viruses with the electron microscope. *Cancer Res.* **20,** 712–730. [380]

Berns, A., and Jaenisch, R. (1976). Increase of AKR-specific sequences in tumor tissues of leukemic AKR mice. *Proc. Natl. Acad. Sci. U.S.* **73,** 2448–2452. [452]

Berns, K. I. (1974). Molecular biology of the adenoassociated viruses. *Curr. Topics Microbiol. Immunol.* **65,** 1–20. [344]

Berns, K. I., and Kelley, T. J., Jr. (1974). Visualization of the inverted terminal repetition in adeno-associated virus DNA. *J. Mol. Biol.* **82**, 267–271. [345]

Berns, K. I., and Rose, J. A. (1970). Evidence for a single-stranded adenovirus-associated virus genome: isolation and separation of complementary single strands. *J. Virol.* **5**, 693–699. [344]

Bernstein, C., and Bernstein, H. (1974). Coiled rings of DNA released from cells infected with bacteriophages T7 or T4 or from uninfected *Escherichia coli. J. Virol.* **13**, 1346–1355. [165]

Bertani, L. E., and Bertani, G. (1971). Genetics of P2 and related phages. *Adv. Genetics* **16**, 199–237. [237]

Bertrand, K., Korn, L., Lee, F., Platt, T., Squires, C. L., Squires, C., and Yanofsky, C. (1975). New features of the regulation of the tryptophan operon. *Science* **189**, 22–26. [72, 73, 219]

Besmer, P., Smotkin, D., Haseltine, W., Fan, H., Wilson, A. T., Paskind, M., Weinberg, R., and Baltimore, D. (1974). Mechanism of induction of RNA tumor viruses by halogenated pyrimidines. *Cold Spring Harbor Symposia Quant. Biol.* **39**, 1103–1107. [386]

Bessman, M. J., Myzyczka, N., Goodman, M. F., and Schnaar, R. L. (1974). Studies on the biochemical basis of spontaneous mutation II. The incorporation of a base and its analogue into DNA by wild-type, mutator and antimutator DNA polymerases. *J. Mol. Biol.* **88**, 409–421. [167]

Best, R. J., and Gallus, H. P. C. (1965). Further evidence for the transfer of character-determinants (recombination) between strains of tomato spotted wilt virus. *Enzymologia* **17**, 207–221. [466]

Best, R. J., and Palk, B. A. (1964). Electron microscopy of strain E of tomato spotted wilt virus and comments on its probable biosynthesis. *Virology* **23**, 445–460. [471]

Beveridge, W. I. B., and Burnet, F. M. (1946). The cultivation of viruses and rickettsiae in the chick embryo. *Med. Res. Council Spec. Rep. No. 256.* [17, 26]

Biggs, P. M., de Thé, G., and Payne, L. N. (1972). Oncogenesis and herpesvirus. International Agency for Research on Cancer, Lyon. [367]

Bishop, J. M., Deng, C.-T., Mahy, B. W. J., Quintrell, N., Stavnezer, E., and Varmus, M. E. (1976). Synthesis of viral RNA in cells infected by avian sarcoma viruses. In *Animal Virology* (ed. D. Baltimore et al.). Plenum Press, New York, pp. 1–20. [386]

Bishop, J. O. (1974). The gene numbers game. *Cell* **2**, 81–85. [255, 257]

Black, D. H., Rowe, W. P., Turner, H. C., and Huebner, R. J. (1963). A specific complement-fixing antigen present in SV40 tumor and transformed cells. *Proc. Natl. Acad. Sci. U.S.* **50**, 1148–1158. [354]

Black, D. R., and Knight, C. A. (1970). Ribonucleic acid transcriptase activity in purified wound tumor virus. *J. Virol.* **6**, 194–198. [462]

Black, L. M. (1945). A virus tumor disease of plants. *Am. J. Botany* **32**, 408–415. [20]

Black, L. M. (1972). Plant tumors of viral origin. In: *Progress in Experimental Tumor Research: Plant Tumor Research*, Vol. XV (ed. A. C. Braun). Konger, Basel. [461, 467]

Black, L. M., Smith, K. M., Hills, G. J., and Markham, R. (1965). Ultrastructure of potato yellow dwarf virus. *Virology* **27**, 446–448.

Blumberg, D. D., and Malamy, M. H. (1974). Evidence for the presence of nontranslated T7 late in RNA in infected F' (P1F⁺) episome-containing cells. *J. Virol.* **13,** 378–393. [201]

Bode, V. C., and Gillin, F. D. (1971). The arrangement of DNA in lambda phage heads. I. Biological consequences of micrococcal nuclease attack on a portion of the chromosome exposed in tailless heads. *J. Mol. Biol.* **62,** 493–502. [216]

Bodian, D. (1955). Emerging concept of poliomyelitis infection. *Science* **122,** 105–108. [417, 422]

Bodo, G., Scheirer, W., Suh, M., Schultze, B., Horak, I., and Jungwirth, C. (1972). Protein synthesis in pox-infected cells treated with interferon. *Virology* **50,** 140–147. [413]

Bollum, F. J. (1975). Mammalian DNA polymerases. *Prog. Nucleic Acid. Res.* **15,** 109–144. [76]

Bolognesi, D. P. (1974). Structural components of RNA tumor viruses. *Adv. Virus Research* **19,** 315–360. [380]

Boone, C. W. (1975). Malignant hemangioendotheliomas produced by subcutaneous inoculation of Ba1b/3T3 cells attached to glass beads. *Science* **188,** 68–70. [439]

Borsa, J., and Graham, A. F. (1968). Reovirus: RNA polymerase activity in purifed virions. *Biochem. Biophys. Res. Commun.* **33,** 895–901. [339]

Borsa, J., Grover, J., and Chapman, J. D. (1970). Prescence of nucleoside triphosphate phosphohydrolase activity in purified virions of reovirus. *J. Virol.* **6,** 295–302. [340]

Borst, P. (1972). Mitochondrial nucleic acids. *Ann. Rev. Biochem.* **41,** 333–376. [485]

Botchan, M., McKenna, G., and Sharp, P. A. (1974). Cleavage of mouse DNA by a restriction enzyme as a clue to the arrangement of genes. *Cold Spring Harbor Symposia Quant. Biol.* **38,** 383–395. [267]

Botchan, M., Topp, W., and Sambrook, J. (1976). The arrangement of simian virus 40 sequences in the DNA of transformed cells. *Cell,* in press. [447]

Both, G. W., Lavi, S., and Shatkin, A. J. (1975a). Synthesis of all the gene products of the reovirus genome *in vivo* and *in vitro*. *Cell* **4,** 173–180. [330]

Both, G. W., Moyer, S. A., and Banerjee, A. K. (1975b). Translation and identification of the viral mRNA species isolated from subcellular fractions of vesicular stomatitis virus-infected cells. *J. Virol.* **15,** 1012–1019. [329]

Both, G. W., Moyer, S. A., and Banerjee, A. K. (1975c). Translation and identification of the mRNA species synthesized *in vitro* by the virion-associated RNA polymerase of vesicular stomatitis virus. *Proc. Natl. Acad. Sci. U.S.* **72,** 274–278. [339]

Botstein, D., and Herskowitz, I. (1974). Properties of hybrids between *Salmonella* phage P22 and coliphage λ. *Nature* **251,** 584–589. [238]

Botstein, D., and Susskind, M. M. (1974). Regulation of lysogeny and the evolution of temperate bacterial viruses. In: *Mechanisms of Virus Disease* (eds. W. S. Robinson and C. F. Fox). W. A. Benjamin, Menlo Park, California, pp. 363–384. [72, 132, 239]

Bouché, J. P., Zechel, K., and Kornberg, A. (1975). DNAG gene product, a Rifampicin-resistant RNA polymerase, initiates the conversion of a single stranded coliphage DNA to its duplex replicative form. *J. Biol. Chem.* **250,** 5995–6001. [75, 76, 85]

Bourguignon, G. J., Tattersall, P. J., and Ward, D. C. (1976). The DNA of minute virus of mice: a self-priming non-permuted single-stranded genome with a 5'-terminal hairpin duplex. *J. Virol.* **20,** 290–306. [345]

Brakke, M. K. (1967). Density-gradient centrifugation. In: *Methods in Virology*, Vol. II (eds. K. Maramorsch and H. Koprowski). Academic Press, New York, pp. 93–117. [104]

Brakke, M. K., Natter, A. E., Black, L. M. (1954). Size and shape of wound-tumor virus. *Brookhaven Nat'l. Lab.* (Upton, N.Y.) *Symp. in Biology* **6,** 137–156. [471]

Brandhorst, B. P., and McConkey, E. H. (1974). Stability of nuclear RNA in mammalian cells. *J. Mol. Biol.* **85,** 451–463. [259]

Brandner, G., and Mueller, N. (1974). Cytosine arabinoside- and interferon-mediated control of polyoma and SV40 genome expression. *Cold Spring Harbor Symposia Quant. Biol.* **39,** 305–308. [413]

Branton, D. (1969). Membrane structure. *Ann. Rev. Plant Physiol.* **20,** 209–233. [262]

Branton, D., and Klug, A. (1975). Capsid geometry of bacteriophage T2: a freeze-etching study. *J. Mol. Biol.* **92,** 559–565. [44, 175]

Bratt, M., and Robinson, W. (1967). Ribonucleic acid synthesis in cells infected with Newcastle disease virus. *J. Mol. Biol.* **23,** 1–21. [329]

Brawerman, G. (1974). Eukaryotic messenger RNA. *Ann. Rev. Biochem.* **43,** 621–642. [311]

Brawerman, G. (1976). Characteristic and significance of the polyadenylate sequence in mammalian mRNA. *Prog. Nucleic Acid. Res.* **17,** 118–148. [78]

Brawerman, G., and Diez, J. (1975). Metabolism of the polyadenylate sequence of nuclear RNA and messenger RNA in mammalian cells. *Cell* **5,** 271–280.

Brenner, S., and Horne, R. W. (1959). A negative staining method for high resolution electron microscopy of viruses. *Biochim. Biophys. Acta* **34,** 103–110. [35]

Brenner, S., Jacob, F., and Meselson, M. (1961). An unstable intermediate carrying information from genes to ribosomes for protein synthesis. *Nature* **190,** 576–581. [65, 173]

Bretscher, M. S., and Raff, M. C. (1975). Mammalian plasma membranes. *Nature* **258,** 43–49. [262]

Brinton-Darnell, M., and Plagemann, P. G. W. (1975). Structure and chemical-physical characteristics of lactate dehydrogenase-elevating virus and its RNA. *J. Virol.* **16,** 420–433. [324]

Britten, R. J. (1969). Repeated DNA and transcription. In: *Problems in Biology: RNA in Development* (ed. W. Hanley). University of Utah Press, Salt Lake City, pp. 187–216.

Britten, R. J., Graham, D. E., and Newfeld, B. R. (1974). Analysis of repeating DNA sequences by reassociation. In: *Methods in Enzymology*, Vol. XXIX (eds. L. Grossman and K. Moldave), Academic Press, New York, pp. 363–418. [99, 102]

Britten, R. J., and Kohne, D. E. (1968). Repeated sequences in DNA. *Science* **161,** 529–540. [101, 266]

Britton, J. R., and Haselkorn, R. (1975). Permeability lesions in male *Esche-*

richia coli infected with bacteriophage T7. *Proc. Natl. Acad. Sci. U.S.* **72,** 2222–2226. [201]

Broker, T. R. (1973). An electron microscopic analysis of pathways for bacteriophage T4 DNA recombination. *J. Mol. Biol.* **81,** 1–16. [96]

Broker, T. R., and Doermann, A. H. (1976). Molecular and genetic recombination of bacteriophage T4. *Ann. Rev. Genetics* **9,** 213–244. [128, 186, 193]

Brown, D. T., and Burlingham, B. T. (1973). Penetration of host cell membranes by Adenovirus 2. *J. Virol.* **12,** 386–396. [298]

Brown, D. T., Waite, M. R. F., and Pfefferkorn, E. R. (1972). Morphology and morphogenesis of Sindbis virus as seen with freeze-etching techniques. *J. Virol.* **10,** 524–536. [336]

Brown, F., Newman, J., Stott, J., Porter, A., Frisby, D., Newton, C., Carey, N., and Fellner, P. (1974). Poly (c) in animal viral RNA's. *Nature* **251,** 342–344. [311]

Brownlee, G. G., Sanger, F., and Barrell, B. G. (1968). The sequence of 5S ribosomal ribonucleic acid. *J. Mol. Biol.* **34,** 379–412. [80, 89]

Brugge, J. S., and Butel, J. S. (1975). Role of simian virus 40 gene A function in maintenance of transformation. *J. Virol.* **15,** 619–635. [449]

Brutlag, D. C., and Peacock, W. J. (1975). Sequences of highly repeated DNA in *Drosophila melanogaster*. In: *The Eukaryote Chromosome*, Vol. XXXXVI (ed. W. J. Peacock and R. D. Brock). Australian National University Press, Canberra, pp. 35–46. [267]

Bruton, C. J., and Kennedy, S. I. T. (1975). Semliki forest virus intracellular RNA: properties of the multi-stranded RNA species and kinetics of positive and negative strand synthesis. *J. Gen. Virol.* **28,** 111–127. [327]

Bujard, H. (1969). Location of single-strand interruptions in the DNA of bacteriophage T5. *Proc. Natl. Acad. Sci. U.S.* **62,** 1167–1174. [197]

Buonassisi, V., Sato, G., and Cohen, A. I. (1962). Hormone-producing cultures of adrenal and pituitary tumor origin. *Proc. Natl. Acad. Sci. U.S.* **48,** 1184–1190. [409]

Burge, B. W., and Strauss, J. H., Jr. (1970). Glycopeptides of the membrane glycoprotein of Sindbis virus. *J. Mol. Biol.* **47,** 449–466. [122, 279]

Burger, M., and Goldberg, A. R. (1967). Identification of a tumor-specific determinant on neoplastic cell surfaces. *Proc. Natl. Acad. Sci. U.S.* **57,** 359–366. [443]

Burgess, A. B. (1969). Studies on the proteins of ϕX174. II. *Proc. Natl. Acad. Sci. U.S.* **64,** 613–617.

Burgess, R. R. (1971). RNA polymerase. *Ann. Rev. Biochem.* **40,** 711–740. [171]

Burgess, R., and Travers, A. A. (1969). Cyclic re-use of the RNA polymerase sigma factor. *Nature* **222,** 537–540.

Burgess, R. R., Travers, A. A., Dunn, J. J., and Bautz, E. K. F. (1969). Factor simulating transcription by RNA polymerase. *Nature* **221,** 43–46. [76]

Burke, W., Newton, C. S., and Fangman, W. L. (1975). Replication initiation in yeast chromosomes. In: *The Eukaryotic Chromosome* (eds. W. J. Peacock and R. D. Brock). Australian National University Press, Canberra, pp. 327–343. [257]

Burnet, F. M. (1929). A method for the study of bacteriophage multiplication in broth. *Brit. J. Exp. Path.* **10,** 109–115. [142]

Burnet, F. M. (1945). *Virus as Organism*. Cambridge, Harvard University Press. [5, 483]

Butler, P. J. G. (1976). Assembly of tobacco mosaic virus. *Philosoph. Trans. Royal Soc. London* B **276**, 151-163. [459, 460]

Butterworth, B. E., Gruenert, R. R., Korant, B. D., Lonberg-Holm, K., and Yin, F. H. (1976). Replication of rhinoviruses. *Archives of Virology* **51**, 169-189. [301]

Cairns, J. (1960). The initiation of vaccinia infection. *Virology* **11**, 603-623. [379, 392]

Caliguiri, L. A., and Tamm, I. (1968). Action of guanidine on the replication of poliovirus RNA. *Virology* **35**, 408-417. [317]

Caliguiri, L. A., and Tamm, I. (1970a). The role of cytoplasmic membranes in poliovirus biosynthesis. *Virology* **42**, 100. [312]

Caliguiri, L. A., and Tamm, I. (1970b). Characterization of poliovirus-specific structures associated with cytoplasmic membranes. *Virology* **42**, 112-122. [312]

Caliguiri, L. A., and Tamm, I. (1973). Guanidine and 2 (α hydroxybenzyl)-benzimidazole (HBB): Selective inhibitors of picornavirus multiplication. In: *Selective Inhibitors of Viral Functions* (ed. W. A. Carter). Chemical Rubber Company Press, Cleveland, Ohio, pp. 257-294. [429]

Campbell, A. (1976). Defective bacteriophages and incomplete prophages. In: *Comprehensive Virology* (eds. H. L. Fraenkel-Conrat and R. Wagner). Plenum Press, New York. [489]

Cancedda, R., Villa-Komaroff, L., Lodish, H. F., and Schlesinger, M. (1975). Initiation sites for translation of Sindbis virus 42S and 26S messenger RNA's. *Cell* **6**, 215-222. [325]

Capecchi, M. (1966). Initiation of *E. coli* proteins. *Proc. Natl. Acad. Sci. U.S.* **55**, 1517-1524. [69]

Carroll, R. B., Hager, L., and Dulbecco, R. (1974). Simian virus 40 T antigen binds to DNA. *Proc. Natl. Acad. Sci. U.S.* **71**, 3754-3757. [354]

Carter, B. J., and Rose, J. A. (1974). Transcription *in vivo* of a defective parvovirus: sedimentation and electrophoretic analysis of RNA synthesized by adenovirus-associated virus and its helper adenovirus. *Virology* **61**, 182-199. [345]

Carter, M. F., Biswal, N., and Rawls, W. E. (1974). Polymerase activity of Pichinde virus. *J. Virol.* **13**, 577-583. [333]

Carter, W. A. (1973). *Selective Inhibitors of Viral Functions*, Chemical Rubber Company Press, Cleveland, Ohio. [429]

Casals, J. (1957). The arthropod-borne group of animal viruses. *Trans. N.Y. Acad. Sci. Series II* **19**, 219-235. [423]

Casals, J. (1971). Arboviruses: incorporation into a general system of virus classification. In: *Comparative Virology* (eds. K. Maramorosch and E. Kurstak). Academic Press, New York, pp. 307-333. [323]

Casjens, S., and King, J. (1975). Virus assembly. *Ann. Rev. Biochem.* **44**, 555-611. [54, 131, 132, 174, 215, 224, 484]

Caspar, D. L. D. (1956). Structure of tobacco mosaic virus. Radial density distribution in the tobacco mosaic virus particle. *Nature* **177**, 928. [39]

Caspar, D. L. D., and Klug, A. (1962). Physical principles in the construction of regular viruses. *Cold Spring Harbor Symposia Quant. Biol.* **27,** 1–24. [14, 41, 43, 54]

Cassells, A. C., and Burke, D. C. (1973). Changes in the constitutive enzymes of chick cells following infection with Semliki forest virus. *J. Gen. Virol.* **18,** 135–141. [416]

Celma, M. L., and Ehrenfeld, E. (1974). Effect of poliovirus double-stranded RNA on viral and host-cell protein synthesis. *Proc. Natl. Acad. Sci. U.S.* **71,** 2440–2444. [402]

Chambon, P. (1975). Eukaryotic nuclear RNA polymerases. *Ann. Rev. Biochem.* **44,** 613–638. [76, 77, 271]

Champness, J. N., Bloomer, A. C., Bricagne, G., Butler, P. J. G., and Klug, A. (1976). The structure of the protein disk of tobacco mosaic virus to $5A^0$ resolution. *Nature* **259,** 20–24. [39, 459]

Champoux, J. J., and Dulbecco, R. (1972). An activity from mammalian cells that untwists superhelical DNA—a possible swivel for DNA replication. *Proc. Natl. Acad. Sci. U.S.* **69,** 143–146. [358]

Chang, L., and Bollum, F. J. (1971). Deoxynucleotide-polymerizing enzymes of calf thymus gland. V. Homogeneous terminal deoxynucleotidyl transferase. *J. Biol. Chem.* **246,** 909–916. [78]

Chao, J., Chao, L., and Speyer, J. F. (1974). Bacteriophage T4 head morphogenesis: host DNA enzymes affect frequency of petite forms. *J. Mol. Biol.* **85,** 41–50. [179]

Chardonnet, Y., and Dales, S. (1970). Early events in the interaction of adenoviruses with HeLa cells. I. Penetration of type 5 and intracellular release of the DNA genome. *Virology* **40,** 462–477. [298]

Chardonnet, Y., and Dales, S. (1970). Early events in the interaction of adenoviruses with HeLa cells. II. Comparative observations on the penetration of types 1, 5, 7 and 12. *Virology* **40,** 478–485. [298]

Chardonnet, Y., and Dales, S. (1972). Early events in the interaction of adenovirus with HeLa cells. III. Relationship between an ATPase activity in nuclear envelopes and transfer of core material: a hypothesis. *Virology* **48,** 342–359. [298]

Charney, J., Machlowitz, R., Tytell, A. A., Sagin, J. F., and Spicer, D. S. (1961). The concentration and purification of poliomyelitis virus by the use of nucleic acid precipitation. *Virology* **15,** 269–280. [105]

Chase, M., and Doermann, A. H. (1958). High negative interference over short segments of the genetic structure of bacteriophage T4. *Genetics* **43,** 332–353. [187]

Chattopadhyay, S. K., Rowe, W. P., Teich, N. M., and Lowy, D. R. (1975). Definitive evidence that the murine C-type virus inducing locus *Aiku*-1 is viral genetic material. *Proc. Natl. Acad. Sci. U.S.* **72,** 906–910. [389]

Chen, M. C. Y., Birkenmeier, E., and Salzman, N. P. (1976). Simian virus 40 DNA replication: characterization of gaps in the termination region. *J. Virol.* **17,** 614–621. [357]

Childs, J. D. (1973). Superinfection exclusion by incomplete genomes of bacteriophage T4. *J. Virol.* **11,** 1–8. [163]

Chiu, R. J., and Black, L. M. (1967). Monolayer cultures of insect cell lines and their inoculation with a plant virus. *Nature* **215,** 1076–1078. [455]

Choppin, P. W. (1964). Multiplication of a myxovirus (SV5) with minimal cytolopathic effects and without interference. *Virology* **23**, 224–233. [253]

Choppin, P. W., and Compans, R. W. (1975a). Reproduction of paramyxovirions. In: *Comprehensive Virology*, Vol. IV (eds. H. Frankel-Conrat and R. R. Wagner). Plenum Press, New York, pp. 95–178. [122, 327, 333]

Choppin, P. W., and Compans, R. W. (1975b). The structure of influenza virus. In: *The Influenza Viruses and Influenza* (ed. E. D. Kilborne). Academic Press, New York, pp. 50–51.

Choppin, P. W., Compans, R. W., Scheid, A., McSharry, J. J., and Lazarowitz, S. G. (1972). Structure and assembly of viral membranes. In: *Membrane Research* (ed. C. F. Fox). Academic Press, New York, pp. 163–186. [335]

Chou, J. Y., and Martin, R. G. (1974). Complementation analysis of simian virus 40 mutants. *J. Virol.* **13**, 1101–1109. [351]

Chou, J. Y., and Martin, R. G. (1975). DNA infectivity and the induction of host DNA synthesis with temperature-sensitive mutants of simian virus 40. *J. Virol.* **15**, 145–151. [355]

Chou, N., and Shatkin, A. J. (1975). Blocked and unblocked 5' termini in reovirus genome RNA. *J. Virol.* **15**, 1057–1064.

Chow, N., and Shatkin, A. J. (1975). Blocked and unblocked 5' termini in reovirus genome RNA. *J. Virol.* **15**, 1057–1064. [340]

Choy, Y.-M., Fehmel, F., Frank, N., and Stirm, S. (1975). *Escherichia coli* capsule bacteriophages, VI. *J. Virol.* **16**, 581–590. [138]

Čiampor, F., and Križanová, O. (1971). Interaction of plasma membranes with influenza virus. III. Electron microscopic study of interactions between influenza virus and isolated plasma membranes. *Acta Virol.* **15**, 361–366.

Citarella, R. V., Muller, R., Schlaboch, A., and Weissbach, A. (1972). Studies on vaccinia virus-directed deoxyribonucleic acid polymerase. *J. Virol.* **10**, 721–729. [377]

Claude, A. (1946). Fractionation of mammalian liver cells by differential centrifugation. *J. Exp. Med.* **84**, 61–89. [259]

Clegg, J. C. S. (1975). Sequential translation of capsid and membrane protein genes of alpha viruses. *Nature* **254**, 454–455. [325]

Clegg, J. C. S., and Kennedy, S. I. T. (1975). Initiation of synthesis of the structural proteins of Semliki Forest virus. *J. Mol. Biol.* **97**, 401–411. [325]

Clowes, R. C. (1972). Molecular study of bacterial plasmids. *Bacteriol. Rev.* **36**, 361–405.

Coffin, J. M., and Billeter, M. A. (1976). A physical map of the Rous Sarcoma virus genome. *J. Mol. Biol.* **100**, 293–318. [383]

Coffin, J., and Temin, H. M. (1972). Hybridization of Rous sarcoma virus DNA polymerase product and RNA from chicken and rat cells infected with Rous sarcoma virus. *J. Virol.* **9**, 766–775. [386]

Cohen, G. H. (1972). Ribonucleotide reductase activity of synchronized K B cells infected with herpes simplex virus. *J. Virol.* **9**, 408–418. [371]

Cohen, L. B., Herner, A. E., and Goldberg, I. H. (1969). Inhibition by pactamycin of the initiation of protein synthesis. Binding of N-acetyl-phenylalanyl transfer ribonucleic acid and polyuridylic acid to ribosomes. *Biochemistry* **8**, 1312–1326. [272]

Cohen, S. N., Chang, A. C. Y., Boyer, H. W., and Helling, R. B. (1973).

Construction of biologically functional bacterial plasmids *in vitro*. *Proc. Natl. Acad. Sci. U.S.* **70**, 3240–3244. [156]

Cohen, S. S. (1968). *Virus-Induced Enzymes,* Vol XXII. Columbia University Press, New York, p. 315. [130, 163]

Colby, C., and Chamberlain, M. (1969). The specificity of interferon induction in chick embryo cells by helical RNA. *Proc. Natl. Acad. Sci. U.S.* **63**, 160–167. [410]

Cole, C. N., and Baltimore, D. (1973a). Defective interfering particles of poliovirus. II. Nature of the defect. *J. Mol. Biol.* **76**, 325–344. [321]

Cole, C. N., and Baltimore, D. (1973b). Defective interfering particles of poliovirus III. Interference and enrichment. *J. Mol. Biol.* **76**, 345–361. [322, 323]

Cole, C. N., Smoler, D., Wimmer, E., and Baltimore, D. (1971). Defective interfering particles of poliovirus. I. Isolation and physical properties. *J. Virol.* **7**, 478–485. [321]

Colonno, R. J., and Banerjee, A. K. (1976). A unique RNA species involved in initiation of vesicular stomatitis virus RNA transcription *in vitro. Cell* **8**, 197–204. [330]

Colter, J. S., Bird, H. H., and Brown, R. A. (1957). Infectivity of ribonucleic acid from Ehrlich ascites tumor cells infected with mengo encephalitis. *Nature* **179**, 859–862. [65, 107, 309]

Colter, J. S., Bird, H. H., Moyer, A. W., and Brown, R. A. (1957). Infectivity of ribonucleic acid isolated from virus-infected tissues. *Virology* **4**, 522–532. [113]

Compans, R. W. (1971). Location of the glycoprotein in the membrane of Sinbis virus. *Nat. New Biol.* **229**, 114–117. [284]

Compans, R. W., and Choppin, P. W. (1974). Reproduction of myxoviruses. In: *Comprehensive Virology*, Vol. IV (eds. H. Fraenkel-Conrat and R. Wagner). Plenum Press, New York, pp. 179–252. [283, 327]

Compans, R. W., Klenk, H. D., Caliguiri, L. A., and Choppin, P. W. (1970). Influenza virus proteins. 1. Analysis of polypeptides of the virion and identification of spike glycoproteins. *Virology* **42**, 880–889. [283]

Conley, M. P., and Wood, W. B. (1975). Bacteriophage T4 wiskers: A rudimentary environment-sensitive device. *Proc. Natl. Acad. Sci. U.S.* **72**, 3701–3705. [159]

Coons, A. H. (1957). The morphological aspects of virus infection of cells as revealed by fluorescent antibody. In: *The Nature of Viruses* (eds. G. E. W. Wolstenholme and E. C. P. Millar). Ciba Fndn. Symposium, Churchill, London, pp. 203–207. [59]

Cooper, P. D. (1969). The genetic analysis of poliovirus. In: *The Biochemistry of Viruses* (ed. H. Levy). Marcel Dekker, New York, pp. 177–218. [323]

Cooper, P. D., Geissler, E., and Tannock, G. A. (1975). Attempts to extend the genetic map of poliovirus temperature sensitive mutants. *J. Gen. Virol.* **29**, 109–120. [323]

Coppo, A., Manzi, A., Pulitzer, J. F., Takahashi, H. (1973). Abortive bacteriophage T4 head assembly in mutants of *Escherichia coli. J. Mol. Biol.* **76**, 61–87. [177]

Cordell-Stewart, B., and Taylor, M. W. (1973). Effect of viral double-stranded RNA on protein synthesis in intact cells. *J. Virol.* **11**, 232–237. [410]

Cormack, D. V., Holloway, A. F., and Pringle, C. R. (1973). Temperature-sensitive mutants of vesicular stomatitis virus: Homology and nomenclature. *J. Gen. Virol.* **19**, 295–300. [332]

Cornblath, M., Randle, P. J., Parmeggiani, A., and Morgan, H. E. (1963). Regulation of glycogenolysis in muscle. *J. Biol. Chem.* **238**, 1592–1597. [409]

Cox, R. F. (1976). Quantitation of elongating form A and B RNA polymerases in chick oviduct nuclei and effects of extradiol. *Cell* **7**, 455–465. [271]

Craig, E. A., Tal, J., Nishimoto, T., Zimmer, S., McGrogan, M., and Raskas, H. J. (1974). RNA transcription in cultures productively infected with adenovirus 2. *Cold Spring Harbor Symposia Quant. Biol.* **39**, 483–493. [364]

Crawford, L. V. (1960). A study of the Rous sarcoma virus by density gradient centrifugation. *Virology* **12**, 143–153. [57]

Crick, F. H. C. (1958). On protein synthesis. *Soc. Exp. Med. Symp. London* **12**, 138–140.

Crick, F. II. C. (1966). Codon-anticodon pairing: the wobble hypothesis. *J. Mol. Biol.* **19**, 548–555. [67]

Crick, F. H. C. (1968). The origin of the genetic code. *J. Mol. Biol.* **38**, 267–279. [66]

Crick, F. H. C., Barnett, L., Brenner, S., and Watts-Tobin, R. J. (1961). General nature of the genetic code for proteins. *Nature* **192**, 1227–1232. [65, 157, 190]

Crick, F. H. C., and Watson, J. D. (1957). The structure of small viruses. *Nature* **177**, 473–475.

Crowell, R. L. (1968). Effect of pH on attachment of enteroviruses to HeLa cells. *Bact. Proc.* **212**, 180.

Crowell, R. L., and Philipson, L. (1971). Specific alterations of coxsackievirus B3 eluted from HeLa cells. *J. Virol.* **8**, 509–515. [284, 298]

Crowther, R. A., and Amos, L. A. (1971). Harmonic analysis of electron microscopic images with rotational symmetry. *J. Mol. Biol.* **60**, 123–130. [35]

Cummings, D. J., and Bolin, R. W. (1975). Head length control in T4 bacteriophage morphogenesis: effect of canavanine on assembly. *Bacteriol. Rev.* **40**, 314–359. [43, 175]

Curtis, M. J., and Alberts, B. (1976). Studies on the structure of intracellular bacteriophage T4 DNA. *J. Mol. Biol.* **102**, 793–816. [165]

Curtiss, P. J., and Weissman, C. (1976). Purification of globin messenger RNA from dimethysulfoxide-induced Friend cells and detection of a putative globin messenger RNA precursor. *J. Mol. Biol.* **106**, 1061–1075. [269]

Dales, S. (1963). Association between the spindle apparatus and reovirus. *Proc. Natl. Acad. Sci. U.S.* **50**, 268–275. [393]

Dales, S. (1973). Early events in cell-animal virus interactions. *Bacteriol. Rev.* **37**, 103–135. [277, 279, 290, 297, 301]

Danna, K., and Nathans, D. (1971). Specific cleavage of simian virus 40 DNA by restriction endonuclease of *Hemophilus Influenzae*. *Proc. Natl. Acad. Sci. U.S.* **68**, 2913–2917. [351]

Danna, K., and Nathans, D. (1972). Bidirectional replication of simian virus 40 DNA. *Proc. Natl. Acad. Sci. U.S.* **69**, 3097–3100. [357, 365]

Danna, K. J., Sack, G. M., Jr., and Nathans, D. (1973). Studies of simian virus

40 DNA. VII. A cleavage map of the SV40 genome. *J. Mol. Biol.* **78,** 363–376. [351]

Darnell, J. E., Jr. (1968). Ribonucleic acids from animal cells. *Bacteriol. Rev.* **32,** 262–290. [84, 107, 268, 270]

Darnell, J. E. (1972). Biochemistry of animal virus reproduction. In: *Viral and Rickettsial Infections of Man* (eds. F. L. Horsfall and I. Tamm). J. B. Lippincott, Philadelphia, pp. 233–266.

Darnell, J. E., Jr. (1975). The origin of mRNA and the structure of the mammalian chromosome. *Harvey Lect.* **69,** 1–47. [65, 78, 84, 107, 266, 268, 269, 270]

Darnell, J. E., Jr., Girard, M., Baltimore, D., Summers, D. F., and Maizel, J. V., Jr. (1967). The synthesis and translation of poliovirus RNA. In: *The Molecular Biology of Viruses* (eds. J. S. Colter and W. Parenchych). Academic Press, New York, pp. 375–401. [316, 404, 406]

Darnell, J. E., and Levintow, L. (1960). Poliovirus protein: source of amino acids and time course of synthesis. *J. Biol. Chem.* **235,** 74–77. [399]

Darnell, J. E., Jr., Levinton, L., Thorén, M. M., and Hooper, J. L. (1961). The time course of synthesis of poliovirus RNA. *Virology* **13,** 271–279.

Darnell, J. E., and Sawyer, T. K. (1960). The basis for variation in susceptibility to poliovirus in HeLa cells. *Virology* **11,** 665–675. [279]

Davenport, F. M. (1971). Killed influenza virus vaccines; present status, suggested use, desirable developments. In: *Viral, Rickettsial, and Bacterial Diseases of Man*. Sci. Publ. No. 226, Pan American Health Organization, Washington, D.C. [427]

Davidson, E. H., and Britten, R. J. (1973). Organization, transcription, and regulation in the animal genome. *Quart. Rev. Biol.* **48,** 565–613. [101, 266, 268]

Davidson, J. N. (1972). The biochemistry of nucleic acids. Charman and Hall, London. [77, 79, 80]

Davis, R. E., and Whitcomb, R. F. (1971). Micoplasmas, rickettsiae, and chlamidiae: possible relation to yellows diseases and other disorders of plants and insects. *Am. Rev. Phytopath.* **9,** 119–154. [455]

Davis, R. W., and Davidson, N. (1968). Electron-microscopic visualization of deletion mutations. *Proc. Natl. Acad. Sci. U.S.* **60,** 243–250. [96]

Davison, P. F., Freifelder, D., Hede, R., and Levinthal, C. (1961). The structural unity of the DNA of T2 bacteriophage. *Proc. Natl. Acad. Sci. U.S.* **47,** 1123–1129. [176]

Davson, H., and Danielli, J. F. (1943). *The permeability of natural membranes*. Cambridge University Press. [262]

De Clerq, E., and Merigan, T. C. (1969). Requirement of a stable secondary structure for the antiviral activity of polynucleotides. *Nature* **222,** 1148–1152. [409]

deDuve, C. (1959). Lysosomes, a new group of cytoplasmic particles. In: *Subcellular particles* (ed. T. Hayashi). Ronald Press, New York, 128–159. [75, 259]

Deinhardt, F. (1976). Hepatitis in primates. *Adv. Virus Research* **20,** 113–158. [418]

Deinhardt, F., and Shranek, G. J. (1969). Immunization against mumps. *Prog. Med. Virol.* **11,** 126–153. [425]

Delbrück, M. (1940). Adsorption of bacteriophages under various physiological conditions of the host. *J. Gen. Physiol.* **23,** 631–642. [146]

Delbrück, M. (1945). Interference between bacterial viruses. III. *J. Bacteriol.* **50,** 151–170. [144]

Delius, H., Howe, C., and Kazinski, A. W. (1971). Structure of the replicating DNA from bacteriophage T4. *Proc. Natl. Acad. Sci. U.S.* **68,** 3049–3053. [164]

Denhardt, D., Eisenberg, S., Bartok, K., and Carter, B. J. (1976). Multiple structures of adeno-associated virus DNA: analysis of terminally labeled molecules with endonuclease R·HaeIII. *J. Virol.* **18,** 672–684. [345]

Derman, E., Goldberg, S., and Darnell, J. E. (1976). HnRNA in HeLa cells: Distribution of transcript sizes estimated from nascent molecule profile. *Cell* **9,** 465–472. [268]

Desmyter, J., Melwick, J. L., and Rawls, W. E. (1968). Defectiveness of interferon production and of rubella virus interference in a line of African Green monkey kidney cells (Vero). *J. Virol.* **2,** 955–961. [411]

Deutscher, M. (1973). Synthesis and functions of the –C–C–A terminus of transfer RNA. *Prog. Nucleic Acid Res.* **13,** 51–94. [78]

Devillers-Thiery, A., Kindt, T., Scheele, G., and Blobel, G. (1975). Homology in amino-terminal sequence of precursors to pancreatic secretory proteins. *Proc. Natl. Acad. Sci. U.S.* **72,** 5016–5020.

Dewey, M. J., Wiberg, J. S., Frankel, F. R. (1974). Genetic control of whisker antigen of bacteriophage T4D. *J. Mol. Biol.* **84,** 625–634. [174]

Dharmalingam, K., and Goldberg, E. B. (1976). Mechanism localization and control of restriction cleavage of phage T4 and λ chromosomes *in vivo*. *Nature* **260,** 406–410. [195]

Dharmalingam, K., and Goldberg, E. B. (1976). Phage-coded protein prevents restriction of unmodified progeny T4 DNA. *Nature* **260,** 454–456. [195]

Diener, T. O. (1972). Viroids. *Adv. Virus Research* **17,** 295–313. [132, 462]

Diener, T. O. (1972). Is the scrapie agent a viroid? *Nature New Biol.* **235,** 218–219. [420, 462]

Diener, T. O. (1974). Viroids: the smallest known agents of infectious disease. *Ann. Rev. Microbiol.* **28,** 23–40.

Diener, T. O., and Smith, D. R. (1975). Potato spindle tuber virus. XIII. *Virology* **63,** 421–427. [462]

Dintzis, H. (1961). Assembly of the peptide chains of hemoglobin. *Proc. Natl. Acad. Sci. U.S.* **47,** 247–261. [69, 316, 357]

Dodgson, J. B., Nes, I. F., Porter, B. W., and Wells, R. D. (1976). Two new genetic assays for noninfectious fragments of ϕX174 DNA. *Virology* **69,** 782–785.

Doerfler, W. (1969). Nonproductive infection of baby hamster kidney cells (BHK21) with adenovirus type 12. *Virology* **38,** 587–606. [106]

Doermann, A. H. (1948). Lysis and lysis inhibition with *Escherichia coli* bacteriophage. *J. Bacteriol.* **55,** 257–276. [177]

Doermann, A. H. (1972). T4 and the rolling circle model of replication. *Ann. Rev. Genetics* **7,** 325–341. [184]

Doermann, A. H., and Boehner, L. (1963). An experimental analysis of bacteriophage T4 heterozygotes. I. *Virology* **21,** 551–567. [187]

Doi, Y., Teranaka, M., Yora, K., and Asuyama, H. (1969). Mycoplasma—or

PLT-like microorganisms found in the phloem elements in plants infected with mulberry dwarf, potato witches' broom, aster yellows, or powlownia witches' broom. *Rev. Plant Protect. Res.* **2,** 84–89. (English translation). [455]

Dorsch-Häsler, K., Yogo, Y., and Wimmer, E. (1976). Replication of picornaviruses. I. Evidence from *in vitro* RNA synthesis that poly(A) of the poliovirus genome is genetically coded. *J. Virol.* **17,** 1512–1527. [312]

Dubbs, D. R., and Kit, S. (1964a). Isolation and properties of vaccinia mutants deficient in thymidine kinase-inducing activity. *Virology* **22,** 214–225. [371]

Dubbs, D. R., and Kit, S. (1964b). Mutant strains of herpes simplex deficient in thymidine kinase-inducing activity. *Virology* **22,** 493–502. [377]

Dubin, D. T., and Stollar, V. (1975). Methylation of Sindbis virus "26S" messenger RNA. *Biochem. Biophys. Res. Commun.* **66,** 1373–1379. [118]

Duffy, J. J., Petrusek, R. L., and Gaidushek, E. P. (1975). Conversion of *Bacillus subtilis* polymerase activity *in vitro* by a protein induced by phage SP01. *Proc. Natl. Acad. Sci. U.S.* **72,** 2366–2370. [171]

Dulbecco, R. (1952). Production of plaques in monolayer tissue cultures by single particles of an animal virus. *Proc. Natl. Acad. Sci. U.S.* **38,** 747–752. [251, 276]

Dulbecco, R., and Vogt, M. (1953). Some problems of animal virology as studied by the plaque technique. *Cold Spring Harbor Symposia Quant. Biol.* **18,** 273–279. [18, 26]

Dunn, J. J. ed. (1974). Processing of RNA. In: *Brookhaven Symposium in Virology*, Vol. XXVI. Brookhaven National Laboratory. [75, 84]

Dunn, J. J., Anderson, C. W., Atkins, J. F., Bartelt, D. C., and Crockett, W. C. (1976). Bacteriophages T7 and T3 as model systems for RNA synthesis and processing. *Proc. Nuc. Acid Res.* **19,** 263–274. [84]

Dunn, J. J., and Studier, F. W. (1973). T7 early RNA's and *E. coli* ribosomal RNA's are cut from large precursor RNA's *in vivo* by RNAse III. *Proc. Natl. Acad. Sci. U.S.* **70,** 3296–3300. [84]

Eagle, H. (1960). Metabolic studies with normal and malignant human cells in culture. *Harvey Lectures* **54,** 156–175. [252]

Easterbrook, K. B. (1962). Interference with the maturation of vaccinia virus by isatin β-thiosemicarbazone. *Virology* **17,** 245–251. [430]

Eaton, B. T., and Faulkner, P. (1972). Heterogeneity in the poly(A) content of the genome of Sindbis virus. *Virology* **50,** 865–873. [324]

Eckhart, W. (1968). Transformation of animal cells by oncogenic DNA viruses. *Physiol. Revs.* **48,** 513–533. [356]

Edelman, G. M. (1976). Surface modulation in cell recognition and cell growth. *Science* **192,** 218–226. [262, 263]

Edelman, G. M., and Yahara, I. (1976). Temperature-sensitive changes in surface modulating assemblies of fibroblasts transformed by mutants of Rous sarcoma virus. *Proc. Natl. Acad. Sci. U.S.* **73,** 2047–2051. [443]

Edgar, R. S., and Wood, W. B. (1966). Morphogenesis of bacteriophage T4 in extracts of mutant-infected cells. *Proc. Natl. Acad. Sci. U.S.* **55,** 498–505. [157, 174]

Edmonds, M., and Winters, M. A. (1976). Polyadenylate polymerases. *Prog. Nucleic Acid Res.* **17,** 149–181. [78]

Egyhazi, E. (1975). Inhibition of Balbiani ring RNA synthesis at the initiation level. *Proc. Natl. Acad. Sci. U.S.* **72**, 947–950. [273]

Eggers, H. J. (1976). Successful treatment of enterovirus infected nicely 2-(α-hydroxybenzyl)-benzimidazole and guanidine. *J. Exp. Med.* **143**, 1367–1381. [430]

Ehrenfeld, E., and Hunt, T. (1971). Double-stranded poliovirus RNA inhibits initiation of protein synthesis by reticulocyte lysates. *Proc. Natl. Acad. Sci. U.S.* **68**, 1075–1078. [402]

Eiserling, F. A., Geiduschek, E. P., Epstein, R. H., and Metter, E. J. (1970). Capsid side and deoxyribonucleic acid length: the petite variant of bacteriophage T4. *J. Virology* **6**, 865–876. [46]

Elford, W. J. (1938). The sizes of virus and bacteriophages and methods for their determination. In: *Handbuch der Virusforschung* (eds. R. Doerr and C. Hallauer). Wien, Springer. [34]

Eliasson, R., Martin, R., and Reichard, P. (1974). Characterization of the RNA initiating the discontinuous synthesis of polyoma DNA. *Biochem. Biophys. Res. Commun.* **59**, 307–313. [85]

Elliott, J., Richter, C., Souther, A., and Bruner, R. (1973). Synthesis of bacteriophage and host DNA in toluene-treated cells prepared from T4-infected *Escherichia coli*: role of bacteriophage gene D2a. *J. Virol.* **12**, 1253–1258. [167]

Ellis, E. L., and Delbrück, M. (1939). The growth of bacteriophage. *J. Gen. Physiol.* **22**, 365–384. [141]

Emerson, S. U., and Yu, Y.-H. (1975). Both NS and L proteins are required for *in vitro* RNA synthesis by vesicular stomatitis virus. *J. Virol.* **15**, 1248–1356. [330]

Enders, J. F., Weller, T. H., and Robbins, F. C. (1949). Cultivation of the Lansing strain of poliomyelitis virus in cultures of various human embryonic tissues. *Science* **109**, 85–87. [251, 426]

Emrich, J. (1968). Lysis of T4-infected bacteria in the absence of lysozyme. *Virology* **35**, 158–165. [177]

Ensinger, M. J., Martin, S. A., Paoletti, E., and Moss, B. (1975). Modification of the 5′-terminus of mRNA by soluble guanylyl and methyl transferases from vaccinia virus. *Proc. Natl. Acad. Sci. U.S.* **72**, 2525–2529. [375, 376]

Epstein, M. A. (1970). Aspects of the EB virus. *Adv. Cancer Res.* **13**, 383–411. [436]

Epstein, R. H., Bolle, A., Steinberg, C. M., Kellenberger, E., Boy de la Tour, E., Chevalley, R., Edgar, R. S., Susman, M., Denhardt, G. H., Lielausis, A. (1963). Physiological studies of conditional lethal mutants of bacteriophage T4D. *Cold Spring Harbor Symposia Quant. Biol.* **28**, 375–394. [127, 157]

Esau, K. (1965). An anatomist's view of virus diseases. *Amer. J. Bot.* **43**, 739–748. [467]

Eskin, B., and Linn, S. (1972). The deoxyribonucleic acid modification and restriction enzymes of *Escherichia coli* B. *J. Biol. Chem.* **247**, 6183–6191. [156]

Etkind, P. R., and Krug, R. M. (1975). Purification of influenza viral complementary RNA: Its genetic content and activity in wheat germ cell-free extracts. *J. Virol.* **16**, 1464–1475. [333]

Everitt, E., Sundquist, B., Pettersson, U., and Philipson, L. (1973). Structural proteins of adenoviruses. X. Isolation and topography of low-molecular weight antigens from the virions of adenovirus type 2. *Virology* **52,** 130–147.

Fan, H. J., and Baltimore, D. (1973). RNA metabolism of murine leukemia virus: detection of virus-specific DNA sequences in infected and uninfected cells and identification of virus-specific messenger RNA. *J. Mol. Biol.* **80,** 93–117. [386]

Fareed, G. C., Garon, C. F., and Salzman, N. P. (1972). Origin and direction of simian virus 40 deoxyribonucleic acid replication. *J. Virol.* **10,** 484–491. [357]

Fareed, G. C., Khoury, G., and Salzman, N. P. (1973). Self-annealing of 4S strands from replicating simian virus 40 DNA. *J. Mol. Biol.* **77,** 457–462. [358]

Farkas, G. L., Kiraly, Z., and Salymosz, F. (1960). Role of oxidative metabolism in the localization of plant viruses. *Virology* **12,** 408. [467]

Fazekos de St. Groth, S. (1948). Viropexis, the mechanism of influenza virus infection. *Nature* **162,** 294–295. [289, 290]

Feemster, R. F., and Wells, W. F. (1933). Experimental and statistical evidence of the particulate nature of the bacteriophage. *J. Exp. Med.* **58,** 385–391. [24]

Fenner, F. (1948). The pathogenesis of the acute exanthems: An interpretation based on experimental investigations with mouse pox (infectious ectromelia of mice). *Lancet* **2,** 915–920. [421]

Fenner, F., McAuslan, B. R., Mims, C. A., Sambrook, J., and White, D. O. (1974). *The Biology of Animal Viruses.* Academic Press, New York. [323]

Fernandez-Tomas, C., and Baltimore, D. (1973). The morphogenesis of poliovirus. II. Demonstration of a new intermediate, the provirion. *J. Virol.* **12,** 1122–1130. [321]

Fernandez-Munoz, R., and Darnell, J. E. (1976). Structural difference between the 5' termini of viral and cellular mRNA in poliovirus-infected cells: Possible basis for the inhibition of host protein synthesis. *J. Virol.* **18,** 719–726. [313, 403]

Feunteun, J., Sompayrac, L., Fluck, M., and Benjamin, T. (1976). Localization of gene functions in polyoma virus. *Proc. Natl. Acad. Sci. U.S.,* in press. [355]

Fey, G., and Hirt, B. (1974). Fingerprints of polyoma virus proteins and mouse histones. *Cold Spring Harbor Symposia Quant. Biol.* **39,** 235–241. [349]

Field, A. K., Tytell, A. A., Lampson, G. P., and Hilleman, M. R. (1968). Inducers of interferon and host resistance. V. *In vitro* studies. *Proc. Natl. Acad. Sci. U.S.* **61,** 340–346. [410]

Fields, B. N., and Joklik, W. K. (1969). Isolation and preliminary genetic and biochemical characterization of temperature-sensitive mutants of reovirus. *Virology* **37,** 335–342. [341]

Fiers, W., Contreras, R., Duerinck, F., Hageman, G., Iserentant, D., Merregaert, J., Min Jou, W., Molemans, F., Raeymaekers, A., Van Den Berghe, A., Volckaert, G., and Ysebaert, M. (1976). Complete nucleotide sequence of bacteriophage MSZ RNA: primary and secondary structure of the replicase gene. *Nature* **260,** 500–506. [64, 80, 206]

Fiers, W., and Sinsheimer, R. L. (1962). The structure of the DNA of bacteriophage øX174. *J. Mol. Biol.* **5,** 408–434. [108]

Finch, J. T., and Klug, A. (1959). Structure of poliomyelitis virus. *Nature* **183,** 1709–1714. [309]

Flamand, A., and Pringle, C. R. (1971). The homologies of spontaneous and induced temperature-sensitive mutants of vesicular stomatitis virus isolated in chick embryos and BHK-21 cells. *J. Gen. Virol.* **11,** 81–85. [332]

Flamm, W. G., Walker, P. M. B., and McCallum, M. (1969). Some properties of the single strands isolated from the DNA of the nuclear satellite of the mouse (*mus musculus*). *J. Mol. Biol.* **40,** 423–443. [267]

Flanegan, J. B., Pettersson, R. F., Ambros, V., Hewlett, M. J., and Baltimore, D. (1977). Covalent linkage of a protein to a defined nucleotide sequence at the 5′ terminus of virion and replicative intermediate RNAs of poliovirus. *Proc. Nat. Acad. Sci. U.S.A.* **74,** 961–965. [309]

Fleishman, R. A., and Richardson, C. C. (1971). Analysis of host range restriction in *Escherichia coli* treated with toluene. *Proc. Natl. Acad. Sci. U.S.* **68,** 2527–2531. [195]

Flint, J. (1977). The topography and transcription of the Adenovirus genome. *Cell* **10,** 153–166. [365]

Flint, S. J., Berget, S. M., and Sharp, P. A. (1976). Adenovirus transcription. VI. Adenovirus 2 transformed cell and "early" mRNA species. *Proc. Natl. Acad. Sci. U.S.,* in press. [364, 448]

Fox, T. D., and Pero, J. (1974). New phage-SP01-induced polypeptides associated with *Bacillus subtilis* RNA polymerase. *Proc. Natl. Acad. Sci. U.S.* **71,** 2761–2765. [171]

Fraenkel, N., and Roizman, B. (1971). Herpes simplex virus: Genome size and redundancy renaturation kinetics. *J. Virol.* **8,** 591–593. [103]

Fraenkel-Conrat, H. (1974). Descriptive catalogue of viruses. In: *Comprehensive Virology*, Vol. I. Plenum Press, New York. [115]

Fraenkel-Conrat, H., and Rueckert, P. R. (1967). Analysis of protein constituents of viruses. In: *Methods in Virology*, Vol. III (eds. K. Maramorosch and H. Kaprowski). Academic Press, New York, pp. 1–35. [107]

Fraenkel-Conrat, H., Singer, B., and Williams, R. C. (1957). Infectivity of viral nucleic acid. *Biochim. Biophys. Acta* **25,** 87–96. [65, 113]

Fraenkel-Conrat, H., and Williams, R. C. (1955). Reconstitution of active tobacco mosaic virus from its inactive protein and nucleic acid components. *Proc. Natl. Acad. Sci. U.S.* **41,** 690–698. [38, 460]

Francis, T., Jr. (1953). Vaccination against influenza. *Bull. World Health Organiz.* **8,** 725. [427]

Francis, T., Jr. (1960). On the doctrine of the original antigenic sin. *Proc. Am. Philosoph. Soc.* **104,** 572–578. [61]

Francis, T., Jr. et al. (1957). Evaluation of the 1954 field trial of poliomyelitis vaccine (final report). University of Michigan Poliomyelitis Vaccine Evaluation Center. [426]

Francki, R. I. B. (1973). Plant rhabdoviruses. *Adv. Virus Research* **18,** 257–345. [463, 471]

Franklin, N. (1967). Extraordinary recombinational events in *Escherichia coli*. Their independence of the *rec*⁺ function. *Genetics* **55,** 699–707.

Franklin, N. C. (1974). Altered reading of genetic signals fused to the N operon of bacteriophage λ: genetic evidence for modification of polymerase by the protein product of the N gene. *J. Mol. Biol.* **89**, 33–48. [217, 234]

Franklin, R. E., Klug, A., and Holmes, K. C. (1957). The structure and morphology of tobacco mosaic virus. In: *The Nature of Viruses* (eds. G. E. W. Wolstenholme and E. C. P. Millar). Churchill, London. [37]

Franklin, R. M. (1962). The significance of lipids in animal viruses. *Prog. Med. Virol.* **4**, 1–53. [306]

Franklin, R. M., and Baltimore, D. (1962). Patterns of macromolecular synthesis in normal and virus-infected mammalian cells. *Cold Spring Harbor Symposia Quant. Biol.* **27**, 175–198. [312, 404, 406]

Franklin, R. M., and Rosner, J. (1962). Localization of RNA synthesis in Mengovirus infected L cells. *Biochim. Biophys. Acta* **55**, 240–241. [404]

Franklin, R. M., Rubin, H., and Davis, C. A. (1957). The production, purification and properties of Newcastle disease virus labeled with radiophosphorus. *Virology* **3**, 96–114.

Franklin, R. M., Wecker, E., and Henri, C. (1959). Some properties of an infectious ribonucleic acid from mouse encephalomyelitis virus (Appendix). *Virology* **7**, 220–235. [25]

Fraser, D. K. (1957). Host range mutants and semitemperate mutants of bacteriophage T3. *Virology* **3**, 527–553. [201]

Frearson, P. M., and Crawford, L. V. (1972). Polyoma virus basic proteins. *J. Gen. Virol.* **14**, 141–155. [349]

Freeman, G. J., Rose, J. K., Clinton, G. M., and Huang, A. S. (1977). Ribonucleic acid synthesis of vesicular stomatitis virus VII. Complete separation of the messenger RNA's of VSV by duplex formation. *J. Virol.* in press.

Freeman, V. J. (1951). Studies on virulence of bacteriophage-infected strains of *Corynebacterium diphtheriae. J. Bacteriol.* **61**, 675–688. [242]

Freese, E. (1961). The molecular mechanism of mutations. In: *Proceedings of the Fifth International Congress of Biochemistry*, Moscow, Vol. I (ed. V. A. Engelhardt). Oxford, Pergamon Press, pp. 204–229. [153]

Freidman, R. M. (1970). Studies on the mechanism of interferon action. *J. Gen. Physiol.* **56**, 149–171. [412]

Frenkel, N., and Roizman, B. (1971). Herpes simplex virus: studies of the genome size and redundancy by renaturation kinetics. *J. Virol.* **8**, 591–593. [368]

Frenkel, N., and Roizman, B. (1972). Separation of the herpesvirus deoxyribonucleic acid on sedimentation in alkaline gradients. *J. Virol.* **10**, 565–572. [371]

Frenkel, N., Silverstein, S., Cassai, E., and Roizman, B. (1973). RNA synthesis in cells infected with herpes simplex virus. VII. Control of transcription and of transcript abundancies of unique and common sequences of herpes simplex 1 and 2. *J. Virol.* **11**, 886–892. [372]

Fried, M. (1965). Cell-transforming ability of a temperature-sensitive mutant of polyoma virus. *Proc. Natl. Acad. Sci. U.S.* **53**, 486–491. [449]

Fry, B. (1959). Conditions for the infection of *Escherichia coli* with lambda phage and for the establishment of lysogeny. *J. Gen. Microbiol.* **21**, 676–684. [226]

Frye, L. D., and Edidin, M. (1970). The rapid intermixing of cell surface antigens after formation of mouse-human heterokaryons. *J. Cell Science* **7**, 319–335. [263]

Fulginiti, V. A., Kemp, C. H., Hathaway, W. E., Pearlman, D. S., Sieber, D. F., Jr., Eller, J. J., and Robinson, A. (1968). Progressive vaccinia in immunologically deficient individuals. In: *Birth Defects: Immunologic Deficiency Diseases in Man* (eds. Bergsiya, D., and Good, R. A.). National Foundation, New York. [425]

Furuichi, Y. (1974). "Methylation-coupled" transcription by virus-associated transcriptase of cytoplasmic polyhedrosis virus containing double-stranded RNA. *Nuc. Acids Res.* **1**, 809–822. [115]

Furuichi, Y., Morgan, M., Muthukrishnan, S., and Shatkin, A. J. (1975a). Reovirus messenger RNA contains a methylated, blocked 5'-terminal structure: $m^7G(5')ppp(5')G^mpCp$. *Proc. Natl. Acad. Sci. U.S.* **72**, 362–366. [340, 458]

Furuichi, Y., Muthukrishnan, S., and Shatkin, A. J. (1975b). 5'-terminal $m^7G(5')ppp(5')G^mp$ *in vivo*: identification in reovirus genome RNA. *Proc. Natl. Acad. Sci. U.S.* **72**, 742–745. [340]

Gadjusek, D. C., Roger, N. G., Basnight, M., Gibbs, C. J., Jr., and Alpers, M. (1969). Transmission experiments with kuru in chimpanzees and the isolation of latent viruses from the explanted tissues of affected animals. *Ann. N. Y. Acad. Sci.* **162**, 529–550. [420]

Galibert, F., Sedat, J., and Ziff, E. B. (1974). Direct determination of DNA nucleotide sequences: Structure of a fragment of bacteriophage ϕX174 DNA. *J. Mol. Biol.* **87**, 377–407. [91]

Gardner, S. D., Field, A. M., Coleman, D. V., and Hulme, B. (1971). New human papovavirus (B. K.) isolated from urine after renal transplantation. *Lancet* **1**, 1253–1257. [349]

Garen, A. (1968). Sense and nonsense in the genetic code. *Science* **160**, 149–159. [69, 70]

Garoff, H., and Simons, K. (1974). Location of the spike glycoproteins in the Semliki forest virus membrane. *Proc. Natl. Acad. Sci. U.S.* **71**, 3988–3992. [122, 336, 337]

Garoff, H., Simons, K., and Renkonen, O. (1974). Isolation and characterization of the membrane proteins of Semliki forest virus. *Virology* **61**, 493–504. [324]

Garon, C. F., Berry, K. W., and Rose, J. A. (1972). A unique form of terminal redundancy in adenovirus DNA molecules. *Proc. Natl. Acad. Sci. U.S.* **69**, 2391–2395. [360]

Garro, A. J., and Marmur, J. (1970). Defective bacteriophages. *J. Cell Physiol.* **76**, 253–263. [245]

Gauntt, C. J., and Graham, A. F. (1969). The reoviruses. In: *The Biochemistry of Viruses* (ed. H. Levy). Marcel Dekker, New York, pp. 259–291. [338]

Gefter, M. (1975). DNA replication. *Ann. Rev. Biochem.* **44**, 45–78. [75, 76, 79, 80, 84]

Gefter, M., Hausmann, R., Gold, M., and Hurwitz, J. (1966). The enzymatic methylation of ribonucleic acid and deoxyribonucleic acid. *J. Biol. Chem.* **241**, 1995–2006. [201]

Gelb, L. D., Kohne, D. E., and Martin, M. A. (1971). Quantitation of simian virus 40 sequences in African green monkey, mouse and virus-transformed cell genomes. *J. Mol. Biol.* **57,** 129–145. [102]

Gellert, M., Mizuuchi, K., O'Dea, M. H., and Nash, H. A. (1976). DNA gyrase: an enzyme that introduces superhelical turns in DNA. *Proc. Natl. Acad. Sci. U.S.* **73,** 3872–3876. [109]

Georgopoulos, C. P., and Herskowitz, I. (1971). *Escherichia coli* mutants blocked in lambda DNA synthesis. In: *The Bacteriophage* λ (ed. A. D. Hershey). Cold Spring Harbor Laboratory.

Gerber, P. (1972). Activation of Epstein-Barr virus by 5'-bromodeoxyuridine in virus free human cells. *Proc. Natl. Acad. Sci. U.S.* **69,** 83–85. [372]

Gerber, P., Whang-Peng, J., and Monroe, J. H. (1969). Transformation and chromosome changes induced by Epstein-Barr virus in normal human leukocyte cultures. *Proc. Natl. Acad. Sci. U.S.* **63,** 740–747. [372]

Germond, J. E., Hirt, J. E., Oudet, P., Gross-Bellard, M., and Chambon, P. (1975). Folding of the DNA double helix in chromatin-like structures from simian virus 40. *Proc. Natl. Acad. Sci. U.S.* **72,** 1843–1847. [350]

Gerry, H. W., Kelly, T. J., Jr., and Berns, K. I. (1973). Arrangement of nucleotide sequences in adeno-associated virus DNA. *J. Virol.* **79,** 207–225. [345]

Gershon, D., Hauser, P., Sachs, L., and Winocour, E. (1965). On the mechanism of polyoma virus-induced synthesis of cellular DNA. *Proc. Natl. Acad. Sci. U.S.* **54,** 1584–1589. [356]

Geshelin, P., and Berns, K. (1974). Characterization and localization of the naturally occurring cross-links in vaccinia virus DNA. *J. Mol. Biol.* **88,** 785–796. [374]

Gey, G. O., Coffman, W. D., and Kubicek, M. T. (1952). Tissue culture studies of the proliferative capacity of cervical carcinoma and normal epithelium. *Cancer Res.* **12,** 264–265. [252]

Gianni, A. M., Hutton, J. R., Smotkin, D., and Weinberg, R. A. (1976). Proviral DNA of Moloney leukemia virus: purification and visualization. *Science* **191,** 569–571. [385]

Gianni, A. M., Smotkin, D., and Weinberg, R. A. (1975). Murine leukemia virus: detection of unintegrated double-stranded DNA forms of the provirus. *Proc. Natl. Acad. Sci. U.S.* **72,** 447–451. [385]

Gibbs, A., Skotnicki, A. H., Gardiner, J. E., and Walker, E. S. (1975). A tobamovirus of a green alga. *Virology* **64,** 571–574. [39]

Gibor, A. (1965). Chloroplast heredity and nucleic acids. *Am. Natural.* **99,** 229–239. [485]

Gibson, W., and Roizman, B. (1971). Compartmentalization of spermine and spermidine in the herpes simplex virion. *Proc. Natl. Acad. Sci. U.S.* **68,** 2818–2821. [122]

Gierer, A. (1960). Recent investigations on tobacco mosaic virus. *Prog. Biophys.* **10,** 299–342. [113]

Gierer, A., and Schramm, G. (1956). Infectivity of ribonucleic acid from tobacco mosaic virus. *Nature* **177,** 702–703. [3, 65, 106, 113]

Gilbert, W., and Dressler, D. (1968). DNA replication—the rolling circle model. *Cold Spring Harbor Symposia Quant. Biol.* **33,** 473–484. [94, 152]

Gilbert, W., Maizels, N., and Maxim, A. (1973). Sequences of controlling regions of the lactose operon. *Cold Spring Harbor Symposia Quant. Biol.* **38,** 845–855. [72]

Gillespie, D., and Spiegelman, S. (1965). A quantitative assay for RNA-DNA hybrids with DNA immobilized on a membrane. *J. Mol. Biol.* **12,** 829–842. [99]

Ginsberg, H. S., Bella, L. J., and Levine, A. J. (1967). Control of biosynthesis of host macromolecules in cells infected with adenovirus. In: *The Molecular Biology of Viruses* (eds. J. S. Colter and W. Parenchych). Academic Press, New York, pp. 547–572. [399, 404]

Girard, M. (1969). *In vitro* synthesis of poliovirus ribonucleic acid: Role of the replicative intermediate. *J. Virol.* **3,** 376–384. [315]

Girard, M., Baltimore, D., and Darnell, J. E., Jr. (1967). The poliovirus replication complex: Site for synthesis of poliovirus RNA. *J. Mol. Biol.* **24,** 59–74. [312]

Girardi, A. J., and Defendi, V. (1970). Induction of SV40 transplantation antigen (TrAg) during the lytic cycle. *Virology* **42,** 688–698. [355]

Goff, C. G. (1974). Chemical structure of a modification of the *Escherichia coli* ribonucleic acid polymerase α polypeptides induced by bacteriophage T4 infection. *J. Biol. Chem.* **249,** 6181–6190. [171]

Gold, M., Hausmann, R., Maitra, U., and Hurwitz, J. (1964). The enzymatic methylation of RNA and DNA. VIII. Effects of bacteriophage infection on the activity of the methylation enzymes. *Proc. Natl. Acad. Sci. U.S.* **52,** 292–297. [77]

Goldberg, S., Weber, J., and Darnell, J. E. (1977). Definition of a large viral transcription unit late in Ad2 infection of HeLa cells: Mapping by effects of ultraviolet irradiation. *Cell* **10,** 617–622. [364]

Gomatos, P. J., and Tamm, I. (1963a). The secondary structure of reovirus RNA. *Proc. Natl. Acad. Sci. U.S.* **49,** 707–714. [339]

Gomatos, P. J., and Tamm, I. (1963b). Animal and plant viruses with double helical RNA. *Proc. Natl. Acad. Sci. U.S.* **50,** 878–885. [114, 471]

Gomatos, P. J., Tamm, I., Dales, S., and Franklin, R. M. (1962). Reovirus type 3: Physical characteristics and interaction with L cells. *Virology* **17,** 441–454. [338]

Gorter, E., and Grendel, F. (1925). On bimolecular layers of lipoids on the chromocytes of the blood. *J. Exp. Med.* **41,** 439–443. [262]

Gottschalk, A. (1959). Chemistry of virus receptors. In: *The Viruses,* Vol. III (eds. F. M. Burnet and W. M. Stanley). Academic Press, New York, pp. 51–61. [278, 283]

Gourlay, R. N., Garwes, D. J., Bruce, J., and Wyld, S. Q. (1973). Further studies on the morphology and composition of Mycoplasmatales Virus-laidlarvii 2. *J. Gen. Virol.* **18,** 127–133. [6]

Grace, T. D. C. (1962). The development of a cytoplasmic polyhedrosis in insect cells grown *in vitro*. *Virology* **18,** 33–42.

Grace, T. D. C. (1969). Insect tissue culture and its use in virus research. *Adv. Virus Res.* **14,** 201–220. [27, 473]

Graessmann, A., Graessman, M., Hoffman, H., Niebel, J., Brandner, G., and Mueller, N. (1974). Inhibition interferon of SV40 tumor antigen formation in cells infected with SV40 cRNA transcribed *in vitro*. *FEBS Letters* **39,** 249–251. [413]

Grafstrom, R. H., Alwine, J. C., Steinhart, W. L., and Hyman, R. W. (1975). The terminal repetition of herpes simplex virus DNA *Virology* **67**, 144–157. [368]

Granboulan, N., and Girard, M. (1969). Molecular weight of poliovirus ribonucleic acid. *J. Virol.* **4**, 475–479. [309]

Gratia, A. (1936). Des relations numériques entre bactéries lysogènes et particules de bactériophage. *Ann. Inst. Pasteur* **57**, 652–694. [17]

Green, D. M. (1964). Infectivity of DNA isolated from *Bacillus subtilis* bacteriophage, SP82. *J. Mol. Biol.* **10**, 438–451.

Green, D. M., and Krieg, D. R. (1961). The delayed origin of mutants induced by exposure of extracellular phage T4 to ethyl methane sulfonate. *Proc. Natl. Acad. Sci. U.S.* **47**, 64–72. [155]

Green, M. (1962). Studies on the biosynthesis of viral DNA. *Cold Spring Harbor Symposia Quant. Biol.* **27**, 219–235.

Green, M. (1970). Oncogenic viruses. *Ann. Rev. Biochem.* **39**, 701–756. [415]

Green, R. G. (1935). On the nature of filterable viruses. *Science* **82**, 443–445. [483]

Greenblatt, J., and Schleif, R. (1971). Arabinose C protein: regulation of the arabinose operon *in vitro. Nature New Biol.* **233**, 166–170. [72]

Grell, R. F. (ed.) (1974). *Mechanisms in Recombination.* Plenum Press, New York. [152]

Griffin, B., Fried, M., and Cowie, A. (1974). Polyoma DNA—a physical map. *Proc. Natl. Acad. Sci. U.S.* **71**, 2077.

Griffith, J., Dieckmann, M., and Berg, P. (1975). Electron microscope localization of a protein bound near the origin of simian virus 40 DNA replication. *J. Virol.* **15**, 167–172. [356]

Griffith, J. D. (1975). Chromatin structure: Deduced from a minichromosome. *Science* **187**, 1202–1203.

Grimes, W. J., and Burge, B. W. (1971). Modification of Sindbis virus glycoprotein by host-specified glycosyl transferases. *J. Virol.* **7**, 309–313. [336]

Grodzicker, T., Arditti, R. R., and Eisen, H. (1972). Establishment of repression by lambdoid phage in catabolite activator protein and adenylate cyclase mutants of *Escherichia coli. Proc. Natl. Acad. Sci. U.S.* **69**, 366–370. [226]

Grollman, A. P. (1966). Structural basis for inhibition of protein synthesis by emetine and cycloheximide based on an analogy between ipecac alkaloids and glutarimide antibiotics. *Proc. Natl. Acad. Sci. U.S.* **56**, 1867–1874. [272]

Groman, N. B. (1959). The relation of bacteriophage to human disease: A review of conversion to toxigenicity in *Corynebacterium diphtheriae*. In: *Immunity and Virus Infection* (ed. V. A. Najjar). John Wiley and Sons, New York, pp. 196–205. [244]

Gross, L. (1970). *Oncogenic Viruses*, 2nd ed. Pergamon Press, Oxford. [380, 435]

Grossman, L., Braun, A., Feldberg, R., and Mahler, I. (1975). Enzymatic repair of DNA. *Ann. Rev. Biochem.* **44**, 19–44. [79, 80, 81]

Groudine, M., and Weintraub, H. (1975). Rous sarcoma virus activates embryonic globin genes in chicken fibroblasts. *Proc. Natl. Acad. Sci. U.S.* **72**, 4464–4468. [446]

Günthert, V., Schweiger, M., Stupp, M., and Doerfler, W. (1976). DNA methylation in adenovirus, adenovirus-transformed cells, and host cells. *Proc. Natl. Acad. Sci. U.S.* **73**, 3923–3927. [113]

Habel, K. (1961). Resistance of polyoma virus immune animals to transplanted polyoma tumors. *Proc. Soc. Exptl. Biol. Med.* **106**, 722–725. [452]

Hall, B. D., Nygaard, A. P., and Green, M. H. (1964). Control of T2-specific RNA synthesis. *J. Mol. Biol.* **9**, 143–153. [100]

Hall, B. D., and Spiegelman, S. (1961). Sequence complementarity of T2-DNA and T2-specific RNA. *Proc. Natl. Acad. Sci. U.S.* **47**, 137–146. [99]

Hall, T. C., Wepprich, R. K., Davies, J. W., Weathers, L. G., and Semancik, J. S. (1974). Functional distinctions between the ribonucleic acids from citrus exocartis viroid and plant viruses: cell-free translation and amino-acylation reactions. *Virology* **61**, 486–492. [462]

Hamilton, R. I. (1974). Replication of plant viruses. *Ann. Rev. Phytopathol.* **12**, 233–245. [456, 462]

Hamkalo, B., and Miller, O. L., Jr. (1973). Electronmicroscopy of genetic activity. *Ann. Rev. Biochem.* **42**, 379–396. [93]

Hamlett, N. V., and Berger, H. (1975). Mutations altering genetic recombination and repair in bacteriophage T4. *Virology* **63**, 539–567. [192]

Hampar, B., Derge, J. G., Martos, L. M., Tagamets, M. A., Chang, S. Y., and Chakrabarty, A. (1973). Identification of a critical period during the S phase for activation of the Epstein-Barr virus by 5-iododeoxyuridine. *Nature New Biol.* **244**, 214–217. [373]

Hampar, B., Tanaka, A., Nonoyama, M., and Derge, J. (1974). Replication of the resident repressed Epstein-Barr Virus genome during early S phase (S-1 period) of nonproducer Raji cells. *Proc. Natl. Acad. Sci. U.S.* **71**, 631–633. [373]

Hamre, D. (1968). Rhinoviruses. In: *Monographs in Virology*, Vol. I. Karger, New York. [299]

Hanafusa, H. (1977). Transformation by RNA tumor virus. In: *Comprehensive Virology* (ed. H. Fraenkel-Conrat and R. R. Wagner). Plenum Press, New York, in press. [448]

Hanafusa, H., Baltimore, D., Smoler, D., Watson, K. F., Yaniv, A., and Spiegelman, S. (1972). Absence of polymerase protein in virions of alpha-type Rous sarcoma virus. *Science* **177**, 1188–1191. [388]

Hand, R., and Tamm, I. (1972). Rate of DNA chain growth in mammalian cells infected with cytocidal RNA viruses. *Virology* **47**, 331–337.

Hand, R., and Tamm, I. (1974). DNA replication and rate of chain growth in mammalian cells. In: *Cell Cycle Controls* (eds. G. M. Padilla, I. L. Cameron, and A. Zimmerman). Academic Press, New York, pp. 273–288. [406, 407]

Hantke, K., and Braun, V. (1973). Covalent binding of lipid to protein. *Europ. J. Biochem.* **34**, 284–296. [137]

Harada, F., Sawyer, R. C., and Dahlberg, J. E. (1975). Primer ribonucleic acid for initiation of *in vitro* Rous sarcoma virus deoxyribonucleic acid synthesis. *J. Biol. Chem.* **250**, 3487–3497. [387]

Hardy, K. G. (1975). Colicinogeny and related phenomena. *Bacteriol. Rev.* **39**, 464–515. [245, 486]

Harriman, P. D., and Stent, G. S. (1964). The effect of radiophosphorus decay on cistron function in bacteriophage T4. Long-range and short-range hits. *J. Mol. Biol.* **10,** 488–507. [193]

Harrison, S. C., Caspar, D. L. D., Camerini-Otero, R. D., and Franklin, R. M. (1971). Lipid and protein arrangement in bacteriophage PM2. *Nature New Biol.* **229,** 197–201. [48]

Harrison, S. C., David, A., Jumblatt, J., and Darnell, J. E. (1971). Lipid and protein organization in Sindbis virus. *J. Mol. Biol.* **60,** 523–528. [336]

Hartman, N., and Zinder, N. D. (1974). The effect of B specific restriction and modification of DNA on linkage relationships in f1 bacteriophage. I. Studies on the mechanism of B restriction *in vivo. J. Mol. Biol.* **85,** 345–356.

Hartwell, L., Vogt, M., and Dulbecco, R. (1965). Induction of cellular DNA synthesis by polyoma. II. Increase in the rate of enzyme synthesis after infection with polyoma virus in mouse embryo kidney cells. *Virology* **27,** 262–272. [352, 355]

Haruna, I., and Spiegelman, S. (1965). Specific template requirements of RNA replicases. *Proc. Natl. Acad. Sci. U.S.* **54,** 1189–1193. [487]

Haselkorn, R., and Rothman-Denes, L. B. (1973). Protein synthesis. *Ann. Rev. Biochem.* **42,** 397–429. [67, 71]

Haseltine, W., Kleid, D. G., Panet, A., Rothenberg, E., and Baltimore, D. (1976). Ordered transcription of RNA tumor virus genomes. *J. Mol. Biol.,* **106,** 109–131. [387]

Hatanaka, M., Augl, C., and Gilden, R. V. (1971). Evidence for a functional change in the plasma membrane of murine sarcoma virus-infected mouse embryo cells. *J. Biol. Chem.* **245,** 714–717. [443]

Hattman, S., and Fukasawa, T. (1963). Host-induced modification of T-even phages due to defective glucosylation of their DNA. *Proc. Natl. Acad. Sci. U.S.* **50,** 297–300. [195]

Hausen, P., and Stein, H. (1970). Ribonuclease H: An enzyme degrading the RNA moiety of DNA-RNA hybrids. *Europ. J. Biochem.* **14,** 278–283. [80]

Hausmann, R. (1973). The genetics of T-odd phages. *Ann. Rev. Microbiol.* **27,** 51–66. [197]

Havel, E. A., and Vilcek, J. (1972). Production of high-titered interferon in cultures of human diploid cells. *Antimicrobiol. Agents and Chemotherapy* **2,** 476–484. [408]

Hay, A. J. (1974). Studies on the formation of the influenza virus envelope. *Virology* **60,** 398–418. [337]

Hay, J., and Subak-Sharpe, J. H. (1976). Mutants of herpes simplex virus types 1 and 2 that are resistant to phosphonoacetic acid induce altered DNA polymerase activities in infected cells. *J. Gen. Virol.* **31,** 145–148. [371]

Hayashi, M., Spiegelman, S., Franklin, N. C., and Luria, S. E. (1963). Separation of the RNA message transcribed in response to a special inducer. *Proc. Natl. Acad. Sci. U.S.* **49,** 729–736. [100]

Hayes, W. (1968). *The Genetics of Bacteria and Their Viruses.* John Wiley and Sons, New York. [129, 131, 135, 145, 198, 486]

Hayflick, L., and Moorhead, P. (1961). The serial cultivation of human diploid cell strains. *Exp. Cell Res.* **25,** 585–621. [436]

Hayward, G. S., Jacob, R. J., Wadsworth, S. C., and Roizman, B. (1975). Anatomy of herpes simplex virus DNA: Evidence for four populations of molecules that differ in the relative orientations of their long and short components. *Proc. Natl. Acad. Sci. U.S.* **72,** 4243–4247. [103, 111, 368]

Hefti, E., Bishop, D. H. L., Dubin, D. T., and Stollar, V. S. (1976). 5'nucleotide sequence of Sindbis viral RNA. *J. Virol.* **17,** 149–159. [324]

Helenius, A., and Söderlund, H. (1973). Stepwise dissociation of the Semliki forest virus membrane with triton X-100. *Biochim. Biophys. Acta* **307,** 287–300. [337]

Heller, E. (1963). Enhancement of Chikungunya virus replication and inhibition of interferon production by actinomycin D. *Virology* **21,** 652–656. [411]

Henle, G., and Henle, W. (1947). The effect of ultraviolet irradiation on various properties of influenza viruses. *J. Exp. Med.* **85,** 347–364. [408]

Henle, W., and Henle, G. (1972). Epstein-Barr virus: The cause of infectious mononucleosis. In: *Oncogenesis and Herpes Viruses* (eds. P. M. Biggs, G. de-Thé, and L. N. Payne). Int. Agency for Canc. Res. Pub. #2, Lyon, France. [418, 419]

Henry, T. J., and Knippers, R. (1974). Isolation and function of the gene A initiator of bacteriophage φX174, a highly specific DNA endonuclease. *Proc. Natl. Acad. Sci. U.S.* **71,** 1549–1553. [202]

Herman, R. C., Williams, J. G., and Penman, S. (1976). Message and non-message sequences adjacent to poly(A) in steady state heterogenous nuclear RNA of HeLa cells. *Cell* **7,** 429–437. [269]

Herriott, R. M., and Barlow, J. L. (1952). Preparation, purification, and properties of *E. coli* virus T2. *J. Gen. Physiol.* **36,** 17–18. [105]

Hershey, A. D. (1946). Spontaneous mutations in bacterial viruses. *Cold Spring Harbor Symposia Quant. Biol.* **11,** 67–77. [149, 157]

Hershey, A. D. (1957). Some minor components of bacteriophage T2 particles. *Virology* **4,** 237–264. [139]

Hershey, A. D., ed. (1971). *The Bacteriophage* λ. Cold Spring Harbor Laboratories. [212, 220]

Hershey, A. D., Burgi, E., and Ingraham, L. (1963). Cohesion of DNA molecules isolated from phage lambda. *Proc. Natl. Acad. Sci. U.S.* **49,** 748–755. [109]

Hershey, A. D., and Chase, M. (1951). Genetic recombination and heterozygosis in bacteriophages. *Cold Spring Harbor Symposia Quant. Biol.* **16,** 471–479. [184]

Hershey, A. D., and Chase, M. (1952). Independent functions of viral protein and nucleic acid in growth of bacteriophage. *J. Gen. Physiol.* **36,** 39–56. [129, 138]

Hershey, A. D., and Dove, W. (1971). Introduction to lambda. In: *The Bacteriophage Lambda* (ed. A. D. Hershey). Cold Spring Harbor Laboratories. [489]

Hershey, A. D., Kamen, M. D., Kennedy, J. W., and Gest, H. (1951). The mortality of bacteriophage containing assimilated radioactive phosphorus. *J. Gen. Physiol.* **34,** 305–319. [193]

Herskowitz, I. (1973). Control of gene expression in bacteriophage lambda. *Ann. Rev. Genetics* **7,** 289–324. [224]

Heston, W. E. (1976). The genetic aspects of human cancer. *Adv. Cancer Res.* **23,** 1–22. [453]

Hewitt, W. B., Raski, D. J., and Goheen, A. C. (1958). Nematode vector of soil borne fanleaf virus of grapevines. *Phytopath.* **48,** 586–595. [471]

Hewlett, M. J., Rose, J. K., and Baltimore, D. (1976). 5′-Terminal structure of poliovirus polyribosomal RNA is pUp. *Proc. Natl. Acad. Sci. U.S.* **73,** 327–330. [313, 403]

Hill, A. K. (1973). Interferon induction by polynucleotides. In: *Selective Inhibitors of Viral Functions* (ed. W. A. Carter). Chemical Rubber Company Press, Cleveland, Ohio, pp. 149–176. [426]

Hill, M., Hillova, J., Dantchev, D., Mariage, R., and Goubin, G. (1974). Infectious viral DNA in Rous sarcoma virus-transformed nonproducer and producer animal cells. *Cold Spring Harbor Symposia Quant. Biol.* **39,** 1015–1026. [384]

Hirsch, I., and Vonka, V. (1974). Ribonucleotides linked to DNA of herpes simplex virus type 1. *J. Virol.* **13,** 1162–1168. [368]

Hirst, G. K., and Pons, M. W. (1973). Mechanism of influenza virus recombination. II. Virus aggregation and its effect on plaque formation by so-called noninfectious virus. *Virology* **56,** 620–631. [335]

Hirst, G. K. (1941). The agglutination of red blood cells by allantoic fluid of chick embryos infected with influenza virus. *Science* **94,** 22–23. [279]

Hirst, G. K. (1950). Receptor destruction by viruses of the mumps-NDV-influenza group. *J. Exp. Med.* **91,** 161–175. [283]

Hirst, G. K. (1962). Genetic recombination with Newcastle disease virus, polioviruses and influenza. *Cold Spring Harbor Symposia Quant. Biol.* **27,** 303–309. [113]

Hirst, G. K. (1965). Cell-virus attachment and the action of antibodies on viruses. In: *Viral and Rickettsial Diseases of Man*, 4th ed. (eds. F. Horsfall and I. Tamm). Lippincott, Philadelphia, p. 216. [278]

Hirt, B. (1967). Selective extraction of polyoma DNA from infected mouse cell cultures. *J. Mol. Biol.* **26,** 365–369. [357, 384]

Hodge, L. D., and Scharff, M. D. (1969). Effect of adenovirus on host cell DNA synthesis in synchronized cells. *Virology* **37,** 554–564. [366, 407]

Hoffman, G. E. (1973). Amandantine HCl and related compounds. In: *Selective Inhibitors of Viral Functions* (ed. W. A. Carter). Chemical Rubber Company Press, Cleveland, Ohio, pp. 199–212. [429]

Hoggan, M. D. (1970). Adenovirus associated viruses. In: *Progress in Medical Virology*, Vol. XII (ed. J. L. Melnick), Karger, Basel, pp. 211–239. [344]

Hoggan, M. D. (1971). Small DNA Viruses. In: *Comparative Virology* (eds. K. Maramorosch and L. E. Kurstak), Academic Press, New York, pp. 43–79. [344]

Hohne, T., and Hohn, B. (1970). Structure and assembly of simple RNA bacteriophages. *Adv. Virus Research* **16,** 43–98. [120]

Holland, J. J. (1963). Depression of host-controlled RNA synthesis in human cells during poliovirus infection. *Proc. Natl. Acad. Sci. U.S.* **49,** 23–27. [285, 406]

Holland, J. J. (1964). Enterovirus entrance into specific host cells, and subsequent alterations of cell protein and nucleic acid synthesis. *Bacteriol. Rev.* **28,** 3–13.

Holland, J. J., and Kiehn, E. D. (1968). Specific cleavage of viral proteins as steps in the synthesis and maturation of enteroviruses. *Proc. Natl. Acad. Sci. U.S.* **60**, 1015–1022. [317]

Holland, J. J., and McLaren, L. C. (1959). The mammalian cell-virus relationship. II. Adsorption, reception, and eclipse of poliovirus by HeLa cells. *J. Exp. Med.* **109**, 487–504. [278]

Holley, R., and Kiernan, J. (1968). "Contact inhibition" of cell division in 3T3 cells. *Proc. Natl. Acad. Sci. U.S.* **60**, 300–304. [439]

Holmes, F. O. (1929). Local lesions in tobacco mosaic. *Bot. Gaz.* **87**, 39–55. [27]

Holmes, F. O. (1931). Local lesions of mosaic in *Nicotiana tabacum* L. *Contrib. Boyce Thompson Inst.* **3**, 163–172. [19]

Holmes, F. O. (1937). Inheritance of resistance to tobacco-mosaic disease in the pepper. *Phytopath.* **27**, 637–642. [19]

Holmes, F. O. (1948). Order Virales: The filterable viruses. In: *Bergey's Manual of Determinative Bacteriology*, 6th ed. Williams and Wilkins, Baltimore.

Holowczak, J. A. (1976). Poxvirus DNA 1. Studies on the structure of the vaccinia genome. *Virology* **72**, 121–133. [111]

Holowczak, J. A., and Joklik, W. K. (1967). Studies of the structural proteins of vaccinia virus. II. Kinetics of the synthesis of individual groups of structural proteins. *Virology* **33**, 726–739. [377]

Honess, R. W., and Roizman, B. (1973). Proteins specified by herpes simplex virus. XI. Identification and relative molar rates of synthesis of structural and non-structural herpes virus polypeptides in the infected cell. *J. Virol.* **12**, 1347–1365. [368]

Honess, R. W., and Roizman, B. (1974). Regulation of herpesvirus macromolecular synthesis. I. Cascade regulation of the synthesis of three groups of viral proteins. *J. Virol.* **14**, 8–19. [371]

Hood, L. E., Wilson, J. H., and Wood, W. B. (1975). *Molecular Biology of Eukaryotic Cells: A Problem Approach*. W. A. Benjamin, Menlo Park, Calif. [102]

Hope-Simpson, R. E. (1965). The nature of herpes zoster: a long term study and a new hypothesis. *Proc. Roy. Soc. Med.* **58**, 9–20. [418]

Horiuchi, K., and Zinder, N. D. (1972). Cleavage of bacteriophage f1 DNA by the restriction enzyme of *Escherichia coli* B. *Proc. Natl. Acad. Sci. U.S.* **69**, 3220–3224. [156]

Horiuchi, K., and Zinder, N. D. (1976). Origin and direction of synthesis of bacteriophage f1 DNA. *Proc. Natl. Acad. Sci. U.S.* **73**, 2341–2345. [202]

Horowitz, N. H. (1945). On the evolution of biochemical syntheses. *Proc. Natl. Acad. Sci. U.S.* **31**, 153–157. [483]

Horvitz, H. R. (1973). Polypeptide bound to the host RNA polymerase is specified by T4 control gene 33. *Nature New Biol.* **244**, 137–140. [171]

Horwitz, M. (1976). Bidirectional replication of adenovirus type 2 DNA. *J. Virol.* **18**, 307–315. [365]

Howard, A., and Pelc, S. (1953). Synthesis of DNA in normal and irradiated cells and its relation to chromosome breakage. *Heredity* **6**, Suppl. 261–273. [264]

Howatson, A. F. (1970). Vesicular stomatitis and related viruses. *Adv. Virus Research* **16**, 195–256. [327]

Howe, M. M., and Bade, E. G. (1975). Molecular biology of bacteriophage Mu. *Science.* **190**, 624–632. [239]

Hoyle, L. (1962). The entry of myxoviruses into the cell. *Cold Spring Harbor Symposia Quant. Biol.* **27**, 113–121. [289, 291]

Hsu, H. T. (1974). Polykaryocytosis induced by potato yellow dwarf virus in monolayer of vector cells. *Proc. Am. Phytopathol. Soc.* **1**, 82. [456]

Hsu, H. T., and Black, L. M. (1974). Multiplication of potato yellow dwarf virus on vector cell monolayers. *Virology* **59**, 331–334. [456]

Hsu, M.-T., Kung, H. J., and Davidson, N. (1973). An electron microscope study of Sindbis virus RNA. *Cold Spring Harbor Symposia Quant. Biol.* **38**, 943–950. [325]

Hu, S., Ohtsubo, E., Davidson, N., and Saedler, H. (1975). Electron microscope heteroduplex studies of sequence relations among bacterial plasmids: identification and mapping of the insertion sequences IS1 and IS2 in F and R plasmids. *J. Bacteriol.* **122**, 764–775.

Huang, A. S. (1973). Defective interfering viruses. *Ann. Rev. Microbiol.* **27**, 101–117. [308]

Huang, A. S., and Baltimore, D. (1970a). Initiation of polyribosome formation in poliovirus-infected HeLa cells. *J. Mol. Biol.* **47**, 275–291. [308]

Huang, A. S., and Baltimore, D. (1970b). Defective viral particles and viral disease processes. *Nature* **226**, 325–327. [312, 317]

Huang, A. S., and Baltimore, D. (1977). Defective interfering animal viruses. In: *Comprehensive Virology* (ed. H. Fraenkel-Conrat and R. R. Wagner). Plenum Press, New York, in press. [308]

Huang, A. S., and Manders, E. (1972). Ribonucleic acid synthesis of vesicular stomatitis virus. IV. Transcription by standard virus in the presence of defective interfering particles. *J. Virol.* **9**, 909–916. [332]

Huang, A. S., and Wagner, R. R. (1966). Defective T particles of vesicular stomatitis virus. II. Biologic role in homologous interference. *Virology* **30**, 173–181. [333]

Huebner, R. J. et al. (1956). Viruses in search of disease. Spec. publication, *N. Y. Acad. Sci.* **67**, 209–446. [416]

Huebner, R. J., and Todaro, G. J. (1969). Oncogenesis of RNA tumor viruses as determinants of cancer. *Proc. Natl. Acad. Sci. U.S.* **64**, 1087–1099. [453]

Hughes, K. M. (1952). Development of the inclusion bodies of a granulosis virus. *J. Bacteriol.* **64**, 375–380. [475]

Humphries, E. M., and Temin, H. M. (1972). Cell cycle-dependent activation of Rous sarcoma virus-infected stationary chicken cells: avian leukosis virus group-specific antigens and ribonucleic acid. *J. Virol.* **10**, 82–87. [389]

Humphries, E. M., and Temin, H. M. (1974). Requirement for cell division for initiation of transcription of Rous sarcoma virus RNA. *J. Virol.* **14**, 531–546. [389]

Hunt, D. M., Emerson, S. U., and Wagner, R. R. (1976). RNA-temperature-sensitive mutants of vesicular stomatitis virus: L-protein thermosensitivity accounts for transcriptase restriction of group I mutants. *J. Virol.* **18**, 596–603. [332]

Hunt, J. M., and Marcus, P. I. (1974). Mechanism of Sindbis virus-induced intrinsic interferon with vesicular stomatitis virus replication. *J. Virol.* **14**, 99–109. [414]

Hunter, T. H., Hunt, T., Knowland, J., and Zimmern, D. (1976). Messenger

RNA for the coat protein of tobacco mosaic virus. *Nature* **260,** 759–764. [326, 458]

Hurlburt, H. S., and Thomas, J. I. (1960). The experimental host range of the arthropod-borne animal viruses in arthropods. *Virology* **12,** 391–407. [423]

Hyman, R. W., Brunovskis, I., and Summers, W. C. (1973). Base sequence homology between coliphages T7 and øII and between T3 and øII as determined by heteroduplex mapping in the electron microscope. *J. Mol. Biol.* **77,** 189–196. [201]

Hynes, R. O. (1974). Role of surface alterations in cell transformation: the importance of proteases and surface proteins. *Cell* **1,** 147–156. [443]

Ikeda, H., and Tomizawa, J. (1968). Prophage P1, an extrachromosomal replication unit. *Cold Spring Harbor Symposia Quant. Biol.* **33,** 791–798.

Ingram, V. M. (1958). Abnormal human haemoglobins. I. The comparison of normal human and sickle-cell haemoglobins by "fingerprinting." *Biochim. Biophys. Acta* **28,** 539–545. [64, 89]

Inman, R. B. (1966). A denaturation map of the λ phage DNA molecule determined by electron microscopy. *J. Mol. Biol.* **18,** 464–476. [151]

Inman, R. B. (1967). Denaturation maps of the left and right sides of the lambda DNA molecule determined by electron microscopy. *J. Mol. Biol.* **28,** 103–116. [96]

Isaacs, A. (1963). Interferon. *Adv. Virus Research* **10,** 1–38. [409, 425]

Isaacs, A., and Lindenmann, J. (1957). Virus interference: I. The interferon. II. Some properties of interferon. *Proc. Royal Soc. London* **147B,** 258–267; 268–273. [408, 409]

Iseki, S., and Sakai, T. (1953). Artificial transformation of O antigens in Salmonella E group. I. Transformation by antiserum and bacterial autolysate. *Proc. Japan. Acad.* **29,** 121–138. [242]

Ishii, K., and Green, H. (1973). Lethality of adenosine for cultured mammalian cells by interference with pyrimidine biosynthesis. *J. Cell. Sci.* **13,** 429–439. [273]

Ishikawa, A., and Di Mayorca, G. (1971). Recombination between two temperate sensitive mutants of polyoma virus. In: *The Biology of Oncogenic Viruses* (ed. L. G. Silvestri). North-Holland Publishing Company, Amsterdam and London, pp. 294–299. [360]

Jackson, G. G., and Muldoon, R. L. (1975). *Viruses Causing Common Respiratory Infection in Man.* The University of Chicago Press, Chicago. [252, 416, 421, 428]

Jacob, F., and Brenner, S., and Cuzin, F. (1963). On the regulation of DNA replication in bacteria. *Cold Spring Harbor Symposia Quant. Biol.* **28,** 329–348. [130, 486]

Jacob, F., and Monod, J. (1961). Genetic regulatory mechanisms in the synthesis of proteins. *J. Mol. Biol.* **3,** 318–356. [65, 72, 131]

Jacob, F., and Wollman, E. L. (1958). *Sexuality and the Genetics of Bacteria.* Aademic Press, New York. [133]

Jacobson, M. F., Asso, J., and Baltimore, D. (1970). Further evidence on the formation of poliovirus proteins. *J. Mol. Biol.* **49,** 657–669. [317, 321]

Jacobson, M., and Baltimore, D. (1968). Morphogenesis of poliovirus. I. Association of the viral RNA with coat protein. *J. Mol. Biol.* **33,** 369–378. [121, 272, 321, 487]

Jacobson, M. F., and Baltimore, D. (1968). Polypeptide cleavages in the formation of poliovirus proteins. *Proc. Natl. Acad. Sci. U.S.* **61,** 77–84. [317, 319, 320]

Jamieson, J. D., and Palade, G. E. (1968). Intracellular transport of secretory proteins in the pancreatic exocrine cell. III. Dissociation of intracellular transport from protein synthesis. *J. Cell. Biol.* **39,** 580–588. [69, 261]

Jamieson, J. D., and Palade, G. E. (1968). Intracellular transport of secretory proteins in the pancreatic exocrine cell. IV. Metabolic requirements. *J. Cell Biol.* **39,** 589–603. [69, 261]

Jarvik, J., and Botstein, D. (1973). A genetic method for determining the order of events in a biological pathway. *Proc. Natl. Acad. Sci. U.S.* **70,** 2046–2050. [128]

Jaspars, E. M. J. (1974). Plant viruses with multipartite genome. *Adv. Virus Research* **19,** 37–149. [461, 467]

Jazwinski, S. M., and Kornberg, A. (1975). DNA replication *in vitro* starting with an intact ϕX174 phage. *Proc. Natl. Acad. Sci. U.S.* **72,** 3863–3867. [202]

Jazwinski, S. M., Lindberg, A. A., and Kornberg, A. (1975a). The lipopolysaccharide receptor for bacteriophages ϕX174 and S13. *Virology* **66,** 268–282. [202]

Jazwinski, S. M., Lindberg, A. A., and Kornberg, A. (1975b). The gene H spike protein of bacteriophages ϕX174 and S13. I. *Virology* **66,** 283–293. [202]

Jelinek, W. (1976). Oligonucleotides that are present in mRNA and HnRNA from HeLa cells. *Science,* in press.

Jenner, E. (1798). An inquiry into the causes and effects of the variolae vaccina, a disease discovered in some of the western counties of England, particularly Gloucestershire, and known by the name of cowpox. Reprinted by Cassell and Co., Ltd., 1896. Available in pamphlet vol. 4232, Army Medical Library, Washington, D.C. [426]

Jensen, D. D. (1959). A plant virus lethal to its insect vector. *Virology* **8,** 164–175.

Jessel, D., Hudson, J., Landau, T., Tenen, D., and Livingston, D. M. (1975). Interaction of partially purified simian virus 40 T antigen with circular viral DNA molecules. *Proc. Natl. Acad. Sci. U.S.* **72,** 1960–1964. [354]

Johnson, F. (1941). Transmission of plant viruses by dodder. *Phytopath.* **31,** 649–656. [469]

Johnson, M. W., and Markham, R. (1962). Nature of the polyamine in plant viruses. *Virology* **17,** 276–281. [123]

Joho, R. H., Billeter, M. A., and Weissman, C. (1975). Mapping of biological functions on RNA of avian tumor viruses: location of regions required for transformation and determination of host range. *Proc. Natl. Acad. Sci. U.S.* **72,** 4772–4776. [383]

Joho, R. H., Stoll, E., Friis, R. R., Billeter, M. A., and Weissmann, C. (1976). A partial genetic map of Rous sarcoma virus RNA: location of polymerase, envelope and transformation markers. In: *Animal Virology* (eds. D. Baltimore et al.). Academic Press, New York, pp. 127–146. [383]

Joklik, W. K. (1964). The intracellular uncoating of poxvirus DNA. II. The molecular basis of the uncoating process. *J. Mol. Biol.* **8,** 277–288. [377]

Joklik, W. K. (1966). The poxviruses. *Bacteriol. Rev.* **30,** 33–66. [376]

Joklik, W. K. (1968). The poxviruses. *Ann. Rev. Microbiol.* **22,** 359–390. [297, 374]

Joklik, W. K. (1972). Studies on the effect of chymotrypsin on reovirions. *Virology* **49,** 700–715. [339]

Joklik, W. K. (1974). Reproduction of reoviridae. In: *Comprehensive Virology* (eds. H. Fraenkel-Conrat and R. Wagner). Plenum Press, New York, pp. 231–334. [301, 338, 341, 342]

Joklik, W. K., and Becker, Y. (1964). The replication and coating of vaccinia DNA. *J. Mol. Biol.* **10,** 452–474. [407]

Joklik, W. K., and Darnell, J. E. (1961). The adsorption and early fate of purified poliovirus in HeLa cells. *Virology* **13,** 439–447. [279, 298]

Joklik, W. K., and Merigan, T. C. (1966). Concerning the mechanism of action of interferon. *Proc. Natl. Acad. Sci. U.S.* **56,** 558–565. [413]

Jolicoeur, P., and Baltimore, D. (1976). Effect of *Fv*-1 gene product on proviral DNA formation and integration in cells infected with murine leukemia viruses. *Proc. Natl. Acad. Sci. U.S.* **73,** 2236–2240. [452]

Joslin, R. (1971). Physiological studies of the *t* gene defect in T4-infected *Escherichia coli. Virology* **44,** 101–107. [177]

Josse, J., Kaiser, A. D., and Kornberg, A. (1961). Enzymatic synthesis of DNA. VIII. Frequencies of nearest neighbor base sequences in deoxyribonucleic acid. *J. Biol. Chem.* **236,** 864–875. [80]

Judd, B. H. (1975). Genes and chromosomes of Drosophila. In: *The Eukaryote Chromosome* (eds. W. J. Peacock and R. D. Brock). Australian National University Press, Canberra, pp. 169–184. [257]

Jungwirth, C., Horak, I., Bodo, G., Lindner, J., and Schultze, B. (1972). The synthesis of poxvirus-specific RNA in interferon-treated cells. *Virology* **48,** 59–70. [413]

Jungwirth, C., and Joklik, W. K. (1965). Studies on "early" enzymes in Hela cells infected with vaccinia virus. *Virology* **27,** 80–93. [377]

Kacian, D. L., Mills, D. R., Kramer, F. R., and Spiegelman, S. (1972). A replicating RNA molecule suitable for a detailed analysis of extracellular evolution and replication. *Proc. Natl. Acad. Sci. U.S.* **69,** 3038–3042. [209]

Kacian, D. L., Watson, K. F., Burry, A., and Spiegelman, S. (1971). Purification of the DNA polymerase of avian myeloblastosis virus. *Biochim. Biophys. Acta* **246,** 365–383. [387]

Kado, C. I., and Agrawal, H. O. (1972). *Principles and Techniques in Plant Virology,* Ch. 9, 10. Van Nostrand Reinhold, New York. [472]

Kado, C. I., and Knight, C. A. (1966). Location of a local lesion gene in tobacco mosaic virus RNA. *Proc. Natl. Acad. Sci. U.S.* **55,** 1276–1283. [466]

Kaiser, D., Syvanen, M., and Masuda, T. (1975). DNA packaging steps in bacteriophage lambda head assembly. *J. Mol. Biol.* **91,** 175–186.

Kamen, R. I. (1975). The structure and function of the QB replicase. In: *RNA Phages* (ed. N. D. Zinder). Cold Spring Harbor. [71]

Kamen, R. I., and Shure, H. (1976). Topography of polyoma messenger RNA molecules. *Cell* **7,** 361–371. [356]

Kao, F. T., and Puck, T. T. (1968). Genetics of somatic mammalian cells. VII. Induction and isolation of nutritional mutants in Chinese hamster cells. *Proc. Natl. Acad. Sci. U.S.* **60,** 1275–1281. [274]

Kaper, J. M., and Steere, R. L. (1959). Infectivity of tobacco ringspot virus nucleic acid preparations. *Virology* **7**, 127–139. [113]

Kapikian, A. Z., Kalica, A. R., Shih, J. W.-K., Cline, W. L., Thornhill, T. S., Wyatt, R. G., Chanock, R. M., Kim, H. W., and Gerin, J. L. (1976). Buoyant density in cesium chloride of the human reoviruslike agent of infantile gastroenteritis by ultracentrifugation, electron microscopy, and complement fixation. *Virology* **70**, 564–569. [338]

Kaplan, A. S. (1957). A study of the herpes simplex virus–rabbit kidney cell system by the plaque technique. *Virology* **4**, 435–437.

Kapuler, A. M., Mendelsohn, N., Klett, H., and Acs, G. (1970). Four base-specific 5'-triphosphatases in the subviral core of reovirus. *Nature* **225**, 1209–1213. [340]

Kassanis, B. K., and Glover, D. A. (1971). New evidence on the mechanism of aphid transmission of potato C and potato ancuba mosaic viruses. *J. Gen. Virol.* **10**, 99–101. [470]

Kassanis, B., and Macfarlane, I. (1965). Interaction of virus strain, fungus isolate, and host species in the transmission of tobacco necrosis virus. *Virology* **26**, 603–612. [472]

Katagiri, S., Aikawa, S., and Hinuma, Y. (1971). Stepwase degradation of poliovirus capsid by alkaline treatment. *J. Gen. Virol.* **13**, 101–109. [299]

Katagiri, S., Hinuma, Y., and Ishida, N. (1968). Relation between the adsorption to cells and antigenic properties in poliovirus particles. *Virology* **34**, 797–799. [298]

Kates, J. (1970). Transcription of the vaccinia virus genome and the occurrence of polyriboadenylic acid sequences in messenger RNA. *Cold Spring Harbor Symposia Quant. Biol.* **35**, 743–752. [375]

Kates, J., and Beeson, J. (1970). Ribonucleic acid synthesis in vaccinia virus. I. The mechanism of synthesis and release of RNA in vaccinia cores. *J. Mol. Biol.* **50**, 1–18. [376]

Kates, J., and McAuslan, B. R. (1967a). Messenger RNA synthesis by a "coated" viral genome. *Proc. Natl. Acad. Sci. U.S.* **57**, 314–320. [119, 297, 376, 377]

Kates, J., and McAuslan, B. R. (1967b). Poxvirus DNA-dependent RNA polymerase. *Proc. Natl. Acad. Sci. U.S.* **58**, 134–141. [119, 297, 375]

Kates, J., and McAuslan, B. R. (1967c). Relationship between protein synthesis and viral DNA synthesis. *J. Virol.* **1**, 110–114. [377]

Kaufman, H. E. (1962). Clinical cure of herpes simplex keratitis by 5'-iodo-2'-deoxyuridine. *Proc. Soc. Exptl. Biol. Med.* **109**, 251–252. [431]

Kaufman, Y., Goldstein, E., and Penman, S. (1976). Poliovirus-induced inhibition of polypeptide initiation *in vitro* on native polyribosomes. *Proc. Natl. Acad. Sci. U.S.* **73**, 1834–1838. [403]

Kavenoff, R., and Zimm, B. H. (1973). Chromosome-sized DNA molecules from *Drosophila. Chromosoma* **41**, 1–27. [257]

Kayajanian, G. (1968). Studies on the genetics of biotin-transducing, defective variants of bacteriophage λ. *Virology* **36**, 30–41.

Kellenberger, E., and Arber, W. (1957). Electron microscopal studies of phage multiplication. I. A method for quantitative analysis of particle suspensions. *Virology* **3**, 245–255. [55]

Keller, W., and Wendel, I. (1974). Stepwise relaxation of supercoiled SV40 DNA. *Cold Spring Harbor Symposia Quant. Biol.* **39,** 199–208. [358]

Kelly, T. J., Jr., Lewis, A. M., Jr., Levine, A. S., and Siegel, S. (1974). Structural studies of two adenovirus 2-SV40 hybrids containing the entire SV40 genome. *Cold Spring Harbor Symposia Quant. Biol.* **39,** 409–417. [367]

Kelly, T. J., Jr., and Smith, H. O. (1970). A restriction enzyme from *Hemophilus influenzae*. II. Base sequence of the recognition site. *J. Mol. Biol.* **51,** 393–409. [81]

Kelner, A. (1949). Effect of visible light on the recovery of Streptomyces griseus conidia from ultraviolet irradiation injury. *Proc. Natl. Acad. Sci. U.S.* **35,** 73–79. [192]

Kennedy, S. I. T. (1976). Sequence relationships between the genome and the intracellular RNA species of standard and defective-interfering Semliki forest virus. *J. Mol. Biol.* **108,** 491–511. [324, 326]

Kennedy, S. I. T., Brzeski, H., and Clegg, J. C. S. (1976). Synthesis of the polymerase polypeptides of Semliki forest virus. In: *Animal Viruses* (ed. D. Baltimore et al.). Academic Press, New York, pp. 677–688. [324, 325]

Kerr, I. M., Olshevsky, U., Lodish, H. F., and Baltimore, D. (1976). The translation of murine leukemia virus ribonucleic acid in cell-free systems from animal cells. *J. Virol.* **18,** 627–635. [386]

King, J., and Mykolajewycz, N. (1973). Bacteriophage T4 tail assembly: Proteins of the sheath, core, and baseplate. *J. Mol. Biol.* **75,** 339–358. [121]

Kingsbury, D. W. (1973). Paramyxovirus replication. *Curr. Topics Microbiol.* **59,** 1–30. [114]

Kissling, R. E. (1960). The arthropod-borne viruses of man and other animals. *Ann. Rev. Microbiol.* **14,** 261–282.

Kit, S. (1968). Viral-induced enzymes and the problem of viral oncogenesis. *Adv. Cancer Res.* **11,** 73–207. [415]

Kleckner, N., Chan, R. K., Tye, B., and Botstein, D. (1975). Mutagenesis by insertion of a drug-resistance element carrying an inverted repetition. *J. Mol. Biol.* **97,** 561–576.

Kleiman, J. H., and Moss, B. (1975a). Purification of a protein kinase and two phosphate acceptor proteins from vaccinia virions. *J. Biol. Chem.* **250,** 2420–2429. [376]

Kleiman, J. H., and Moss, B. (1975b). Characterization of a protein kinase and two phosphate acceptor proteins from vaccinia virions. *J. Biol. Chem.* **250,** 2430–2437.

Klein, G. (1972). Herpesviruses and oncogenesis. *Proc. Natl. Acad. Sci. U.S.* **69,** 1056–1064. [452]

Kleinschmidt, A. K. (1968). Monolayer techniques in electron microscopy of nucleic acid molecules. In: *Methods in Enzymology*, Vol. XIIB (eds. L. Grossman and K. Moldave). Academic Press, New York, pp. 361–377. [92]

Klenk, H. D., and Choppin, P. W. (1970). Plasma membrane lipids and parainfluenza virus assembly. *Virology* **40,** 939–947. [336]

Klimenko, S. M., Tikchonenko, T. I., and Andreev, V. M. (1967). Packing of DNA in the head of bacteriophage T2. *J. Mol. Biol.* **23,** 523–533. [176]

Klug, A., Longley, W., and Leberman, R. (1966). Arrangement of protein subunits and the distribution of nucleic acid in turnip yellow mosaic virus. I. X-ray diffraction studies. *J. Mol. Biol.* **15,** 315–343. [43]

Klug, A. (1972). Assembly of tobacco mosaic virus. *Federation Proc.* **31,** 30–42. [3, 459]

Klug, A., and Caspar, D. L. D. (1960). The structure of small viruses. *Adv. Virus Res.* **7,** 225–325. [37]

Knight, C. A. (1946). Precipitin reactions of highly purified influenza viruses and related materials. *J. Exp. Med.* **83,** 281–294.

Knight, E., Jr. (1974). Interferon-sepharose: Induction of the antiviral state. *Biochem. Biophys. Res. Commun.* **56,** 860–864. [412]

Knight, E., Jr. (1975). Heterogenity of purified mouse interferons. *J. Biol. Chem.* **250,** 4139–4144. [409]

Knight, E., Jr. (1976). Interferon: Purification and initial characterization from human diploid cells. *Proc. Natl. Acad. Sci. U.S.* **73,** 520–523. [409]

Knipe, D., Rose, J. K., and Lodish, H. F. (1975). Translation of the individual species of vesicular stomatitis virus in RNA. *J. Virol.* **15,** 1004–1011. [329]

Knipe, D. M., Lodish, H. F., and Baltimore, D. (1977a). Localization of two cellular forms of the vesicular stomatitis viral glycoprotein. *J. Virol.,* in press.

Knipe, D., Baltimore, D., and Lodish, H. F. (1977b). Separate pathways of maturation of the major structure proteins of vesicular stomatitis virus. *J. Virol.,* in press. [336, 337]

Knipe, D. M., Lodish, H. F., and Baltimore, D. (1977c). Maturation of viral proteins in cells infected with temperature-sensitive mutants of vesicular stomatitis virus. *J. Virol.,* in press. [337]

Knowland, J. (1974). Protein synthesis directed by the RNA from a plant virus in a normal animal cell. *Genetics* **78,** 383–394. [458]

Knowland, J., Hunter, T., Hunt, T., and Zimmern, D. (1976). Translation of tobacco mosaic virus RNA and isolation of the messenger for TMV coat protein. *Proc. INSERMEMBO Coll.* on *"In vitro* transcription of viral genomes,"* INSERM **47,** 211–216. [458]

Koch, A. L. (1960). Encounter efficiency of colliphage-bacterium interaction. *Biochim. Biophys. Acta* **39,** 311–318. [277]

Koch, A. S., and Feher, G. (1973). The possible nature of the chance-event in initiation of virus infections at the cellular level. *J. Gen. Virol.* **18,** 319–327. [277]

Koczot, F. J., Carter, B. J., Garon, C F., and Rose, J. A. (1973). Self-complementarity of terminal sequences within plus or minus strands of adenovirus-associated virus DNA. *Proc. Natl. Acad. Sci. U.S.* **70,** 215–219. [110, 345]

Koprowski, H., Jervis, G. A., and Norton, T. W. (1952). Immune responses in human volunteers upon oral administration of a rodent adapted strain of poliomyelitis virus. *Am. J. Hyg.* **55,** 108–126. [427]

Korant, B. D. (1972). Cleavage of viral precursor proteins *in vivo* and *in vitro*. *J. Virol.* **10,** 751–759. [432]

Korant, B. D. (1975). Regulation of animal virus replication by protein cleavage. In: *Proteases and Biological Control* (eds. E. Reich, D. Rifkind, and E. Shaw). Cold Spring Harbor Laboratory, pp. 621–644. [121]

Korant, B. D., and Butterworth, B. E. (1976). Inhibition by zinc of rhinovirus protein cleavage: interaction of zinc with capsid polypeptides. *J. Virol.* **18,** 298–306. [319]

Korant, B. D., Kauer, J. C., and Butterworth, B. E. (1974). Zinc ions inhibit replication of rhinoviruses. *Nature* **248,** 588–590. [432]

Korant, B. D., and Lonberg-Holm, K. (1974). Zonal electrophoresis and isoelectric focusing of proteins and virus particles in density gradients of small volume. *Anal. Biochem.* **59,** 75–82. [106]

Korant, B. D., Lonberg-Holm, K., Noble, J., and Stasny, J. T. (1972). Naturally occurring and artificially produced components of three rhinoviruses. *Virology* **48,** 71–86. [284, 299]

Kornberg, A. (1974). *DNA Synthesis.* W. H. Freeman, San Francisco. [75, 79, 84, 111]

Kornberg, R. D. (1974). Chromatin structure: a repeating unit of histones and DNA. *Science* **184,** 868–871. [257]

Koschel, K. (1974). Poliovirus infection and poly(A) sequences of cytoplasmic cellular RNA. *J. Virol.* **13,** 1061–1066. [403]

Kozinski, A. W. (1961). Fragmentary transfer of P³²-labeled parental DNA to progeny phage. *Virology* **13,** 124–134. [164, 186]

Kozloff, L. M., Lute, M., and Crosby, L. (1975a). Bacteriophage T4 baseplate components. I. *J. Virol.* **16,** 1391–1400. [174]

Kozloff, L. M., Crosby, L. K., Lute, M., and Hall, D. H. (1975b). Bacteriophage T4 baseplate components. II. *J. Virol.* **16,** 1401–1408. [174]

Kozloff, L. M., Crosby, L. K., and Lute, M. (1975c). Bacteriophage T4 baseplate components. III. *J. Virol.* **16,** 1409–1419. [174]

Kozloff, L. M., Verses, C., Lute, M., and Crosby, L. K. (1970). Bacteriophage tail components. II. Dihydrofolate reductase in T4D bacteriophage. *J. Virol.* **5,** 740–753. [122]

Krisch, H. M., Bolle, A., and Epstein, R. H. (1974). Regulation of the synthesis of bacteriophage T4 gene 32 protein. *J. Mol. Biol.* **88,** 89–104. [167]

Krueger, D. H., Presber, W., Hansen, S., and Rosenthal, H. A. (1975). Biological functions of the bacteriophage T3 SAMase gene. *J. Virol.* **16,** 453–455. [201]

Krugman, S. (1971). Present status of measles and rubella immunization in the U.S.: A medical progress report. *J. Pediatrics* **78,** 1–16. [425, 428]

Krywienczyk, J., and Bergold, G. H. (1960). Serological relationships of viruses from some lepidopterous and hymenopterous insects. *Virology* **10,** 308–315. [474]

Kuhn, C. M. (1964). Cellular susceptibility to enteroviruses. *Bacteriol. Rev.* **28,** 382–390. [285, 286]

Kung, H. J., Hu, S., Bender, W., Bailey, J. M., Davidson, N., Nicholson, M. O., and McAllister, R. M. (1976). RD-114, Baboon and wooly monkey viral RNAs compared in size and structure. *Cell* **7,** 609–620.

Kurtz-Fritsch, C., and Hirth, L. (1972). Uncoating of two spherical plant viruses. *Virology* **47,** 385–396.

Labedan, B., Crochet, M., Legault-Demare, J., and Stevens, B. J. (1973). Location of the first step transfer fragment and single-strand interruptions in T5stO bacteriophage DNA. *J. Mol. Biol.* **75,** 213–234. [198]

Lachmi, B. E., Glanville, N., Keränen, S., and Kääriäinen, L. (1975). Tryptic peptide analysis of nonstructural and structural precursor proteins from Semliki Forest virus mutant-infected cells. *J. Virol.* **16**, 1615-1629. [325]

Lachmi, B. E., and Kääriäninen, L. (1976). Sequential translation of nonstructural proteins in Semliki Forest virus mutant infected cells. *Proc. Natl. Acad. Sci. U.S.* (in press).

Laemmli, U. K. (1970). Cleavage of structural proteins during the assembly of the head of bacteriophage T4. *Nature* **227**, 680-685. [121]

Laemmli, U. K., and Quittner, S. F. (1974). Maturation of the head of bacteriophage T4. IV. *Virology* **62**, 485-499. [174, 176]

Laemmli, U. K., Teaff, N., and D'Ambrosia, J. (1974). Maturation of the head of bacteriophage T4. III. DNA packaging into preformed heads. *J. Mol. Biol.* **88**, 749-765. [176]

Lafay, F. (1974). Envelope proteins of vesicular stomatitis virus virus: effect of temperature-sensitive mutations in complementation groups III and V. *J. Virol.* **14**, 1220-1228. [332]

Lafferty, K. L., and Ortelis, S. (1963). The interaction between virus and antibody. III. Examination of virus-antibody complexes with the electron microscope. *Virology* **21**, 91-99. [60]

Lai, C. J., and Nathans, D. (1974). Mapping the genes of simian virus 40. *Cold Spring Harbor Symposia Quant. Biol.* **39**, 53-60. [351]

Lai, M.-H. T., Wérenne, J. J., and Joklik, W. K. (1973). The preparation of reovirus top component and its effect on host DNA and protein synthesis. *Virology* **54**, 237-244.

Lam, S. T., Stahl, M. M., McMilin, K. D., and Stahl, F. W. (1974). Rec-mediated recombinational hot spot activity in bacteriophage lambda. II. A mutation which causes hot spot activity. *Genetics* **77**, 425-433. [222]

La Montagne, J. R., and McDonald, W. C. (1972). A bacteriophage of *Bacillus subtilis* which forms plaques only at temperatures above 50 C. III. *J. Virology* **9**, 659-663. [137]

Lampson, G. P., Tytell, A. A., Field, A. K., Nemes, M. M., and Hilleman, M. R. (1967). Inducers of interferon and host resistance 1. Double-stranded RNA from extracts of *Penicillium funiculosum*. *Proc. Natl. Acad. Sci. U.S.* **58**, 782-989. [409]

Lane, L. C. (1974). The bromoviruses. *Adv. Virus Research* **19**, 152-213. [119]

Langridge, R., and Gomatos, P. J. (1963). The structure of RNA. *Science* **141**, 694-698. [339]

Lanni, Y. T. (1968). First-step-transfer deoxyribonucleic acid of bacteriophage T5. *Bacteriol. Rev.* **32**, 227-242. [198]

Lark, K. (1969). Initiation and control of DNA synthesis. *Am. Rev. Biochem.* **38**, 569-604.

Lauffer, M. A., and Price, W. C. (1945). Infection by viruses. *Arch. Biochem.* **8**, 449-468. [24, 27]

Lavelle, G., Patch, C., Khoury, G., and Rose, J. (1975). Isolation and characterization of single-stranded adenoviral DNA produced during synthesis of adenovirus type 2 DNA. *J. Virol.* **16**, 775-782. [385]

Laver, W. G. (1973). The polypeptides of influenza virus. *Adv. Virus Research* **18**, 57-104. [106, 279, 283]

Laver, W. G., Wrigley, N. G., and Pereira, H. G. (1969). Removal of pentons from particles of adenovirus type 2. *Virology* **39**, 599-605. [284]

Lawrence, C., and Thach, R. E. (1974). Encephalomyocarditis virus infection of mouse plasmacytoma cells I. Inhibition of cellular protein synthesis. *J. Virol.* **14**, 598–610. [403]

Lawrence, C., and Thach, R. E. (1975). Identification of a viral protein involved in post-translational maturation of the encephalomyocarditis virus capsid. *J. Virol.* **15**, 918–928. [321]

Lazarowitz, S. G., and Choppin, P. W. (1975). Enhancement of the infectivity of influenza A and B viruses by proteolytic cleavage of the hemagglutinin polypeptide. *Virology* **68**, 440–454. [338]

Le Bouvier, G. L., and McCollum, R. W. (1970). Australia (Hepatitis-associated) antigen: Physicochemical and immunological characteristics. *Adv. Virus Research* **16**, 357–396. [418]

Lebowitz, P., Kelly, T. J., Jr., Nathans, D., Lee, T. N. H., and Lewis, A. M., Jr. (1974). A colinear map relating the simian virus 40 (SV40) DNA segments of six adenovirus-SV40 hybrids to the DNA fragments produced by restriction endonuclease cleavage of SV40 DNA. *Proc. Natl. Acad. Sci. U.S.* **71**, 441–445. [367]

Leder, P., and Nirenberg, M. (1964). RNA codewords and protein synthesis. II. Nucleotide sequence of a valine RNA codeword. *Proc. Natl. Acad. Sci. U.S.* **52**, 420–427. [65]

Lederberg, J. (1952). Cell genetics and hereditary symbiosis. *Physiol. Revs.* **32**, 403–430. [133]

Ledinko, N. (1963). Genetic recombination with poliovirus type I. Studied of crosses between a normal horse serum resistant mutant and several guanidine resistant mutants of the same strain. *Virology* **20**, 117–119. [323]

Ledinko, N. (1968). Enhanced deoxyribonucleic acid polymerase activity in human embryonic kidney cultures infected with adenovirus 2 or 12. *J. Virol.* **2**, 89–98. [366]

Ledinko, N. (1972). Nucleolar ribosomal precursor RNA and protein metabolism in human embryo kidney cultures infected with adenovirus 12. *Virology* **49**, 79–89. [406]

Ledinko, N., and Hirst, G. (1961). Mixed infection of Hela cells with poliovirus types 1 and 2. *Virology* **14**, 207–219. [323]

Lee, F. Y., Nomoto, A., Detjen, B. M., and Wimmer, E. (1977). A protein covalently linked to poliovirus genome RNA. *Proc. Natl. Acad. Sci. U.S.* **74**, 59–63. [309]

Lehman, I. R. (1974). DNA ligase: structure, mechanism, and function. *Science* **186**, 790–797. [79]

Lehman, I. R., and Uyemura, D. G. (1976). DNA polymerase. I. Essential replication enzyme. *Science* **193**, 963–968. [76]

Leibowitz, R., and Penman, S. (1971). Regulation of protein synthesis in HeLa cells. III. Inhibition during poliovirus infection. *J. Virol.* **8**, 661–668. [403, 406]

Lenard, J., and Compans, R. W. (1974). The membrane structure of lipid-containing viruses. *Biochim. Biophys. Acta* **344**, 51–94. [335]

Lengyel, P. (1974). The process of translation: a birdseye view. In: *Ribosomes* (eds. M. Nomura, A. Tissieres, and P. Lengyel). Cold Spring Harbor Laboratory, pp. 13–52. [67, 69, 71]

Lerner, A. M., Bailey, E. J., and Tillotson, J. R. (1965). Enterovirus hemagglutination. Inhibition by several enzymes and sugars. *J. Immunol.* **95,** 1111–1115. [289]

Lerner, A. M., Gelb, L. D., Tillotson, J. R., Carruthers, M. M., and Bailey, E. J. (1966). Enterovirus hemagglutination: Inhibition by aldoses and a possible mechanism. *J. Immunol.* **96,** 629–636.

Lerner, A. M., and Miranda, Q. R. (1968). Cellular interactions of several enteroviruses and a reovirus after treatment with sodium borohydride or carbohydrases. *Virology* **36,** 277–285. [289]

Levin, D. H., Mendelsohn, N., Schonberg, M., Klett, H., Silverstein, S. C., Kapuler, A. M., and Acs, G. (1970). Properties of RNA transcriptase in reovirus subviral particles. *Proc. Natl. Acad. Sci. U.S.* **66,** 890–897. [339]

Levin, J. G., Grimley, P. M., Ramseur, J. M., and Berekesky, I. K. (1974). Deficiency of 60 to 70S RNA in murine leukemia virus particles assembled in cells treated with actinomycin D. *J. Virol.* **14,** 152–161. [387]

Levine, A. J. (1976). SV40 and adenovirus early functions involved in DNA replication and transformation. *BBA Revs. Cancer,* in press. [349]

Levine, A. J., van der Vliet, P. C., Rosenwirth, B., Rabek, J., Frenkel, G., and Ensinger, M. (1974). Adenovirus-infected, cell-specific, DNA-binding proteins. *Cold Spring Harbor Symposia Quant. Biol.* **39,** 559–566. [365]

Levinthal, C., and Visconti, N. (1953). Growth and recombination in bacterial viruses. *Genetics* **38,** 500–511. [184]

Levintow, L. (1974). The reproduction of picornaviruses. In: *Comprehensive Virology,* Vol. II (eds. H. Fraenkel-Conrat and R. R. Wagner). Plenum Press, New York, pp. 109–169. [309]

Levintow, L., and Darnell, J. E. (1960). A simplified procedure for purification of large amounts of poliovirus: Characterization and amino acid analysis of Type I poliovirus. *J. Biol. Chem.* **235,** 70–73. [278]

Levintow, L., and Eagle, H. (1961). Biochemistry of cultured mammalian cells. *Ann. Rev. Biochem.* **30,** 605–640. [258]

Levitt, N. H., and Crowell, R. L. (1967). Comparative studies of the regeneration of HeLa cell receptors for poliovirus T1 and Coxsackie-virus B3. *J. Virol.* **1,** 693–700. [289]

Lewandovsky, L. J., and Millward, S. (1971). Characterization of the genome of cytoplasmic polyhedrosis virus. *J. Virology* **7,** 434–437. [475]

Lewis, A. M., Jr., Levine, A. S., Crumpacker, C. S., Levin, M. J., Samaha, J., and Henry, P. H. (1973). Studies on nondefective adenovirus 2-simian virus 40 hybrid viruses. V. Isolation of additional hybrids which differ in their simian virus 40-specific biological properties. *J. Virol.* **11,** 655–664. [367]

Lewis, A. M., and Rowe, W. P. (1971). Studies on nondefective Adenovirus-simian virus 40 hybrid viruses. I. A newly characterized SV40 antigen induced by Ad2$^+$ND genome. *J. Virol.* **7,** 189–197. [355]

Lewis, J. B., Atkins, J. F., Anderson, C. W., Baum, P. R., and Gesteland, R. F. (1975). Mapping of late adenovirus genes by cell-free translation of RNA selected by hybridization to specific DNA fragments. *Proc. Natl. Acad. Sci. U.S.* **72,** 1344–1348. [364]

Lewis, J. B., Atkins, J. F., Baum, P. R., Solem, R., Gesteland, R. F., and

Anderson, C. W. (1976). Location and identificaton of the genes for adenovirus type 2 early polypeptides. *Cell* **7**, 141–151. [364]

L'Héritier, Ph. (1951). The CO_2 sensitivity problem in Drosophila. *Cold Spring Harbor Symposia Quant. Biol.* **16**, 99–111. [478]

L'Héritier, Ph. (1958). The hereditary virus of Drosophila. *Adv. Virus Res.* **5**, 195–245. [478]

Lhoas, P. (1972). Mating pairs of *Sacchararomyus cerevisiae* infected with double stranded RNA viruses from *Aspergillus niger. Nature New Biol.* **236**, 86–87.

Lilly, F., and Pincus, T. (1973). Genetic control of murine viral leukemogenesis. *Adv. Cancer Res.* **17**, 231–278. [452]

Lindahl, T., Adams, A., Bjursell, G., Bornkamm, G. W., Kaschka-Dierich, C., and Jehn, U. (1976). Covalently closed circular duplex DNA of Epstein-Barr virus in a human lymphoid cell line. *J. Mol. Biol.* **102**, 511–530. [373, 447]

Lindberg, A. A. (1973). Bacteriophage receptors. *Ann. Rev. Microbiol.* **27**, 205–241.

Lister, R. M. (1969). Tobacco rattle, NETV, viruses in relation to functional heterogeneity in plant viruses. *Federation Proc.* **28**, 1875–1889. [461]

Littlefield, J. W. (1966). The use of drug-resistant markers to study the hybridization of mouse fibroblasts. *Exp. Cell Res.* **41**, 190–196. [274]

Litvak, S., Tarragó, A., Tarragó-Litvak, L., and Allende, J. E. (1973). Elongation factor-viral genome interaction dependent on the aminoacylation of TYMV and TMV RNAs. *Nature New Biol.* **241**, 88–90. [458]

Lazzardi, P. (1976). Biogenesis of silk fibroin RNA: An example of extremely fast processing. *Prog. Nucleic Acid Res.* **24**, in press. [269]

Lockart, R. Z., Jr. (1964). The necessity for cellular RNA and protein synthesis for viral inhibition resulting from interferon. *Biochem. Biophys. Res. Commun.* **15**, 513–518. [412]

Lockart, R. Z., Jr. (1970). Interferon induction by viruses. *J. Gen. Physiol.* **56**, 3_s–12_s. [410]

Lockart, R. Z., Jr., Bayliss, N. L., Toy, S. T., and Yin, F. H. (1968). Viral events necessary for the induction of interferon in chick embryo cells. *J. Virol.* **2**, 962–965. [410]

Lodish, H. F. (1975). Regulation of *in vitro* protein synthesis by bacteriophage RNA by RNA tertiary structure. In: *RNA Phages* (ed. N. D. Zinder). Cold Spring Harbor Laboratory. [206]

Lonberg-Holm, K. (1975). The effects of concanavalin A on the early events of infection by rhinovirus type 2 and poliovirus type 2. *J. Gen. Virol.* **28**, 313–327. [299, 301]

Lonberg-Holm, K., and Butterworth, B. E. (1976). Investigation of the structure of polio and human rhinovirions through the use of selective chemical reactivity. *Virology* **71**, 207–216. [284, 298]

Lonberg-Holm, K., Crowell, R. L., and Philipson, L. (1976). Unrelated animal viruses share receptors. *Nature* **259**, 679–681. [277, 286]

Lonberg-Holm, K., Gossner, L. B., and Kauer, J. C. (1975). Early alteration of poliovirus in infected cells and its specific inhibition. *J. Gen. Virol.* **27**, 329–342. [301]

Lonberg-Holm, K., and Korant, B. D. (1972). Early interaction of rhinoviruses with host cells. *J. Virol.* **9**, 29–40. [278, 298]

Lonberg-Holm, K., and Noble-Harvey, J. (1973). Comparison of *in vitro* and cell-mediated alteraton of a human rhinovirus and its inhibition by SDS. *J. Virol.* **12**, 819–826.

Lonberg-Holm, K., and Philipson, L. (1969). Early events of virus-cell interaction in an adenovirus system. *J. Virol.* **4**, 323–338. [284, 298]

Lonberg-Holm, K., and Philipson, L. (1974). *Early Interactions Between Viruses and Cells.* Karger, New York. [277, 278, 279, 284, 290]

Lonberg-Holm, K., and Yin, F. H. (1973). Antigenic determinants of infective and inactivated human rhinovirus type 2. *J. Virol.* **12**, 114–123. [285, 299]

Losick, R. (1972). *In vitro* transcription. *Ann. Rev. Biochem.* **41**, 409–446. [76]

Losick, R., and Sonenshein, A. L. (1968). Change in the template sensitivity of RNA polymerase during sporulation of *Bacillus subtilis. Nature* **224**, 35–37. [135]

Louie, A. J. (1974). The organization of proteins in polyoma and cellular chromatin. *Cold Spring Harbor Symposia Quant. Biol.* **39**, 259–266. [349, 350]

Lowy, D. R., Rowe, W. P., Teich, N., and Hartley, J. M. (1971). Murine leukemia virus: high frequency activation *in vitro* by 5-iododeoxyuridine. *Science* **174**, 155–156. [389]

Lucké, B. (1938). Carcinoma in the leopard frog: its probable causation by a virus. *J. Exp. Med.* **68**, 457–468. [7]

Lundquist, R. E., Ehrenfeld, E., and Maizel, J. V. (1974). Isolation of a viral polypeptide associated with poliovirus RNA polymerase. *Proc. Natl. Acad. Sci. U.S.* **71**, 4773–4777. [312]

Luria, S. E. (1945). Mutations of bacterial viruses affecting their host range. *Genetics* **30**, 84–99. [157]

Luria, S. E. (1951). The frequency distribution of spontaneous bacteriophage mutants as evidence for the exponential rate of phage reproduction. *Cold Spring Harbor Symposia Quant. Biol.* **16**, 463–470. [153]

Luria, S. E. (1953a). *General Virology*, 1st ed. John Wiley and Sons, New York. [2, 4]

Luria, S. E. (1953b). Host-induced modifications of viruses. *Cold Spring Harbor Symposia Quant. Biol.* **18**, 237–244.

Luria, S. E. (1959). Viruses as infective genetic materials. In: *Immunity and Virus Infection* (ed. V. Najjar). John Wiley and Sons, New York. [2]

Luria, S. E., and Darnell, J. E., Jr. (1967). *General Virology*, 2nd ed. John Wiley and Sons, New York. [2]

Luria, S. E., and Delbrück, M. (1943). Mutations of bacteria from virus sensitivity to virus resistance. *Genetics* **28**, 491–511. [157]

Luria, S. E., Delbrück, M., and Anderson, T. F. (1943). Electron microscope studies of bacterial viruses. *J. Bacteriol.* **46**, 57–77. [147]

Luria, S. E., and Dulbecco, R. (1949). Genetic recombination leading to production of active bacteriophage from ultraviolet inactivated bacteriophage particles. *Genetics* **34**, 93–125. [187]

Luria, S. E., and Human, M. L. (1952). A non-hereditary host-induced variation of bacterial viruses. *J. Bacteriol.* **64**, 557–569. [155]

Luria, S. E., Williams, R. C., and Backus, R. C. (1951). Electron micrographic counts of bacteriophage particles. *J. Bacteriol.* **61**, 179–188. [25, 142]

Luzzati, D. (1970). Regulation of λ exonuclease synthesis: role of the N gene product and λ repressor. *J. Mol. Biol.* **49**, 515–519. [221]

Luzzati, V., and Husson, J. (1962). The structure of liquid crystalline phases of lipid-water systems. *J. Cell. Biol.* **12**, 207–219. [262]

Lwoff, A. (1943). *L'évolution physiologique, Études des pertes de fonctions chez les microorganismes.* Hermann, Paris. [483]

Lwoff, A. (1957). The concept of virus. *J. Gen. Microbiol.* **17**, 239–253. [1, 3, 4]

Lwoff, A. (1959). Factors influencing the evolution of viral diseases at the cellular level and in the organism. *Bacteriol. Rev.* **23**, 109–124. [424]

Lwoff, A., Dulbecco, R., Vogt, M., and Lwoff, M. (1955). Kinetics of the release of poliomyelitis virus from single cells. *Virology* **1**, 128–139. [306]

Lwoff, A., Horne, R., and Tournier, P. (1962). A system of viruses. *Cold Spring Harbor Symposia Quant. Biol.* **27**, 51–55. [14]

MacLeod, R., Black, L. M., and Moyer, F. H. (1966). The fine structure and intracellular localization of potato yellow dwarf virus. *Virology* **29**, 540–552. [463]

Macpherson, I. (1970). The characteristics of animal cells transformed *in vitro*. *Adv. Cancer Res.* **13**, 169–215. [435]

Macpherson, I., and Montagnier, L. (1964). Agar suspension culture for the selective assay of cells transformed by polyoma virus. *Virology* **23**, 291–294. [439]

Maden, B. E. H., and Salim, M. (1974). The methylated nucleotide sequences in HeLa cell ribosomal RNA and its precursors. *J. Mol. Biol.* **88**, 133–164. [77]

Maizel, J. V. (1969). Acrylamide gel electrophoresis of proteins and nucleic acids. In: *Fundamental Techniques in Virology* (eds. K. Habel and N. P. Salzman). Academic Press, New York, pp. 334–362. [120]

Malamy, M. H., Fiandt, M., and Szibalski, W. (1972). Electron microscopy of polar insertions in the *lac* operon of *Escherichia coli*. *Mol. Gen. Gen.* **119**, 207–222.

Manaker, R. A., and Groupé, V. (1956). Discrete foci of altered chicken embryo cells associated with Rous sarcoma virus in tissue cultures. *Virology* **2**, 838–840. [18]

Mandel, B. (1962). Early stages of virus-cell interaction as studied by using antibody. *Cold Spring Harbor Symposia Quant. Biol.* **27**, 123–136. [309]

Mandel, M., and Higa, A. (1970). Calcium-dependent bacteriophage DNA infection. *J. Mol. Biol.* **53**, 159–162. [139]

Mandel, P. (1964). Free nucleotides in animal tissue. *Prog. Nucleic Acid Res.* **3**, 299–334. [259]

Maniatis, T., Ptashne, M., Bachman, K., Kleid, D., Flashman, S., Jeffrey, A., and Maurer, R. (1975). Sequences of repressor binding sites in the operators of bacteriophage λ. *Cell* **5**, 109–113. [72]

Marchesi, V. T., Tillack, T. W., Jackson, R. L., Segrest, J. P., and Scott, R. E. (1972). Chemical characterization and surface orientation of the major glycoprotein of the human erythrocyte membrane. *Proc. Natl. Acad. Sci. U.S.* **69**, 1445–1449. [263]

Marchesi, V. T., Tillack, T. W., and Scott, R. E. (1971). The role of glycoproteins in red cell membrane structure. In: *Glycoproteins of Blood Cells and Plasma* (eds. Jamieson and Greenwalt). Lippincott, Philadelphia, pp. 94–105. [289]

Marcu, K., and Dudock, B. (1975). Methylation of TMV RNA. *Biochem. Biophys. Res. Commun.* **62,** 798–807. [458]

Marcus, P. I., Engelhardt, D., Hunt, J. M., and Sekellick, M. J. (1971). Interferon action: inhibition of vesicular stomatitis virus RNA synthesis induced by virion-bound polymerase. *Science* **174,** 593–598. [413]

Marcus, P. I., and Hirsch, D. (1963). Renewal of a normal surface component. Cellular receptors for virus attachment. *J. Cell Biol.* **19,** 47A. [289]

Marcus, P. J., and Carver, D. H. (1967). Intrinsic interference: A new type of viral interference. *J. Virol.* **1,** 334–343. [414]

Marcus, P. I., Terry, T. M., and Levine, S. (1975). Interferon action II membrane-bound alkaline ribonuclease activity in chick embryo cells manifesting interferon-mediated interference. *Proc. Natl. Acad. Sci. U.S.* **72,** 182–186. [413]

Margulis, L. (1970). *Origin of Eukaryotic Cells.* Yale University Press, New Haven. [485]

Mark, D. F., and Richardson, C. C. (1976). *Escherichia coli* thioredoxin: a subunit of bacteriophage T7 DNA polymerase. *Proc. Natl. Acad. Sci. U.S.* **73,** 780–784. [198]

Markham, R., Hitchborn, J. H., Hills, G. J., and Frey, S. (1964). The anatomy of the tobacco mosaic virus. *Virology* **22,** 342–359. [39, 459]

Markham, R., and Smith, K. M. (1949). Studies on the virus of turnip yellow mosaic. *Parasitology* **39,** 330–342. [53]

Marmur, J., Brandon, C., Neubort, S., Ehrlich, M., Mandel, M., and Konvicka, J. (1972). Unique properties of nucleic acid from *Bacillus subtilis* phage SP-15. *Nature New Biol.* **239,** 68. [113]

Marmur, J., Rownd, R., and Schildkraut, C. L. (1963). Denaturation and renaturation of DNA. *Prog. Nucleic Acid Res.* **1,** 231–300. [98]

Martin, G. S. (1970). Rous sarcoma virus: a function required for the maintenance of the transformed state. *Nature* **227,** 1021–1023. [448]

Martin, R. G., Chou, J. Y., Avila, J., and Saral, R. (1974). The semiautonomous replicon: a molecular model for the oncogenicity of SV40. *Cold Spring Harbor Symposia Quant. Biol.* **39,** 17–24. [354]

Martin, S. A., and Moss, B. (1975). Modification of RNA by mRNA guanylyltransferase and mRNA (guanine-7-)-methyltransferase from vaccinia virions. *J. Biol. Chem.* **250,** 9330–9335. [375]

Martin, S. A., Paoletti, E., and Moss, B. (1975). Purification of mRNA guanylyltransferase and mRNA (guanine-7-)-methyltransferase from vaccinia virions. *J. Biol. Chem.* **250,** 9322–9329. [375, 376]

Marvin, D. A., and Wachtel, E. J. (1975). Structure and assembly of filamentous bacterial viruses. *Nature* **253,** 19–23. [205]

Matheson, A. T., and Thomas, C. A., Jr. (1960). The indirect effect accompanying P^{32} suicide of bacteriophage. *Virology* **11,** 289–291. [192, 194]

Mathews, C. K. (1971). *Bacteriophage Biochemistry*, Vol. VIII. Van Nostrand Reinhold, New York. [130]

Matsumoto, S. (1970). Rabies virus. *Adv. Virus Research* **16,** 257–302. [393]

Matsumara, T., Stollar, V., and Schlessinger, R. W. (1971). Studies on the nature of Dengue viruses. V. Structure and development of Dengue virus in Vero cells. *Virology* **46**, 344–355. [393]

Maurer, R., Maniatis, T., and Ptashne, M. (1974). Promoters are in the operators in phage lambda. *Nature* **249**, 221–224.

Maxam, A. M., and Gilbert, W. (1977). A new method for sequencing DNA. *Proc. Natl. Acad. Sci. U.S.* **74**, 560–564. [92]

May, E., Maizel, J. V., and Salzman, N. P. (1976). Mapping of the transcription sites of SV40-specific late 16S and 10S mRNA by electron microscopy. *Proc. Natl. Acad. Sci. U.S.*, in press.

Maynard-Smith, S., and Symonds, N. (1973). Involvement of bacteriophage T4 genes in radiation repair. *J. Mol. Biol.* **74**, 33–44. [192]

McAllister, P. E., and Wagner, R. R. (1976). Differential inhibition of host protein synthesis in L cells infected with RNA temperature-sensitive mutants of vesicular stomatitis virus. *J. Virol.* **18**, 550–558. [404]

McAuslan, B. R. (1963a). Control of induced thymidine kinase activity in the Poxvirus-infected cell. *Virology* **20**, 162–168. [377, 378, 410]

McAuslan, B. R. (1963b). The induction and repression of thymidine kinase in the poxvirus infected HeLa cell. *Virology* **21**, 383–389. [377, 378, 410]

McAuslan, B. R. (1969). The biochemistry of poxvirus replication. In: *The Biochemistry of Viruses* (ed. Hilton B. Levy). Marcel Dekker, New York.

McAuslan, B. R., and Kates, J. R. (1967). Poxvirus-induced acid deoxyribonuclease: regulation of synthesis; control of activity *in vivo*; purification and properties of the enzyme. *Virology* **33**, 709–716. [377]

McAuslan, R. R., and Armentrout, R. W. (1974). The biochemistry of icosahedral cytoplasmic deoxyviruses. *Curr. Topics Microbiol. Immunol.* **68**, 77–105. [374]

McClelland, L., and Hare, R. (1941). The adsorption of influenza virus by red cells and a new *in vitro* method of measuring antibodies for influenza virus. *Canad. Publ. Health J.* **32**, 530–538. [279]

McCormick, W., and Penman, S. (1967). Inhibition of RNA synthesis in HeLa and L cells by mengovirus. *Virology* **31**, 135–141. [405]

McCorguodale, D. J. (1975). The T-odd bacteriophages. *CRC Critical Reviews in Microbiology*, Dec. 1975, 101–159. [197]

McDowell, M. J., Joklik, W. K., Villa-Komaroff, L., and Lodish, H. F. (1972). Translation of reovirus messenger RNAs synthesized *in vitro* into reovirus polypeptides by several mammalian cell-free extracts. *Proc. Natl. Acad. Sci. U.S.* **69**, 2649–2653. [339]

McGeoch, D. J., Crawford, L. V., and Follett, E. A. C. (1970). The DNAs of three parvoviruses. *J. Gen. Virol.* **6**, 33–40. [344]

McGregor, S., Hall, L., and Rueckert, R. R. (1975). Evidence for the existence of protomers in the assembly of encephalomyocarditis virus. *J. Virol.* **15**, 1107–1120. [321]

McLaren, L. C., and Holland, J. J. (1974). Defective interfering particles from poliovirus vaccine and vaccine reference strains. *Virology* **60**, 579–583. [323]

McLaren, L. C., Holland, J. J., and Syverton, J. T. (1959). The mammalian cell-virus relationship. I. Attachment of poliovirus to cultivated cells of primate and nonprimate origin. *J. Exp. Med.* **109**, 475–485. [285]

McLaren, L. C., Holland, J. J., and Syverton, J. T. (1960). The mammalian cell-virus relationship. V. Susceptibility and resistance of cells *in vitro* to infection by Coxsackie A9 virus. *J. Exp. Med.* **112,** 581–594. [286]

MacLeod, R., Black, L. M., and Moyer, F. H. (1966). The fine structure and intracellular localization of potato yellow dwarf virus. *Virology* **29,** 540–552.

McNicol, L. A., and Goldberg, E. B. (1973). An immunochemical characterization of glycosylation in bacteriophage T4. *J. Mol. Biol.* **76,** 285–301. [164]

McQuillen, K., Roberts, R. B., and Britten, R. J. (1959). Synthesis of nascent protein by ribosomes in *E. coli. Proc. Natl. Acad. Sci. U.S.* **45,** 1437–1447. [69]

McSharry, J. J., Compans, R. W., and Choppin, P. W. (1971). Proteins of vesicular stomatitis virus and of phenotypically mixed vesicular stomatitis virus-simian virus 5 virions. *J. Virol.* **8,** 722–729.

Medrano, L., and Green, H. (1973). Picornavirus receptors and picornavirus multiplication in human-mouse hybrid cell lines. *Virology* **54,** 515–524. [287]

Merigan, T. C. (1964). Purified interferons: Physical properties and species specificity. *Science* **145,** 811–813. [408]

Meselson, M., and Weigle, J. J. (1961). Chromosome breakage accompanying genetic recombination in bacteriophage. *Proc. Natl. Acad. Sci. U.S.* **47,** 857–868.

Meselson, M., and Yuan, R. (1968). DNA restriction enzyme from E. coli. *Nature* **216,** 1110–1114. [81]

Metzenberg, R. L. (1972). Genetic regulatory systems in *Neurospora. Ann. Rev. Genetics* **6,** 111–132. [265]

Meynell, E. W. (1961). A phage, $\phi\chi$, which attacks mobile bacteria. *Virology* **47,** 743–752.

Milcarek, C., Price, R. P., and Penman, S. (1974). The metabolism of a poly(A) minus mRNA fraction in HeLa cells. *Cell* **3,** 1–10. [273]

Miller, D. A., Miller, O. J., Dev, V. G., Hashmi, S., Tantravahi, R., Medrano, L., and Green, H. (1974). Human chromosome 19 carries a poliovirus receptor gene. *Cell* **1,** 167–173. [287]

Miller, G., Robinson, J., and Heston, L. (1974). Immortalizing and nonimmortalizing laboratory strains of Epstein-Barr virus. *Cold Spring Harbor Symposia Quant. Biol.* **39,** 773–781. [372]

Miller, O. L., Jr., and Beatty, B. R. (1969). Visualization of nucleolar genes. *Science* **164,** 955–957. [93, 96]

Miller, S. L., and Orgel, L. E. (1974). *The Origins of Life on the Earth.* Prentice-Hall, Englewood Cliffs, New Jersey. [483]

Min Jou, W., Haegeman, G., Ysebaert, M., and Fiers, W. (1972). Nucleotide sequence of the gene coding for the bacteriophage MS2 coat protein. *Nature* **237,** 82–88. [64]

Model, P., and Zinder, N. (1974). *In vitro* synthesis of bacteriophage f1 proteins. *J. Mol. Biol.* **83,** 231–251. [205]

Moelling, K. (1975). Reverse transcriptase and RNase H: present in a murine virus and in both subunits of an avian virus. *Cold Spring Harbor Symposia Quant. Biol.* **39,** 969–973. [387]

Moffett, H. L., and Cramblett, H. G. (1962). Viral isolations and illnesses in young infants attending a well-baby clinic. *New England J. Med.* **267,** 1213–1218. [421]

Molling, K., Bolognesi, D. P., Bauer, H., Busen, W., Plassmann, H. W., and Hausen, P. (1971). Association of viral reverse transcriptase with an enzyme degrading the RNA moiety of RNA:DNA hybrids. *Nature New Biol.* **234,** 240–243. [80]

Montagnier, L., and Sanders, F. K. (1963). The replicative form of encephalomyocarditis virus ribonucleic acid. *Nature* **199,** 664–667. [312, 313]

Moody, M. F. (1965). The shape of the T-even bacteriophage head. *Virology* **26,** 567–576. [44]

Moody, M. F. (1973). Sheath of bacteriophage T4. III. *J. Mol. Biol.* **80,** 613–636. [161]

Morgan, C., and Howe, C. (1968). Structure and develoment of viruses as observed in the electron microscope. IX. Entry of parainfluenza I (Sendai) virus. *J. Virol.* **2,** 1122–1132. [291]

Morgan, C., Rifkind, R. A., and Rose, H. M. (1962). The use of ferritin-conjugated antibodies in electron microscopic studies of influenza and vaccinia viruses. *Cold Spring Harbor Symposia Quant. Biol.* **27,** 57–65. [50]

Morgan, C., and Rose, H. M. (1968). Structure and development of viruses as observed in the electron microscope. VIII. Entry of influenza virus. *J. Virol.* **2,** 925–936. [291]

Morgan, C., Rose, H. M., and Mednis, B. (1968a). Electron microscopy of herpes simplex virus. I. Entry. *J. Virol.* **2,** 507–516. [297]

Morgan, C., Rosenkranz, H. S., and Mednis, B. (1968b). Structure and development of viruses as observed in the electron microscope. X. Entry and uncoating of adenovirus. *J. Virol.* **4,** 777–796. [298]

Morgenson, K. E., and Cantell, K. (1974). Human leukocyte interferon: a role for disulphide bonds. *J. Gen. Virol.* **22,** 95–103. [408]

Morris, C. F., Sinha, N. K., and Alberts, B. M. (1975). Recombination of bacteriophage T4 DNA replication apparatus from purified components. *Proc. Natl. Acad. Sci. U.S.* **72,** 4800–4804. [164, 168]

Morrison, T. G., Stampfer, M., Lodish, H. F., and Baltimore, D. (1975). *In vitro* translation of vesicular stomatitis virus messenger RNA's and the existence of a 40S "plus" strand. In: *Negative Strand Viruses*, Vol. I. (eds. B. W. J. Mahy and R. D. Barry). Academic Press, New York, pp. 293–300. [331]

Morrow, T., and Berg, P. (1972). Cleavage of simian virus 40 DNA at a unique site by a bacterial restriction enzyme. *Proc. Natl. Acad. Sci. U.S.* **69,** 3365–3369. [351]

Mosig, G. (1963). Genetic recombination in bacteriophage T4 during replication of DNA fragments. *Cold Spring Harbor Symposia Quant. Biol.* **28,** 35–42. [25, 167, 175]

Mosig, G. (1968). A map of distances along the DNA molecule of bacteriophage T4. *Genetics* **59,** 137–151. [179]

Mosig, G., Bowden, D. W., and Bock, S. (1972). *E. coli* DNA polymerase I and other host functions participate in T4 DNA replication and recombination. *Nature New Biol.* **240,** 12–16.

Mosig, G., Carnigan, J. R., Bibring, J. B., Cole, R., Bock, H. O., and Bock, S. (1972). Coordinate variation in lengths of deoxyribonucleic acid molecules and head lengths in morphological variants of bacteriophage T4. *J. Virol.* **9**, 857–871. [46]

Mosig, G., and Werner, R. (1974). On the replication of incomplete chromosomes of phage T4. *Proc. Natl. Acad. Sci. U.S.* **64**, 747–754. [167]

Moss, B. (1968). Inhibition of HeLa cell protein synthesis by the vaccinia virion. *J. Virol.* **2**, 1028–1037.

Moss, B. (1973). Ansamycins: (A) Rifamycin SV derivations. In: *Selective Inhibitors of Viral Functions* (ed. W. A. Carter). Chemical Rubber Company Press, Cleveland, Ohio, pp. 313–328. [379, 431, 432]

Moss, B. (1974). Reproduction of poxviruses. In: *Comprehensive Virology*, Vol. III (eds. H. Fraenkel-Conrat and R. R. Wagner). Plenum Press, New York, pp. 405–474. [374]

Moss, B., Rosenblum, E. N., and Garon, C. F. (1973). Glycoprotein synthesis in cells infected with vaccinia virus. III. Purification and biosynthesis of the virion glycoprotein. *Virology* **55**, 143–156. [375]

Moss, B., Rosenblum, E. N., and Gershowitz, A. (1975). Characterization of a polyriboadenylate polymerase from vaccinia virions. *J. Biol. Chem.* **250**, 4722–4729. [375]

Mosser, A. G., Caliguiri, L. A., and Tamm, I. (1972). Incorporation of lipid precursors into cytoplasmic membranes of poliovirus-infected HeLa cells. *Virology* **47**, 39–47. [415]

Motulsky, A. (1974). Genetics of human hemoglobins: An overview. *Ann. N.Y. Acad. Sci.* **241**, 7–11. [64]

Moulder, J. W. (1962). *The Biochemistry of Intracellular Parasitism.* University of Chicago Press, Chicago. [484]

Moulder, J. W. (1964). *The Psittacosis Group as Bacteria.* John Wiley and Sons, New York. [484]

Mousset, S., and Gariglio, P. (1976). Isolation and preliminary characterization of SV40 transcription complex. In: *Animal Virology* (ed. D. Baltimore et al.). Academic Press, New York, pp. 55–66. [357]

Mowshowitz, D. (1970). Transfer RNA synthesis in HeLa cells. II. Formation of tRNA from a precursor *in vitro* and formation of pseudouridine. *J. Mol. Biol.* **50**, 143–151. [270]

Moyer, S. A., and Banerjee, A. K. (1975). Messenger RNA species synthesized *in vitro* by the virion-associated RNA polymerase of vesicular stomatitis virus. *Cell* **4**, 37–43. [330]

Mulder, C., Arrand, J. R., Delius, H., Keller, W., Pettersson, U., Roberts, R. J., and Sharp, P. A. (1974). Cleavage maps of DNA from adenovirus types 2 and 5 by restriction endonucleases *ECO*RI and *Hpa*I. *Cold Spring Harbor Symposia Quant. Biol.* **39**, 397–400. [363]

Muller, H. J. (1967). The genetic material as the initiator and the organizing basis of life. In: *Heritage from Mendel* (ed. R. A. Brink). University of Wisconsin Press, Madison, Wisconsin, pp. 419–447. [257]

Müller, U. R., and Marchin, G. L. (1975). Temporal appearance of bacteriophage T4-modified valyl tRNA synthetase in *Escherichia coli. J. Virol.* **15**, 238–243. [173]

Mundry, K. W. (1965). Interactions between plant viruses and host cells. *Symp. Soc. Developmental Biol.* **24,** 65–94. [465]

Munyon, W., Paoletti, E., and Grace, J. T., Jr. (1967). RNA polymerase activity in purified infectious vaccinia virus. *Proc. Natl. Acad. Sci. U.S.* **58,** 2280–2287. [119, 297, 375]

Munyon, W. H., and Kit, S. (1966). Induction of cytoplasmic RNA synthesis in vaccinia-infected LM cells during inhibition of protein synthesis. *Virology* **29,** 303–309. [376]

Murphy, J. R., Pappenheimer, A. M., Jr., and Tayart de Borms, S. (1974). Synthesis of diphtheria *tox*-gene products in *Escherichia coli* extracts. *Proc. Natl. Acad. Sci. U.S.* **71,** 11–15. [244]

Naficy, K., and Nategh, R. (1972). Measles vaccine and its use in developing countries. *Adv. Virus Research* **17,** 279–294. [252, 417]

Naora, H., and Whitelam, J. M. (1975). Presence of sequences by hybridisable to dsRNA in cytoplasmic mRNA molecules. *Nature* **256,** 756–759. [269]

Narkhammer, M., and Magnusson, G. (1976). DNA polymerase activities induced by polyoma virus infection of 3T3 mouse fibroblasts. *J. Virol.* **18,** 1–6. [356]

Nash, H. A. (1975). Integrative recombination of bacteriophage lambda DNA *in vitro*. *Proc. Natl. Acad. Sci. U.S.* **72,** 1072–1075. [234]

Nathans, D., Notani, G., Schwartz, J. H., and Zinder, N. D. (1962). Biosynthesis of the coat protein of coliphage *f2* by *E. coli* extracts. *Proc. Natl. Acad. Sci. U.S.* **48,** 1424–1431. [65]

Nathans, D., and Smith, H. O. (1975). Restriction endonucleases in the analysis and restructuring of DNA molecules. *Ann. Rev. Biochem.* **44,** 273–293. [81, 86]

Neurath, A. R., Hartzell, R. W., and Rubin, B. A. (1970). Partial characterization of the complementary sites involved in the reaction between adenovirus type 7 and erythrocyte receptors. *Virology* **42,** 789–793. [278]

Ng, M. H., and Vilcek, J. (1972). Interferons: Physicochemical properties and control of cellular synthesis. *Adv. in Protein Chem.* **26,** 173–234. [408, 410]

Ngan, J. S. C., Holloway, A. F., and Cormack, D. V. (1974). Temperature-sensitive mutants of vesicular stomatitis virus: comparison of the *in vitro* RNA polymerase defects of group I and group IV mutants. *J. Virol.* **14,** 765–772. [332]

Nicholson, G. L. (1976). Transmembrane control of the receptors on normal and tumor cells. 1. Cytoplasmic influence over cell surface components. *Biochim. Biophys. Acta.* **457,** 57–108. [336, 443]

Nicolson, G. L. (1971). Difference in topology of normal and tumour cell membranes shown by different surface distributions of ferritin-conjugated concanavalin A. *Nature New Biol.* **233,** 244–246. [263]

Nicolson, G. L. (1974). The interaction of lectins with animal cell surfaces. *Intern. Rev. Cytology* **39,** 89–190. [60]

Nieman, P. F. (1972). Rous sarcoma virus nucleotide sequences in cellular DNA: Measurement by RNA DNA hybridization. *Science* **178,** 750–753. [384]

Nirenberg, M., and Matthaei, H. (1961). The dependence of cell-free protein synthesis in *E. coli* upon naturally occurring or synthetic polyribonucleotides. *Proc. Natl. Acad. Sci. U.S.* **47,** 1588–1602. [65]

Nishimura, S., Jones, D. S., and Khorama, S. (1965). Studies on polynucleotides. XVVIII. *J. Mol. Biol.* **13**, 302–324. [65]

Noll, H. and Stutz, E. (1968). The use of sodium and lithium dodecyl sulfate in nucleic acid isolation. In: *Methods in Enzymology*, Vol. XII (eds. L. Grossman and K. Moldave). Academic Press, New York, pp. 129–155. [106]

Nomoto, A., Lee, Y. F., and Wimmer, E. (1976). The 5′ end of poliovirus mRNA is not capped with m⁷G(5′)ppp(5′)Np. *Proc. Natl. Acad. Sci. U.S.* **73**, 375–380. [313, 403]

Nomura, M., and Held, W. A. (1974). Reconstitution of ribosomes: Studies of ribosome structure, function and assembly. In: *Ribosomes* (eds. M. Nomura, A. Tissieres, and P. Lengyel). Cold Spring Harbor Laboratory. pp. 193–224. [71]

North, A. C. T., and Rich, A. (1961). X-ray diffraction studies of bacterial viruses. *Nature* **191**, 1242–1245. [43]

Norrick, A., and Szilard, L. (1951). Virus strains of identical phenotype but different genotype. *Science* **113**, 34–35.

Notani, G. W., Engelhardt, D. L., Konigsberg, W., and Zinder, N. D. (1965). Suppression of a coat protein mutant of the bacteriophage f². *J. Mol. Biol.* **12**, 439–447. [71]

Novick, A., and Szilard, L. (1951). Virus strains of identical phenotype but different genotype. *Science* **113**, 34–35. [131, 176]

Novikoff, A., and Holtzman, E. (1975). *Cells and Organelles*, 2nd ed. Holt, Rinehart and Winston, New York. [254, 259]

Öberg, B., and Philipson, L. (1971). Replicative structures of poliovirus RNA *in vivo*. *J. Mol. Biol.* **58**, 725–737. [313, 431]

Öberg, B., and Philipson, L. (1972). Binding of histidine to tobacco mosaic virus RNA. *Biochem. Biophys. Res. Comm.* **48**, 927–932. [458]

O'Callaghan, D. J., Hyde, J. M., Gentry, G. A., and Randall, C. C. (1968). Kinetics of viral deoxyribonucleic acid, protein, and infectious particle production and alteration in host macromolecular syntheses in equine abortion (herpes) virus-infected cells. *J. Virol.* **2**, 793–804. [371]

Oda, K., and Joklik, W. K. (1967). Hybridization and sedimentation studies on "early" and "late" vaccinia messenger RNA. *J. Mol. Biol.* **27**, 395–419. [376, 377, 378, 379]

O'Farrell, P. Z., and Gold, L. M. (1973a). Bacteriophage T4 gene expression. *J. Biol. Chem.* **248**, 5502–5511. [170]

O'Farrell, P. Z., and Gold, L. M. (1973b). Transcription and translation of prereplicative bacteriophage T4 genes *in vitro*. *J. Biol. Chem.* **248**, 5512–5519. [168, 171]

O'Farrell, P. H. (1975). High resolution two-dimensional electrophoresis of proteins. *J. Biol. Chem.* **250**, 4007–4021. [71]

Ogston, A. G. (1963). On uncertainties inherent in the determination of the efficiency of collision between virus particles and cells. *Biochim. Biophys. Acta* **66**, 279–281. [277]

Ohe, K., and Weissman, S. M. (1970). Nucleotide sequence of an RNA from cells infected with adenovirus type 2. *Science* **167**, 879–881. [363]

Okada, Y. (1958). The fusion of Ehrlich's tumor cells caused by HVJ virus *in vitro*. *Biken J.* **1**, 103–110. [291]

Okada, Y. (1972). Fusion of cells by HVJ (Sendai virus). In: *Membrane Research* (ed. C. F. Fox). Academic Press, New York, pp. 371-382. [333]

Okazaki, R., Okazaki, T., Sakabe, K., Sugimoto, K., and Sugino, A. (1968). Mechanism of DNA chain growth. 1. Possible discontinuity and unusual secondary structure of newly synthesized chains. *Proc. Natl. Acad. Sci. U.S.* **59,** 598-605. [358]

Okubo, S., Strauss, B., and Stodolsky, M. (1964). The possible role of recombination in the infection of competent *Bacillus subtilis* by bacteriophage nucleic acids. *Virology* **24,** 552-562. [25]

Oldstone, M. B. A., and Dixon, F. J. (1970). Persistent lymphocytic choriomeningitis viral infection. III. Virus-anti-viral antibody complexes and associated chronic disease followed transplacental infection. *J. Immunol.* **105,** 829-837. [419]

O'Neill, F. J., and Rapp, F. (1971). Synergistic effect of herpes simplex virus and cytosine arabinoside on human chromosomes. *J. Virol.* **7,** 692-695. [393, 407]

Osborn, M., and Weber, K. (1974). SV40: T antigen, the A function and transformation. *Cold Spring Harbor Symposia Quant. Biol.* **39,** 267-276. [354]

Oxman, M., and Levin, M. J. (1971). Interferon and transcription of virus-specific RNA in cells infected with simian virus 40. *Proc. Natl. Acad. Sci. U.S.* **68,** 255-302. [413]

Padgett, B. L., Walker, D. L., ZuRhein, G. M., Echroade, R. J., and Dessel, B. H. (1971). Cultivation of papova-like virus from human brain with progressive multifocal leukoencephalopathy. *Lancet* **1,** 1257-1260. [349]

Padmanabhan, R., Padmanabhan, R., and Green, M. (1976). Evidence for palindromic sequences near the termini of adenovirus 2 DNA. *Biochem. Biophys. Res. Commun.* **69,** 860-867. [111, 360]

Painter, R. B., and Schaeffer, A. W. (1969). Rate of synthesis along replicons of different kinds of mammalian cells. *J. Mol. Biol.* **45,** 467-479.

Palese, P., and Schulman, J. L. (1976). Mapping of the influenza virus genome: identification of the hemagglutinin and the neuraminidase genes. *Proc. Natl. Acad. Sci. U.S.*, in press. [333]

Palma, E. L., Perlman, S. M., and Huang, A. S. (1974). Ribonucleic acid synthesis of vesicular stomatitis virus VI. Correlation of defective particle RNA synthesis with standard RNA replication. *J. Mol. Biol.* **85,** 127-136. [333]

Panet, A., Haseltine, W. A., Baltimore, D., Peters, G., Harada, F., and Dahlberg, J. E. (1975a). Specific binding of tryptophan transfer RNA to avian myeloblastosis virus reverse transcriptase. *Proc. Natl. Acad. Sci. U.S.* **72,** 2535-2539. [387]

Panet, A., Baltimore, D., and Hanafusa, T. (1975b). Quantitation of avian RNA tumor virus reverse transcriptase by radioimmunoassay. *J. Virol.* **115,** 146-152. [387]

Paoletti, E., and Moss, B. (1974). Two nucleic acid-dependent nucleoside triphosphate phosphohydrolases from vaccinia virus: Nucleotide substrate and polynucleotide cofactor specificities. *J. Biol. Chem.* **249,** 3281-3286. [376]

Paoletti, E., Rosemond-Hornbeak, H., and Moss, B. (1974). Two nucleic acid-dependent nucleoside triphosphate phosphohydrolases from vaccinia virus. *J. Biol. Chem.* **249**, 3273–3280. [376]

Pardee, A. B. (1974). A restriction point for control of normal animal cell proliferation. *Proc. Natl. Acad. Sci. U.S.* **71**, 1286–1290. [437]

Pardue, M. L., and Gall, J. G. (1970). Chromosomal localization of mouse satellite DNA. *Science* **168**, 1356–1358. [267]

Paskind, M. P., Weinberg, R. A., and Baltimore, D. (1975). Dependence of Moloney MuLV production on cell growth. *Virology* **67**, 242–248. [389]

Pasternak, G. (1969). Antigens induced by mouse leukemia viruses. *Adv. Cancer Res.* **12**, 1–99. [452]

Pasteur, L. (1885). Méthode pour prevenir la rage après morsure. *C. r. Acad. Sci.* **101**, 765–772. [423]

Paterson, R., and Knight, C. A. (1975). Protein synthesis in tobacco protoplasts infected with tobacco mosaic virus. *Virology* **64**, 10–22. [457]

Penman, S. (1965). Stimulation of the incorporation of choline in poliovirus-infected cells. *Virology* **25**, 148–152. [415]

Penman, S., Becker, Y., and Darnell, J. E. (1964). A cytoplasmic structure involved in the synthesis and assembly of poliovirus components. *J. Mol. Biol.* **8**, 541–555. [312]

Penman, S., Scherrer, K., Becker, Y., and Darnell, J. E. (1963). Polyribosomes in normal and poliovirus infected HeLa cells and their relationship to messenger RNA. *Proc. Natl. Acad. Sci. U.S.* **49**, 654–662.

Penman, S., and Summers, D. (1965). Effects on host cell metabolism following synchronous infection with poliovirus. *Virology* **27**, 614–620. [402]

Pennington, T. H., and Follett, E. A. C. (1974). Vaccinia virus replication in enucleate BSC-1 cells: particle production and synthesis of viral DNA and proteins. *J. Virol.* **13**, 488–493. [374]

Perlin, M., and Phillips, B. A. (1973). *In vitro* assembly of polioviruses III. Assembly of 14 S particles into empty capsids by poliovirus-infected HeLa cell membranes. *Virology* **53**, 107–114. [321]

Perlman, S., Phillips, C., and Bishop, J. O. (1976). A study of foldback DNA. *Cell* **8**, 33–42. [268]

Perlman, S. M., and Huang, A. S. (1973). RNA synthesis of vesicular stomatitis virus. V. Interaction between transcription and replication. *J. Virol.* **12**, 1395–1400. [332]

Perlman, S. M., and Huang, A. S. (1973/4). Virus-specific RNA specified by the group I and IV temperature-sensitive mutants of vesicular stomatitis virus. *Intervirology* **2**, 312–325. [332]

Pero, J., Nelson, J., and Losick, R. (1975). *In vitro* and *in vivo* studies of transcription by vegetative and sporulating *Bacillus subtilis*. In: *Spores*, Vol. VI., pp. 202–212. [76]

Perol-Vauchez, Y., Tournier, M. P., and Lwoff, M. (1961). Attenuation de la virulence du virus de l'encephalomyocardite de la souris par culture à basse temperature. Influence de l'hypo-et de l'hy; erthermie sur l'evolution de la maladie experimentale. *C. r. Acad. Sci.* **253**, 2164–2166. [424]

Perrault, J., and Holland, J. J. (1972). Absence of transcriptase activity or transcription-inhibiting ability in defective interfering particles of vesicular stomatitis virus. *Virology* **50**, 159–170. [332]

Perry, R. P. (1965). The nucleolus and the synthesis of ribosomes. *National Cancer Inst. Monograph* **18,** 325–340. [273]

Perry, R. P., and Kelley, D. E. (1974). Existence of methylated messenger RNA in mouse L cells. *Cell* **1,** 37–42. [115]

Pestka, S. (1969). Translocation, aminoacyloligonucleotides, and antibiotic action. *Cold Spring Harbor Symp. Quant. Biol.* **34,** 395–410. [272]

Peters, G., Harada, F., Dahlberg, J. E., Haseltine, W. A., Panet, A., and Baltimore, D. (1977). The low molecular weight RNAs of Moloney murine leukemia virus: identification of the primer for RNA-directed DNA synthesis. *J. Virol.,* in press. [387]

Peterson, R. F., Kievitt, K. D., and Eunis, H. L. (1972). Membrane protein synthesis after infection of *Escherichia coli* B with phage T4: The rII B protein. *Virology* **50,** 520–527. [190]

Pettersson, U., and Philipson, L. (1974). Synthesis of complementary RNA during productive adenovirus infection. *Proc. Natl. Acad. Sci. U.S.* **71,** 4887–4891. [365]

Pettersson, U., Tibbetts, C., and Philipson, L. (1976). Hybridization maps of early and late messenger RNA sequences on the adenovirus 2 genome. *J. Mol. Biol.* **101,** 479–501.

Pfefferkorn, E. R., and Burge B. W. (1967). Genetics and biochemistry of arbovirus temperature-sensitive mutants. In: *The Molecular Biology of Viruses* (ed. J. Colter). Academic Press, New York, pp. 403–426. [327]

Pfefferkorn, E. R., and Clifford, R. L. (1964). The origin of the protein of Sindbis virus. *Virology* **23,** 217–223. [337]

Pfefferkorn, E. R., and Shapiro, D. (1974). Reproduction of Togaviruses. In: *Comprehensive Virology* (eds. Fraenkel-Conrat, H. and Wagner, R.). Plenum Press, New York, pp. 17–230. [323, 415, 423]

Philipson, L. (1967a). Chromatography and membrane separation. In: *Methods in Virology*, Vol. II (eds. K. Maramorosch and H. Koprowski). Academic Press, New York, pp. 179–235. [105]

Philipson, L. (1967b). Water-organic solvent phase systems. In: *Methods in Virology*, Vol. II (eds. K. Maramorosch, and H. Koprowski). Academic Press, New York, pp. 236–244. [105]

Philipson, L., and Lindberg, U. (1974). Reproduction of adenoviruses. In: *Comprehensive Virology* (eds. H. Fraenkel-Conrat and R. R. Wagner). Plenum Press, New York, pp. 143–227. [360]

Philipson, L., Lonberg-Holm, K., and Petterson, U. (1968). Virus-receptor interaction in an adenovirus system. *J. Virol.* **2,** 1064–1075. [284, 289]

Philipson, L., Pettersson, U., and Lindberg, U. (1975). *Molecular Biology of Adenoviruses.* Springer-Verlag, New York. [360]

Philipson, L., Pettersson, U., Lindberg, U., Tibbetts, C., Vennström, B., and Persson, T. (1974). RNA synthesis and processing in adenovirus-infected cells. *Cold Spring Harbor Symposia Quant. Biol.* **39,** 447–456. [406]

Piaget, U., Eliasson, R., and Reichard, P. (1974). Replication of polyoma DNA in isolated nuclei. III. The nucleotide sequence at the RNA-DNA junction of nascent strands. *J. Mol. Biol.* **84,** 197–216. [360]

Pollack, R., and Goldman, R. (1973). Synthesis of infective poliovirus in BSC-1 monkey cells enucleated with cytochalasin B. *Science* **179,** 915–916. [312]

Pollack, R., Osborn, M., and Weber, K. (1975). Patterns of organization of actin and myosin in normal and transformed cells. *Proc. Natl. Acad. Sci. U.S.* **72,** 994–998. [443]

Pollack, R., Risser, R., Conlon, S., and Rifkin, D. (1974). Plasminogen activator production accompanies loss of anchorage regulation in transformation of primary rat embryo cells by simian virus 40. *Proc. Natl. Acad. Sci. U.S.* **71,** 4792–4796. [445]

Pollard, T. D., and Weihing, R. R. (1974–75). Actin and myosin and cell movement. In: *Critical Reviews of Biochemistry,* Vol. 2 (ed. G. D. Fasman). Chemical Rubber Company Press, Cleveland, Ohio, pp. 1–65. [261]

Pons, M. U., Schulze, I. T., and Hirst, G. K. (1969). Isolation and characterization of the ribonucleoprotein of influenza virus. *Virology* **39,** 250–259. [333]

Pontén, J. (1971). *Spontaneous and Virus-Induced Transformation in Cell Culture.* Springer-Verlag, New York. [435]

Pope, J. H., and Rowe, W. P. (1964). Immunofluorescent studies of adenovirus 12 tumors and of cells transformed or infected by adenoviruses. *J. Exp. Med.* **120,** 577–588. [363]

Porter, K., Claude, A., and Fullan, E. F. (1945). A study of tissue culture cells by electron microscopy. *J. Exp. Med.* **81,** 233–246. [259]

Poste, G. (1970). Virus-induced polykaryocytosis and the mechanism of cell fusion. *Adv. Virus Research* **16,** 303–356. [415]

Prage, L., Petterson, U., Höglund, S., Lonberg-Holm, K., and Philipson, L. (1970). Structural proteins of adenoviruses. IV. Sequential degradation of the adenovirus type 2 virion. *Virology* **42,** 341–358. [284]

Prescott, D. M., Kates, J., and Kirkpatrick, J. B. (1971). Replication of vaccinia virus DNA in enucleated L-cells. *J. Mol. Biol.* **59,** 505–508. [374]

Price, R., and Penman, S. (1972). A distinct RNA polymerase activity, synthesizing 5.5S, 5S and 4S RNA in nuclei from adenovirus 2 infected HeLa cells. *J. Mol. Biol.* **70,** 435–450. [363]

Price, W. C. (1940). Acquired immunity from plant virus diseases. *Quart. Rev. Biol.* **15,** 338–361. [468, 469]

Printz, P. (1973). Relationship of sigma virus to vesicular stomatitis virus. *Adv. Virus Research* **18,** 143–147. [479]

Prives, C., Aviv, H., Gilboa, E., Winocour, E., and Revel, M. (1975). *The Cell-Free Translation of Early and Late Classes of SV40 Messenger RNA in* in vitro *Transcription and Translation of Viral Genomes* (eds. A. L. Haenni and G. Beaud). INSERM, Paris, pp. 305–312. [356]

Ptashne, M., Bachman, K., Humazun, M. Z., Jeffrey, A., Maurer, R., Mayer, B. and Sauer, R. T. (1976). Autoregulation and function of repressor in bacteriophage lambda. *Science* **194,** 156–161. [217, 226]

Ptashne, K., and Cohen, S. N. (1975). Occurrence of insertion sequences (IS) regions on plasmid deoxyribonucleic acid as direct and inverted nucleotide sequence duplications. *J. Bacteriol.* **122,** 776–781.

Puck, T. T., Marcus, P. I., and Cieciura, S. J. (1956). Clonal growth of mammalian cells *in vitro.* Growth characteristics of colonies from single HeLa cells with and without a feeder layer. *J. Exp. Med.* **103,** 273–284. [252]

Puck, T. T., Garen, A., and Cline, J. (1951). The mechanism of virus attachment to host cells. *J. Exp. Med.* **103,** 65–88. [146, 278]

Pulleyblank, D. E., Shure, M., Tang, D., Vinograd, J., and Vosberg, H. P. (1975). Action of nicking-closing enzyme on supercoiled and nonsupercoiled closed circular DNA: Formation of a Boltzmann distribution of topological isomers. *Proc. Natl. Acad. Sci. U.S.* **72**, 4280–4284. [81, 86, 109]

Quersin-Thiry, L. (1961). Interaction between cellular extracts and animal viruses. I. Kinetic studies and some notes on the specificity of the interaction. *Acta Virol. Praha* **5**, 141–152. [286]

Quersin-Thiry, L., and Nihoul, E. (1961). Interaction between cellular extracts and animal viruses. II. Evidence for the presence of different inactivators corresponding to different viruses. *Acta Virol. Praha* **5**, 283–293. [286]

Randall, J., and Disbrey, C. (1965). Evidence for the presence of DNA at basal body sites in *Tetrahymena pyriformis*. *Proc. Royal Soc. B.* **162**, 473–491. [486]

Ranki, M., and Pettersson, R. F. (1975). Uukuniemi virus contains an RNA polymerase. *J. Virol.* **16**, 1420–1425. [333]

Raskas, H. J., Thomas, D. C., and Green, M. (1970). Biochemical studies on adenovirus multiplication XVII. Ribosome synthesis in uninfected and infected KB cells. *Virology* **40**, 893–902. [405]

Rawls, W. E. (1968). Congenital rubella: The significance of virus persistance. *Prog. Med. Virol.* **10**, 238–285. [253, 424]

Reed, S. I., Ferguson, J., Davis, R., and Stark, G. R. (1975). T antigen binds to Simian Virus 40 DNA at the origin of replication. *Proc. Natl. Acad. Sci. U.S.* **72**, 1605–1609. [97, 354]

Reed, W., and Carroll, J. (1902). The etiology of yellow fever; a supplementary note. *Am. Med.* **3**, 301–305. [423]

Reed, W., Carroll, J., Agramonte, A., and Lazear, J. W. (1911). *Yellow fever: a compilation of various publications.* U.S. 61st Congress, Docum. No. 822, Washington, D.C. [423]

Reedman, B. M., and Klein, G. (1973). Cellular localization of an Epstein-Barr virus (EBV) associated complement fixing antigen in producer and non producer lymphoblastoid cell lines. *Int. J. Cancer* **11**, 499–520. [372]

Rees, M. W., Short, M. N., and Kassanis, B. (1970). The amino acid composition, antigenicity and other characteristics of the satellite viruses of tobacco necrosis virus. *Virology* **40**, 448–461. [115]

Reich, E., Franklin, R. M., Shatkin, A. J., and Tatum, E. L. (1961). Effect of actinomycin D on cellular nucleic acid synthesis and virus production. *Science* **134**, 556–557. [273]

Reichardt, L. (1975). Control of bacteriophage lambda repressor synthesis: regulation of the maintenance pathway by the *cro* and *cI* products. *J. Mol. Biol.* **93**, 289–309.

Reichmann, M. E. (1964). The satellite tobacco necrosis virus: a single protein and its genetic code. *Proc. Natl. Acad. Sci. U.S.* **52**, 1009–1017. [115, 130, 462]

Reid, M. S., and Matthews, R. E. F. (1966). On the origin of the mosaic induced by turnip yellow mosaic virus. *Virology* **28**, 563–570. [463]

Revel, H. R., and Luria, S. E. (1970). DNA-glucosylation in T-even phage: genetic determination and role in phage-host interaction. *Ann. Rev. Genetics* **4**, 177–192. [163, 195]

Revel, M., Bash, D., and Ruddle, F. H. (1976). Antibodies to a cell-surface component coded by human chromosome 21, inhibit action of interferon. *Nature* **260**, 139–141. [412]

Rhoades, M., and Rhoades, E. A. (1972). Terminal repetition in the DNA of bacteriophage T5. *J. Mol. Biol.* **69**, 187–200. [198]

Rhode, S. L., III. (1976). Replication process of the parvovirus H-1. V. Isolation and characterization of the temperature-sensitive H-1 mutants defective in progeny DNA synthesis. *J. Virol.* **17**, 659–667. [349]

Rhodes, D. P., and Banerjee, A. K. (1976). 5'-terminal sequences of vesicular stomatitis virus mRNAs synthesized *in vitro*. *J. Virol.* **17**, 33–42. [329]

Rhodes, D. P., Moyer, S., and Banerjee, A. K. (1974). *In vitro* synthesis of methylated messenger RNA by the virion-associated RNA polymerase of vesicular stomatitis virus. *Cell* **3**, 327–333. [330]

Richards, K. E., and Williams, R. C. (1973). Assembly of tobacco mosaic virus rods *in vitro*. Elongation of partially assembled rods. *Biochemistry* **12**, 4574–4581. [460]

Richards, K. E., and Williams, R. C. (1976). Assembly of tobacco mosaic virus *in vitro*. In: *Comprehensive Virology*, Vol. 6 (eds. H. Fraenkel-Conrat and R. R. Wagner). Plenum Press, New York, pp. 1–37. [120]

Rifkind, D., Goodman, N., and Hill, R. B. (1967). The clinical significance of cytomegalovirus infection in renal transplant recipients. *Ann. Int. Med.* **66**, 1116–1128. [418]

Risser, R., and Pollack, R. (1974). A nonselective analysis of SV40 transformation of mouse 3T3 cells. *Virology* **59**, 477–489. [436]

Ritchey, M. B., Palese, P., and Kilbourne, E. D. (1976). RNA's of influenza A, B and C viruses. *J. Virol.* **18**, 738–744. [335]

Ritchie, D. A., Thomas, C. A., Jr., MacHattie, L. A., and Wensink, P. C. (1967). Terminal repetition in non-permuted T3 and T7 bacteriophage DNA molecules. *J. Mol. Biol.* **23**, 365–376. [199]

Riva, S., Cascino, A., and Geiduschek, E. P. (1970). Coupling of late transcription to viral replication in bacteriophage T4 development. *J. Mol. Biol.* **54**, 85–102. [171]

Riva, S., Cascino, A., and Gaiduschek, E. P. (1970). Uncoupling of late transcription from DNA replication in bacteriophage T4 development. *J. Mol. Biol.* **54**, 103–119. [171]

Rivers, T. M. (1937). Viruses and Koch's postulates. *J. Bacteriol.* **33**, 1–12. [14]

Robb, J. A., and Martin, R. G. (1972). Genetic analysis of simian virus 40. III. Characterization of a temperature-sensitive mutant blocked at an early stage of productive infection in monkey cells. *J. Virol.* **9**, 956–968. [354]

Robbins, P. W., and Uchida, T. (1962). Determinants of specificity in Salmonella: changes in antigenic structure mediated by bacteriophage. *Federation Proc.* **21**, 702–710. [242]

Robbins, P. W., Wickus, G. G., Branton, P. E., Gaffney, B. J., Hirschberg, C. B., Fuchs, P., and Blumberg, P. M. (1974). The chick fibroblast cell surface after transformation by Rous sarcoma virus. *Cold Spring Harbor Symposia Quant. Biol.* **39**, 1173–1180. [443]

Robbins, P. W., Wright, A., and Dankert, M. (1966). Polysaccharide biosynthesis. *J. Gen. Physiol.* **49**, 331–346. [482]

Roberts, B. E., Matthews, M. B., and Bruton, C. J. (1973). Tobacco mosaic virus RNA directs the synthesis of a coat protein peptide in a cell-free system from wheat. *J. Mol. Biol.* **80,** 733-742. [457]

Roberts, J. W. (1975). Transcription termination and late control in phage lambda. *Proc. Natl. Acad. Sci. U.S.* **72,** 3300-3304. [224]

Roberts, J. W., and Roberts, C. W. (1975). Proteolytic cleavage of bacteriophage lambda repressor in induction. *Proc. Natl. Acad. Sci. U.S.* **72,** 147-151. [212]

Roberts, J. W., and Steitz, J. A. (1967). The reconstitution of infective bacteriophage R17. *Proc. Natl. Acad. Sci. U.S.* **58,** 1416-1421. [205]

Roberts, R. J., Arrand, J. R., and Keller, W. (1974). The length of the terminal repetition in adenovirus-2 DNA. *Proc. Natl. Acad. Sci. U.S.* **71,** 3829-3833. [360]

Robinson, A. J., and Bellett, A. J. D. (1974). A circular DNA-protein complex from adenoviruses and its possible role in DNA replication. *Cold Spring Harbor Symposia Quant. Biol.* **39,** 523-531. [363]

Rochow, W. F. (1972). The role of mixed infections in the transmission of plant viruses by aphids. *Ann. Rev. Phytopathol.* **10,** 101-124.

Rodbell, M. (1964). Metabolism of isolated fat cells. *J. Biol. Chem.* **239,** 375-380. [409]

Roeder, R. G., and Rutter, W. J. (1970). Specific nucleolar and nucleoplasmic RNA polymerases. *Proc. Natl. Acad. Sci. U.S.* **65,** 675-682. [271, 273]

Roizman, B. (1965). An inquiry into the mechanisms of recurrent herpes infections of man. *Perspectives in Virol.* **4,** 283-301. [418]

Roizman, B., and Furlong, D. (1974). The replication of herpes viruses. In: *Comprehensive Virology*, Vol. 3 (eds. H. Fraenkel-Conrat and R. R. Wagner). Plenum Press, New York, pp. 229-403. [367, 371]

Roizman, B., and Heine, J. W. (1972). Modification of human cell membranes by herpes virus. *Proceedings of the First California Membrane Conference,* Vol. 3 (ed. C. F. Fox). Academic Press, New York, pp. 203-237. [393, 415]

Roizman, B., and Kieff, E. D. (1975). Herpes simplex and Epstein-Barr viruses in human cells and tissues: a study in contrasts. In: *Cancer*, Vol. 2 (ed. F. F. Becker). Plenum Press, New York. [367]

Roizman, B., and Spear, P. G. (1971). The role of herpesvirus glycoproteins in the modification of membranes of infected cells. In: *Nucleic Acid-Protein Interactions—Nucleic Acid Synthesis in Viral Infections* (ed. D. W. Ribbons et al.). North-Holland, Amsterdam, 1971, pp. 435-460. [372]

Rose, J. A. (1974). Parvovirus reproduction. In: *Comprehensive Virology*, Vol. 3 (eds. H. Fraenkel-Conrat and R. R. Wagner). Plenum Press, New York, pp. 1-62. [344]

Rose, J. K. (1975). Heterogeneous 5'-terminal structures occur on vesicular stomatitis virus mRNA's. *J. Biol. Chem.* **250,** 8098-8104. [109, 329]

Rose, J. K., and Knipe, D. (1975). Nucleotide sequence complexities, molecular weights, and poly(A) content of the vesicular stomatitis virus mRNA species. *J. Virol.* **15,** 994-1003. [329]

Rosemond-Hornbeak, H., Paoletti, E., and Moss, B. (1974). Single-stranded DNA-specific nuclease V_1 from vaccinia virus: Purification and characterization. *J. Biol. Chem.* **249,** 3287-3291. [376]

Rosenberg, M., Kramer, R. A., and Steitz, J. A. (1974). T7 early messenger

RNA's are the direct products of ribonuclease III cleavage. *J. Mol. Biol.* **89,** 777–782.

Rosenberg, N., and Baltimore, D. (1976). A quantitative assay for transformation of bone marrow cells by Abelson murine leukemia virus. *J. Exp. Med.* **143,** 1453–1463. [449]

Ross, Jr. (1976). A precursor of globin messenger RNA. *J. Mol. Biol.* **106,** 403–420. [269]

Rous, P. (1911). A sarcoma of the fowl transmissible by an agent separable from the tumor cells. *J. Exp. Med.* **13,** 397–411. [435]

Roux, L., and Kolakofsky, D. (1975). Isolation of RNA transcripts from the entire Sendai viral genome. *J. Virol.* **16,** 1426–1434. [333]

Roy, P., and Bishop, D. H. L. (1970). Isolation and properties of poliovirus minus strand ribonucleic acid. *J. Virol.* **6,** 604–609.

Rubin, H. (1960). An analysis of the assay of Rous sarcoma *in vitro* by the infective center technique. *Virology* **10,** 29–49. [26]

Rueckert, R. (1971). Picornaviral architecture. In: *Comprehensive Virology* (eds. K. Maramorosch and E. Kurstak). Academic Press, New York, pp. 225–306. [309]

Ruiz-Gomez, J., and Isaacs, A. (1963). Optimal temperature for growth and sensitivity to interferon among different viruses. *Virology* **19,** 1–7. [412]

Rymo, L., Parsons, J. T., Coffin, J. M., and Weissmann, C. (1974). *In vitro* synthesis of Rous sarcoma virus-specific RNA is catalyzed by a DNA-dependent RNA polymerase. *Proc. Natl. Acad. Sci. U.S.* **71,** 2782–2786. [386]

Sabelnikov, A. G., Avdeeva, A. V., Ilijushenko, B. N. (1975). Enhances uptake of donor DNA by Ca^{++}-treated *Escherichia coli*. *Mol. Gen. Gen.* **138,** 351–358. [139]

Sabin, A. B. (1952). Nature of inherited resistance to viruses affecting the nervous system. *Proc. Natl. Acad. Sci. U.S.* **38,** 540–546. [18]

Sabin, A. (1957). Properties of attenuated poliovirus and their behavior in human beings. *Spec. Publ. N.Y. Acad. Sci.* **5,** 113–127. [427]

Saborio, J. L., Pong, S. S., and Koch, G. (1974). Selective and reversible inhibition of initiation of protein synthesis in mammalian cells. *J. Mol. Biol.* **85,** 195–211. [320]

Safferman, R. S., and Morris, M.-E. (1963). Algal virus: isolation. *Science* **140,** 679–680. [6]

Safferman, R. S., Schneider, I. R., Steere, R. L., Morris, M. E., and Diener, T. O. (1969). Phycovirus SM-1: A virus infecting unicellular blue-green algae. *Virology* **47,** 105–113.

Sager, R. (1972). *Cytoplasmic Genes and Organelles*, Vol. XIV. Academic Press, New York. [134]

Sager, R., and Kitchin, R. (1975). Selective silencing of eukaryotic DNA. *Science* **189,** 426–433. [155]

Sagik, B. P., and Levine, S. (1957). The interaction of Newcastle disease virus (NDV) with chicken erythrocytes: attachment, elution, and hemolysis. *Virology* **3,** 401–416. [278]

Saigo, K., and Uchida, H. (1974). Connection of the righthand terminus of DNA to the proximal end of the tail in bacteriophage lambda. *Virology* **61,** 524–536. [216]

Sakuma, S., and Watanabe, Y. (1972). Incorporation of *in vitro* synthesized reovirus double-stranded ribonucleic acid into virus corelike particles. *J. Virol.* **10**, 943–950. [341]

Salk, J. E. with the collaboration of Bennett, B. L., Lewis, L. J., Ward, E. N., and Youngner, J. S. (1953). Studies in human subjects on active immunization against poliomyelitis. I. A preliminary report of experiments in progress. *J. Am. Med. Assoc.* **151**, 1081–1098. [426]

Salser, W. (1974). DNA sequencing techniques. *Ann. Rev. Biochem.* **43**, 923–966. [89]

Salser, W., Gesteland, R. F., and Bolle, A. (1967). *In vitro* synthesis of bacteriophage lysozyme. *Nature* **215**, 588–591.

Salzman, N. P. (1960). The rate of formation of vaccinia DNA and vaccinia virus. *Virology* **10**, 150–152.

Salzman, N., and Khoury, G. (1974). Reproduction of papovaviruses. In: *Comprehensive Virology*, Vol. III. (eds. H. Fraenkel-Conrat and R. R. Wagner). Plenum Press, New York, pp. 63–141. [349]

Salzman, N. P., Sebring, E. D., and Radnovich, M. (1973). Unwinding of parental strands during simian virus 40 DNA replication. *J. Virol.* **12**, 669–676. [81]

Sambrook, J. (1972). Transformation by polyoma virus and simian virus 40. *Adv. Cancer Res.* **16**, 141–180. [349]

Sambrook, J., Sugden, B., Keller, W., and Sharp, P. A. (1973). Transcription of simian virus 40. III. Orientation of RNA synthesis and mapping of "early" and "late" species of viral RNA extracted from lytically infected cells. *Proc. Natl. Acad. Sci. U.S.* **70**, 3711–3715. [356]

Sambrook, J., and Shatkin, A. J. (1969). Polynucleotide ligase activity in cells infected with simian virus 40, polyoma virus or vaccinia virus. *J. Virol.* **4**, 719–726. [377]

Sambrook, J. F., Westphal, H., Srinivasan, P. R., and Dulbecco, R. (1968). The integrated state of viral DNA in SV40 transformed cells. *Proc. Natl. Acad. Sci. U.S.* **60**, 1288–1295. [253]

Sanford, K. K., Earle, W. R., and Likely, G. D. (1948). The growth *in vitro* of single isolated tissues cells. *J. Natl. Cancer Inst.* **9**, 229–246. [252]

Sanford, K. K., Merwin, R. M., Hobbs, G. L., Young, J. M., and Earle, W. R. (1959). Clonal analysis of variant cell lines transformed to malignant cells in tissue culture. *J. Natl. Cancer Inst.* **23**, 1035–1069. [252]

Sanger, F., Air, G. M., Barrell, B. G., Brown, N. L., Coulson, A. R., Fiddes, J. C., Hutchison, C. A., III, Slocombe, P. M., and Smith, M. (1977). Nucleotide sequence of bacteriophage ϕX174 DNA. *Nature* **265**, 687–695. [92]

Sanger, F., and Coulson, A. R. (1975). A rapid method for determining sequences in DNA by primed synthesis with DNA polymerase. *J. Mol. Biol.* **94**, 441–448. [91]

Sänger, H. L. (1968). Characteristics of tobacco rattle virus. I. *Mol. Gen. Gen.* **101**, 346–367. [467]

Sänger, H. L., Klotz, G., Riesner, D., Gross, H. J., and Kleinschmidt, A. K. (1976). Viroids are single-stranded covalently closed circular RNA molecules existing as highly base-paired rod-like structures. *Proc. Natl. Acad. Sci. U.S.* **73**, 3852–3856. [462]

Sarabhai, A. S., Stretton, A. O. W., and Brenner, S. (1964). Co-linearity of the gene with the polypeptide chain. *Nature* **201,** 13-17. [64, 69, 176]

Sarov, I., and Joklik, W. K. (1972a). Studies on the nature and location of the capsid polypeptides of vaccinia virions. *Virology* **50,** 579-592. [375]

Sarov, I., and Joklik, W. K. (1972b). Characterization of intermediates in the uncoating of vaccinia virus DNA. *Virology* **50,** 593-602. [375]

Sarov, I., and Joklik, W. K. (1973). Isolation and characterization of intermediates in vaccinia virus morphogenesis. *Virology* **52,** 223-233. [379]

Sawicki, S., Jelinek, W., and Darnell, J. E. (1977). 3' terminal addition to HeLa cell nuclear and cytoplasmic poly(A). *J. Mol. Biol.,* in press. [78, 268, 274]

Scandella, D., and Arber, W. (1974). An *Escherichia coli* mutant which inhibits the injection of phage λ DNA. *Virology* **58,** 504-513. [216]

Schade, S. Z., Adler, J., and Ris, H. (1967). How bacteriophage λ attacks motile bacteria. *J. Virol.* **1,** 599-609.

Schaefer, T. W., and Lockart, R. Z., Jr. (1970). Interferon required for viral resistance induced by poly I·poly C. *Nature* **226,** 449-450. [411]

Schaffer, F. L., and Hackett, A. J. (1963). Early events in poliovirus-HeLa cell interaction: Acridine orange photosensitization and detergent extraction. *Virology* **21,** 124-126. [298]

Scharff, M., and Levintow, L. (1963). Quantitative study of the formation of poliovirus antigens in infected HeLa cells. *Virology* **19,** 491-500.

Scheible, P. P., and Rhoades, M. (1975). Heteroduplex mapping of heat-resistant deletion mutants of bacteriophage T5. *J. Virol.* **15,** 1276-1280. [197]

Scheid, A., Caliguiri, L. A., Compans, R. W., and Choppin, P. W. (1972). Isolation of paramyxovirus glycoproteins. Association of both hemagglutinating and neuraminidase activities with the larger SV5 glycoprotein. *Virology* **50,** 640-652. [284, 333]

Scheid, A., and Choppin, P. W. (1974). Identification of biological activities of paramyxovirus glycoproteins. Activation of cell fusion, hemolysis, and infectivity by proteolytic cleavage of an inactive precursor protein of Sendai virus. *Virology* **57,** 475-490. [291, 333, 338]

Schekman, R., Weiner, A., and Kornberg, A. (1974). Multienzyme systems of DNA replication. *Science* **186,** 987-993. [202]

Schekman, R., Wickner, W., Westergaard, O., Brutlag, D., Geider, K., Bertsch, L. L., and Kornberg, A. (1972). Initiation of DNA synthesis: synthesis of ϕX174 replicative form requires RNA synthesis resistant to rifamycin. *Proc. Natl. Acad. Sci. U.S.* **69,** 2691-2695. [221]

Scher, C. D., Handenschild, C., and Klagsbrun, M. (1976). The chick chorioallantoic membrane as a model system for the study of tissue invasion by viral transformed cells. *Cell* **8,** 373-382. [446]

Scherer, W. F., Syverton, J. T., and Gey, G. O. (1953). Studies on the propagation *in vitro* of poliomyelitis viruses. IV. Viral multiplication in a stable strain of human malignant epithelial cells (strain HeLa) derived from an epidermoid carcinoma of the cervix. *J. Exp. Med.* **97,** 695-709. [252]

Schlesinger, M. J., and Schlesinger, S. (1973). Large-molecular-weight precursors of Sindbis virus proteins. *J. Virol.* **11,** 1013-1016. [325]

Schlesinger, R. W. (1969). Adenoviruses: The nature of the virion and of controlling factors in productive or abortive infection and tumorogenesis. *Adv. Virus Research* **14**, 1–61.

Schlesinger, S., and Schlesinger, M. J. (1972). Formation of Sindbis virus proteins: identification of a precursor for one of the envelope proteins. *J. Virol.* **10**, 925–932. [325]

Schlesinger, W. (1959). Interference between animal viruses. In: *The Viruses*, Vol. III (eds. F. M. Burnet and W. M. Stanley). Academic Press, New York, pp. 157–194. [407, 414, 418]

Schlessinger, D. (1975). *Microbiology—1974*. Amer. Soc. Microbiol. Washington, D.C. [247, 249, 485, 486, 487]

Schmeiger, H., and Backhaus, H. (1973). The origin of DNA in transducing particles in P22⁻ mutants with increased transduction-frequencies (HT-mutants). *Mol. Gen.* **120**, 181–190. [239]

Schmitt, F. O., Bear, R. S., and Clark, G. L. (1955). X-ray diffraction studies on nerve. *Radiology* **25**, 131–151. [262]

Schneider, I. R., Hall, R., and Markham, R. (1972). Multidense satellite of tobacco ringspot virus: a regular series of components of different densities. *Virology* **47**, 320–330. [456, 462]

Schnos, M., and Inman, R. B. (1971). Starting point and direction of replication in P2 DNA. *J. Mol. Biol.* **55**, 31–38.

Schonberg, M., Silverstein, S. C., Levin, D. H., and Acs, G. (1971). A synchronous synthesis of the complementary strands of the reovirus genome. *Proc. Natl. Acad. Sci. U.S.* **68**, 505–508. [341]

Schulman, J. L., Khakpour, M., and Kilbourne, E. D. (1968). Protective effects of specific immunity to viral neuraminidase on influenza virus infection of mice. *J. Virol.* **2**, 778–786. [283, 429]

Schwerdt, C. E., and Fogh, J. (1957). The ratio of physical particles per infectious unit observed for poliomyelitis viruses. *Virology* **4**, 41–52. [25]

Scott, J. R. (1970). A defective P1 prophage with a chromosomal location. *Virology* **40**, 144–151. [241]

Sebring, E. D., Kelly, T. J., Jr., Thoren, M. M., and Salzman, N. P. (1971). Structure of replicating simian virus 40 deoxyribonucleic acid molecules. *J. Virol.* **8**, 478–490. [81, 358]

Séchaud, J., Streisinger, G., Emrich, J., Newton, J., Lanford, H., Reinhold, H., and Stahl, F. M. (1965). Chromosome structure in phage T4. II. Terminal redundancy and heterozygosis. *Proc. Natl. Acad. Sci. U.S.* **54**, 1333–1339. [176, 187]

Seecof, R. L. (1964). Deleterious effects on *Drosophila* development associated with the sigma virus infection. *Virology* **22**, 142–148. [478]

Sefton, B. M., Wickus, G. G., and Burge, B. W. (1973). Enzymatic iodination of Sindbis virus proteins. *J. Virol.* **11**, 730–735. [338]

Sehgal, P., Derman, E., Molloy, G., Darnell, J. E., and Tamm, I. (1976a). 5,6-Dichloro-1-β-D-ribofuranosyl benzimidazole inhibits the initiation of hnRNA chains in HeLa cells. *Science* **192**, in press. [410, 273]

Sehgal, P. B., Tamm, I., Vilček, J. (1976b). On the mechanism of enhancement of human interferon production by actinomycin D and cycloheximide. *Virology* **70**, 256–259. [410]

Semancik, J. S., Morris, T. J., Weathers, L. G., Rodorf, B. F., and Kearns, D. R. (1975). Physical properties of a minimal infectious agent (viroid) associated with the exocortis disease. *Virology* **63,** 160–167. [462]

Semancik, J. S., Vidaver, A. K., and Van Etten, J. L. (1973). Characterization of a segmental double-helical RNA from bacteriophage φ6. *J. Mol. Biol.* **78,** 617–626. [139, 205]

Sen, G. C., Gupta, S. L., Brown, G. E., Lebleu, B., Rebella, M. A., and Lengyel, P. (1976). Interferon treatment of Ehrlich ascites tumor cells: effects on exogenous mRNA translation and tRNA inactivation in the cell extract. *J. Virol.* **17,** 191–203. [413]

Sen, G. C., Lebleu, B., Brown, G. E., Rebello, M. A., Furuichi, Y., Morgan, M., Shatkin, A. J., and Lengyel, P. (1975). Inhibition of reovirus mRNA methylation in extracts of interferon-treated Ehrlich ascites cells. *Biochem. Biophys. Res. Commun.* **65,** 427–434. [413]

Shapiro, L., Agabian-Keshishian, N., and Bendis, I. (1971). The cell cycle of *Caulobacter* is used as a model system for studying the molecular basis of differentiation. *Science* **173,** 884–892. [135]

Sharp, D. G. (1965). Quantitative use of the electron microscope in virus research. Methods and recent results of particle counting. *Lab. Invest.* **14,** 831–863. [54]

Sharp, J. D., Donta, S., and Freifelder, D. (1971). Lack of polarity of DNA injection by *Escherichia coli* phage λ. *Virology* **43,** 176–184. [216]

Sharp, P. A., Gallimore, P. H., and Flint, S. J. (1974). Mapping of adenovirus 2 RNA sequences in lytically infected cells and transformed cell lines. *Cold Spring Harbor Symposia Quant. Biol.* **39,** 457–474. [102, 363, 365]

Sharp, P. A., Hsu, M., and Davidson, N. (1972). Note on the structure of prophage λ. *J. Mol. Biol.* **71,** 499–501.

Sharp, P. A., Pettersson, U., and Sambrook, J. (1974). Viral DNA in transformed cells. 1. A study of the sequences of Adenovirus 2 DNA in a line of transformed rat cells using specific fragments of the viral genome. *J. Mol. Biol.* **86,** 709–726.

Shatkin, A. J. (1963). Actinomycin D and vaccinia virus infection of Hela cells. *Nature* **199,** 357–358. [379]

Shatkin, A. J. (1971). Viruses with segmented RNA genomes: multiplication of influenza versus reovirus. *Bacteriol. Rev.* **35,** 250–266. [114, 115, 461]

Shatkin, A. J. (1974). Methylated messenger RNA synthesis *in vitro* by purified reovirus. *Proc. Natl. Acad. Sci. U.S.* **71,** 3204–3207. [115, 341]

Shatkin, A. J. (1974). Animal RNA viruses: Genome structure and function. *Ann. Rev. Bochem.* **43,** 643–665. [118]

Shatkin, A. J. (1976). Capping of eukaryotic mRNA's. *Cell* **9,** 645–653. [116, 268, 403]

Shatkin, A. J. (1976). 5′ Terminal caps in eukaryotic mRNA's. *New Scientist* (London), (in press). [117]

Shatkin, A. J., and Both, G. W. (1976). Reovirus mRNA: Transcription and translation. *Cell* **7,** 305–313. [338]

Shatkin, A. J., and Sipe, J. D. (1968). RNA polymerase activity in purified reoviruses. *Proc. Natl. Acad. Sci. U.S.* **61,** 1462–1469. [339]

Sheldrick, P., and Berthelot, N. (1974). Inverted repetitions in the chromosome of herpes simplex virus. *Cold Spring Harbor Symposia Quant. Biol.* **39**, 667–678. [368]

Sheldrick, P., Laithier, M., Londo, D., and Ryhiner, M. L. (1973). Infectious DNA from herpes simplex virus: infectivity of double-stranded and single-stranded molecules. *Proc. Natl. Acad. Sci. U.S.* **70**, 3621–3625. [370]

Shelokov, A., Vogel, J. E., and Chi, L. (1958). Hemadsorption (adsorption-hemagglutination) test for viral agents in tissue culture with special reference to influenza. *Proc. Soc. Exptl. Biol. Med.* **97**, 802–809. [26, 283]

Shenk, T. E., Rhodes, C., Rigby, P. W. J., and Berg, P. (1974). Mapping of mutational alterations in DNA with S nuclease: the location of deletions, insertions and temperature-sensitive mutations in SV40. *Cold Spring Harbor Symposia Quant. Biol.* **39**, 61–67. [351]

Shepherd, R. J. (1976). DNA viruses of higher plants. *Adv. Virus Research* **20**, 305–339. [455]

Shih, D. S., and Kaesberg, P. (1973). Translation of brome mosaic viral ribonucleic acid in a cell-free system derived from wheat embryo. *Proc. Natl. Acad. Sci. U.S.* **70**, 1799–1803. [326]

Shih, D. S., Lane, L. C., and Kaesberg, P. (1972). Origin of the small component of brome mosaic virus RNA. *J. Mol. Biol.* **64**, 353–362. [115]

Shimada, K., and Campbell, A. (1974). Int-constitutive mutants of bacterio-phage λ. *Proc. Natl. Acad. Sci. U.S.* **71**, 237–241. [217]

Shimada, K., Weisberg, R. A., and Gottesman, M. E. (1973). Prophage lambda at unusual chromosomal locations. II. Mutations induced by bacterio-phage lambda in *Escherichia coli* K12. *J. Mol. Biol.* **80**, 297–314. [231]

Shimojo, H., Shiroki, K., and Yamaguchi, K. (1974). The viral DNA replication machinery of adenovirus type 12. *Cold Spring Harbor Symposia Quant. Biol.* **39**, 533–538. [365]

Shimojo, H., and Yamashita, T. (1968). Induction of DNA synthesis by adenoviruses in contact-inhibited hamster cells. *Virology* **36**, 422–433. [366]

Shin, S.-I., Freedman, U. H., Risser, R., and Pollack, R. (1975). Tumorigenic-ity of virus-transformed cells in *nude* mice is correlated specifically with anchorage dependence of growth *in vitro*. *Proc. Natl. Acad. Sci. U.S.* **72**, 4435–4439. [439]

Showe, M. H., and Kellenberger, E. (1975). Control mechanisms in virus assembly. *Symp. Soc. Gen. Microbiol.* **25**, 407–438. [43, 132]

Siden, E. J., and Hayashi, M. (1974). Role of the gene B product in bacterio-phage φX174 development. *J. Mol. Biol.* **89**, 1–16. [204]

Siegel, A., Zaitlin, M., and Sehgal, O. P. (1962). The isolation of defective tobacco mosaic virus strains. *Proc. Natl. Acad. Sci. U.S.* **48**, 1845–1851. [466]

Siegel, A. (1971). Pseudovirions of tobacco mosaic virus. *Virology* **46**, 50–59.

Siegel, P. J., and Schaechter, M. (1973). Bacteriophage T4 head maturation: release of progeny DNA from the host cell membrane. *J. Virol.* **11**, 359–367. [165]

Siegl, G., and Gautschi, M. (1976). Multiplication of parvovirus LuIII in a synchronized culture system. III. Replication of viral DNA. *J. Virol.* **17**, 841–853. [345]

Siev, M., Weinberg, R., and Penman, S. (1969). The selective interruption of nucleolar RNA synthesis in HeLa cells by cordycepin. *J. Cell. Biol.* **41**, 510-520. [274]

Silverstein, J. L., and Goldberg, E. B. (1976a). T4 DNA injection: I. Growth cycle of a gene 2 mutant. *Virology* **72**, 195-211. [163]

Silverstein, J. L., and Goldberg, E. B. (1976b). T4 DNA injection: II. Protection of entering DNA from host endonuclease V. *Virology* **72**, 212-223. [163]

Silverstein, S. C., Astell, C., Levin, D. H., Schonberg, M., and Acs, G. (1972). The mechanisms of reovirus uncoating and gene activation *in vivo*. *Virology* **47**, 797-806. [342]

Silverstein, S. C., and Dales, S. (1968). The penetration of reovirus RNA and initiation of its genetic function in L-strain fibroblasts. *J. Cell. Biol.* **36**, 197-230. [342]

Simmons, D. T., and Strauss, J. H., Jr. (1972a). Replication of Sindbis virus. I. Relative size and genetic content of 26S and 49S RNA. *J. Mol. Biol.* **71**, 599-614. [324]

Simmons, D. T., and Strauss, J. H., Jr. (1972b). Replication of Sindbis virus. II. Multiple forms of double-stranded RNA isolated from infected cells. *J. Mol. Biol.* **71**, 615-632. [327]

Simmons, D. T., and Strauss, J. H. (1974a). Translation of Sindbis virus 26S DNA and 49S RNA in lysates of rabbit reticulocytes. *J. Mol. Biol.* **86**, 397-409. [326]

Simmons, D. T., and Strauss, J. H. (1974b). Replication of Sindbis virus. V. Polyribosomes and mRNA in infected cells. *J. Virol.* **14**, 552-559. [325]

Simon, E. H. (1961). Evidence for the nonparticipation of DNA in viral RNA synthesis. *Virology* **13**, 105-118. [312]

Simon, L. D., and Anderson, T. F. (1967). The infection of *Escherichia coli* by T2 and T4 bacteriophages as seen in the electron microscope. I. II. *Virology* **32**, 279-305. [158]

Simons, K., Garoff, H., Helenius, A., Kääriäinen, L., and Renkonen, O. (1974). Structure and assembly of virus membranes. In: *Perspectives in Membrane Biology* (eds. S. Estrada and C. Gitler). Academic Press, New York, pp. 45-77. [335, 337]

Simpson, R. W., and Hirst, G. K. (1968). Temperature-sensitive mutants of influenza A virus: Isolation of mutants and preliminary observations on genetic recombination and complementation. *Virology* **35**, 41-49. [335]

Singer, S. J. (1971). The molecular organization of biological membranes. In: *Structure and Function of Biological Membranes* (ed. L. Rothfield). Academic Press, New York, pp. 145-222. [336, 337]

Singer, S. J. (1974). The molecular organization of membranes. *Ann. Rev. Biochem.* **43**, 805-833. [337]

Singer, S. J., and Nicolson, G. L. (1972). The fluid mosaic model of the structure of cell membranes. *Science* **175**, 720-731. [262]

Sinha, R. C. (1968). Recent work on leafhopper-transmitted viruses. *Adv. Virus Research* **13**, 181-220. [469]

Sinha, R. C., and Black, L. M. (1962). Studies on the smear technique for detecting virus antigens in an insect vector by use of fluorescent antibodies. *Virology* **17**, 582-587. [471]

Sinsheimer, R. L. (1959a). A single-stranded DNA from bacteriophage ϕX174. *Brookhaven Symp. in Biol.* **12,** 27-34. [107]

Sinsheimer, R. L. (1959b). Purification and properties of bacteriophage ϕX174. *J. Mol. Biol.* **1,** 37-42. [105, 107]

Sippel, A., and Hartmann, G. (1968). Mode of action of rifamycin on the RNA polymerase reaction. *Biochim. Biophys. Acta* **157,** 218-219. [273]

Sjögren, H. O., Hellstrom, I., and Klein, G. (1961). Transplantation of polyoma virus induced tumors in mice. *Cancer Res.* **21,** 329-337. [452]

Sjöström, J.-E., and Philipson, L. (1974). Role of the ϕ11 phage genome in competence of *Staphylococcus aureus*. *J. Bacteriol.* **119,** 19-32.

Skalka, A., and Enquist, L. W. (1974). Overlapping pathways for replication, recombination and repair in bacteriophage lambda. In: *Mechanisms of DNA Replication* (eds. A. R. Kolber and M. Kohiyama). Plenum Press, New York, pp. 181-200. [222]

Skehel, J. J., and Joklik, W. K. (1969). Studies on the *in vitro* transcription of reovirus RNA catalyzed by reovirus cores. *Virology* **39,** 822-831. [339]

Smith, H. O., and Nathans, D. (1973). A suggested nomenclature for bacterial host modification and restriction systems and their enzymes. *J. Mol. Biol.* **81,** 419-423. [156]

Smith, H. O., and Wilcox, K. W. (1970). A restriction enzyme from *Hemophilus influenzae*. I. Purification and general properties. *J. Mol. Biol.* **51,** 379-391. [81]

Smith, J. A., and Martin, L. (1973). Do cells cycle? *Proc. Natl. Acad. Sci. U.S.* **70,** 1263-1267. [264]

Smith, J. D. (1976). Transcription and processing of transfer RNA precursors. *Prog. Nucleic Acid Res.* **16,** 25-74. [75, 77, 84]

Smith, K. M. (1959). The insect viruses. In: *The Viruses*, Vol. 3 (eds. F. M. Burnet and W. M. Stanley). Academic Press, New York, pp. 369-392. [473]

Smith, K. M., and Brown, R. M., Jr. (1965). A study of the long virus rods associated with insect granuloses. *Virology* **27,** 512-519. [476]

Smith, P. F. (1964). Comparative physiology of pleuropneumonia-like and L-type organisms. *Bacteriol. Rev.* **28,** 97-125. [3]

Smith, P. F. (1971). *The Biology of Mycoplasmas*. Academic Press, New York. [484]

Smith, S. H., and Schlegel, D. E. (1965). The incorporation of ribonucleic acid precursors in healthy and virus-infected plants. *Virology* **26,** 180-189. [463]

Smotkin, D., Gianni, A. M., Rozenblatt, S., and Weinberg, R. A. (1975). Infectious viral DNA of murine leukemia virus. *Proc. Natl. Acad. Sci. U.S.* **72,** 4910-4913. [384, 385]

Smotkin, D., Yoshimura, F. K., and Weinberg, R. A. (1976). Infectious linear unintegrated DNA of Moloney murine leukemia virus. *J. Virol.,* **20,** 621-626. [385]

Snustad, D. P., and Conroy, L. M. (1974). Mutants of bacteriophage T4 deficient in the ability to induce nuclear disruption. *J. Mol. Biol.* **89,** 663-673.

Snustad, D. P., Parson, K. A., Warner, H. R., Tutas, D. J., Wehner, J. M , and

Koerner, J. F. (1974). Mutants of bacteriophage T4 deficient in the ability to induce nuclear disruption. II. *J. Mol. Biol.* **89**, 675–687. [177]

Snyder, L. R., and Montgomery, D. L. (1974). Inhibition of T4 growth by an RNA polymerase mutant of *Escherichia coli*: physiological and genetic analysis of the effects during phage development. *Virology* **62**, 184–196. [177]

Sobell, H. N. (1973). The stereochemistry of actinomycin binding to DNA and its implications in molecular biology. *Prog. Nucleic Acid Res.* **13**, 153–190. [273]

Soderlund, H., Pettersson, U., Vennstrom, B., Philipson, L., and Mathews, M. (1976). A new species of virus-coded low molecular weight RNA from cells infected with adenovirus type 2. *Cell* **7**, 585–593. [364]

Soeiro, R., Vaughan, M. H., Warner, J. R., and Darnell, J. E., Jr. (1968). The turnover of nuclear DNA-like RNA in HeLa cells. *J. Cell. Biol.* **39**, 112–118. [271]

Sogo, J. M., Koller, T., and Diener, T. O. (1973). Potato spindle tuber viroid. X. *Virology* **55**, 70–80.

Sommer, S., Salditt-Georgieff, M., Bachenheimer, S., Darnell, J. E., Furuichi, Y., Morgan, M., and Shatkin, A. J. (1976). The methylation of adenovirus-specific nuclear and cytoplasmic RNA. *Nucleic Acids. Res.* **3**, 749–766. [363, 406]

Sonnabend, J. A., Kerr, I. M., and Martin, E. M. (1970). Development of the antiviral state in response to interferon. *J. Gen. Physiol.* **56**, 172$_s$–183$_s$.

Sonneborn, T. M. (1967). The evolutionary integration of the genetic material into genetic systems. In: *Heritage from Mendel* (ed. R. A. Brink). University of Wisconsin Press, Madison, Wisconsin, pp. 375–401. [482]

Soria, M. H., Little, S. P., and Huang, A. S. (1974). Characterization of vesicular stomatitis virus nucleocapsids. I. Complementary 40S RNA molecules in nucleocapsids. *Virology* **61**, 270–280. [328, 332]

Souther, A., Bruner, R., and Elliott, J. (1972). Degradation of *Escherichia coli* chromosome after infection by bacteriophage T4: role of bacteriophage gene D2a. *J. Virol.* **10**, 979–984.

Southern, E. M. (1970). Base sequence and evolution of guinea-pig α-satellite DNA. *Nature* **227**, 794–798. [267]

Spector, D. H., and Baltimore, D. (1974). Requirement of 3'-terminal polyadenylic acid for the infectivity of poliovirus RNA. *Proc. Natl. Acad. Sci. U.S.* **71**, 2983–2987. [311]

Spector, D. H., and Baltimore, D. (1975a). Polyadenylic acid in poliovirus RNA. II. Poly(A) on intracellular RNA's. *J. Virol.* **15**, 1418–1431. [309, 313, 430]

Spector, D. H., and Baltimore, D. (1975b). Poly(A) on Mengovirus RNA. *J. Virol.* **16**, 1081–1084. [311]

Spendlove, R. S., Lennette, E. H., and Jehn, A. C. (1963). The role of the mitotic apparatus in the intracellular location of reovirus antigen. *J. Immunol.* **90**, 554–560. [393]

Spiegelman, S., and Haruna, I. (1966). A rationale for an analysis of RNA replication. *Proc. Natl. Acad. Sci. U.S.* **55**, 1539–1554. [130]

Spizizen, J., Reilly, B. E., and Evans, A. H. (1966). Microbial transformation and transfection. *Ann. Rev. Microbiol.* **20**, 371–400. [139]

Sreevalsan, T., and Lockart, R. Z., Jr. (1962). A comparison of methods for interferon assay. *Virology* **17**, 207–208. [408]

Stahl, F. W., and Stahl, M. M. (1974). Red-mediated recombination in bacteriophage lambda. In: *Mechanisms in Recombination* (ed. R. Grell). Plenum Press, New York, pp. 407–419. [223]

Stampfer, M., Baltimore, D., and Huang, A. S. (1971). Absence of interference during high-multiplicity infection by clonally purified vesicular stomatitis virus. *J. Virol.* **7**, 409–411. [333]

Stanier, R. Y. (1964). Toward a definition of bacteria. In: *The Bacteria*, Vol. 5 (eds. I. C. Gunsalus and R. Y. Stanier). Academic Press, New York. [6]

Stanley, W. M. (1935). Isolation of a crystalline protein possessing the properties of the tobacco-mosaic virus. *Science* **81**, 644–645. [3, 6, 105]

Starlinger, P., and Saedler, H. (1972). Insertion mutations in microorganisms. *Biochimie* **54**, 177–185. [250]

Steck, F., and Rubin, H. (1966). The mechanism of interference between an avian leukosis virus and Rous sarcoma virus. II. Early steps of infection by RSV of cells under conditions of interference. *Virology* **29**, 642–653. [414]

Stehelin, D., Guntaka, R. V., Varmus, H. E., and Bishop, J. M. (1976a). Purification of DNA complementary to sequences required for neoplastic transformation of fibroblasts by avian sarcoma viruses. *J. Mol. Biol.* **101**, 349–365. [451]

Stehelin, D., Varmus, H. E., Bishop, J. M., and Vogt, P. K. (1976b). DNA related to the transforming genes of avian sarcoma viruses is present in normal avian DNA. *Nature* **260**, 170–173. [451]

Steinberg, C., and Stahl, F. (1961). The clone-soze distribution of mutants arising from a steady-state pool of vegetative phage. *J. Theoret. Biol.* **1**, 488–497. [154]

Steiner-Pryor, A., and Cooper, P. D. (1973). Temperature sensitive poliovirus mutants defective in repression of host protein synthesis are also defective in structural protein. *J. Gen. Virol.* **21**, 215–225. [402]

Steinhaus, E. A. (1946). *Insect Microbiology*. Comstock, Ithaca. [484]

Steinhaus, E. A., and Leutenegger, R. (1963). Icosahedral virus from a scarab (Sericesthis). *J. Insect Pathol.* **5**, 266–270. [476]

Steitz, J. A. (1975). Ribosome recognition of initiator regions in the RNA bacteriophage genome. In: *RNA Phages* (ed. N. Zinder). Cold Spring Harbor, New York, pp. 319–352. [206]

Stent, G. S., and Fuerst, C. R. (1960). Genetic and physiological effects of the decay of incorporated radioactive phosphorus in bacterial viruses and bacterial. *Adv. Med. Biol. Phys.* **7**, 1–75. [194]

Stent, G. S., and Wollman, E. L. (1950). Studies on activation of T4 bacteriophage by cofactor. II. *Biochim. Biophys. Acta* **6**, 307–316. [159]

Stevens, J. G. (1975). Latent herpes simplex virus and the nervous system. *Curr. Topics Microbiol. Immunol.* **70**, 31–50. [373]

Stevens, A. (1972). New small polypeptides associated with DNA-dependent RNA polymerase of *Escherichia coli* after infection with bacteriophage T4. *Proc. Natl. Acad. Sci. U.S.* **69**, 603–607. [171]

Stewart II, W. E., and Desmyter, J. (1975). Molecular heterogeneity of human leukocyte interferon: two populations differing in molecular weights, requirements for renaturation, and cross-species antiviral activity. *Virology* **67,** 68–73. [408, 409]

Stewart II, W. E., and Lockart, R. Z. (1970). Relative antiviral resistance induced in homologous and heterologous cells by cross-reacting interferons. *J. Virol.* **6,** 795–799. [408]

Stoker, M. (1968). Abortive transformation by polyoma virus. *Nature* **218,** 234–238. [446]

Stoltzfus, C. M., Shatkin, A. J., and Banerjee, A. K. (1973). Absence of polyadenylic acid from reovirus messenger RNA. *J. Biol. Chem.* **248,** 7993–7998. [340]

Stoltzfus, C. M., and Snyder, P. N. (1975). Structure of B77 sarcoma virus RNA: stabilization of RNA after packaging. *J. Virol.* **16,** 1161–1170. [386]

Stoltzfus, C. M., and Rueckert, R. (1972). Capsid polypeptides of mouse Elberfeld virus. I. Amino acid compositions and molar rations in the virions. *J. Virol.* **10,** 347–355. [309]

Stone, J. D. (1948). Prevention of virus infection with enzyme of *V. cholerae*. I. Studies with viruses of mumps-influenza group in chick embryos. II. Studies with influenza virus in mice. *Austral. J. Exptl. Biol. and Med.* **26,** 49–64; 287–298. [283, 429]

Strathern, A., and Herskowitz, I. (1975). Defective prophage in *Escherichia* K12 strains. *Virology* **67,** 136–143. [246]

Straus, S. E., Sebring, E. D., and Rose, J. A. (1976). Concatemers of alternating plus and minus strands are intermediates in adenovirus-associated virus DNA synthesis. *Proc. Natl. Acad. Sci. U.S.* **73,** 743–746. [345, 347]

Strauss, J. H., Jr., Burge, B. W., and Darnell, J. E., Jr. (1969). Sindbis virus infection of chick and hamster cells: synthesis of virus-specific proteins. *Virology* **37,** 367–376. [399]

Streisinger, G., Edgar, R. S., and Denhardt, G. H. (1964). Chromosome structure in phage T4. I. Circularity of the linkage map. *Proc. Natl. Acad. Sci. U.S.* **51,** 775–779. [109, 176, 181]

Streisinger, G., Okada, Y., Emrich, J., Newton, J., Tsugita, A., Terzaghi, E., and Inouye, M. (1966). Frame shift mutations and the genetic code. *Cold Spring Harbor Symposia Quant. Biol.* **31,** 77–84. [191]

Studier, F. W. (1972). Bacteriophage T7: Genetic and biochemical analysis of this simple phage gives information about basic genetic processes. *Science* **176,** 367–376. [198]

Sugar, J., and Kaufman, H. E. (1973). Halogenated pyrimidines in antiviral therapy. In: *Selective Inhibitors of Viral Functions* (ed. W. A. Carter). Chemical Rubber Company Press, Cleveland, Ohio, pp. 295–312. [431]

Sugino, A., and Okazaki, R. (1973). RNA-linked fragments *in vitro*. *Proc. Natl. Acad. Sci. U.S.* **70,** 88–92. [75, 85]

Sugiyama, T. (1966). Tobacco mosaic viruslike rods formed by "mixed reconstitution" between MS2 ribonucleic acid and tobacco mosaic virus protein. *Virology* **28,** 488–492. [38]

Suhadolnik, R. J. (1970). *Nucleoside Antibiotics.* John Wiley and Sons, New York. [273, 274]

Summers, D. F., and Maizel, J. V., Jr. (1968). Evidence for large precursor proteins in poliovirus synthesis. *Proc. Natl. Acad. Sci. U.S.* **59,** 966–971. [317]

Summers, D. F., and Maizel, J. V., Jr. (1971). Determination of the gene sequence of poliovirus with pactamycin. *Proc. Natl. Acad. Sci. U.S.* **68**, 2852-2856. [320]

Summers, D. F., Maizel, J. V., and Darnell, J. E. (1965). Evidence for virus-specific noncapsid proteins in poliovirus-infected HeLa cells. *Proc. Natl. Acad. Sci. U.S.* **54**, 505-513. [121, 298, 309, 317, 402]

Summers, D. F., Maizel, J. V., Jr., and Darnell, J. E., Jr. (1967). The decrease in size and synthetic activity of poliovirus polysomes late in the infectious cycle. *Virology* **31**, 427-435.

Summers, W. C., Brunovskis, I., and Hyman, R. W. (1973). The process of infection with coliphage T7. VII. *J. Mol. Biol.* **74**, 291-300.

Sussman, R., and ben Zeev, H. (1975). Proposed mechanism of bacteriophage lambda induction: acquisition of binding sites for lambda repressor by DNA of the host. *Proc. Natl. Acad. Sci. U.S.* **72**, 1973-1976.

Sutton, W. D. (1971). A crude nuclease preparation suitable for use in DNA reassociation experiments. *Biochim. Biophys. Acta* **240**, 522-531. [79]

Suzuki, E., and Shimojo, H. (1971). A temperature-sensitive mutant of adenovirus 31, defective in viral deoxyribonucleic acid replication. *Virology* **43**, 488-494. [366]

Svehag, S., and Mandel, B. (1964). The formation and properties of poliovirus-neutralizing antibody. *J. Exp. Med.* **119**, 1-39. [425]

Sydiskis, R. J., and Roizman, B. (1967). The disaggregation of host polyribosomes in productive and abortive infection with herpes simplex virus. *Virology* **32**, 678-686. [399, 404]

Szilágyi, J. F., and Uryvayev, L. (1973). Isolation of an infectious ribonucleoprotein from vesicular stomatitis virus containing an active RNA transcriptase. *J. Virol.* **11**, 279-286. [328]

Szybalski, W. (1974). Bacteriophage lambda. In: *Handbook of Genetics* Vol. 1 (ed. R. C. King). Plenum Press, New York, pp. 309-322. [226]

Taber, R., Rekosh, D. M., and Baltimore, D. (1971). Effect of pactamycin on synthesis of poliovirus proteins: A method for genetic mapping. *J. Virol.* **8**, 395-401. [320]

Takahashi, S. (1974). The rolling-circle replicative structure of a bacteriophage λ DNA. *Biochem. Biophys. Res. Commun.* **61**, 607-613.

Takashashi, T., and Diener, T. O. (1975). Potato spindle tuber viroid. XIV. *Virology* **64**, 106-114. [462]

Takebe, I. (1975). The use of protoplasts in plant virology. *Ann. Rev. Phytopathol.* **13**, 105-125. [456]

Takeda, Y., Matsubara, K., and Ogata, K. (1975). Regulation of early gene expression in bacteriophage lambda: effect of *tof* mutations on strand-specific transcription. *Virology* **65**, 374-384. [218]

Talbot, P., and Brown, F. (1972). A model for foot-and-mouth disease virus. *J. Gen. Virol.* **15**, 163-170. [284]

Tamm, I. (1975). Cell injury with viruses. *Am. J. Path.* **81**, 163-177. [393, 406, 407]

Tamm, I., and Eggers, H. J. (1963). Specific inhibition of replication of animal viruses. *Science* **142**, 24-33. [317]

Tan, K. B., Sambrook, J. F., and Bellett, A. J. D. (1969). Semliki forest virus temperature-sensitive mutants: isolation and characterization. *Virology* **38**, 427–439. [406]

Tan, Y. H., Armstrong, J., Ke, Y. H., and Ho, M. (1970). Regulation of cellular interferon production: Enhancement by antimetabolites. *Proc. Natl. Acad. Sci. U.S.* **67**, 464–471. [410]

Tan, Y. H., Creagan, R. P., and Ruddle, F. H. (1974). The somatic cell genetics of human interferon: assignment of human interferon loci to chromosomes 2 and 5. *Proc. Natl. Acad. Sci. U.S.* **71**, 2251–2555. [410]

Tattersall, P. (1972). Replication of the parvovirus MVM. I. Dependence of virus multiplication and plaque formation on cell growth. *J. Virol.* **10**, 586–590. [344]

Tattersall, P., Crawford, L. V., and Shatkin, A. J. (1973). Replication of the parvovirus MVM. II. Isolation and characterization of intermediates in the replication of the viral deoxyribonucleic acid. *J. Virol.* **12**, 1446–1456. [345]

Tattersall, P., Cawte, P. J., Shatkin, A. J., and Ward, D. C. (1976). Three structural polypeptides coded for by minute virus of mice, a parvovirus. *J. Virol.* **20**, 273–289. [344]

Tattersall, P., and Ward, D. C. (1976). Rolling hairpin model for replication of parvovirus and linear chromosomal DNA. *Nature* **263**, 106–109. [347]

Taylor, J. H. (1968). Rates of chain growth and units of replication in DNA of mammalian chromosomes. *J. Mol. Biol.* **31**, 579–594.

Taylor, J. M., and Illmensee, R. (1975). Site on the RNA of an avian sarcoma virus at which primer is bound. *J. Virol.* **16**, 553–558. [387]

Taylor, R. B., Duffus, P. H., Raff, M. C., and de Petris, S. (1971). Redistribution and pinocytosis of lymphocyte surface immunoglobulin molecules induced by antiimmunoglobulin antibody. *Nature New Biol.* **223**, 225–229. [263]

Teakle, D. S. (1962). Transmission of tobacco necrosis virus by a fungus, *Olpidium brassicae*. *Virology* **18**, 224–231.

Tegtmeyer, P. (1972). Simian virus 40 deoxyribonucleic acid synthesis: the viral replicon. *J. Virol.* **10**, 591–598. [355]

Tegtmeyer, P., Schwartz, M., Collins, J. K., and Rundell, K. (1975). Regulation of tumor antigen synthesis by simian virus 40 gene A. *J. Virol.* **16**, 168–178. [354]

Temin, H. (1963). The effects of actinomycin D on growth of Rous sarcoma virus *in vitro*. *Virology* **20**, 577–582. [386]

Temin, H. M. (1971). Mechanism of cell transformation by RNA tumor viruses. *Ann. Rev. Microbiol.* **25**, 609–648. [384]

Temin, H. M. (1973). The cellular and molecular biology of RNA tumor viruses, especially avian leukosis-sarcoma viruses and their relatives. *Adv. Cancer Res.* **19**, 47–104. [380]

Temin, H. M., and Baltimore, D. (1972). RNA-directed DNA synthesis and RNA tumor viruses. *Adv. Virus Research* **17**, 129–186. [76, 102, 380, 387]

Temin, H., and Mizutani, S. (1970). RNA-dependent DNA polymerase in virions of Rous sarcoma virus. *Nature* **226**, 1211–1213. [130, 384, 387]

Temin, H. M., and Rubin, H. (1958). Characteristics of an assay for Rous sarcoma virus and Rous sarcoma cells in tissue culture. *Virology* **6**, 669–688. [435]

Tenen, D. G., Baygell, P., and Livingston, D. M. (1975). Thermolabile T (tumor) antigen from cells transformed by a temperature-sensitive mutant of simian virus 40. *Proc. Natl. Acad. Sci. U.S.* **72,** 4351–4355. [354]

Terzaghi, E., Okada, Y., Streisinger, G., Emrich, J., Inouye, M., and Tsugita, A. (1966). Change of a sequence of amino acids in phage T4 lysozyme by acridine-induced mutations. *Proc. Natl. Acad. Sci. U.S.* **56,** 500–507. [177]

Thach, S. S., Dobbertin, D., Lawrence, C., Golini, F., and Thach, R. E. (1974). The mechanism of viral replication. Structure of replication complexes of encephalomyocarditis virus. *Proc. Natl. Acad. Sci. U.S.* **71,** 2549–2553. [313]

Thomas, C. A., Jr. (1966). The arrangement of information in DNA molecules. *J. Gen. Physiol.* **49,** (No. 6, Part 2), 143–169. [109]

Thomas, M., Cameron, J. R., and Davies, R. W. (1974). Viable molecular hybrids of bacteriophage lambda and eukaryotic DNA. *Proc. Natl. Acad. Sci. U.S.* **11,** 4579–4583. [233]

Thomas, M., White, R. L., and Davis, R. W. (1976). Hybridization of RNA to double-stranded DNA; Formation of R-loops. *Proc. Natl. Acad. Sci. U.S.* **73,** 2294–2298. [97]

Thompson, A. D. (1961). Interaction between plant viruses. I. Appearance of new strains after mixed infection with potato virus X strains. *Virology* **13,** 507–514. [466]

Thompson, R. L., Minton, S. A., Officer, J. E., and Hitchings, G. H. (1953). Effect of heterocyclic and other thiosemicarbazones on vaccinia infection in the mouse. *J. Immunol.* **70,** 229–235. [430]

Thormar, H., and Palsson, P. A. (1967). Visna and maedi—two slow infections of sheep and their etiological agents. *Perspectives in Virol.* **5,** 291–308. [420]

Thurm, P., and Garro, A. J. (1975). Isolation and characterization of prophage mutants of the defective *Bacillus subtilis* bacteriophage PBSX. *J. Virol.* **16,** 184–191. [245]

Tissières, A., Watson, J. D., Schlessinger, D., and Hollingworth, B. R. (1959). Ribonucleoprotein particles from *E. coli. J. Mol. Biol.* **1,** 221–233. [69]

Todaro, G., and Green, H. (1963). Quantitative studies on the growth of mouse embryo cells in culture and their development into established lines. *J. Cell. Biol.* **17,** 299–313. [436]

Todaro, G. J., Benveniste, R., and Sherr, C. J. (1976). Interspecies transfer of RNA tumor virus genes: implications for the search for "human" type C viruses. In: *Animal Virology* (ed. D. Baltimore et al.). Plenum Press, New York, in press. [389]

Tolmach, L. J. (1957). Attachment and penetration of cells by viruses. *Adv. Virus Research* **4,** 63–110. [146]

Tomasi, T. B., Jr., and Bienenstock, J. (1968). Secretory immunoglobulins. *Adv. Imunol.* **9,** 1–96. [425]

Tooze, J. (1973). *The Molecular Biology of Tumour Viruses.* Cold Spring Harbor Laboratory. [349, 380, 435]

Tseng, B. Y., and Goulian, M. (1975). Evidence for covalent association of RNA with nascent DNA in human lymphocytes. *J. Mol. Biol.* **99,** 339–344. [85, 406]

Tsukagoshi, N., Peterson, M. H., and Franklin, R. M. (1975). Structure and synthesis of a lipid-containing bacteriophage. *Virology* **66,** 206–216. [48]

Tye, B., Chan, R. K., and Botstein, D. (1974a). Packaging of an oversize transducing genome by phage P22. *J. Mol. Biol.* **85,** 485–500. [239]

Tye, B., Huberman, J., and Botstein, D. (1974b). Nonrandom circular permutation of phage P22 DNA. *J. Mol. Biol.* **85,** 501–532. [238]

Uetake, H., Luria, S. E., and Burrows, J. W. (1958). Conversion of somatic antigens in *Salmonella* by phage infection leading to lysis or lysogeny. *Virology* **5,** 68–91. [242]

Uhlenhopp, E. L., Zimm, B. H., and Cummings, D. J. (1974). Structural aberrations in T-even bacteriophage. VI. Molecular weight of DNA from giant heads. *J. Mol. Biol.* **89,** 689–702. [165, 176]

Unkeless, J. C., Tobia, A., Ossowski, L., Quigley, J. P., Rifkin, D. B., and Reich, E. (1973). An enzymatic function associated with transformation of fibroblasts by oncogenic viruses. I. *J. Exp. Med.* **137,** 85–111. [444]

Utermann, G., and Simons, K. (1974). Studies on the amphipathic nature of the membrane proteins in Semliki forest virus. *J. Mol. Biol.* **85,** 569–582. [122, 336]

Vago, C. (1951). The phenomenon of latency in a disease caused by the ultravirus of insects. *Rev. Canad. Biol.* **10,** 299–307. [477]

Vago, C., and Bergoin, M. (1968). Viruses of invertebrates. *Adv. Virus Research* **13,** 247–303. [473]

Valentine, R. C., and Allison, A. C. (1959). Virus particle adsorption. I. Theory of adsorption and experiments on the attachment of particles to non-biological surfaces. *Biochim. Biophys. Acta* **34,** 10–23. [277]

Valentine, R. C., and Pereira, H. G. (1965). Antigens and structure of the adenovirus. *J. Mol. Biol.* **13,** 13–20. [43]

Varmus, H. E., Bishop, J. M., and Vogt, P. K. (1973a). Appearance of virus-specific DNA in mammalian cells following transformation by Rous sarcoma virus. *J. Mol. Biol.* **74,** 613–626. [386]

Varmus, H. E., Vogt, P. K., and Bishop, J. M. (1973b). Integration of deoxyribonucleic acid specific for Rous sarcoma virus after infection of permissive and nonpermissive hosts. *Proc. Natl. Acad. Sci. U.S.* **70,** 3067–3071. [386]

Varmus, H. E., Guntaka, R. V., Fan, W. J. W., Heasley, S., and Bishop, J. M. (1974a). Synthesis of viral DNA in the cytoplasm of duck embryo fibroblasts and in enucleated cells after infection by avian sarcoma virus. *Proc. Natl. Acad. Sci. U.S.* **71,** 3874–3878. [384, 385]

Varmus, H., Guntaka, R. V., Deng, C. T., and Bishop, J. M. (1974b). Synthesis, structure and function of avian sarcoma virus—specific DNA in permissive and nonpermissive cells. *Cold Spring Harbor Symposia Quant. Biol.* **39,** 987–996. [388]

Varmus, H. E., Heasley, S., Linn, J., and Wheeler, K. (1976). Use of alkaline sucrose gradients in a zonal rotor to detect integrated and unintegrated avian sarcoma virus-specific DNA in cells. *J. Virol.* **18,** 574–585. [385]

Vaughan, M. H., and Hansen, B. S. (1973). Control of initiation of protein synthesis in human cells. Evidence for a role of uncharged transfer ribonucleic acid. *J. Biol. Chem.* **248,** 7087–7096. [272]

Vengris, V. E., Stollar, B. D., and Pitha, P. M. (1975). Interferon externalization by producing cell before induction of antiviral state. *Virology* **65,** 410–417. [411]

Verma, I. M., Mason, W. S., Drost, S. D., nd Baltimore, D. (1974). DNA polymerase activity from two temperature-sensitive mutants of Rous sarcoma virus is thermolabile. *Nature* **251**, 27–31. [388]

Verwoerd, D. W. (1970). Diplornaviruses: A newly recognized group of double-stranded RNA viruses. *Prog. Med. Virol.* **12**, 192–210. [338]

Vilček, J., and Havel, E. A. (1973). Stabilization of interferon messenger RNA activity by treatment of cells with metabolic inhibitors and lowering of the incubation temperature. *Proc. Natl. Acad. Sci. U.S.* **70**, 3909–3913. [410]

Villa-Komaroff, L., Guttman, N., Baltimore, D., and Lodish, H. F. (1975). Complete translation of poliovirus RNA in a eukaryotic cell-free system. *Proc. Natl. Acad. Sci. U.S.* **72**, 4157–4161. [321]

Vinograd, J. (1963). Sedimentation equilibrium in a buoyant density gradient. In: *Methods in Enzymology*, Vol. VI (eds. J. P. Colowik and N. O. Kaplan). Academic Press, New York, 854–887. [104]

Vinograd, J., Bruner, R., Kent, R., and Weigle, J. (1963). Band-centrifugation of macromolecules and viruses in self-generating density gradients. *Proc. Natl. Acad. Sci. U.S.* **49**, 902–910. [57]

Vinograd, J., and Lebowitz, J. (1966). Physical and topological properties of circular DNA. *J. Gen. Physiol.* **49**, 103–126. [350]

Vinograd, J., Lebowitz, J., Radloff, R., Watson, R., and Laipis, P. (1965). The twisted circular form of polyoma viral DNA. *Proc. Natl. Acad. Sci. U.S.* **53**, 1104–1110. [108, 350]

Visconti, N., and Delbrück, M. (1953). The mechanism of genetic recombination in phage. *Genetics* **38**, 5–33. [183]

Vogt, M., Dulbecco, R., and Smith, B. (1966). Induction of cellular DNA synthesis by polyoma virus. III. Induction in productively infected cells. *Proc. Natl. Acad. Sci. U.S.* **55**, 956–960. [356]

Vog, P. K. (1971). Genetically stable reassortment of markers during mixed infection with avian tumor viruses. *Virology* **46**, 947–952. [383]

Vogt, P. K. (1976). The genetics of RNA tumor viruses. In: *Comprehensive Virology* (eds. H. Fraenkel-Conrat and R. R. Wagner). Plenum Press, New York, in press. [380, 383, 449]

Vogt, V. M., and Eisenman, R. (1973). Identification of a large polypeptide precursor of avian oncornavirus proteins. *Proc. Natl. Acad. Sci. U.S.* **70**, 1734–1739. [381, 386]

Volkin, E., and Astrachan, L. (1956). Phosphorus incorporation in *Escherichia coli* ribonucleic acid after infection with bacteriophage T2. *Virology* **2**, 149–161. [65, 99, 173]

Wadsworth, S., Hayward, G. S., and Roizman, B. (1976). Anatomy of herpes simplex virus DNA. V. Terminally repetitive sequences. *J. Virol.* **17**, 503–512. [368]

Wadsworth, S., Jacob, R. J., and Roizman, B. (1975). Anatomy of herpes simplex virus DNA. II. Size, composition, and arrangement of inverted terminal repetitions. *J. Virol.* **15**, 1487–1497. [368]

Wagner, E. K., and Roizman, B. (1969). Ribonucleic acid synthesis in cells infected with herpes simplex virus. I. Patterns of RNA synthesis in productively infected cells. *J. Virol.* **4**, 36–46. [297, 405, 406]

Wagner, R. R. (1963). Cellular resistance to viral infection with particular reference to endogenous interferon. *Bacteriol. Rev.* **27**, 72–86. [408, 425]

Wagner, R. R. (1975). Reproduction of rhabdoviruses. In: *Comprehensive Virology*, Vol. IV (eds. H. Fraenkel-Conrat and R. R. Wagner). Plenum Press, New York, pp. 1–94. [327]

Wais, A. C., and Goldberg, E. B. (1969). Growth and transformation of phage T4 in *Escherichia coli* B/4, *Salmonella, Aerobacter, Proteus* and *Serratia*. *Virology* **39**, 153–161. [163]

Walker, D. L. (1968). Persistent viral infection in cell culture. In: *Medical and Applied Virology* (eds. M. Sanders and E. H. Lennette). W. Green, St. Louis, Missouri. [253, 419]

Waller, J. P. (1963). The NH_2-terminal residues of the proteins from cell-free extracts of *E. coli*. *J. Mol. Biol.* **7**, 483–496. [69]

Wang, J. C. (1971). Interaction between DNA and an *Escherichia coli* protein ω. *J. Mol. Biol.* **55**, 523–533. [81]

Wang, L.-H., Duesberg, P. H., Kawai, S., and Hanafusa, H. (1976a). The location of envelope-specific and sarcomaspecific oligonucleotides on the RNA of Schmidt-Ruppin Rous sarcoma virus. *Proc. Natl. Acad. Sci. U.S.* **73**, 447–451. [383]

Wang, L.-H., Duesberg, P., Mellon, P., and Vogt, P. K. (1976b). Distribution of envelope-specific and sarcoma-specific nucleotide sequences from different parents in the RNAs of avian tumor virus recombinants. *Proc. Natl. Acad. Sci. U.S.* **73**, 1073–1077. [383]

Warner, H. R., Snustad, D. P., Korner, J. F., and Childs, J. D. (1972). Identification and genetic characterization of mutants of bacteriophage T4 defective in the ability to induce exonuclease A. *J. Virol.* **9**, 399–407. [167]

Watanabe, Y., Prevec, C., and Graham, A. (1967). Specificity in transcription of the reovirus genome. *Proc. Natl. Acad. Sci. U.S.* **58**, 1040–1046. [339]

Watson, J. D. (1950). The properties of X-ray inactivated bacteriophage. I. *J. Bacteriol.* **60**, 697–718.

Watson, J. D. (1972). Origin of concatemeric T7 DNA. *Nature New Biol.* **239**, 197–201. [152, 165, 199]

Watson, J. D., and Littlefield, J. W. (1960). Some properties of DNA from Shope papilloma virus. *J. Mol. Biol.* **2**, 161–165. [349]

Weber, H., Kamen, R., Meyer, F., and Weissman, C. (1974). Interactions between Qβ replicase and Qβ RNA. *Experientia* **30**, 711 [206]

Webster, R. E., Engelhardt, D. L., and Zinder, N. D. (1966). *In vitro* protein synthesis: chain initiation. *Proc. Natl. Acad. Sci. U.S.* **55**, 155–161. [69]

Webster, R. G., Campbell, C. H., and Granoff, A. (1973). The *in vivo* production of new influenza viruses. III. Isolation of recombinant influenza viruses under simulated conditions of natural transmission. *Virology* **51**, 149. [335]

Wecker, E. (1959). The extraction of infectious virus nucleic acid with hot phenol. *Virology* **7**, 241–243. [107, 113, 324]

Wei, C.-M., Gershowitz, A., and Moss, B. (1975). N^6, $O^{2'}$-di-methyladenosine, a novel methylated ribonucleoside next to the 5′-terminal of animal cell and virus mRNA's. *Nature* **257**, 251–253. [363]

Wei, C.-M., and Moss, B. (1974). Methylation of newly synthesized viral messenger RNA by an enzyme in vaccinia virus. *Proc. Natl. Acad. Sci. U.S.* **71,** 3014–3018. [115]

Wei, C.-M., and Moss, B. (1975). Methylated nucleotides block 5′-terminus of vaccinia virus messenger RNA. *Proc. Natl. Acad. Sci. U.S.* **72,** 318–322. [375]

Weidel, W., and Pelzer, H. (1964). Bagshaped macromolecules—a new outlook on bacterial cell walls. *Adv. Enzymology* **26,** 193–232. [482]

Weil, R., Michel, M. R., and Ruschman, G. K. (1965). Induction of cellular DNA synthesis by polyoma virus. *Proc. Natl. Acad. Sci. U.S.* **53,** 1468–1475. [416]

Weinberg, R. A., and Newbold, J. E. (1974). Mapping of SV40 mRNA species on the viral genome. *Cold Spring Harbor Symposia Quant. Biol.* **39,** 161–164. [355]

Weinberg, R. A., Warnaar, S. O., and Winocour, E. (1972). Isolation and characterization of Simian virus 40 RNA. *J. Virol.* **10,** 193–201. [356]

Weiner, A. M., and Weber, K. (1971). Natual read-through at the UGA termination signal of Qφ coat protein cistron. *Nature New Biol.* **234,** 206–209. [206]

Weinman, R., Brendler, T. G., Raskas, H. J., and Roeder, R. G. (1976). Low molecular weight viral RNA's transcribed by RNA polymerase III during adenovirus 2 infection. *Cell* **7,** 557–566. [363]

Weintraub, H., and Gourdine, M. (1976). Chromosomal subunits in active genes have an altered conformation. *Science* **193,** 848–856. [257]

Weissmann, C., Parsons, J. T., Coffin, J. W., Rymo, L., Billeter, M. A., and Hofstetter, H. (1974). Studies on the structure and synthesis of Rous sarcoma virus RNA. *Cold Spring Harbor Symposia Quant. Biol.* **39,** 1043–1056. [381]

Wellauer, P. K., and David, I. B. (1973). Secondary structure maps of RNA: Processing of HeLa ribosomal RNA. *Proc. Natl. Acad. Sci. U.S.* **70,** 2827–2831. [96]

Wensink, P. C., Finnegan, D. J., Donelson, J. E., and Hogness, D. S. (1974). A system for mapping DNA sequences in the chromosomes of *Drosophila melanogaster*. *Cell.* **3,** 315–325.

Wessels, N. K., Sponer, B. S., Ash, J. F., Bradley, M. O., Luduena, M. A., Taylor, E. L., Wrenn, J. T., and Yamada, K. M. (1971). Microfilaments in cellular and developmental processes. *Science* **171,** 135–143. [261]

Westphal, H. (1970). SV40 DNA strand selection by *Escherichia coli* RNA polymerase. *J. Mol. Biol.* **50,** 407–420. [356]

Westphal, H., Meyer, J., and Maizel, J. V. (1976). Mapping of adenovirus mRNA by electron microscopy. *Proc. Natl. Acad. Sci. U.S.* **73,** 2069–2071.

Wetmur, J. G., and Davidson, N. (1968). Kinetics of renaturation of DNA. *J. Mol. Biol.* **31,** 349–370. [101]

Wheelock, E. F., and Tamm, I. (1959). Mitosis and division in HeLa cells infected with influenza or Newcastle disease virus. *Virology* **8,** 532–536. [399, 407]

Wheelock, E. F., and Tamm, I. (1961). Biochemical basis for alterations in structure and function of HeLa cells infected with Newcastle disease virus. *J. Exp. Med.* **114,** 617–622.

Wiberg, J. S., Mendelsohn, S., Warner, V., Hercules, K., Aldrich, C., and

Munro, J. L. (1973). SP62, a viable mutant of bacteriophage T4D defective in regulation of phage enzyme synthesis. *J. Virol.* **12,** 775–792. [171]

Wieland, T. (1968). Poisonous principles of mushrooms of the genus *Amanita*. *Science* **159,** 946–952. [273]

Wiebe, M. R., and Joklik, W. K. (1975). The mechanism of inhibition of reovirus replication by interferon. *Virology* **66,** 229–240. [412, 413]

Wildy, P. (1971). Classification and nomenclature of viruses. *Monographs in Virology*, Vol. V. S. Karger, Basel, pp. 1–81. [14]

Wilgus, G. S., Mural, R. J., Friedman, D. I., Fiandt, M., and Szybalski, W. (1973). λ*imm*λ434: a phage with a hybrid immunity region. *Virology* **56,** 46–53.

Wilkie, N. M., Clements, J. B., Macnab, J. C. M., and Subak-Sharpe, J. H. (1974). The structure and biological properties of herpes simplex virus DNA. *Cold Spring Harbor Symposia Quant. Biol.* **39,** 657–666. [368]

Wilkins, M. H. F., Stockes, A. R., Seeds, W. E., and Oster, G. (1950). Tobacco mosaic virus crystals and tri-dimensional microscopic vision. *Nature* **166,** 127–139. [463]

Willems, M., and Penman, S. (1966). The mechanism of host cell protein synthesis inhibition by poliovirus. *Virology* **30,** 355–367.

Williams, J. F., Grodzicker, T., Sharp, P., and Sambrook, J. (1975). Adenovirus recombination: physical mapping of crossover events. *Cell* **4,** 113–119. [364]

Williams, J. F., Young, C. S. H., and Austin, P. E. (1974). Genetic analysis of human adenovirus type 5 in permissive and nonpermissive cells. *Cold Spring Harbor Symposia Quant. Biol.* **39,** 427–437. [364]

Williams, R. C. (1953). A method of freeze-drying for electron microscopy. *Exptl. Cell. Res.* **4,** 188–201. [35]

Williams, R. C., and Backus, R. C. (1949). Macromolecular weights determined by direct particle counting. I. *J. Am. Chem. Soc.* **71,** 4052–4057. [54]

Williams, R. C., and Wyckoff, R. W. G. (1946). Applications of metallic shadowcasting to microscopy. *J. Appl. Phys.* **17,** 23–33. [35]

Wilson, D. A., and Thomas, C. A., Jr. (1974). Palindromes in chromosomes. *J. Mol. Biol.* **84,** 115–144. [268]

Wilson, J. H., Luftig, R. B., and Wood, W. B. (1970). The interaction of bacteriophage T4 tail components with purified *E. coli* lipopolysaccharide. *J. Mol. Biol.* **51,** 423–434. [161]

Wilson, J. N., and Cooper, P. D. (1962). Photodynamic demonstration of two stages in the growth of poliovirus. *Virology* **17,** 195–196. [298]

Wilson, J. T. (1963). Function of the bacteriophage T4 transfer RNA's. *J. Mol. Biol.* **74,** 753–757. [173]

Winocur, E., and Robbins, E. (1970). Histone synthesis in polyoma- and SV40-infected cells. *Virology* **40,** 307–315. [356, 416, 417]

Wittman, H. G. (1974). Purification and identification of *E. coli* ribosomal proteins. In: *Ribosomes* (eds. M. Nomura, A. Tissieres and P. Lengyel). Cold Spring Harbor Laboratory, pp. 93–114. [71]

Wold, W. S. M., Green, M., and Büttner, W. (1976). Adenoviruses. In: *The Molecular Biology of Animal Viruses* (ed. D. D. Nayak). Marcel Dekker, New York, in press. [360]

Wolfson, J., and Dressler, D. (1972). Adenovirus-2 DNA contains an inverted terminal repetition. *Proc. Natl. Acad. Sci. U.S.* **69**, 3054–3057. [110, 360]

Wolfson, J., Dressler, D., and Magazin, M. (1972). Bacteriophage T7 DNA replication: a linear replicating intermediate. *Proc. Natl. Acad. Sci. U.S.* **69**, 499–504. [199]

Wong, K., and Paranchych, W. (1976). The polarity of penetration of phage R17 RNA. *Virology* **73**, 489–497. [206]

Wood, H. A. (1973). Viruses with double-stranded RNA genome. *J. Gen. Virol.* **20** (suppl.), 61–85.

Wood, W. B., and Revel, H. R. (1976). The genome of bacteriophage T4. *Bacteriol. Rev.* **40**, 847–868. [157, 179]

Woodroofe, G. M., and Fenner, F. (1962). Serological relationships within the poxvirus group: an antigen common to all members of the group. *Virology* **16**, 334–341. [374]

Woodson, B. (1967). Vaccinia mRNA synthesis under conditions which prevent uncoating. *Biochem. Biophys. Res. Commun.* **27**, 169–175. [376]

Woodson, B., and Joklik, W. K. (1965). The inhibition of vaccinia virus multiplication by isatin β-thiosemicarbazone. *Proc. Natl. Acad. Sci. U.S.* **54**, 946–953. [430]

Worcel, A., and Burgi, E. (1972). On the structure of the folded chromosome of *Escherichia coli. J. Mol. Biol.* **71**, 127–147. [165]

Wu, M., and Davidson, N. (1973). A technique for mapping transfer RNA genes by electron microscopy of hybrids of ferritin-labeled transfer RNA and DNA: The $\phi80hpsu_{III}^{+-}$ system, *J. Mol. Biol.* **78**, 1–21. [97]

Wu, M., and Davidson, N. (1975). Use of gene 32 protein staining of single-strand polynucleotides for gene mapping by electron microscopy: Application to the $\phi80d_3ilvsu^{+}7$ system. *Proc. Natl. Acad. Sci. U.S.* **72**, 4506–4510. [97]

Wurster, D. H., and Benirschke, K. (1970). Indian Muntjac, Muntiacus muntjac: A deer with a low diploid chromosome number. *Science* **168**, 1364–1366. [257]

Wyatt, G. R., and Cohen, S. S. (1953). The bases of the nucleic acids of some bacterial and animal viruses: The occurrence of 5-hydroxymethylcytosine. *Biochem. J.* **55**, 774–782. [111]

Xeros, N. (1954). A second virus disease of the leather-jacket, *Tipula paludosa. Nature* **170**, 1073. [476]

Xeros, N. (1962). Deoxynucleoside control and synchronization of mitosis. *Nature* **194**, 682–683. [274]

Yamafuji, K. (1964). Metabolic virogens having mutagenic action and chromosomal previruses. *Enzymologia* **27**, 217–274. [477]

Yamashita, T., and Shimojo, H. (1969). Induction of cellular DNA synthesis by adenovirus type 12 in human embryo kidney cells. *Virology* **38**, 351–355. [366]

Yankofsky, S. A., and Spiegelman, S. (1963). Distinct cistrons for the two ribosomal RNA components. *Proc. Natl. Acad. Sci. U.S.* **49**, 538–544. [100]

Yanofsky, C., Drapeau, G. R., Guest, J. R., and Carlton, B. C. (1967). The complete amino acid sequence of the tryptophan synthetase A protein (α subunit) and its colinear relationship with the genetic map of the A gene. *Proc. Natl. Acad. Sci. U.S.* **57**, 296–298. [64]

Yogo, Y., and Wimmer, E. (1972). Polyadenylic acid at the 3'-terminus of poliovirus RNA. *Proc. Natl. Acad. Sci. U.S.* **68,** 1877–1882. [309]

Young, C. S. H., Pringle, C. R., and Follett, E. A. C. (1975). Action of interferon in enucleated cells. *J. Virol.* **15,** 428–429. [412]

Young, C. W., Schochetman, G., Hodas, S., and Balis, M. E. (1967). Inhibition of DNA synthesis by hydroxyurea: structure-activity relationships. *Cancer Res.* **27,** 535–540. [274]

Young, E. T., and Sinsheimer, R. L. (1964). Novel intra-cellular form of lambda DNA. *J. Mol. Biol.* **10,** 562–564. [109]

Younger, J. S., and Hallum, J. V. (1968). Interferon production in mice by double-stranded polynucleotides: Induction or release? *Virology* **35,** 177–179. [408]

Younghusband, H. B., and Inman, R. B. (1974). The electron-microscopy of DNA. *Ann. Rev. Biochem.* **43,** 605–619. [92, 94, 96]

Yutsudo, M., and Okamoto, K. (1973). Immediate-early expression of the gene causing superinfection breakdown in bacteriophage T4B. *J. Virol.* **12,** 1628–1630. [163]

Zaitlin, M., and Beachy, R. N. (1974). The use of protoplasts and separated cells in plant virus research. *Adv. Virus Research* **19,** 1–32. [455]

Zaitlin, M., Duda, C. T., and Petti, M. A. (1973). Replication of tobacco mosaic virus. V. Properties of the bound and solubilized replicase. *Virology* **53,** 300–311. [456]

Zamecnik, P. C. (1969). An historical account of protein synthesis with current overtones—A personalized view. *Cold Spring Harbor Symposia Quant. Biol.* **34,** 1–16. [67]

Zamecnik, P. C. (1960). Historical and current aspects of the problem of protein synthesis. *Harvey Lect.* **54,** 256–281. [67]

Zimmerman, E. F., Heeter, M., and Darnell, J. E. (1963). RNA synthesis in poliovirus infected cells. *Virology* **19,** 400–408. [404]

Zimmern, D. (1975). The 5' end group of tobacco mosaic virus RNA is m^{7}G$^{5'}$ppp$^{5'}$Gp. *Nucleic Acid Res.* **2,** 1189–1202. [458]

Zimmern, D. (1976). The region of tobacco mosaic virus RNA involved in the nucleation of assembly. *Philosoph. Trans. Royal Soc. London,* B, **276,** 189–204. [461]

Zinder, N. D., ed. (1975). *RNA Phages.* Cold Spring Harbor Laboratory, Cold Spring Harbor, New York, 428 pp. [113, 205]

Zinder, N. D., and Lederberg, J. (1952). Genetic exchange in Salmonella. *J. Bacteriol.* **64,** 679–699. [239]

Zubay, G., Schwartz, D., and Beckwith, J. (1970). Mechanism of activation of catabolite-sensitive genes: A positive control system. *Proc. Natl. Acad. Sci. U.S.* **66,** 104–110. [73]

Zweerink, H. J. (1974. Multiple forms of SS→DS RNA polymerase activity in reovirus-infected cells. *Nature* **247,** 313–315. [341]

Zweig, M., and Cummings, D. J. (1973). Cleavage of head and tail protein during bacteriophage T5 assembly: selective host involvement in the cleavage of a tail protein. *J. Mol. Biol.* **80,** 505–518. [197]

Zylber, E., Vesco, C., and Penman, S. (1969). Selective inhibition of the synthesis of mitochondria-associated RNA by ethidium bromide. *J. Mol. Biol.* **44,** 195–204. [273]

Index